Sources and Studies in the History of Mathematics and Physical Sciences

More information about this series at http://www.springer.com/series/4142

Sources and Studies in the History of Mathematics and Physical Sciences was inaugurated as two series in 1975 with the publication in Studies of Otto Neugebauer's seminal three-volume History of Ancient Mathematical Astronomy, which remains the central history of the subject. This publication was followed the next year in Sources by Gerald Toomer's transcription, translation (from the Arabic), and commentary of Diocles on Burning Mirrors. The two series were eventually amalgamated under a single editorial board led originally by Martin Klein (d. 2009) and Gerald Toomer, respectively two of the foremost historians of modern and ancient physical science. The goal of the joint series, as of its two predecessors, is to publish probing histories and thorough editions of technical developments in mathematics and physics, broadly construed. Its scope covers all relevant work from pre-classical antiquity through the last century, ranging from Babylonian mathematics to the scientific correspondence of H. A. Lorentz. Books in this series will interest scholars in the history of mathematics and physics, mathematicians, physicists, engineers, and anyone who seeks to understand the historical underpinnings of the modern physical sciences.

Venketeswara Pai • K. Ramasubramanian
M.S. Sriram • M.D. Srinivas

Karaṇapaddhati of Putumana Somayājī

Springer
HINDUSTAN BOOK AGENCY

Venketeswara Pai
Department of Humanities
and Social Sciences
Indian Institute of Science
Education and Research
Pune, Maharashtra, India

K. Ramasubramanian
Department of Humanities
and Social Sciences
Indian Institute of Technology Bombay
Mumbai, Maharashtra, India

M.S. Sriram
Prof. K.V. Sarma Research Foundation
Chennai, Tamil Nadu, India

M.D. Srinivas
Centre for Policy Studies
Chennai, Tamil Nadu, India

This work is a co-publication with Hindustan Book Agency, New Delhi, licensed for sale in all countries in electronic form, in print form only outside of India. Sold and distributed in print within India by Hindustan Book Agency, P-19 Green Park Extension, New Delhi 110016, India. ISBN: 978-93-86279-65-1 © Hindustan Book Agency 2018.

ISSN 2196-8810 ISSN 2196-8829 (electronic)
Sources and Studies in the History of Mathematics and Physical Sciences
ISBN 978-981-13-3896-0 ISBN 978-981-10-6814-0 (eBook)
https://doi.org/10.1007/978-981-10-6814-0

Softcover re-print of the Hardcover 1st edition 2018

Printed on acid-free paper

This Springer imprint is published by the registered company Springer Nature Singapore Pte Ltd. part of Springer Nature.
The registered company address is: 152 Beach Road, #21-01/04 Gateway East, Singapore 189721, Singapore

Transliteration scheme

The transliteration scheme followed in this book to represent *Devanāgarī* script is given below.

Table of vowels

अ	आ	इ	ई	उ	ऊ	ऋ	ॠ	ऌ	ए	ओ	ऐ	औ	अं	अः
a	*ā*	*i*	*ī*	*u*	*ū*	*ṛ*	*ṝ*	*ḻ*	*e*	*o*	*ai*	*au*	*aṃ*	*aḥ*

Table of consonants

क	ख	ग	घ	ङ	च	छ	ज	झ	ञ
ka	*kha*	*ga*	*gha*	*ṅa*	*ca*	*cha*	*ja*	*jha*	*ña*
ट	ठ	ड	ढ	ण	त	थ	द	ध	न
ṭa	*ṭha*	*ḍa*	*ḍha*	*ṇa*	*ta*	*tha*	*da*	*dha*	*na*
प	फ	ब	भ	म					
pa	*pha*	*ba*	*bha*	*ma*					
य	र	ल	व	श	ष	स	ह	ळ	
ya	*ra*	*la*	*va*	*śa*	*ṣa*	*sa*	*ha*	*ḷa*	

Contents

7   ग्रहस्फुटानयनम्
     Obtaining the planetary longitudes 177

List of Figures

List of Tables

Foreword

In 1790, a little more than a century after the publication of Isaac Newton's *Principia*, the Scottish mathematician John Playfair wrote a fascinating paper on The *astronomy* of the *Brahmins* in the Transactions of the Royal Society of Edinburgh. He was astonished at the consistently high accuracy of Indian predictions of planetary and sun / moon positions and solar / lunar eclipses, and at how competitive these predictions were with what had just been achieved in Europe. And he wondered how Indians could achieve such high precision with the very simple but elegant rules and tables that they had invented for making their calculations – 'with wonderful certainty and expedition' (but apparently without the aid of any 'principles'), he wrote. Now classical Indian astronomical literature has various *siddhāntas* which formulate algorithms based on observations of planetary motions. Although the implied models were effectively geocentric, Āryabhata had already proposed that the earth rotates around its axis and Nīlakaṇṭha considered that the inner planets moved around the sun but their little inner solar system still revolved around the earth. While such views of planetary motion continued to be discussed, the *panchāngakāras* (calendar-makers) sought the simplest possible 'rules' that would give results very close to those of the siddhāntic calculations / algorithms. The books that describe these rules were known as *karaṇas*. However, how the rules in the *karaṇa* texts were discovered or *derived* – especially as no 'principles' were stated – was a matter of mystery to scholars like Playfair and his French counterparts.

The present book, called *Karaṇa-paddhati* (KP), is not chiefly about the algorithms themselves; rather it is about the *procedure* by which the simple but extraordinarily accurate rules that the *panchāngakāra* could use were derived and constructed. If the *karaṇas* and the formulae and tables they gave could be called 'Astronomical Calculations Made Easy', *Karaṇa-paddhati* considers how simple but accurate rules for astronomical algorithms and calculations could be derived. The authors of this book have produced a scholarly work on the *Karaṇa-paddhati* of Putumana Somayāji. His date is not precisely known, and proposals have varied from the 15th century to the 18th. The authors make a convincing case that Putamana worked in the 16th century, and suggest that he was a junior contemporary of Nīlakaṇṭha. KP itself never mentions the name of its author: both the author's identity and his date have to be inferred from what Putumana says elsewhere and what others have said about his book.

As is generally the case in Sanskrit astronomy KP also describes algorithms in poetry, using the *katapayādi* system for referring to numbers. Here the numbers are so coded that they can be written in words which can in turn be strung into lines of poetry forming complete sentences. This *Vākya* system (as it is called) produced not only ingenious mathematics, but also presented longitudes and various other parameters coded in the form of meaningful sentences, and sometimes even in the form of poetry – generally in praise or support of wisdom and morality. This seems to have been a popular method of exposition particularly in South India. KP thus provides a derivation of each approximation it proposes, many of them ingenious and original, using various different metres (including some rare ones) for composing the *ślokās*. One of the remarkable results that KP has is a new series for π where successive terms $a_n \to 0$ like n^{-4} as $n \to \infty$. Interestingly KP realizes that there might still be errors in the work, or that errors might develop over time (as indeed they did in the siddhāntic algorithms themselves). KP therefore emphasizes the need for correcting them with the aid of continual observation. Clearly therefore KP is about *deriving* and constantly improving rules from siddhāntic algorithms *and* from observations.

The publication of this work is an important event, because it answers the questions that have puzzled many scholars about the mathematics and the algorithms that lay behind those surprisingly accurate Indian calculations. All this was achieved without the use of anything like Newton's laws but with careful observations over centuries, discerning cycles and other patterns in the data, and discovery of algorithms which could describe those observations and therefore also enable predictions. It is clear that this book now before the reading public is a great work of scholarship, and has been a labour of love for the authors – who incidentally provide every *ślokā* in Devanagari, an English transliteration and an English translation, all followed by very interesting commentary. The greatest contribution of this book would, in my view, be the light it sheds on the unique way that ancient Indic astronomers thought and acted. They followed Bhartṛhari's dictum that there are constraints only on the end results (*upeya*), but none on the means (*upāya*) of arriving there; and achieved successes that, till the early 19th century, excited the puzzled – even bewildered – wonder of astronomers like Playfair elsewhere in the world.

All those interested in the history of Indic astronomy must be grateful to the authors of this work for the big step they have taken in revealing *Karaṇa-paddhati's* secrets. My warm congratulations to Dr V Pai, Prof K Ramasubramanian, Prof M S Sriram and Prof M D Srinivas.

Roddam Narasimha

About the Authors

Venketeswara Pai is an assistant professor at the Department of Humanities and Social Sciences, Indian Institute of Science Education and Research (IISER), Pune. He has completed his Ph.D. from the Indian Institute of Technology (IIT) Bombay in 2011. Subsequently, he was a faculty member at the SASTRA University, Thanjavur, Tamil Nadu, India, before joining IISER, Pune. His research interests include Indian astronomy and mathematics. He is a recipient of the Young Historian of Science Award of the Indian National Science Academy (INSA) in 2014. He is also a member of the Indian National Young Academy of Science (INYAS).

K. Ramasubramanian is professor at the Cell for Science and Technology in Sanskrit, Indian Institute of Technology (IIT) Bombay. He worked on nonlinear dynamics to obtain his Ph.D. from the University of Madras in 2001. He has been working in the area of Indian astronomy and mathematics for more than 25 years and is also a scholar in Advaita-Vedānta. He was awarded the Maharṣi Bādarāyaṇa Vyas Samman by the President of India in 2008.

M.S. Sriram worked on field theory and obtained his Ph.D. in physics in 1978 at Indian Institute of Technology (IIT) Kanpur. He was a faculty member at the Allahabad University for 5 years. Subsequently, he joined the Department of Theoretical Physics, University of Madras, and retired as the head of the department in 2011. He has worked in the areas of high energy physics and nonlinear dynamics. He has been engaged in research in Indian astronomy and mathematics for over 25 years. Presently he is associated with Prof. K. V. Sarma Research Foundation, Chennai.

M.D. Srinivas did his Ph.D. in the theoretical physics at the University of Rochester, U.S.A. He was a faculty member at the Department of Theoretical Physics, University of Madras, for two decades and worked on the conceptual and mathematical foundations of quantum mechanics. He has been associated with the Centre for Policy Studies, Chennai, for over two decades and has been working on various aspects of Indian sciences, society and polity.

Introduction

Traditionally, the texts of Indian astronomy have been classed under the heads of *siddhānta* and *karaṇa*.[1] Of these, the *siddhānta* texts start with an ancient date for the epoch and present in detail the theoretical framework apart from giving the computational procedures for calculating the planetary positions, occurrence of eclipses, etc. The *Sūryasiddhānta*, the *Brāhmapshuṭasiddhānta* of Brahmagupta and the *Siddhāntaśiromaṇi* of Bhāskarācārya II are well-known examples of *siddhānta* texts. The *karaṇa* texts are essentially manuals in that they choose an epoch close to the date of composition of the text, and display their ingenuity in coming up with simplified algorithms for computing the planetary positions etc., to a desired level of accuracy. Many a times, they even dispense with the use of *jyās* (Rsines) and other functions, and instead limit themselves to the use of tables or approximate formulae. The *Khaṇḍakhādyaka* of Brahmagupta, *Laghumānasa* of Muñjāla, *Karaṇakutūhala* of Bhāskarācārya II and *Dṛggaṇita* of Parameśvara are well known examples of *karaṇa* texts.

Karaṇapaddhati: A unique text in the *vākya* tradition of Indian astronomy

The *Karaṇapaddhati* of Putumana Somayājī is a special text in that it does not come under the categories mentioned above. More crucially, it is not a *karaṇa* text. As the author declares right at the beginning of the text (Chapter 1,

[1] Subbarayappa and Sarma 1985, pp. 2-3. The general convention seems to be that a *siddhānta* text should choose the beginning of the current *kalpa* or an earlier date for its epoch. If the text employs the beginning of current *caturyuga* or the current *kaliyuga* as its epoch, then it is sometimes referred to as a *tantra*. Also, usually such *tantra* texts do not offer any explanations for the computational procedures, but present the algorithms for doing the computations in detail. The *Śiṣyadhīvṛddhida* of Lalla and the *Tantrasaṅgraha* of Nīlakaṇṭha Somayājī are well known texts of the *tantra* category.

verse 2), he is expounding on the *paddhati* or procedure for preparing an accurate *karaṇa* text for any given epoch. As explained by K. V. Sarma:[2]

> The *Karaṇapaddhati* ... is not a manual prescribing computations; rather it enunciates the rationale behind such manuals. Towards the beginning of the work, the author states that he composed the book to teach how the several multipliers, divisors, and Rsines pertaining to the different computations and the like are to be derived. Thus, the work is addressed not to the almanac maker but to the manual maker. All the topics necessary to make the daily almanac are not treated in *Karaṇapaddhati*, whereas several other items not pertaining to manuals are dealt with.

Another important feature of *Karaṇapaddhati* is that it deals with the computational procedures which are employed in a *karaṇa* text following the so called *vākya* system. The term *vākya* literally means a sentence consisting of one or more words. In the context of astronomy, it refers to a phrase or a string of letters in which the numerical values associated with various physical quantities are encoded. The *vākyas* are composed using the *kaṭapayādi* scheme,[3] which is one of the commonly employed systems to represent numbers in South-Indian texts.The strings used in composing *vākyas* are chosen so that they not only represent numerical values, but form beautiful meaningful sentences that convey worldly wisdom and moral values.

The *vākya* method of finding the true longitudes of the planets is quite different from the methods outlined in the various *siddhāntas*. As per the *siddhāntas*, we first find the mean longitudes of the planets and then apply a series of corrections (*saṃskāras*) to get their true positions.[4] On the other hand, the *vākya* method, by making use of certain *vākyas*—meaningful sentences composed in Sanskrit, which when decoded using *kaṭapayādi* system yield certain numbers—directly gives the true longitudes of the planets at certain intervals. The direct calculation of true longitudes involves making use of auxiliary epochs (*khaṇḍas, śodhyadinas*), chosen to be close to the desired date, and specified longitude values (*dhruvas*) which represent either the true longitude at the chosen epoch or the amounts of change in the true longitude over chosen periods (*maṇḍalas*). These *khaṇḍas*, *śodhyadinas* and *maṇḍalas* are chosen appropriately depending on the mean rates of motion of the planet, its *mandocca, śīghrocca* etc. Since the *vākya* method provides a simple and elegant method for computing the true longitudes without resorting to the normal procedure of calculating various corrections involving sine functions etc., it became very popular in South India where it seems to have originated.[5]

---

[2] Sarma 2008, p. 1837.

[3] For a brief explanation on the *kaṭapayādi* system, see Section 1.2.

[4] The *mandasaṃskāra* is to be applied in the case of the Sun and the Moon, whereas the *mandasaṃskāra* and the *śīghrasaṃskāra* are to be applied in the case of the other five planets. These are explained in Appendix B.

[5] A detailed introduction to the *vākya* system is presented in Appendices D, E. For further details on the *vākya* system see {CV 1948}, {VK 1962}, {SC 1973}, Hari 2001,

Ancient Indian astronomers were aware of the various kinds of periodicities in the motions of celestial bodies. One such periodicity is the 248-day cycle during which the Moon's anomaly completes nearly 9 revolutions.[6] Precisely this cycle was used in India, from early times, to find the true longitudes of the Moon at the sunrise for each day of the cycle and express them in the form of *vākyas*. These are the 248 *Vararuci-vākyas* which are attributed to Vararuci, who is also credited with the invention of the *kaṭapayādi* scheme of notation. He probably hailed from Kerala sometime prior to the 4th century CE.

The canonical text of the *Parahita* system, the *Grahacāranibandhana* of Haridatta, the famous seventh century astronomer from Kerala, introduces *vākyas* for the *manda* and *śīghra* corrections. A fully developed *vākya* system is outlined in the famous *karaṇa* text of 13th century, the *Vākyakaraṇa*, which gives methods for directly computing the true longitudes of the Sun, the Moon and the planets by using *vākyas*. Manuscripts of the work are available in various manuscript libraries of South India, especially Tamilnadu. Kuppanna Sastri and K. V. Sarma estimate that it was composed between 1282 and 1316 CE. The author of this work is not known, but probably hailed from the Tamil speaking region of South India. It has a commentary called *Laghuprakāśikā* by Sundararāja who hailed from Kāñcī near Chennai. The work is based on *Mahābhāskarīya* and *Laghubhāskarīya* of Bhāskara I belonging to the Āryabhaṭa School, and the *Parahita* system of Haridatta.

Mādhava of Saṅgamagrāma (c. 1360-1420), the legendary founder of the Kerala School of Astronomy, composed two works, *Veṇvāroha* and *Sphuṭacandrāpti*, which discuss the *vākya* method for computing the true longitude of the Moon, which shows the maximum variation in the course of a day. Mādhava composed a more accurate set of 248 *vākyas*, which give the true longitudes correct to a second, and also presented an ingenious method by which the these *vākyas* can be used to determine the true longitudes of the Moon at nine instants during the course of a day, thereby reducing the error in calculating the true longitude at any instant using interpolation. Mādhava was also the originator of the *agaṇita* system of computation of planetary positions. The *Agaṇitagrahacāra* of Mādhava is yet to be edited, though manuscripts of this work are available.

The *Vākyakaraṇa* and the above works of Mādhava only present the lists of *vākyas* and the computational procedures for obtaining the longitudes of the planets using these *vākyas*. It is the *Karaṇapaddhati* of Putumana Somayājī which explains the rationale behind them. In particular, it explains in detail the procedures for arriving at suitable *khaṇḍas* and *śodhyadinas* close to any epoch, as also the appropriate *maṇḍalas*, *dhruvas* etc., which have been used in texts such as the *Vākyakaraṇa* and the *Agaṇitagrahacāra*. It is only in *Karaṇapaddhati* that we find a detailed discussion of the procedure for arriving

2003, Madhavan 2012, Pai 2011, 2013, Pai et al 2009, 2015, 2016, Sastri 1989, Sriram 2014, Sriram and Pai 2012.

[6] This cycle has also been noticed by the Babylonians and the Greeks (Jones 1983).

at an accurate set of *vākyas* for the Sun, Moon and the planets, for any suitable epoch.

Karaṇapaddhati extensively employs the method of *vallyupasaṃhāra*, which is essentially the technique of continued fraction expansion of a ratio of two large numbers. The second chapter of the text is devoted to a detailed discussion of the *vallyupasaṃhāra* method, which summarises many of the important properties of the convergents of a continued fraction. This includes the well-known relation between successive convergents and an interesting relation (which we refer to as the remainder theorem) involving the convergents and the remainders obtained in the mutual division of two numbers. The convergents are of course used to arrive at optimal approximations to the rates of motions of planets and their anomalies. The remainder theorem and other properties of the convergents are used to arrive at ingenious algorithms for obtaining *khaṇḍas*, *śodhyadinas*, etc.

In short, the purpose served by *Karaṇapaddhati*[7] may be stated as:

करणकरणे पद्धतिः ।
karaṇakaraṇe paddhatiḥ
[Outlining] the procedure for composing *karaṇa* [texts].

Putumana Somayājī: The author

The author does not give any information about himself in *Karaṇapaddhati* except for the following brief mention at the end of the text (verse 10.12).

इति शिवपुरनामग्रामजः कोऽपि यज्वा...

iti śivapuranāmagrāmajaḥ ko'pi yajvā ...

Thus someone hailing from the village by name *Śivapura*, and who has performed sacrifices (*yajvā*)...

Unfortunately he stops with this and does not provide his name, or lineage, etc. However Putumana (new-house), the name of his house (*illam-peru* in Malayalam) appears in the introduction of one of the commentaries of *Karaṇapaddhati*:[8]

"नूतनगृह" सोमसुता रचितायाः करणपद्धतेर्विदुषा ।
भाषां विलिखति कश्चित् बालानां बोधनार्थमल्पधियाम् ॥

"nūtanagṛha" somasutā racitāyāḥ karaṇapaddhatervidușā |
bhāṣāṃ vilikhati kaścit bālānāṃ bodhanārthamalpadhiyām ||

[7] The word *Karaṇapaddhati* can be derived as: *karaṇe paddhatiḥ karaṇapaddhatiḥ; saptamyāḥ viṣayatvamarthaḥ; viṣayatvaṃ ca karaṇaparam.*
[8] {KP 1956}, p. xxv.

For the benefit of those who are novices to the field, and those who are dull-witted, here is an attempt by someone to explain in the local language (Malayalam) the [content of] *Karaṇapaddhati*, written by the [great] scholar Somayājī belonging to *nūtanagṛha* (*Putumana*).

In yet another work, *Jātakādeśamārga*, a highly popular work on astrology, Putumana Somayājī does refer to his house-name as also the title Somayājī.[9]

नवालयवनाख्येन धीमता सोमयाजिना ।
कृतं प्रकरणं ह्येतद् दैवज्ञजनतुष्टये ॥

navālayavanākhyena dhīmatā somayājinā |
kṛtaṃ prakaraṇaṃ hyetad daivajñajanatuṣṭaye ||

This short treatise (*prakaraṇa*) has been composed by the stalwart known as Putumana (navālayavana) Somayājī in order to please the community of astrologers.

As regards *Śivapura*, we may cite the following points made by S.K.Nayar:[10]

The term Śivapura is generally taken to refer to modern Trichur (Tṛ-ś-Śivapura). But unfortunately Nampūtiri brahmins of the latter village are not privileged to perform sacrifices on account of a curse or royal degradation and hence a native of this village cannot become *'Somayājis'*(sic). There is, however, a neighbouring village Covvaram (Śukapuram) which is also Sanskritised as Śivapuram and it may be taken that our author belonged to this Śivapura where even today there is a house named Putumana whose members are traditional astronomers.

Thus, all that we know about Putumana Somayājī is that he is reputed to have performed *Somayāga* and that his house bore the title Putumana. This is perhaps a unique case of a famous Indian astronomer, whose actual name is not found mentioned anywhere either in his works or in the commentaries.

Works of Putumana Somayājī

Putumana Somayājī seems to have authored several works, not only on astronomy, but also on astrology and *dharmaśāstra*. Unfortunately, apart from *Karaṇapaddhati*, none of his other works have been published, except for the very popular work on astrology, *Jātakādeśamārga*.[11] Works of Putumana Somayājī are identified by the notings on their manuscripts by the scribes and often also by the presence of the following "signature verse" which is the starting invocatory verse in *Karaṇapaddhati* (verse 1.1):

9 {KP 1956}, p. xxii and Pingree 1981, p. 208.

10 {KP 1956}, p. xxv. Based on this information, an attempt was made a few years ago by the authors of the present volume to identify this house of Putumana Somayājī in Covvaram. Unfortunately the efforts did not turn to be successful. Given the fact that Nayar was writing sixty years ago, and the fact that not much importance has been/is being given to maintain such heritage sites, it is not clear whether future efforts may yield any success in locating the *illam "Putumana"* of the author.

11 {JAM, 1930}, {JAM, 1942}, {JAM 1962}, {JAM, 1971}.

मदीयहृदयाकाशे चिदानन्दमयो गुरुः ।
उदेतु सततं सम्यगज्ञानतिमिरारुणः ॥

madīyahṛdayākāśe cidānandamayo guruḥ |
udetu satataṃ samyagajñānatimirāruṇaḥ ||

May the *guru*, an embodiment of intelligence and bliss keep ever rising in the space of my heart, like the Sun dispelling the darkness of ignorance.

The following passage cited from an article of K. V. Sarma gives a succinct summary of the works of Putumana Somayājī other than *Karaṇapaddhati*:[12]

Somayājī was a profile writer, mainly on astronomy and astrology, his only work in a different discipline being *Bahvṛcaprāyaścitta*, a treatise which prescribes expiations (*prāyaścitta*), for lapses in the performance of rites and rituals by *Bahvṛca* (Ṛgvedic) Brahmins of Kerala. In addition to his major work, Somayājī is the author of several other works. In *Pañca-bodha* (Treatise on the Five), he briefly sets out computations at the times of *Vyatipāta* (an unsavory occasion), *Grahaṇa* (eclipse), *Chāyā* (Measurements based on the gnomonic shadow), *Śṛṅgonnati* (Elongation of the moon's horns), and *Mauḍhya* (Retrograde motion of the planets), all of which are required for religious observances. His *Nyāyaratna* (Gems of Rationale), available in two slightly different versions, depicts the rationale of eight astronomical entities: true planet, declination, gnomonic shadow, reverse shadow, eclipse, elongation of the moon's horns, retrograde motion of the planets, and *Vyatipāta*. Three short tracts on the computation of eclipses, including a *Grahaṇāṣṭaka* (Octad on Eclipses), are ascribed to Somayājī. He also composed a work called *Veṇvārohāṣṭaka* (Octad of the Ascent on the Bamboo), which prescribes methods for the computation of the accurate longitudes of the moon at very short intervals. A commentary in the Malayalam language on the *Laghumānasa* of Muñjāla is also ascribed to him. On horoscopy, Somayājī wrote a *Jātakādeśamārga* (methods of making predictions on the basis of birth charts), which is very popular in Kerala.

The time of composition of *Karaṇapaddhati*

Karaṇapaddhati is among the four works that were mentioned in the famous 1834 article of Charles M. Whish on the Kerala School of astronomy.[13] In this article, having given an account of some of the infinite series given in Chapter 6 of *Karaṇapaddhati*, Whish mentions that the grandson of the author "is now alive in his seventieth year." Whish also makes reference to the last verse of *Karaṇapaddhati* as containing a chronogram giving the *kali-ahargaṇa* of the date of composition of the text. The second half of this verse (verse 10.12) reads

व्यधितगणितमेतत् सम्यगालोक्य सन्तः
कथितमिह विदन्तः सन्तु सन्तोषवन्तः ।

[12] Sarma 2008, pp. 1836-7.

[13] *Tantrasaṅgraha, Yuktibhāṣā* and *Sadratnamālā* are the other three works mentioned in this article (Whish 1834).

vyadhitagaṇitametat samyagālokya santaḥ
kathitamiha vidantaḥ santu santoṣavantaḥ |

By assiduously going through this mathematical work and understanding whatever has been set out here, may the noble ones become happy.

According to Whish:[14]

The last verse [of *Karaṇapaddhati*] ...contains its date in numerical letters in terms of the days of the caliyuga ...The words *Ganita Metutsamyac*, written in numbers amounts to 1765653 in terms of the days of the present age; which is found to agree with AD 1733.

Though the above chronogram is somewhat unusual, in that it is embedded in the middle of a *pāda* of the verse, it has been accepted by many scholars, including K. V. Sarma,[15] as giving the date of composition of *Karaṇapaddhati*. According to Sarma, Putumana Somayājī could have been born around 1660 and might have been active till 1740. In any case, the two published Malayalam commentaries on *Karaṇapaddhati* employ dates around 1749-52 for the *khaṇḍadinas*, and thus set an upper limit to the date of *Karaṇapaddhati*.[16]

S. K. Nayar has cited the following verse of Puruṣottama in a commentary of an astrological work called *Praśnāyana* to argue that Putumana Somayājī should have lived around 1700 CE:[17]

मध्यारण्यद्विजास्याद् विगलित इह यः प्रश्नमार्गो मरादौ
कोलम्बे छात्रशिष्योऽस्य च विरचितवान् जातकादेशरत्नम् ।
तच्छिष्यस्यानुमत्या स्थलिपुरनिलयस्यायनप्रश्नतोऽसौ
सिप्रादेशादिवासी व्यरचयमहमप्युत्तमः पूरुषाणाम् ॥

madhyāraṇyadvijāsyād vigalita iha yaḥ praśnamārgo marādau
kolambe chātraśiṣyo'sya ca viracitavān jātakādeśaratnam |
tacchiṣyasyānumatyā sthalipuranilayasyāyanapraśnato'sau
siprādeśādivāsī vyaracayamahamapyuttamaḥ pūruṣāṇām ||

The text *praśnamārga* emanated from the mouth *madhyāraṇyadvija* (Iṭakkāḍ Nampūtiri) in the Kollam year 825 (*Marādau Kolambe*) (A.D. 1650). His student disciple wrote the gem of a work *Jātakādeśa*. With the permission of his student I, Puruṣottama (*uttamaḥ pūruṣāṇām*) belonging to the place called *Siprādeśa*, have authored a commentary on the work called *Praśnāyana*.

The argument presented by Nayar, based on the above verse, in arriving at the conclusion that Putumana Somayājī should have lived around 1700 CE runs as follows:

---

[14] Whish 1834, p. 522.

[15] Sarma 2008. Sarma however notes that the *kali-aharagaṇa* 1765653 corresponds to 1732 CE, whereas it actually corresponds to April 4, 1733 CE.

[16] {KP 1956}, p. xxx.

[17] {KP 1956}, p.xxix-xxx.

This verse states that in the Kollam year 825 (*Marādau Kolambe*) (A.D. 1650) *madhyāraṇyadvija* (Iṭakkāḍ Nampūtiri) wrote his *Praśnamārga* and a pupil of his pupil wrote the *Jātakādeśa*. At the instance of a pupil of this author Puruṣottama wrote his commentary. It may be noted that like *Karaṇapaddhati*, *Jātakādeśa* also extracts verses from other works. The *Praśnamārga* is one such source book. This too points to the fact that *Jātakādeśa* is later and corroborates Puruṣottama's statments quoted above. Since our author is the pupil's pupil of the author of *Praśnamārga* which was composed in 1605 A.D., he must be forty or fifty years younger to him and might have lived about 1700 A.D.

The above argument is also not really convincing as Puruṣottama is referring to a work called *Jātakādeśaratna* (not *Jātakādeśamārga*) and does not make any direct reference to Putumana Somayājī.

In the first volume of his famous *Keralīya Saṃskṛta Sāhitya Caritram* (1937), Vatakkumkur Rajaraja Varma cited the following verse communicated to him by a friend and was supposedly a part of a work *Gaṇitasūcikā* of Govinda Bhaṭṭa, which mentioned that Putumana Somayājī composed his *Karaṇapaddhati* in the *Śaka* year 1353 (1431 CE):[18]

नवीनविपिने महीमखभुजां मणिः सोमया-
ज्युदारगणकोऽत्र यः समभवच्च तेनामुना ।
व्यलेखि सुदृगुत्तमा करणपद्धतिः संस्कृता
'त्रिपञ्चशिखिभू'मित-प्रथित-शाकसंवत्सरे ॥

navīnavipine mahīmakhabhujāṃ maṇiḥ somayā-
jyudāragaṇako'tra yaḥ samabhavac ca tenāmunā |
vyalekhi sudṛguttamā karaṇapaddhatiḥ saṁskṛtā
'tripañcaśikhibhū'mita-prathita-śākasaṁvatsare ||

This well refined [text] *Karaṇapaddhati* that is held high among the works belonging to *dṛk* system, was composed in the śaka year 1353 (*tripañcaśikhibhū*) by that Somayājī, who was a gem amongst the brahmins (*mahīmakhabhuks*)[19], who is an exalted mathematician and was born in a house (*illam*) known as putumana (*navīnavipina*)[20]

This verse, which still remains a hearsay, cannot be given any credence since *Karaṇapaddhati* cites both the *Agaṇitagrahacāra* of Mādhava (composed around 1417-18) and also the *Dṛggaṇita* of Parameśvara which was composed in 1431 CE.

To us the most compelling evidence for the date of Putumana Somayājī seems to be provided by the following initial verse of his *Veṇvārohāṣṭaka*, which is as yet unpublished:[21]

[18] Cited from {KP 1956}, p.xxvii.

[19] The word *makha* is a synonym of *yāga*. Hence, *makhabhuk* = *deva/sura*, and therefore *mahīmakhabhuk* = *bhūsura*, a term often employed to refer to a brahmin.

[20] According to many scholars, it seems to be a common practice in Kerala to translate the Malayalam word *mana* into *vana* or its synonyms such as *vipina* (which refer to forest) in Sanskrit.

[21] Cited from Manuscript No 430, deposited at the K. V. Sarma Research Foundation, Chennai.

रूक्षोऽयं क्रुद्धितास्योनितकलिदिवसात् प्रस्थगोराज्यभक्तात्
कालानङ्गैर्दिवीन्द्रैरपि कलिदिवसं तत्र शिष्टोनमाहुः ...

rūkṣo'yaṃ kruddhitāsyonitakalidivasāt prasthagorājyabhaktāt
kālānaṅgairdivīndrairapi kalidivasaṃ tatra śiṣṭonamāhuḥ...

When 1692162 (*rūkṣo'yaṃ kruddhitāsya*) is subtracted from the *kali-ahargaṇa* and [the remainders are successively] divided by 12372 (*prasthagorājya*), 3031 (*kālānaṅga*) and 248 (*divīndra*), the remaining number of kali days are stated to be ...

The above verse prescribes a procedure for computing the longitude of the Moon, similar to that employed in *Vākyakaraṇa*,[22] where the *khaṇḍadina* number 1692162 is subtracted from the given *kali-ahargaṇa* and the remainders are successively divided by the following number of days corresponding to complete anomalistic cycles: 12372, 3031 and 248. The number 1692162, regarded as a *kali-ahargaṇa*, corresponds to January 8, 1532. It can then be argued that the date of composition of the work should be between the *kali-ahargaṇas* 1692162 and 1692162 + 12372, for otherwise the latter day would have been a better choice for the *khaṇḍadina*. Hence, it follows that the work should have been composed sometime during the period 1532–1566.[23]

From a study of *Karaṇapaddhati*, it is clear that Putumana Somayājī is fully conversant with all the works of Mādhava and the *Dṛk* system of Parameśvara. As regards the *Tantrasaṅgraha* of Nīlakaṇṭha Somayājī, it seems that Putumana Somayājī is well aware of it, especially since there are some verses in common.[24] Further, according to one of the commentaries of *Karaṇapaddhati*, by using the word *nijamadhyama* for the interior planets in verse 7.25, Putumana Somayājī is implying that the equation of centre for those planets is to be applied to their so called *śīghrocca*—a procedure, which was first introduced by Nīlakaṇṭha in *Tantrasaṅgraha*, and is an important feature of his revised planetary model.[25] However, there are many aspects of Nīlakaṇṭha's revised model that are not taken note of in the *Karaṇapaddhati*. Similarly, for many of the problems in spherical astronomy, *Karaṇapaddhati* adopts ideas and techniques, which are very elegant, but totally different from those employed in *Tantrasaṅgraha*.[26] It is therefore reasonable to conclude that Putumana Somayājī was a junior contemporary of Nīlakaṇṭha, living in the sixteenth century. This is in accordance with the above evidence from the *Veṇvārohāṣṭaka*.[27]

---

22 This procedure is discussed in Section D.5 of Appendix D.

23 Precisely the same argument was used by Kuppanna Sastri and K. V. Sarma to conclude that *Vākyakaraṇa* was composed sometime during the period 1282–1316 CE ({VK, 1963}, p. ix).

24 For instance verses 6.19, 9.9 and 9.10 of *Karaṇapaddhati* are the same as the verses 2.16, 6.4 and 6.5, respectively of *Tantrasaṅgraha* {TS, 2011}.

25 For details, see the discussion in Section 7.19.

26 See for instance the discussion in sections 8.2 and 8.13.

27 We may also note that if it were to be assumed that Putumana Somayājī was living in the late 17th or early eighteenth century, then the first subtractive or *khaṇḍadina* chosen

Manuscripts and editions of *Karaṇapaddhati*

Pingree in his *Census of the Exact Sciences in Sanskrit* lists about thirty manuscripts of *Karaṇapaddhati*, most of which are in the Malayalam script, but some are in Grantham and Telugu scripts.[28] Most of the manuscripts are in the Manuscripts Library of the Trivandrum University, but some are in the libraries of Chennai and Baroda. There is also a paper manuscript of the work in the Whish collection of the Royal Asiatic Society, London, which is not listed in the Pingree list.

The first edition of *Karaṇapaddhati* was issued as Vol. 126 of the Trivandrum Sanskrit Series way back in the year 1937. This edition brought out by K. Sāmbaśiva Śāstrī, merely contains the original text in *Devanāgarī* script along with a very brief introduction of one and half pages. However, the meticulous effort that must have been put by Śāstrī in bringing out this largely error-free edition is indeed remarkable. In his acknowledgement, Śāstrī has mentioned that he had made use of five manuscripts (most of which were from private collections) in preparing this edition. Regarding the manuscripts that were available to him, he also states that:[29]

> A commentary on this work in Malayalam was obtained, but it was so worn out and unreadable that we had to give up the idea of getting it printed. We are sparing no pains in unearthing a complete manuscript of the same.

Fortunately, more manuscripts were identified at a later date. In the year 1953, P. K. Koru brought out an edition of the text with his own scholarly annotations and detailed mathematical notes in Malayalam. The title given by Koru for his commentary is *Yuktiprakāśikā* (that which throws light on the rationale). While acknowledging the source material that was available to him, Koru observes:[30]

> In September 1927, in Kannur, a gentleman from the place Cembiloṭṭaṃśaṃkoyoṭṭu (place near Kāṭāccira), named C. Kuññambu Paṇikkar gave me a (hand written) copy of *Karaṇapaddhati*, which also contained an old Malayalam commentary whose authorship is not known. The examples and the explanations contained in that book were found extremely useful to understand the import of the verses.

Besides presenting the original verses along with their import (which Koru rightly calls as *sāram*, as it cannot be considered as translation) he has also worked out several numerical examples. These examples contribute a great

---

in *Veṇvārohāṣṭaka* should have been at least 1741650 (*amitayavotsuka*). This *kalidina*, which corresponds to July 16, 1667, happens to be a well known *khaṇḍadina* employed by the later Kerala astronomers, as has been mentioned by Kunhan Raja in his introduction to *Haricarita* ({HC, 1948}, p. xxxi).

28 Pingree 1981, p. 207.

29 {KP, 1937} p. 1.

30 {KP, 1953}, p. ii (translated from the original passage in Malayalam).

deal to our understanding of the mathematical principles enunciated in the text and thereby add considerable value to this edition.

In 1956, Dr. S. K. Nayar of Madras University published an invaluable critical edition of *Karaṇapaddhati* along with two Malayalam commentaries.[31] Both these commentaries are dated to the middle of eighteenth century, and as Nayar notes:[32]

> Both the commentaries are elaborate and besides the meanings of the verses in detail they give numerous examples worked out in the traditional method which is of great value to a modern student of Indian astronomy. The commentaries are important also from another point of view. In the course of the explanation they quote several tables of *jyās* and other astronomical constants in *vākyas* couched in the *kaṭapayādi* notation, which are also useful to a student of Kerala astronomy.

Nayar mentions that he based his edition of the commentaries on the basis of six manuscripts each. The text itself was edited on the basis of fourteen manuscripts and the published version edited by Sāmbaśiva Śāstrī. Nayar has also written a scholarly introduction to the work, at the end of which he has acknowledged the enormous help and encouragement that he received from K. V. Sarma in bringing out this edition, which is worth reproducing here:[33]

> It now remains for me only to express my indebtedness to friends who enabled me to bring out this publication. The editing of this work was originally assigned to Dr. C. Acyuta Menon, Reader in Malayalam in this University, but his untimely demise prevented him from taking up the work. When I was asked to do it, I felt it a burden on account of my little knowledge of the subject. Luckily for me, my friend Sri K .V. Sarma of the Sanskrit Department of our University came to my rescue and I had the confidence that with this help and advice a good edition of the work could be put through. The original idea was only to print the text and commentary I as available in the Madras manuscript No. M.D.218. I owe to Sri Sarma the expansion of the scheme to bring out a critical edition of the text, to include in the critical apperatus (*sic*) Grantha and Telugu manuscripts, to include the second and better commentary and to edit the commentaries also on the basis of several manuscripts; and I am thankful to him for carrying out the scheme successfully. I am also thankful to him for the manifold new information given in this Introduction and for the several points dealt with therein, and also for adding the tables, appendix etc., which have enriched the value of this publication.

In the present edition, we have relied on the editions of Sāmbaśiva Śāstrī and Nayar for the text of *Karaṇapaddhati*, though at a few places we have indicated and corrected some errors that seem to have crept in.[34] We have made full use of the two Malayalam commentaries, and the Malayalam notes of P. K. Koru and the examples given therein, for understanding the exact import of the verses of the text. We have also cited most of the *vākyas* that have been presented in the commentaries.

---

[31] Nayar also refers to two Tamil commentaries of the text, whose manuscripts were said to be incomplete.

[32] {KP, 1956}, p. xiv.

[33] {KP, 1956}, p. xxxii.

[34] See for instance verses 7.4, 9.12 and 9.13.

The present edition gives the original verses of *Karaṇapaddhati* both in Devanāgarī and in transliterated form, along with English translation and detailed mathematical notes. Each of the ten chapters have been further divided into several sections with an appropriate title given to those sections both in Sanskrit and English. In our notes, we have tried to explain the various algorithms and procedures given in the text by means of equations, diagrams and tables. We have also explained the rationale behind most of the ingenious algorithms presented in the text, and also worked out several illustrative examples. For the benefit of the reader, we have also included a few Appendices, which serve to provide an introduction to important topics such as: the relation between the *vallyupasaṃhāra* technique and continued fraction expansion, the epicyclic and eccentric models of planetary motion, and the *vākya* system of computing the longitudes of the Sun, Moon and the planets.[35]

An outline of the contents of *Karaṇapaddhati*

Karaṇapaddhati is a comparatively short text with 214 verses, which are divided into ten chapters. It is largely based on the *Parahita* system initiated by Haridatta in seventh century, which adopts all the revolution numbers given in the *Āryabhaṭīya* after applying a correction known as the *śakābda-saṃskāra* which is an accumulating correction taking effect from 522 CE.

The first chapter of *Karaṇapaddhati* is similar to the chapter on *Madhyamādhikāra* in most of the astronomical works, and deals with the computation of mean longitudes. Here, soon after the invocation, the author presents the number of revolutions of the planets in a *mahāyuga*. This is followed by a discussion on the procedure for obtaining the solar, lunar and sidereal months, days, etc. Then he outlines the method for finding the *kali-ahargaṇa* (the number of civil days elapsed since the beginning of *Kaliyuga*), and the mean positions of the planets. Following this, there is an elaborate discussion on the application the *śakābda*-correction to the mean planets and also to the rates of motion. The author also presents interesting algorithms which make the calculation of the mean planets easier by making use of suitable intermediate epochs (*khaṇḍas*).

Chapter 2 of the text is devoted to a discussion of the method *vallyupasaṃhāra* which is essentially the same as the technique of continued fraction expansion of a ratio of two integers.[36] This method, also known as the *kuṭṭaka*

[35] Appendices A,B and D. While deriving the *Candravākyas* of Mādhava, by direct computation of the true longitude using the basic parameters of the *parahita* system, we found that there were a few errors in the edited versions of these *vākyas* ({VR, 1956}, {SC, 1973}). The details of our computation and the corrected version of the *Candravākyas* of Mādhava are presented in Appendix E.

[36] An introduction to the continued fraction expansion of a rational number is presented in Appendix A.

method, has been widely used in Indian astronomical literature, since the time of Āryabhaṭa, for solving linear indeterminate equations. *Karaṇapaddhati* explains how this method can be used to obtain optimal approximations to the ratio, say H/G, of two large numbers. It gives recursive algorithms for obtaining the convergents H_i/G_i in the continued fraction expansion of the above ratio. This method is used repeatedly in the text to give optimal approximations for the rates of motion of planets and their anomalies. *Karaṇapaddhati* also reveals a very sophisticated understanding of important properties of the convergents, such as,

$$H_iG_{i+1} - H_{i+1}G_i = (-1)^i.$$

Further, it presents an interesting "remainder theorem", which may be expressed in the form

$$HG_i - GH_i = (-1)^{i-1}r_i,$$

where $\{r_i\}$ are the remainders obtained in the mutual division of H and G. These relations are repeatedly made use of in the later chapters of the text to arrive at very ingenious algorithms for the *khaṇḍas*, *śodhyadinas*, etc.

Chapter 3 deals with the *vākya* method of computing the true longitude of the Moon. It commences with the use of *vallyupasaṃhāra* method for approximating the true rate of motion of the anomaly by ratios of smaller numbers such as 9/248, 110/3031, 449/12372, 6845/188611, etc., which give the various approximate anomalistic cycles (such as 9 cycles completed in 248 days etc). The denominators of the above ratios are known as *kendrahārakas*. The text presents an algorithm for obtaining a *khaṇḍadina* (a day on which the lunar anomaly is zero at sunrise), which is close to a given *ahargaṇa*. This ingenious algorithm (which seems to have been arrived at by making use of the relations between successive convergents mentioned above) can be used to generate several *khaṇḍadinas* based on the choice of the *kendrahāraka*. The text then describes the computation of the *dhruvas*, which are the changes in the true longitude over different anomalistic cycles, and introduces other notions such as *kendraphala*, *dhruva-saṃskāra-hāraka*, which help in simplifying the computations. Finally there is a brief discussion of the procedure for obtaining the *Candravākyas*.

Chapter 4 deals with the *vākya* method of computing the true longitudes of planets. It first gives the method for finding the *manda-khaṇḍa* and *śīghra-khanda* (which are the days close to the given *ahargaṇa* on which the *mandakendra* and the *śīghrakendra* vanish respectively) as given in the *nirgaṇita* system, which is perhaps the same as the *agaṇita* system of Mādhava. Then the text gives the algorithm for finding a set of *maṇḍalas* and the associated *dhruvas* for any planet, as used in the *Vākyakaraṇa*. *Maṇḍala* is a period in which a planet and its *śīghrocca* are in conjunction and nearly return to the *mandocca*, starting from a time when they are in conjunction. Then the text gives the method for finding a set of *śodhyadinas* and the associated *dhruvas* for any planet. The *śodhyadina* is a day close to the given *ahargaṇa* such that the planet and its *śīghrocca* are in conjunction and close to the planet's *mandocca*. The text then gives the method for obtaining a set of *agaṇita-hārakas* (divisors giving the revolutions made by a planet in a solar year) used in the *agaṇita* system. This is followed by the method of computing the *śodhyābda*, which gives the solar year close to the given *ahargaṇa* such that, at *Meṣādi* or the beginning of the year, the mean planet and its *śīghrocca* are in conjunction. The chapter finally gives a method to compute the *grahaṇakhaṇḍas*, which are the days of the occurrence of eclipses close to a given *ahargaṇa*. In this case, the text specifically prescribes the use of the parameters of the *Dṛk* system (for the Sun, Moon and the node) of Parameśvara, instead of those of the *Parahita* system.

In Chapter 5, the text emphasises the importance of correcting the revolution numbers etc., on the basis of observations. The chapter begins with the verse:

ग्रहणग्रहयोगाद्यैः ये ग्रहाः सुपरीक्षिताः ।
दृक्समास्तत्समाः कल्पे कल्प्या वा भगणादयः ॥

grahaṇagrahayogādyaiḥ ye grahāḥ suparīkṣitāḥ |
dṛksamāstatsamāḥ kalpe kalpyā vā bhagaṇādayaḥ ||

Based on a thorough examination of the position of the planets at the time of eclipses, planetary conjunctions, and so on, the revolution numbers etc. [of the planets] in a *kalpa* have to be proposed for achieving concordance with observations.

Having highlighted the importance of the need for correction when the parameters employed do not accord with observation, the text also outlines various methods for this purpose. It also explains how the longitude values which have been assumed for the epoch, such as the beginning of *Kaliyuga* etc., can be altered by using the so called *kalyādi-dhruvas*, so that the computed values are in accordance with observations. Alternative methods are suggested (such as correcting the revolution numbers, etc.) for achieving the same end. Finally, the text emphasises that there can indeed be a multiplicity of theoretical models with different durations (1000 or 1008 *mahāyugas*) being assigned to a *kalpa*, and other theoretical constructs. In this context, verse 15 of this chapter quotes the celebrated dictum from the *Vākyapadīya* of Bhartṛhari, that there are no constraints on the theoretical models (*upāya* or means), but only on the observed results (*upeya*, the end).[37]

कल्पादीनां प्रमाणं तु बहुधा कल्प्यते बुधैः ।
उपेयस्यैव नियमो नोपायस्येति यत् ततः ॥

kalpādīnāṃ pramāṇaṃ tu bahudhā kalpyate budhaiḥ |
upeyasyaiva niyamo nopāyasyeti yat tataḥ ||

[37] {VP, 1980}, p. 79. The same passage has been quoted by Nīlakaṇṭha in his *Āryabhaṭīya-bhāṣya* in a very similar context ({ABB, 1931}, p. 31.)

> The duration of a *kalpa* has been conceived differently by different scholars. This is due to the fact that there are no constraints on the means (*upāya*), but only on the end result (*upeya*).

Chapter 6 presents a succinct summary of the results on the *paridhi-vyāsa-sambandha* (relation between the circumference and the diameter of a circle) and *jyānayana* (computation of the Rsines). As regards the first topic, the text presents all the important fast convergent series given by Mādhava for $\frac{\pi}{4}$, which have been cited in the *Gaṇitayuktibhāṣā* of Jyeṣṭhadeva or *Yuktidīpikā* of Śaṅkara Vāriyar. In addition, it also presents the following interesting series, which has not been mentioned either by Jyeṣṭhadeva or by Śaṅkara, and is perhaps due to Putumana Somayājī himself (verse 6.4):

वर्गैर्युजां वा द्विगुणैर्निरेकैः वर्गीकृतैर्वर्जितयुग्मवर्गैः ।
व्यासं च षड्घ्नं विभजेत् फलं स्वं व्यासे त्रिनिघ्ने परिधिस्तदा स्यात् ॥

vargairyujāṃ vā dviguṇairnirekaiḥ
vargīkṛtairvarjitayugmavargaiḥ |
vyāsaṃ ca ṣaḍghnaṃ vibhajet phalaṃ svaṃ
vyāse trinighne paridhistadā syāt ||

> Or, from the square of even numbers multiplied by two, subtract one, and from the square [of that] subtract the square of the even numbers. Divide the diameters multiplied by six by the above [quantities]. When [the sum of] these are added to three times the diameter, the result will be the circumference.

The series presented by the above verse is,

$$C = 3D + \frac{6D}{(2.2^2-1)^2-2^2} + \frac{6D}{(2.4^2-1)^2-4^2} + \frac{6D}{(2.6^2-1)^2-6^2} + \ldots,$$

which may be written in the form

$$C = 3D + \sum_{n=1}^{\infty} \frac{6D}{(2.(2n)^2-1)^2-(2n)^2}.$$

This is a fast convergent series where the successive terms decrease like the fourth power of $2n$.

As regards the computation of sines, an interesting feature of *Karaṇapaddhati* is that it presents both the Rsine and Rcosine series of Mādhava in terms of just one and a half verses (verses 6.12, 6.13). Finally the text discusses the relation between the arc and the Rsine for small angles.

Chapter 7 begins with a discussion of the (variable) dimensions of the *manda* and *śīghra* epicycles and proceeds to present some interesting formulae for the calculation of the *manda* and *śīghra* corrections (*phalas*). It then gives the formula for the inverse hypotenuse (*vyasta-karṇa*) for the case of the Sun. This formula, which was given by Mādhava to calculate the iterated *manda*-hypotenuse (*asakṛt-manda-karṇa*)[38], is used here to calculate the

[38] See for instance the discussion in {TS, 2011}, pp. 494-497.

mean longitude of the Sun from the true longitude. This, in turn, is used to determine the exact times for the true Sun to enter a *rāśi, nakṣatra* etc., and these are expressed in terms of the *saṅkrāntivākyas*, *nakṣatravākyas*, etc. The text also explains the derivation of the *yogyādivākyas*, which give an extremely simple way to calculate the true longitude of the Sun for any given day. The text then gives the methods of combining the *manda* and *śīghra* corrections to arrive at the true longitudes of different planets. Here, at least according to one of the Malayalam commentaries, the text seems to follow the revised planetary model of Nīlakaṇṭha according to which the *manda* correction or equation of centre should be applied not to the mean Sun, but to (what is traditionally referred to as) the *śīghrocca* of the planets. The chapter ends with a discussion on planetary distances, as well as heliacal rising and setting of planets.

Chapter 8, dealing with the shadow problems, is indeed a fairly long chapter as in other texts of Indian astronomy. Here, as well as in the next two chapters, Putumana Somayājī displays his high felicity in dealing with problems of spherical astronomy. The chapter begins with the determination of the latitude from the midday-shadow and the corrections to be applied to take into account the parallax and the finite size of the solar disc. These latter corrections are found in *Tantrasaṅgraha* also, but they are formulated somewhat differently here. After discussing the standard relations between the declination, altitude and the azimuth, longitude etc, the text presents several ways of calculating the *prāṇakalāntara* (the difference between the right ascension and the longitude of the Sun), and the *carajyā* (ascensional difference). The text then goes on to present a very interesting method for the determination of the declination of the Moon, which is actually more accurate than the method outlined in *Tantrasaṅgraha*. Finally the text deals with *lambana* (parallax) and gives the method for computing some of the *lambanahārakas* and *lambanajyās*, which are used in the computation of eclipses. These have been tabulated as *vākyas*, for the latitude around $10°50'$, which is close to Shoranur or Alattur.

Chapter 9 commences with a discussion of the longitudes and latitudes of junction stars and the method for calculating the declination of a celestial body with non-zero latitude. The text then considers the problem of finding the *lagna* (the ascendant or the rising point of the ecliptic). Here again we notice that Putumana Somayājī introduces concepts and methods that are not found in other texts such as *Tantrasaṅgraha*. The notions of *kālalagna* (the time interval between the rise of the vernal equinox and that of the given body on the ecliptic) and *madhyāhna-kālalagna* (the *kālalagna* when the body is on the meridian) are employed in *Karaṇapaddhati* instead of *madhyakāla* (right ascension of the point of the equator on the meridian) and *madhyāhnalagna* (longitude of the meridian ecliptic point) which have been used in *Tantrasaṅgraha*. Both texts, however, present exact results based on a careful analysis of the properties of spherical triangles.

Chapter 10 begins by highlighting the importance of obtaining accurate values of the longitudes and latitudes of celestial objects by a careful examination of their shadows etc., by means of instruments. This is a brief chapter which deals essentially with the important issue of relating one coordinate system to another. For instance, it explains how *natakāla* (right ascension) may be obtained from the altitude and azimuth, and how the celestial latitude and longitude may be determined from the right ascension and the hour angle, etc.

This brief outline of the contents of *Karaṇapaddhati* should be sufficient to show that the following declaration made by the author at the end of the work is entirely justified (verse 10.11):

गणितमिदमशेषं युक्तियुक्तं पठन्तः भुवि गणितजनानाम् अग्रगण्या भवेयुः ।

gaṇitamidamaśeṣaṃ yuktiyuktaṃ paṭhantaḥ
bhuvi gaṇitajanānām agragaṇyā bhaveyuḥ |

Those who study and comprehend (*paṭhantaḥ*) all the mathematical principles supported by rationales (*yuktiyuktaṃ*) enunciated here, would become the foremost leaders in the community of mathematicians in this world.

The metres employed in the text

Not only has Putumana Somayājī demonstrated his originality in devising ingenious algorithms, he has also demonstrated his exceptional skills as a poet by composing verses in a variety of metres. The 214 verses comprising the work have been composed in 20 different metres. The names of the metres along with the number of verses composed in them are listed in Table 1. It may be noted that Somayājī has employed both *varṇavṛttas* and *mātrāvṛttas*.

Some of the long metres that appear at the end of the table are not quite easy to handle particularly when dealing with topics such as mathematics and astronomy. In the classical *kāvya* literature it may be common to see such metres, but it is rare to find them in scientific literature where the author does not have as large a basket of words to choose from in order to convey a particular thought or idea.

Notwithstanding this constraint, it is remarkable that Putumana Somayājī has handled such metres with great felicity in *Karaṇapaddhati*. In this regard, he can be compared with Bhāskaracārya-II—the author of the famous works *Līlāvatī, Bījagaṇita* and *Siddhāntaśiromaṇi*—who is highly acclaimed for his poetical skills including the use of a variety of metres. It may be recalled that even renowned mathematicians and astronomers like Āryabhaṭa, Brahmagupta, Mādhava, or Nīlakaṇṭha generally stick to one particular metre—*āryā, anuṣṭubh*, etc, that does not impose serious constraints, and gives much leeway in employing words.

No.	Name of *vṛtta* in		No. of syllables	Type of *vṛtta* in		No. of verses
	Devanagari	Transliteration		Devanagari	Transliteration	
1	आर्या	*āryā*	-	मात्रावृत्तम्	*mātrāvṛttam*	13
2	अनुष्टुभ्	*anuṣṭubh*	8	वर्णवृत्तम्	*varṇavṛttam*	76
3	प्रमाणिका	*pramāṇikā*	8	"	"	1
4	मुरली	*muralī*	10-11-10-11	अर्धसमवृत्तम्	*ardhasamavṛttam*	1
5	इन्द्रवज्रा	*indravajrā*	11	वर्णवृत्तम्	*varṇavṛttam*	23
6	रथोद्धता	*rathoddhatā*	11	"	"	3
7	शालिनी	*śālinī*	11	"	"	2
8	स्वागता	*svāgatā*	11	"	"	1
9	उपजातिः	*upajātiḥ*	11	"	"	23
10	उपेन्द्रवज्रा	*upendravajrā*	11	"	"	3
11	मालभारिणी	*mālabhāriṇī*	11-12-11-12	अर्धसमवृत्तम्	*ardhasamavṛttam*	3
12	भुजङ्गप्रयातम्	*bhujaṅgaprayātam*	12	वर्णवृत्तम्	*varṇavṛttam*	2
13	द्रुतविलम्बितम्	*drutavilambitam*	12	"		1
14	वंशस्था	*vaṃśasthā*	12	"	"	1
15	वसन्ततिलका	*vasantatilakā*	14	"	"	23
16	मालिनी	*mālinī*	15	"	"	2
17	मन्दाक्रान्ता	*mandākrāntā*	17	"	"	2
18	शिखरिणी	*śikhariṇī*	17	"	"	1
19	शार्दूलविक्रीडितम्	*śārdūlavikrīḍitam*	19	"	"	21
20	स्रग्धरा	*sragdharā*	21	"	"	12

Table 1 The different metres employed in *Karaṇapaddhati*.

The choice of different metres, at least from the view point of readers, has the following advantages:

- In olden days when the mode of learning was primarily oral, it would have largely facilitated student in recalling the verses dealing with any topic in terms of the metres in which they are composed.
- For describing processes involving several steps, choice of long metres such as *sragdharā* greatly facilitates in remembering the series of operations that needs to be sequentially carried out just by memorizing a single verse.
- It is a general convention that the poet changes the metres of the last verse in a chapter, to alert the reader regarding the change of topic of discussion.
- Also, the change of the rhythm in rendering the verses helps, to a large extent, in reducing the boredom (an inevitable hazard) while reading or memorising the text.

Acknowledgements

Initial studies on the *Karaṇapaddhati* were made by Venketeswara Pai when he started his research career by working in some of projects—that were undertaken by Prof. K. Ramasubramanian—sponsored by the Industrial Research and Consultancy Centre (IRCC), IIT Bombay, and the National Academy of Sciences India (NASI), Allahabad. A critical study of certain chapters of *Karaṇapaddhati* was the topic of his doctoral dissertation, supervised by Prof. K. Ramasubramanian at IIT Bombay, for the award of Ph.D. degree in the year 2011. Both Pai and Ramasubramanian would like to express their sincere gratitude to IRCC and NASI for graciously supporting such studies by way of sanctioning sponsored projects.

Dr. Pai is grateful to IISER Pune for the extraordinary academic freedom which made it possible for him to meet the collaborators frequently to finalize the book without any delay. He would also like to thank Prof. K. N. Ganesh and Prof. L. S. Shashidhara for their support.

During his postdoctoral research, when Dr. Pai was working with Prof. M. S. Sriram on *Karaṇapaddhati*, he received support from an INSA project undertaken by Prof. K. Ramasubramanian. For extending this support, both of them would like to place on record their thanks to the History of Science Division, INSA. Dr. Pai would also like to gratefully acknowledge the support and encouragement he received from the authorities of the SASTRA University, Thanjavur and IISER, Pune during the course of this work.

In the process of finalizing the text for publication, the manuscript went through several rounds of revisions which had to be carefully implemented. The authors would like to express their gratitude to Sri G. Periyasamy and Dr. Dinesh Mohan Joshi for enthusiastically assisting them in this regard.

They would also like to thank the anonymous referee for carefully going through the manuscript and making several valuable comments and suggestions. They would also like to sincerely acknowledge the generous support from the MHRD received through the project on "Science and Heritage Initiative" (SandHI) at IIT Bombay. Finally, the authors would like to thank the Hindustan Book Agency, New Delhi for graciously coming forward to publish this volume as a part of their series on Culture and History of Mathematics.

हेमलम्बि-ज्येष्ठकृष्णद्वादशी
कल्यब्द ५११९

Wednesday
June 21, 2017

Venketeswara Pai
IISER Pune

K Ramasubramanian
IIT Bombay

M S Sriram
K V Sarma Research Foundation, Chennai

M D Srinivas
Centre for Policy Studies, Chennai

Chapter 1
मध्यग्रहः शकाब्दसंस्कारश्च
Mean planets and the *śakābdasaṃskāra*

१.१ मङ्गलाचरणम्

1.1 Invocation

मदीयहृदयाकाशे चिदानन्दमयो गुरुः।
उदेतु सततं सम्यगज्ञानतिमिरारुणः॥ १ ॥

मार्ताण्डादीन् ग्रहान् नत्वा तत्प्रसादात् विलिख्यते।
गुणहारगुणादीनां करणे कापि पद्धतिः॥ २ ॥

madīyahṛdayākāśe cidānandamayo guruḥ |
udetu satataṃ samyagajñānatimirāruṇaḥ || 1 ||

mārtāṇḍādīn grahān natvā tatprasādāt vilikhyate |
guṇahāraguṇādīnāṃ karaṇe kāpi paddhatiḥ || 2 ||

May the *guru*, an embodiment of intelligence and bliss, keep ever rising in the space of my heart, like the Sun dispelling the darkness of ignorance.

Having paid my obeisances to all the planets beginning with the Sun, a novel (*kāpi*) procedure for the construction of multipliers (*guṇa*), divisors (*hāra*), Rsines (*guṇa*) etc. is being written down [by me] by their grace.

Commencing the work with a *maṅgalācaraṇa* (performance of an auspicious act) is a common feature that can be seen in almost all the compositions pertaining to different disciplines in the Indian tradition. *Maṅgalācaraṇa* generally falls under one of the following types:

(i) Offering veneration to the teacher (*guruvandanam*).
(ii) Offering prayers to the desired deity (*iṣṭadevatānamaskāraḥ*).
(iii) Stating the purpose of the text (*vastunirdeśaḥ*). [1]

[1] Generally *maṅgalācaraṇa* will be of the form (i) or (ii) or a combination of both. However, this third type of a *maṅgalācaraṇa* though rare is found in Kālidāsa's *Kumārasambhava*, which begins with the verse *astyuttarasyāṃ diśi devatātmā...*

V. Pai et al., *Karaṇapaddhati of Putumana Somayājī*, Sources and Studies in the History of Mathematics and Physical Sciences, https://doi.org/10.1007/978-981-10-6814-0_1

In the above verses (composed in *anuṣṭubh* metre), it is interesting to note that all the three forms of *maṅgalācarana* have been incorporated by the author. He first extolls his *guru* as an embodiment of intelligence and bliss, then pays his obeisances to all the *grahas*, and finally also briefly mentions the purpose of the text.

In the next few verses the author presents all the parameters that form the basis for the computation of longitudes of the planets.

१.२ महायुगे ग्रहपर्ययाः

1.2 Revolutions of the planets in a *mahāyuga*

नानाज्ञानप्रगल्भस्तिलबलमसुसूक्ष्मं धयेद्राजदम्भो
भद्रोदन्तोधरेन्द्रो निरनुसृगधिसौख्यं वरिष्ठोऽभिषङ्गः।
दोर्दण्डाग्रेऽद्रिनाथो विषमितविपिनं चन्द्ररेखाम्बुखिन्ने-
त्यर्कादेः पर्ययाः स्युः क्षितिदिनमनृशंसःकळार्थीसमर्त्यः॥ ३ ॥

nānājñānapragalbhastilabalamasusūkṣmaṃ dhayedrājadambho
bhadrodantodharendro niranusṛgadhisaukhyaṃ variṣṭho'bhiṣaṅgaḥ |
dordaṇḍāgre'drinātho viṣamitavipinaṃ candrarekhāmbukhinne-
tyarkādeḥ paryayāḥ syuḥ kṣitidinamanṛśaṃsaḥ kaḷārthīsamartyaḥ || 3 ||

The number of revolutions of [the planets] Sun etc. are: 4320000 (*nānājñānapragalbhaḥ*), 57753336 (*tilabalamasusūkṣmaṃ*), 488219 (*dhayedrājadambha*), 2296824 (*bhadrodantodharendra*), 17937020 (*niranusṛgadhisaukhyaṃ*), 364224 (*variṣṭho'bhiṣaṅga*), 7022388 (*dordaṇḍāgre'drinātha*), 146564 (*viṣamitavipina*), 232226 (*candrarekhāmbukhinna*). And the number of civil days (*kṣitidina*) [in a *mahāyuga*] is 1577917500 (*anṛśaṃsaḥkaḷārthīsamartya*).

The above verse is composed in *sragdharā* metre, one of the longest metres with 21 syllables per quarter. The choice of this metre is quite understandable as it facilitates the author to concisely present as much information as possible. In fact, the author has specified the number of revolutions made by *all* the planets in a *mahāyuga*, as well as the total number of civil days, in one single verse. For specifying these numbers, Putumana Somayājī has adopted the *kaṭapayādi* system[2] of numeration—the most preferred choice of the Kerala astronomers starting at least from the time of Haridatta (7th cent.). Table 1.1 presents these numbers along with their Sanskrit equivalents. It may be mentioned here that the revolution numbers as well as the number of civil days in a *mahāyuga* specified in Table 1.1 are the same as in *Āryabhaṭīya*.

An interesting feature of the works of Kerala astronomers is that while specifying the numbers using *kaṭapayādi* system, they try to make the phrases

[2] The following table presents the numbers associated with the consonants *ka*, *kha* etc. in this system:

Planet	Revolutions in a *mahāyuga*	
	in *Kaṭapayādi*	in numerals
Sun	नानाज्ञानप्रगल्भः (Illustrious with various kinds of knowledge)	4320000
Moon	तिलबलमसुसूक्ष्मम् (Power of sesamum [seed] is not subtle)	57753336
Moon's apogee	धयेद्राजदम्भः (Hypocrisy of king)*	488219
Mars	भद्रोदन्तोधरेन्द्रः (King about whom the narratives are good)	2296824
Mercury**	निरनुसृगधिसौख्यम् (Comfort without repeated inarticulate sound)	17937020
Jupiter	वरिष्ठोऽभिषङ्गः (Humiliated senior)	364224
Venus**	दोर्दण्डाग्रेऽद्रिनाथः (Lord of mountains [who is] upholding the pole by [his] arm)	7022388
Saturn	विषमितविपिनम् (Impassable forest)	146564
Moon's node	चन्द्ररेखाम्बुखिन्ना (Ray of the Moon distressed by water)	232226
Number of civil days	अनृशंसः कळार्थीसमर्त्यः (Along with a man is not cruel)*	1577917500

*Literal meaning of the words "*dhayed*" and "*kaḷārthī*" is not clear.
**The revolution numbers are those of the associated *śīghroccas*.

Table 1.1 Revolutions made by the planets in a *mahāyuga*.

Number	1	2	3	4	5	6	7	8	9	0
Consonants	*k*	*kh*	*g*	*gh*	*ṅ*	*c*	*ch*	*j*	*jh*	*ñ*
used	*ṭ*	*ṭh*	*ḍ*	*ḍh*	*ṇ*	*t*	*th*	*d*	*dh*	*n*
to represent	*p*	*ph*	*b*	*bh*	*m*	–	–	–	–	–
numbers	*y*	*r*	*l*	*v*	*ś*	*ṣ*	*s*	*h*	*ḷ*	–

In the case of conjunct consonants, only the last consonant is to be considered. Vowels which are separate, represent the number zero.

Let us consider the phrase *nānājñānapragalbhaḥ* as an example. The table below gives the encoded numbers corresponding to each syllable.

nā	*nā*	*jñā*	*na*	*pra*	*ga*	*lbhaḥ*
0	0	0	0	2	3	4

By reversing the order, we have the number 4320000 which is encoded in the phrase *nānājñānapragalbhaḥ*.

meaningful also. It is ensured that the words that get into these phrases are familiar and commonly employed words, thereby facilitating the reader to pronounce and remember them easily. The literal meanings of the phrases used in the verse above are given in Table 1.1 in parenthesis.

१.३ महायुगे सौरचान्द्रमासाः अधिमासाश्च

1.3 The number of solar, lunar and intercalary months in a *mahāyuga*

रूपाहतार्कभगणाः खलु सौरमासाः
मासा रवीन्दुभगणान्तरमेव चान्द्राः ।
चन्द्रार्कमासविवरं च युगाधिमासाः
मासाः पुनर्नगहता दिवसस्वरूपाः ॥ ४ ॥

rūpāhatārkabhagaṇāḥ khalu sauramāsāḥ
māsā ravīndubhagaṇāntarameva cāndrāḥ |
candrārkamāsavivaraṃ ca yugādhimāsāḥ
māsāḥ punarnagahatā divasasvarūpāḥ || 4 ||

The number of revolutions of the Sun multiplied by 12 (*rūpa*) are indeed the solar months. The difference between the number of revolutions of the Sun and that of the Moon is itself the number of lunar [months]. The difference between the number of lunar months and the solar months gives the number of *adhimāsas* (intercalary months) in a *yuga*. These months multiplied by 30 (*naga*) give the number of [solar and lunar] days.

If M_s represents the number of solar months in a *mahāyuga*, and R_s the number of revolutions of the Sun, then

$$\begin{aligned} M_s &= R_s \times 12 \\ &= 4320000 \times 12 = 51840000. \end{aligned} \tag{1.1}$$

A lunar month, by definition, is the time interval between two successive conjunctions of the Sun and the Moon (two *amāvāsyās* or new Moons). Hence the total number of lunar months (M_m) in a *mahāyuga* will be equal to the total number of new Moons which in turn will be equal to the difference between the number of lunar and the solar revolutions. That is,

$$M_m = 57753336 - 4320000 = 53433336. \tag{1.2}$$

The number of lunar months that are in excess of the number of solar months in a *mahāyuga* are referred to as the intercalary months or *adhimāsas* (M_a), and are given by

$$\begin{aligned} M_a &= M_m - M_s \\ &= 53433336 - 51840000 = 1593336. \end{aligned} \tag{1.3}$$

In the last quarter of the above verse, it has been stated that the number of months multiplied by 30 gives the number of days. As this is a generic prescription, the number of solar months in a *mahāyuga* multiplied by 30 gives the number of solar days in a *mahāyuga*, the number of lunar months multiplied by 30 gives the number of lunar days and so on. If D_s, D_m and D_a be the number of solar days, lunar days, and their difference in a *mahāyuga*, respectively, then they are given by

$$\begin{aligned} D_s &= 30 \times 51840000 \\ &= 1555200000, \end{aligned} \tag{1.4}$$
$$\begin{aligned} D_m &= 30 \times 53433336 \\ &= 1603000080, \end{aligned} \tag{1.5}$$
$$\begin{aligned} D_m - D_s = D_a &= 30 \times 1593336 \\ &= 47800080. \end{aligned} \tag{1.6}$$

Here it may be mentioned that a solar day is purely a theoretical entity and is different from the civil day (24 hrs) that is commonly used in practice. The solar day corresponds to the time taken by the Sun to cover 1° of the ecliptic, which on an average would be slightly larger than a civil day. An average lunar day/*tithi* is shorter than a civil day.

१.४ क्षयतिथिनाक्षत्रदिनानां संख्या

1.4 The number of omitted *tithis* and sidereal days

चान्द्रमासा नगाभ्यस्ता भूदिनोनास्तिथिक्षयाः ।
भूदिनाढ्यार्कभगणाः नाक्षत्रदिवसाः स्मृताः ॥ ५ ॥

cāndramāsā nagābhyastā bhūdinonāstithikṣayāḥ |
bhūdināḍhyārkabhagaṇāḥ nākṣatradivasāḥ smṛtāḥ || 5 ||

The number of lunar months multiplied by 30 [and] diminished by the number of civil days gives the *tithikṣayas* (the number of omitted *tithis*). The number of solar revolutions added to the number of civil days (*bhūdina*) is stated to be the number of *nākṣatradivasas* (sidereal days).

This verse essentially presents the procedure for obtaining the total number of omitted *tithis* and sidereal days in a *mahāyuga* from the number of civil days. We know that the number of *tithis* (lunar days) in a *mahāyuga* is obtained by multiplying the number of lunar months by 30. And, the number of *kṣayatithis* (omitted *tithis*) is the difference between the number of *tithis* and the number of civil days in a *mahāyuga*.

Let M_m and D_c represent the total number of lunar months and civil days in a *mahāyuga*. Then the number of *kṣayatithis* (T_k) is given by

$$\begin{aligned} T_k &= M_m \times 30 - D_c \\ &= 53433336 \times 30 - 1577917500 = 25082580. \end{aligned} \tag{1.7}$$

It may be recalled that the total number of civil days in a *mahāyuga* represents the number of sunrises in it. Similarly, the number of sidereal days (*nākṣatradinas*) represents the number of star-rises in a *mahāyuga*. Since the stars do not have any eastward motion of their own, whereas the Sun completes one full revolution in a sidereal year, the number of sidereal days in a solar year will be greater than the number of civil days by exactly one unit. Hence, the total number of sidereal days or *nākṣatradivasas* (D_n) in a *mahāyuga* will exceed the total number of civil days exactly by the number of solar years, which is the same as the number of revolutions of the Sun (R_s) in the background of stars. That is,

$$\begin{aligned} D_n &= D_c + R_s \\ &= 1577917500 + 4320000 = 1582237500. \end{aligned} \tag{1.8}$$

Having specified the revolution numbers corresponding to a *mahāyuga*, the text proceeds to state the values corresponding to a *kalpa* whose period is stated to be 1008 times that of a *mahāyuga*.

१.५ कल्पे ग्रहपर्ययाः

1.5 Revolutions of the planets in a *kalpa*

एवं युगोक्ता भगणादयस्ते दिनानयघ्नास्तु भवन्ति कल्पे ।
चतुर्दश स्युर्मनवोऽत्र तेषां युगानि रासप्रमितानि यस्मात् ॥ ६ ॥

evaṃ yugoktā bhagaṇādayaste dinānayaghnāstu bhavanti kalpe |
caturdaśasyurmanavo'tra teṣāṃ yugāni rāsapramitāni yasmāt || 6 ||

The revolutions (thus stated) for a *caturyuga*, when multiplied by 1008 (*dinānaya*) would become the revolutions in a *kalpa*. In this [*kalpa*] there would be 14 *Manus* since their period is 72 [*catur*]*yugas*.

In the verse above, the word *yuga* has been employed to refer to a *caturyuga* or a *mahāyuga*, whose period is known to be 4320000 years. Also, the period of a *kalpa* is defined in terms of a *mahāyuga*.

Planet	Revolutions in a *kalpa*
Sun	4354560000
Moon	58215362688
Moon's apogee	492124752
Mars	2315198592
Mercury	18080516160
Jupiter	367137792
Venus	7078567104
Saturn	147736512
Moon's node	234083808

Table 1.2 The number of revolutions of the planets in a *kalpa*.

It has been stated that there are 14 *manvantaras* in a *kalpa*.[3] As there are 72 *caturyugas* in a *manvantara*, there are 1008 (72×14) *mahāyugas* in a *kalpa*.

The above verse also mentions that the number of revolutions of the planets in a *kalpa* is obtained by multiplying the revolutions made by them in a *mahāyuga* by 1008. That is,

$$R\ (kalpa) = 1008 \times R\ (yuga).$$

The revolution numbers for a *kalpa* are listed in Table 1.2.

१.६ वर्तमानकल्पे अतीतकालः

1.6 Period elapsed in the present *kalpa*

कृतत्रेताद्वापराख्यः कलिश्चैते युगाङ्घ्रयः ।
युगाङ्घ्रयस्तु कल्पेऽस्मिन् धिगादित्यमिता गताः ॥ ७ ॥

kṛtatretādvāparākhyaḥ kaliścaite yugāṅghrayaḥ |
yugāṅghrayastu kalpe'smin dhigādityamitā gatāḥ || 7 ||

The [four] quarters (*yugāṅghris*) of the [*mahā*]*yuga* are known as *kṛta*, *tretā*, *dvāpara* and *kali*. In the present *kalpa* the number of elapsed *yugāṅghris* (quarter of a *yuga*) are 1839.

The *mahāyuga* that was referred to in the previous verse is conceived to be made up of four parts. The first one is called the *kṛtayuga*, the second one the *tretā*, the third one the *dvāpara* and the fourth one the *kali*. Here it is stated that, in the present *kalpa*, the number of such quarters of *yuga*[4] elapsed is equal to 1839. It may be noted that,

[3] A *Manu* lives for one *manvantara*.

[4] Here, *yuga* refers to a *caturyuga* or *mahāyuga*.

$$\frac{1839}{4} = 459 + \frac{3}{4}.$$

This means that apart from 459 *caturyugas*, three-fourths of a *caturyuga* is over. If we further divide 459 by 72, we obtain 6 as the quotient and 27 as the remainder. Thus, as per this prescription, in the present *kalpa*, six *manvantaras* have already elapsed. And in the seventh *manvantara* 27 *mahāyugas* have elapsed.

We are currently in the twenty-eighth *mahāyuga*. In this *mahāyuga* too, three quarters have elapsed and we are in the fourth one, namely the *kali*. Therefore, the total number of quarters elapsed since the beginning of the *kalpa* is

$$(72 \times 6 + 27) \times 4 + 3 = 1839.$$

१.७ वर्तमानकलियुगे गताब्दगणानयनम्

1.7 Obtaining the number of elapsed years since the beginning of the present *kaliyuga*

गतवर्षान्तकोलम्बवर्षास्तरळगान्विताः ।
कल्यब्दा धीस्थकालाढ्यशकाब्दा वा भवन्ति ते ॥ ८ ॥

gatavarṣāntakolambavarṣāstaraḷagānvitāḥ |
kalyabdā dhīsthakālāḍhyaśakābdā vā bhavanti te || 8 ||

Adding 3926 (*taraḷaga*) to the number of elapsed *kollam* years gives the number of *kali* years elapsed. These [*kali* years] can also be obtained by adding 3179 (*dhīsthakāla*) to the elapsed *śaka* years (*śakābdas*).

The above verse presents the relations among the three popular eras that are currently in vogue in various parts of India, namely the *kalyabda*, the *śakābda* and the *kollam*. Of them, the first two are used throughout India, whereas the last one is employed only in Kerala.

If y_k, y_s and y_{ko} represent the *kalyabda*, *śakābda* and *kollam* years respectively, then the content of the above verse may be expressed as

$$\begin{aligned} y_k &= y_{ko} + 3926, \\ \text{and,} \quad y_k &= y_s + 3179. \end{aligned} \tag{1.9}$$

Also,

$$y_s = y_k - 3179 = y_{ko} + 747. \tag{1.10}$$

१.८ अहर्गणानयनम्

1.8 Procedure for obtaining the *ahargaṇa*

कल्यब्दतः प्रियहताद् गतमासयुक्ताः
चान्द्राख्यमासगुणिताद्रविमासलब्धः ।
नागाहतस्तिथियुतः क्षितिवासरघ्नः
चान्द्रैर्दिनैरपहृतो द्युगणोऽच्छवारात् ॥ ९ ॥

kalyabdataḥ priyahatād gatamāsayuktāḥ
cāndrākhyamāsaguṇitādravimāsalabdhaḥ |
nāgāhatastithiyutaḥ kṣitivāsaraghnaḥ
cāndrairdinairapahṛto dyugaṇo'cchavārāt || 9 ||

The number of years elapsed since the beginning of *kaliyuga*, multiplied by twelve (*priya*) and added to the months elapsed [in the present year], is multiplied by the number of lunar months [in a *yuga*]. The quotient obtained, while dividing this by the solar months [in a *yuga*], is to be multiplied by thirty (*nāga*) and to that the (number of) *tithis* elapsed [in the present month] is added. [The result], multiplied by the civil days in a *yuga* and divided by the total number of lunar days (*cāndradina*) in a *yuga*, will be the *ahargaṇa*, commencing with Friday.

The term *ahargaṇa* refers to the number of civil days that have elapsed since a given epoch. To obtain this, we need to first find the number of solar months that have elapsed since the epoch, which is taken to be the beginning of *kaliyuga*. The number of (solar) months (m_s) elapsed since the beginning of *kaliyuga* is given by

$$m_s = 12 \times y_k + m,$$

where y_k represents the *kalyabda* as defined in the previous section, and m is the number of months elapsed in the present year. The number of lunar months that have elapsed since the beginning of *kaliyuga* till the desired date is obtained by employing the following rule of three:

$$\begin{aligned} &\text{solar months in a } mah\bar{a}yuga\ (M_s) : \text{lunar months in a } mah\bar{a}yuga\ (M_l) \\ &\quad :: \ \text{elapsed solar months } (m_s) : \text{elapsed lunar months } (m_l)\ ? \end{aligned} \tag{1.11}$$

Now,

$$m_l = \frac{m_s \times M_l}{M_s} = \frac{m_s \times 53433336}{51840000} \tag{1.12}$$

$$= \frac{m_s \times 2226389}{2160000}. \tag{1.13}$$

To find the number of *tithis* that have elapsed since the beginning of *kaliyuga* till the desired date, we need to multiply m_l by 30, and add to that the number of *tithis* that have elapsed in the current lunar month. If t be the number of *tithis* that have elapsed in the present month, then the total number of *tithis*

t_e that have elapsed since the beginning of *kaliyuga* till the desired date, is given by

$$t_e = (m_l \times 30) + t. \tag{1.14}$$

Now from these *tithis*, the *ahargaṇa* A, which represents the total number of civil days elapsed from the epoch till date, is found by employing the rule of three. If the total number of civil days and *tithis* in a *mahāyuga* be D_c and T respectively, then

$$T : D_c \quad :: \quad t_e : A\,? \tag{1.15}$$

Therefore,

$$\begin{aligned} A &= \frac{t_e \times D_c}{T} \\ &= \frac{t_e \times 1577917500}{1603000080} \qquad (1.16) \\ &= \frac{t_e \times 26298625}{26716668}. \qquad (1.17) \end{aligned}$$

In most texts, the beginning of the *kaliyuga* is taken to be the mean sunrise of February 18, 3102 BCE, which happens to be a Friday. Thus, when A is divided by 7, if the remainder is 0, 1, 2, ..., 6, it means that the day for which the *ahargaṇa* A has been computed should be Friday, Saturday, ..., Thursday. If the actual weekday differs from the computed one, then the *ahargaṇa* is corrected generally by adding ± 1 to A. Such differences between the computed and the actual values are not uncommon, as the calculational procedure involves the use of rule of three and rounding off to the nearest integer. This rounding off occurs both in the calculation of m_l and A.

It may also be mentioned here, that the procedure for finding the *ahargaṇa* in *Karaṇapaddhati* is slightly different from the one presented in *Tantrasaṅgraha* and many other texts, wherein the number of *adhimāsas* and *kṣayatithis* are separately computed and added. On the other hand, here the number of lunar months and the number of *tithis* elapsed are directly computed. Having obtained them, by multiplying the number of *tithis* elapsed with the *yugasāvanadina* and dividing it by the number of *tithis* in a *yuga*, we obtain the *ahargaṇa*. We shall now illustrate the above procedure with a couple of examples.

Illustrative examples

Example 1: Find the *kalyahargaṇa* corresponding to *nija-āṣāḍha-kṛṣṇa-navamī*,[5] *Śaka* 1891 (August 6, 1969 CE).

[5] Ninth day of the dark fortnight correspondding to the actual (*nija*) lunar month *Āṣāḍha*. Actual because of the occurrence of an intercalary month preceding this.

Number of *kali* years elapsed, y_k $= 1891 + 3179$
$= 5070.$

Number of solar months elapsed in the present year, m $= 3.$

Total number of solar months elapsed, m_s $= (5070 \times 12) + 3$
$= 60843.$

Number of lunar months elapsed (including *adhimāsas*), m_l $= \dfrac{60843 \times 53433336}{51840000}$
$= 62713.04904.$

Since we are interested in the integral part, we take m_l $= 62713.$

Number of *tithis* elapsed in the present lunar month t $= 15 + 8 = 23.$

Total no. of *tithis* elapsed since the beginning of *kali* (including *kṣayatithis*), t_e $= (62713 \times 30) + 23$
$= 1881413.$

Number of civil days since the beginning of *kali* (corresponding to t_e) $= \dfrac{1881413 \times 1577917500}{1602999600}$
$= 1851974.57156.$

We round off the above fraction and take the *kalyahargaṇa* A, to be $= 1851975$
$= (264567 \times 7) + 6.$

The remainder 6 implies that the day has to be a Thursday. But August 6, 1969 happens to be a Wednesday. Hence the computed value of the *ahargaṇa* is incorrect by a day. This error is evidently due to the error in rounding off. This kind of error is very likely to occur particularly when the fractional value obtained in finding A from t_e is close to 0.5. By leaving out the fractional part, we get the actual *ahargaṇa* as 1851974. Thus the number of civil days elapsed since the beginning of *kaliyuga* till *nija-āṣāḍha-kṛṣṇa-navamī, Śaka* 1891 is 1851974.

It may also be noted that the use of rule of three for finding the elapsed lunar months from the solar months implicitly assumes that the *adhimāsas* occur uniformly. However, this is not true, and the variation could be quite significant depending upon the occurrence or absence of true *saṅkrānti* in a lunar month. We need to be especially careful when the value of m_l is close to an integer. If there is an error in the choice of m_l, the *ahargaṇa* would differ from the actual value by nearly 30 days. On the other hand, as seen earlier, we may also go wrong in the *ahargaṇa* by one day.

These errors can be easily fixed from the knowledge of the occurrence or otherwise of an *adhimāsa* near the desired date, and the day of the week, respectively. We shall now illustrate an example where the value m_l is to be chosen carefully.

Example 2: Find the *kalyahargaṇa* corresponding to *phālguna-kṛṣṇa-trayodaśī*,[6] *Śaka* 1922 (March 22, 2001 CE).

Number of *kali* years elapsed, y_k $= 1922 + 3179 = 5101.$

Number of solar months elapsed in the present year, m $= 11.$

Total number of solar months elapsed, m_s $= (5101 \times 12) + 11 = 61223.$

Number of lunar months elapsed (including *adhimāsas*), m_l

$$= \frac{61223 \times 53433336}{51840000} = 63104.72859.$$

Since we are interested in the integral part, we take m_l $= 63104.$

Number of *tithis* elapsed in the present lunar month t $= 15 + 12 = 27.$

Total number of *tithis* elapsed since the beginning of *kali* (including *kṣayatithis*), t_e $= (63104 \times 30) + 23 = 1893147.$

Number of civil days since the beginning of *kali* (corresponding to t_e)

$$= \frac{1893147 \times 1577917500}{1602999600} = 1863524.96992.$$

We round off the above fraction take the *kalyahargaṇa* A, to be $= 1863525 = (266217 \times 7) + 6.$

The remainder 6 implies that the day has to be a Thursday. March 22, 2001 happens to be a Thursday, and hence the computed value of the *ahargaṇa* is correct. Thus the number of civil days elapsed since the beginning of the *kaliyuga* till *phālguna-kṛṣṇa-trayodaśī, Śaka* 1922 (March 22, 2001 CE) is 1863525.

[6] Thirteenth day of the dark fortnight correspondding to the lunar month *Phālguna*.

Note: In this example, rounding off the value of $m_l = 63104.72859$ to the nearest integer (that is, by considering $m_l = 63105$) would have led to the value of $A = 1863555$. This value differs from the actual value by 30 days.

१.९ गणनलाघवोपायः

1.9 Technique for simplifying the mathematical operations

गुणहारान्तरगुणितं गुण्यं हाराहृतं तु वा गुण्ये ।
गुणकाधिकाल्पकत्वे स्वमृणं कुर्यात् फलस्य संसिद्ध्यै ॥ १० ॥

guṇahārāntaraguṇitaṃ guṇyaṃ hārāhṛtaṃ tu vā guṇye |
guṇakādhikālpakatve svamṛṇaṃ kuryāt phalasya saṃsiddhyai || 10 ||

The difference between the multiplier and the divisor (*guṇa-hārāntara*) multiplied by the multiplicand (*guṇya*) and divided by the divisor, has to be added to or subtracted from the multiplicand, depending on whether the multiplier is greater or smaller [than the *hāra* (divisor)] respectively.

Consider the product $a\left(\frac{c}{b}\right)$. Here, c is the multiplier (*guṇaka* or simply *guṇa*), a the multiplicand (*guṇya*) and b the divisor (*hāra*). Clearly there are two possible cases: (i) $c > b$ and (ii) $c < b$. It is easily seen that the given product in these two cases can be written respectively as

$$a\frac{c}{b} = a + a\frac{(c-b)}{b}, \tag{1.18}$$

$$\text{and} \qquad a\frac{c}{b} = a - a\frac{(b-c)}{b}. \tag{1.19}$$

It is precisely these two relations that are given in the above verse.

१.१० मध्यग्रहानयनम्

1.10 Obtaining the mean longitudes of planets

अहर्गणात् खेचरपर्ययघ्नाद् धरादिनाप्ता भगणादिखेटाः ।
त्रिभान्वितं तत्र भवेद् विधूच्चं विधुन्तुदश्चक्रदलाद्विशुद्धः ॥ ११ ॥

ahargaṇāt khecaraparyayaghnād
dharādināptā bhagaṇādikheṭāḥ |
tribhānvitaṃ tatra bhaved vidhūccaṃ
vidhuntudaścakradalādviśuddhaḥ || 11 ||

> The *ahargaṇa* multiplied by the revolution number of planets and divided by the *bhūdinas* (civil days) gives the elapsed *bhagaṇas* of the planets. The [longitude of the] Moon's apogee is obtained by adding three *rāśis* [to the calculated result], and that of the *Rāhu* (Moon's node) by subtracting it from six *rāśis*.

If A be the *ahargaṇa*, R the number of revolutions made by the planet in a *mahāyuga* and D_c the total number of civil days. Then the mean longitude θ_0 of the planet, is given by

$$\begin{aligned}\theta_0 &= \frac{A \times R}{D_c} \\ &= \frac{A \times R}{1577917500} = I_1 + f_1,\end{aligned} \qquad (1.20)$$

where I_1 represents the integral part of θ_0, and f_1 the fractional part. The integral part gives the number of revolutions that have been completed by the planet since the beginning of the present *kaliyuga*. It is from the fractional part f_1 that the *rāśis* etc., covered by the planet, are obtained. To find this, the fractional part f_1 is first multiplied by 12.

$$f_1 \times 12 = I_2 + f_2.$$

Here, the integral part I_2 gives the number of *rāśis* that the planet has covered in the present revolution. The fractional part f_2 when multiplied by 30 gives the number of degrees covered in the present *rāśi*. Let

$$f_2 \times 30 = I_3 + f_3,$$

where I_3 gives the number of degrees covered by the planet in the present *rāśi*. f_3 multiplied by 60 gives the number of minutes covered in the present degree. Continuing along similar lines, we may further express the product of f_3 and 60 as

$$f_3 \times 60 = I_4 + f_4.$$

In the above equation, I_4 represents the integral number of minutes covered and f_4 the fractional part of it. Thus the mean longitude of the planet may be expressed as

$$\theta_0 = (I_2)^r + (I_3)^\circ + (I_4)'. \qquad (1.21)$$

After giving the general prescription for obtaining the longitudes from *ahargaṇa* for all the planets, which includes Moon's apogee and nodes, the verse specifically mentions that in the case of *Tuṅga* (Moon's apogee) three signs are to be added, and in the case of *Rāhu*, the obtained value has to be subtracted from six *rāśis*. These specific prescriptions point to the fact that the Moon's apogee at the beginning of *kaliyuga* was at 90° and that of *Rāhu* was at 180°. It may also be recalled here that the motion of *Rāhu* is retrograde. We shall now illustrate the above procedure through an example.

Illustrative example

We calculate the mean longitude of the Moon corresponding to the *ahargaṇa* 1754000 (9th May 1701). We have chosen this *ahargaṇa* because the Malayalam Commentary II presents a series of verses giving the *dhruvas* of all the planets corresponding to this *ahargaṇa*. The verses commence with the specification of *ahargaṇa* as follows:[7]

अनूनुवर्णसाध्योनात् सोमवाराद्यहर्गणः ।
anūnuvarṇasādhyonāt somavārādyahargaṇaḥ |

By subtracting 1754000, the *ahargaṇa* commencing from Monday is obtained.

Here the number encoded in the string "*anūnuvarṇasādhya*" is 1754000. The same *ahargaṇa* has been used for illustrating other computational procedures in the following chapters as well.

From (1.20), the longitude of the Moon corresponding to the *ahargaṇa* $A = 1754000$ is given by

$$\theta_0 \text{ (Moon)} = \frac{1754000 \times 57753336}{1577917500} = 64198.1290809. \tag{1.22}$$

The integral part 64198 represents the number of revolutions completed and does not contribute to the longitude. It is from the fractional part 0.12908089 we get the number of *rāśis* etc., covered by the Moon. We have

$$0.12908089 \times 12 = 1 + 0.54897068.$$

This shows that the Moon has covered one *rāśi* and is in the second one, namely *Vṛṣabha*. To get the degrees etc., we multiply the fractional part by 30. Thus we have,

$$0.54897068 \times 30 = 16 + 0.4691204.$$

This means that the mean Moon has covered 16 degrees in the *Vṛṣabha-rāśi*. The fractional part of the above expression further multiplied by 60 gives 28.147224 minutes. The fractional part of this can be further multiplied by 60 to get the seconds etc. Thus the mean longitude of the Moon corresponding to the *ahargaṇa* 1754000 is

$$\theta_0 \text{ (Moon)} = 1^r 16^\circ 28' 8''. \tag{1.23}$$

The mean longitudes[8] of all the planets for this *ahargaṇa*, obtained in a similar manner, are given in Table 1.5.

---

[7] {KP 1956}, pp. 56-57.

[8] The mean longitudes of the planets presented here correspond to the mean sunrise of an observer situated on the meridian passing through Ujjayinī. In the case of *Rāhu*, the value obtained using the above procedure has been subtracted from six signs, and in the case of Moon's apogee, three signs have been added to obtain the mean longitude.

Planet	Mean longitude (θ_0)					
	sign (*rāśi*)	degrees (°)	minutes (′)	seconds (″)	thirds (‴)	fourths (⁗)
Sun	0	27	24	55	48	22
Moon	1	16	28	8	50	14
Moon's apogee	11	12	4	5	22	26
Mars	1	16	58	48	13	7
Mercury	7	21	20	5	40	10
Jupiter	10	12	36	55	33	11
Venus	0	10	9	48	2	31
Saturn	11	0	57	26	50	16
Moon's node	4	9	25	18	54	23

Table 1.5 Mean longitudes corresponding to $A = 1754000$.

It was mentioned earlier that the revolution numbers for planets presented in Table 1.1 are not different from those given in *Āryabhaṭīya*. However, in the *Parahita* system some corrections have been introduced to get more accurate values of the mean longitudes of the planets. These corrections (*saṃskāras*) have been prescribed to be applied to the mean longitude of the planets starting from the *śaka* year 444,[9] and are called *śakābdasaṃskāras*.

१.११ मध्यग्रहे शकाब्दसंस्कारः

1.11 *Śakābdasaṃskāra* for mean planets

वाग्भावोनाच्छकाब्दात् धनशतलयहात् मन्दवैलक्ष्यरागैः
आप्ताभिर्लिप्तिकाभिः विरहिततनवः चन्द्रतत्तुङ्गपाताः ।
शोभानीरूढसंवित् गणकनरहतात् मागराप्ताः कुजाद्याः
संयुक्ता ज्ञारसौराः सुरगुरुभृगुजौ वर्जितौ भानुवर्जम् ॥ १२ ॥

vāgbhāvonācchakābdāt dhanaśatalayahāt mandavailakṣyarāgaiḥ
āptābhirliptikābhiḥ virahitatanavaḥ candratattuṅgapātāḥ |
śobhānīrūḍhasaṃvit gaṇakanarahatāt māgarāptāḥ kujādyāḥ
saṃyuktā jñārasaurāḥ suragurubhṛgujau varjitau bhānuvarjam || 12 ||

Subtracting 444 (*vāgbhāva*) from the *śakābda* multiply it by 09 (*dhana*), 65 (*śata*) and 13 (*laya*) and divide by 85 (*manda*), 134 (*vailakṣya*) and 32 (*rāga*) [respectively].

[9] Note that the *śaka* year 444 or 522 CE happens to be 23 years after the composition of the *Āryabhaṭīya*. In the *Parahita* system the correction term becomes zero in this year.

The results thus obtained, in minutes, are subtracted from the longitudes of Moon, its apogee and its node respectively. [Similarly, subtracting 444 (*vāgbhāva*) from the *śakābda* and] multiplying by 45 (*śobhā*), 420 (*nīrūḍha*), 47 (*saṃvit*), 153 (*gaṇaka*) and 20 (*nara*) and dividing by 235 (*māgara*), the results obtained are applied to the planets Mars etc. [Here again] the quantities thus obtained have to be added to the mean longitudes of the Mars, Mercury and Saturn and subtracted from that of the Jupiter and Venus, omitting [any correction in the case of] the Sun.

This verse (in *sragdharā* metre) essentially presents the magnitude of the *śakābda-saṃskāra* that is to be applied to the planets including Moon's apogee. This is done by specifying multipliers (*guṇakāras, g*) and divisors (*hārakas, h*) for each planet which will be used in conjunction with the number of years elapsed since the epoch, namely the *śaka* year 444. Sundararāja, the 16th century commentator of *Vākyakaraṇa* quotes this verse and mentions that this *śakābda* correction was introduced in the *Parahita* system (c. 683) of Haridatta. However, the edited version of the *Grahacāranibandhana* of Haridatta does not have this verse or this correction. However, later works such as the *Grahacāranibandhasaṅgraha* (c. 932) give these *guṇas* and *hārakas* of the *śakābda* correction.[10]

The *śakābda* correction, denoted by c_1, prescribed in the verse is given by

$$c_1 = \frac{(y_s - 444) \times g}{h}. \tag{1.24}$$

This has to be applied to the mean longitudes of the planets. It is seen from the above expression, that if $y_s = 444$, c_1 reduces to zero. Thus, (1.24) implies that the mean longitudes obtained by using the parameters given in *Āryabhaṭīya* were accurate around the *śaka* year 444 (= 522 CE). It may be noted that this year incidentally happens to be 23 years after the date of composition of *Āryabhaṭīya*.

The *śakābdaguṇakāras* (g) and *śakābdahāras* (h) that are given in the verse for different planets, are listed in Table 1.6. Also, as has been explicitly mentioned in the verse, the result (c_1) obtained would be in minutes (*āptābhirliptikābhiḥ*) and has to be applied to the mean longitude obtained by (1.21) either positively or negatively depending upon the planet. Thus, the corrected mean longitude at this stage may be represented as

$$\theta_1 = \theta_0 \pm c_1. \tag{1.25}$$

Note: In (1.25), the '+' sign is to be chosen in the case of Mercury, Mars and Saturn, and '−' in the case of Venus, Jupiter as also the Moon, its apogee and the node.

---

10 {GCN 1954}, pp. ix-x.

Planet	*śakābdaguṇakāra* (g) in		*śakābdahāraka* (h) in		$\frac{g}{h}$	Nature of correction (+/−)
	kaṭapayādi	no.	*kaṭapayādi*	no.		
Sun	No *śakābda* correction for Sun					
Moon	*dhana*	09	*manda*	85	0.10588	−
Moon's apogee	*śata*	65	*vailakṣya*	134	0.48507	−
Mars	*śobhā*	45	*māgara*	235	0.19149	+
Mercury	*nīrūḍha*	420	*māgara*	235	1.78723	+
Jupiter	*saṃvit*	47	*māgara*	235	0.20000	−
Venus	*gaṇaka*	153	*māgara*	235	0.65106	−
Saturn	*nara*	20	*māgara*	235	0.08511	+
Moon's node	*laya*	13	*rāga*	32	0.40625	−

Table 1.6 *Śakābdaguṇahāras* of the planets.

Illustrative example

In Table 1.5 we have listed the mean longitudes of all the planets corresponding to $A = 1754000$. Considering the mean longitude of Venus, and expressing it in minutes we have

$$\theta_0 \text{ (Venus)} = 609.80070'. \tag{1.26}$$

The *śakābda-saṃskāras* corresponding to $A = 1754000$ for different planets are given in Table 1.7. In arriving at these values we have taken the number of *śaka* years elapsed y_s to be 1623. It is noted from the table that the *śakābda* correction for Venus is -767.60426. Applying this to the mean longitude given by (1.26), we have

$$\begin{aligned}\theta_1 \text{ (Venus)} &= 609.80070 - 767.604255 \\ &= -157.80355 \qquad \text{(in min)}.\end{aligned} \tag{1.27}$$

The *śakābda* corrected values of mean longitudes of the other planets corresponding to $y_s = 1623$ are listed in Table 1.8.

It may be noted from (1.24) that the expression given for c_1 takes into account only that part of the correction corresponding to the complete number of years that have elapsed. It does not include the correction that is to be considered for the time elapsed in the current year. The next verse describes the procedure for obtaining this part of the correction.

Planet	*śakābdasaṃskāra* ($\pm c_1$) in minutes ($'$)
Sun	No *saṃskāra*
Moon	−124.83529
Moon's apogee	−571.90299
Mars	+225.76596
Mercury	+2107.14894
Jupiter	−235.80000
Venus	−767.60426
Saturn	+100.34043
Moon's node	−478.96875

Table 1.7 *Śakābdasaṃskāras* for different planets for $y_s = 1623$.

Planet	*śakābda* corrected mean planet θ_1 in min. ($'$)
Moon	2663.31200
Moon's apogee	19952.18658
Mars	3044.56960
Mercury	15987.24343
Jupiter	18521.12588
Venus	−157.80355
Saturn	19957.78772
Moon's node	7286.34636

Table 1.8 *Śakābda* corrected mean longitudes of the planets for $y_s = 1623$.

१.१२ वर्तमानाब्दगतदिनेषु शकाब्दसंस्कारः

1.12 *Śakābdasaṃskāra* for the number of days elapsed in the present year

नाकाहतं भागितभानुमध्यं हत्वा धनाद्यैर्गुणकैरिहोक्तैः ।
मन्दादिहारैर्विभजेदवाप्ताः कार्याः शशाङ्कादिषु तत्पराद्याः ॥ १३ ॥

nākāhataṃ bhāgitabhānumadhyaṃ
hatvā dhanādyairguṇakairihoktaiḥ |
mandādihārairvibhajedavāptāḥ
kāryāḥ śaśāṅkādiṣu tatparādyāḥ || 13 ||

Having multiplied 10 (*nāka*) times the mean longitude of the Sun, expressed in degrees (*bhāgita*), by the *śakābdaguṇakāras*, 09 (*dhana*) etc. stated here, and dividing the same by the *śakābdahārakas*, 85 (*manda*) etc., the results obtained, which are in thirds, should be applied to the Moon etc. [respectively].

If θ_{s0} is the mean longitude of the Sun expressed in degrees, then the *śakābda* correction corresponding to the number of days elapsed in the current year as given in the verse is of the form

$$c_2 = 10 \times \theta_{s0} \times \frac{g}{h}. \tag{1.28}$$

In the above expression, though the mean longitude θ_{s0} is expressed in degrees, it has been explicitly mentioned that the final result obtained is in thirds (*tatparās*). This quantity needs to be applied positively or negatively to the mean planet to obtain the mean longitudes of the planets for the given *ahargaṇa*.

We now explain the rationale behind the expression given in (1.28). From (1.24), it is clear that the *śakābdasaṃskāra* to be applied to the mean planet for one year is $\frac{g}{h}$ (in min.) or $\frac{g}{h} \times 3600$ (in thirds). Since this is the magnitude of correction to be applied corresponding to a period in which the mean Sun moves 360°, the magnitude of correction for a given number of solar days elapsed in the present year can be obtained simply by applying the rule of three, and is given by

$$c_2 = \left(\frac{g}{h} \times 3600\right) \times \frac{\theta_{s0}\ (\text{in deg.})}{360}, \tag{1.29}$$

which is the same as (1.28).

Note: As indicated earlier, in the case of Mars, Mercury and Saturn, the correction is to be added and in the case of the others, it has to be subtracted from their mean *śakābda* longitudes.

Illustrative example

It may be noted from Table 1.8, that the mean longitude of Venus after the first stage of *śakābdasaṃskāra* is θ_1 (Venus) $= -157.80355$. The mean longitude of the Sun for the given *ahargaṇa* (from Table 1.5) when expressed in degrees is

$$\theta_{s0} = 27.41550°. \tag{1.30}$$

Using (1.30) and the corresponding value of $\frac{g}{h}$ for Venus in (1.28), we get the magnitude of the the second correction for Venus to be

$$c_2\ (\text{Venus}) = 178.49242\ \ (\text{in thirds}).$$

The second correction including the sign for all the planets are listed in Table 1.9.

Since c_2 is subtractive for Venus, the corrected mean longitude θ_2 is given by

Planet	*śakābdasaṃskāra* ($\pm c_2$) in thirds ($'''$)
Sun	No *saṃskāra*
Moon	−29.02818
Moon's apogee	−132.98564
Mars	52.49777
Mercury	489.97918
Jupiter	−54.83100
Venus	−178.49242
Saturn	23.33234
Moon's node	−111.37548

Table 1.9 *Śakābdasaṃskāras* corresponding to the number of solar days elapsed in the current year for *ahargaṇa* $A = 1754000$.

$$\begin{aligned} \theta_2 &= \theta_1 - c_2 \\ &= -157.80355 \times 3600 - 178.49242 \\ &= -568271.28254 \text{ (in thirds).} \end{aligned} \tag{1.31}$$

In the above equation, since c_2 is in thirds and θ_1 is in minutes, the value of the latter has been multiplied by 3600 in order to express that also in thirds.

Usually in the Indian astronomical tradition the longitudes are expressed in *rāśis*, degrees, minutes etc. Hence, we divide (1.31) first by 21600 in order to get the result in degrees, and then further divide it by 30 in order to express the corrected mean longitude in terms of *rāśi* etc. Now,

$$\frac{-568271.28256}{216000 \times 30} = -0.0876961856.$$

Since the result is negative, we add 12 *rāśis* to it. Thus

$$\theta_2 = 12 - 0.087696186 = 11.9123038.$$

Here the integral part represents the *rāśi*. Thus, Venus has completed 11 *rāśis* and is in the 12th *rāśi*. By multiplying the fractional part by 30, we will get the degrees etc. In this way, the corrected mean longitude of Venus is found to be

$$\theta_2 = 11^r 27^\circ 22' 8'' 48''' 43''''.$$

In a similar manner the corrected mean longitudes of all the planets corresponding to $A = 1754000$ have been computed and the results are listed in Table 1.10.

Planet	*śakābda* corrected mean longitude (θ_2)					
	sign (*rāśi*)	deg. (°)	min. (′)	sec. (″)	thirds (‴)	fourths (⁗)
Sun	0	27	24	55	48	22
Moon	1	14	23	18	14	9
Moon's apogee	11	2	32	8	58	42
Mars	1	20	44	35	3	3
Mercury	8	26	27	22	46	19
Jupiter	10	8	41	6	38	21
Venus	11	27	22	8	48	43
Saturn	11	2	37	47	39	8
Moon's node	4	1	26	18	55	30

Table 1.10 *Śakābda* corrected mean longitudes of the planets for the *ahargaṇa* $A =$ 1754000.

१.१३ खण्डध्रुवयोरानयनम्

1.13 Obtaining the *khaṇḍas* and the *dhruvas*

It is evident from the procedure for the computation of mean longitudes of planets from the *kalyahargaṇa*, outlined in the previous sections, that the process involves handling huge numbers such as the *ahargaṇa*, the revolution numbers and the number of civil days in a *mahāyuga*. It is therefore very important to arrive at procedures that would simplify the arithmetical operations. One of the main objectives of this text *Karaṇapaddhati* is to describe methods by which such a simplification can be achieved, and in this connection the author introduces several new concepts and techniques that are not found in the *siddhāntic* texts. In particular, the notions of *khaṇḍas* and *dhruvas* are quite important and play a key role in simplifying many of the calculations.

The next couple of verses outline the procedures for obtaining the *khaṇḍas* and *dhruvas*.

१.१३.१ खण्डानयनप्रकारः

1.13.1 Procedure for obtaining the khaṇḍas

धीभावाङ्गैः कलिदिनगणाल्लभ्यते गुण्यसंज्ञः
तच्छेषोनः[11] कलिदिनगणः खण्डसंज्ञो ग्रहाणाम् ।

[11] तच्छेषोनः = धीभावाङ्गैः विभज्य यः शेषः लब्धः, तेन ऊनः कलिदिनगणः ।
taccheṣonaḥ = dhībhāvāṅgaiḥ vibhajya yaḥ śeṣaḥ labdhaḥ, tena ūnaḥ kalidinagaṇaḥ |

गुण्यात् तत्तद्भगणगुणितात् ऊनमूर्च्छाशुभाप्ताः
खेटास्तुङ्गस्त्रिभवनयुतः षड्भशुद्धोऽत्र राहुः ॥ १४ ॥

dhībhāvāṅgaiḥ kalidinagaṇāllabhyate guṇyasaṃjñaḥ
taccheṣonaḥ kalidinagaṇaḥ khaṇḍasaṃjño grahāṇām |
guṇyāt tattadbhagaṇaguṇitāt ūnamūrcchāśubhāptāḥ
kheṭāstuṅgastribhavanayutaḥ ṣaḍbhaśuddho'tra rāhuḥ || 14 ||

The quotient obtained by dividing the *ahargaṇa* (*kalidinagaṇa*) by 3449 (*dhībhāvāṅga*) is *guṇya*; the remainder subtracted from the *ahargaṇa* is known as the *khaṇḍa* of the planets. The results obtained by multiplying the *guṇya* by respective revolution numbers (*bhagaṇas*) [of the planets] and dividing by 457500 (*ūnamūrcchāśubha*) are [the uncorrected *dhruvas* of] the planets. In the case of *Tuṅga* (Moon's apogee) three *rāśis* are added to the result and in the case of *Rāhu* (Moon's node) the result has to be subtracted from six *rāśis*.

The above verse essentially outlines the procedure for obtaining *khaṇḍas*. The term *khaṇḍa* literally means 'a part'. However, in the above verse, it refers to the *ahargaṇakhaṇḍa*, a certain number of civil days that have already elapsed since a given epoch. The mean longitudes at the end of the *khaṇḍa* are called *khaṇḍāntyadhruvas*, or simply *dhruvas* of the planets. We now proceed to explain the procedure to obtain the uncorrected mean longitude (without the "*śakābda-saṃskāra*") as outlined in the verse.

Let A represent the *ahargaṇa* for the desired day. When this is divided by 3449 (*dhībhāvāṅga*) we get,

$$\frac{A}{3449} = G + \frac{r}{3449}. \tag{1.32}$$

Here the quotient G is called *guṇya*, and the remainder r the *śeṣa*. Now, the *khaṇḍa* (K) is defined as

$$K = A - r = 3449 \times G. \tag{1.33}$$

From the above equation it is evident that *khaṇḍa* represents a certain large chunk of *ahargaṇa* that is an integral multiple of 3449. If R represents the revolutions made by the planet in a *mahāyuga*, then the uncorrected mean longitude of the planet at the end of the *khaṇḍa* is stated to be

$$\theta_{k0} = \frac{G \times R}{457500}. \tag{1.34}$$

The number 457500 appearing in the above equation is obtained by dividing the total number of civil days in a *mahāyuga* D_c by 3449.

$$\frac{D_c}{3449} = \frac{1577917500}{3449} = 457500. \tag{1.35}$$

Using (1.35) and (1.33) in (1.34) we have

$$\theta_{k0} = \frac{K \times R}{D_c}, \tag{1.36}$$

which is the standard expression for obtaining the mean longitudes. It should be kept in mind that what we are considering here is only the uncorrected mean longitude.

Illustrative example

We shall now find the *khaṇḍa* and *dhruvas* of the planets corresponding to the *ahargaṇa* $A = 1754000$ for which the mean longitudes were computed earlier.

Dividing A by 3449, we get the *guṇya* G to be 508. That is,

$$\frac{1754000}{3449} = 508 + \frac{1908}{3449}. \tag{1.37}$$

As per the prescription, the remainder $r = 1908$ when subtracted from A gives the *khaṇḍa*

$$K = 1754000 - 1908 = 1752092.$$

We know that the revolution made by the Moon in a *Mahāyuga* is 57753336. From (1.34), the mean longitude of the Moon at the end of the *khaṇḍa* 1752092 may be obtained as follows:

$$\begin{aligned}\theta_{k0}\ (\text{Moon}) &= \frac{508 \times 57753336}{457500} \\ &= 64128.294399.\end{aligned} \tag{1.38}$$

This shows that at the end of the *khaṇḍa*, Moon has completed 64128 revolutions and 3 signs ($0.294399 \times 12 = 3.532799$). By multiplying the fractional part by 30 etc., we obtain the uncorrected mean longitude of the Moon at *khaṇḍāntya* to be

$$\theta_{k0}\ (\text{Moon}) = 3^r 15^\circ 59' 2'' 23''' 59''''.$$

It may be noted that the mean longitude θ_{k0} of the planet at the end of the *khaṇḍa* is obtained by making use of the multiplier ($G = 508$) and the divisor (457500) whose magnitudes are considerably smaller compared to the magnitude of the *khaṇḍa* (1752092) and *bhūdina* (1577917500). This is indeed an advantage from the computational view point.

The uncorrected mean longitudes of all the other planets at the end of the *khaṇḍa* are listed in Table 1.11. The next verse explains the *śakābdasaṃskāra* to be applied to this mean planet in order to find the *dhruva*, D.

Planet	Mean longitude (θ_{k0}) at the end of the *khaṇḍa* $K = 1752092$					
	sign (*rāśi*)	deg. (°)	min. (′)	sec. (″)	thirds (‴)	fourths (⁗)
Sun	10	6	53	6	53	6
Moon	3	15	59	2	23	59
Moon's apogee	4	9	32	33	12	23
Mars	4	7	9	15	41	54
Mercury	11	13	12	28	19	40
Jupiter	5	4	3	57	0	35
Venus	6	13	15	43	47	24
Saturn	8	27	9	25	8	27
Moon's node	7	20	30	42	15	20

Table 1.11 The uncorrected *khaṇḍāntyadhruvas* of the planets.

१.१३.२ ध्रुवानयने शकाब्दसंस्कारः

1.13.2 Applying śakābdasaṃskāra *to find the* dhruvas

गुण्यात् तिथीशगुणिताद् गिरितुङ्गनिघ्नं पोतं त्यजेदथ धनादिगुणाहतं तत् ।
मन्दादिहारहतपोतहृतं यथोक्तं कुर्याद् विहङ्गमकलादिषु ते ध्रुवाः स्युः ॥ १५ ॥

guṇyāt tithīśaguṇitād girituṅganighnaṃ
potaṃ tyajedatha dhanādiguṇāhataṃ tat |
mandādihārahatapotahṛtaṃ yathoktaṃ
kuryād vihaṅgamakalādiṣu te dhruvāḥ syuḥ || 15 ||

[The number] 61 (*pota*) multiplied by 3623 (*girituṅga*) has to be subtracted from 576 (*tithīśa*) multiplied by the *guṇya*. [The result thus obtained] should be multiplied by [the *śakābdaguṇakāras*], 09 (*dhana*) etc. and divided by the product of 61 (*pota*) and the *śakābdahārakas*, 85 (*manda*) etc. By applying [the results] in minutes to the mean planets, the *dhruvas* can be obtained.

The above verse prescribes the *śakābdasaṃskāra* that is to be applied to the uncorrected mean longitude to obtain the *dhruva* D. If G represents the *guṇya*, and g and h the *śakābdaguṇakāra* and *hāraka* of the planet respectively, then the *śakābda* correction is given by

$$c_d = \frac{(G \times 576 - 3623 \times 61) \times g}{h \times 61}. \tag{1.39}$$

Here it has been specifically stated that the result obtained should be applied to the longitude of the planet expressed in minutes (*vihaṅgamakalādiṣu*). This implies that the correction given by (1.39) is in minutes. Multiplying the fractional part by 60, seconds can be obtained. This correction c_d has to be applied positively in the case of Mercury, Mars and Saturn and negatively in the case of others. That is,

$$D = \theta_{k0} + c_d \text{ (for Mercury, Mars and Saturn),} \tag{1.40a}$$

and

$$D = \theta_{k0} - c_d \text{ (for other planets).} \tag{1.40b}$$

We now explain the rationale for the above correction (1.39). It may be recalled that while obtaining mean planets from A, two corrections c_1 and c_2, given by (1.24) and (1.28), were to be applied. However, here only one correction term is given. It can easily be seen that this single correction term given by (1.39) is equivalent to the sum of the two correction terms given by (1.24) and (1.28). Recalling the two expressions and finding their sum we have

$$\begin{aligned} c_1 + c_2 &= \frac{(y_s - 444) \times g}{h} + \frac{\theta_{s0} \times g \times 10}{h \times 3600} \\ &= \left(y_k - 3623 + \frac{\theta_{s0}}{360}\right) \times \frac{g}{h}. \end{aligned} \tag{1.41}$$

Since c_1 is in minutes and c_2 given in (1.28) is in thirds (*tatparas*), we have divided the latter by 3600 in order to express that also in minutes in the above expression.

In (1.41), θ_{s0} is the number of solar days elapsed in the current year and 360 is the number of solar days in a year. Therefore, $\frac{\theta_{s0}}{360}$, say y_f, represents fraction of the current year elapsed. Rewriting (1.41),

$$c_1 + c_2 = ((y_k + y_f) - 3623) \times \frac{g}{h}. \tag{1.42}$$

Now $(y_k + y_f)$ is the number of solar years elapsed corresponding to the *khaṇḍa* K. Since 4320000 is the total number of the number of solar years consisting of 1577917500 days, the number of solar years corresponding to K can be expressed using the rule of three as

$$y_k + y_f = \frac{K \times 4320000}{1577917500}. \tag{1.43a}$$

Dividing 4320000 and 1577917500 by 7500 (which is their G.C.D.), we get 576 and 210389. Thus the above equation becomes

$$y_k + y_f = \frac{K \times 576}{210389} = \frac{K \times 576}{3449 \times 61}. \tag{1.43b}$$

Substituting (1.43*b*) in (1.42), we get

$$\begin{aligned} c_1 + c_2 &= \left(\frac{K \times 576}{3449 \times 61} - 3623 \right) \times \frac{g}{h} \\ &= \left(\frac{\frac{K}{3449} \times 576 - 3623 \times 61}{61} \right) \times \frac{g}{h} \\ &= \frac{(G \times 576 - 3623 \times 61) \times g}{h \times 61}, \end{aligned} \tag{1.44}$$

which is the same as the correction (1.39), prescribed in the text.

Illustrative example

It was shown in the previous section that the uncorrected mean longitude of the Moon corresponding to $K = 1752092$ is

$$\theta_{k0} = 3^r 15^\circ 59' 2'' 23''' 59'''' = 6359.04'.$$

Substituting appropriate values of G from (1.37) and also the values of g and h from (Table 1.6) for the Moon in (1.39), we get the *śakābda* correction to be

$$c_d\,(\text{Moon}) = 124.290260'. \tag{1.45}$$

Planet	*śakābdasaṃskāra* (c_d) in minutes ($'$)
Sun	No *Saṃskāra*
Moon	−124.29026
Moon's apogee	−569.40604
Mars	+224.78026
Mercury	+2097.94908
Jupiter	−234.77049
Venus	−764.25288
Saturn	+99.90233
Moon's node	+476.87756

Table 1.12 *Śakābdasaṃskāra* for the *dhruvas* of the planets.

Subtracting c_d of Moon from θ_{k0}, we get the *dhruva*

$$D\,(\text{Moon}) = 6359.04 - 124.29026 = 6234.74974'.$$

Converting the above into signs, degrees etc. we have

$$D\text{ (Moon)} = 3^r 13^\circ 54' 44'' 59''' 3''''.$$

The *śakābdasaṃskāras* for the other planets obtained in a similar manner are listed in Table 1.12. These corrections when applied to their mean longitudes give the *dhruvas* or the *śakābda*-corrected mean longitudes of all the planets at the end of the *khaṇḍa* (see Table 1.13).

Planet	*śakābda*-corrected mean planet (*dhruva* D) corresponding to *khaṇḍa* $K = 1752092$					
	sign (*rāśi*)	deg. (°)	min. (′)	sec. (″)	thirds (‴)	fourths (⁗)
Sun	10	6	53	6	53	6
Moon	3	13	54	44	59	3
Moon's apogee	4	0	3	8	50	38
Mars	4	10	54	2	30	49
Mercury	12	18	10	25	16	20
Jupiter	5	0	9	10	46	49
Venus	6	0	31	28	37	3
Saturn	8	28	49	19	16	52
Moon's node	7	12	33	49	36	7

Table 1.13 *Dhruvas* corresponding to the *khaṇḍa* $K = 1752092$.

While Table 1.13 presents the *dhruvas* corresponding to a date that goes back by three centuries, the *dhruvas* corresponding to a recent date are listed in Table 1.14. Here we take the *ahargaṇa* $A = 1851974$ corresponding to *nija-āṣāḍha-kṛṣṇa-navamī*, *Śaka* 1891 (August 6, 1969 CE). Then we obtain the corresponding *khaṇḍa* $K = 1848664$. The *dhruvas* for this *khaṇḍa* are listed in Table 1.14.

१.१४ खण्डध्रुवगणनायां प्रकारान्तरम्

1.14 An alternative method for obtaining the *khaṇḍas* and *dhruvas*

भूदिनाद् द्युगणेनाप्तो हारस्तेन स्वपर्ययात् ।
भगणाद्या ग्रहास्तत्र शश्युच्चे भत्रयं क्षिपेत् ॥ १६ ॥

पाते तु मण्डलाच्छुद्धे चक्रार्धमपि योजयेत् ।
हारेण भूदिनाल्लब्धो द्युगणः खण्डसंज्ञकः ॥ १७ ॥

Planet	*śakābda*-corrected mean planet (*dhruva* D) corresponding to *khaṇḍa* $K = 1848664$					
	sign (*rāśi*)	deg. (°)	min. (′)	sec. (″)	thirds (‴)	fourths (⁗)
Sun	2	28	31	28	31	28
Moon	11	0	23	23	42	30
Moon's apogee	2	14	44	24	27	36
Mars	11	7	10	41	33	48
Mercury	10	8	35	17	42	32
Jupiter	8	14	8	29	13	58
Venus	3	10	26	2	1	23
Saturn	8	18	24	37	31	57
Moon's node	4	24	11	15	31	47

Table 1.14 *Dhruvas* corresponding to the *khaṇḍa* $K = 1848664$.

तत्राधिकमुतोनं वा क्रमादृणधनात्मकम् ।
स्वस्वमध्यमभुक्तिघ्नमेतद् भास्करपर्यये ॥ १८ ॥

हारघ्नगिरितुङ्गोने धनादिस्वगुणाहते ।
मन्दादिस्वहरेणाप्ते स्वर्णं साम्यभिदावशात् ॥ १९ ॥

क्रमाद् धनमृणं कृत्वा ततो हारेण संहृतम् ।
कुर्याद् ग्रहेषु लिप्ताद्यं तदा ते स्युर्ध्रुवा इह ॥ २० ॥

bhūdinād dyugaṇenāpto hārastena svaparyayāt |
bhagaṇādyā grahāstatra śaśyucce bhatrayaṃ kṣipet || 16 ||

pāte tu maṇḍalācchuddhe cakrārdhamapi yojayet |
hāreṇa bhūdinállabdho dyugaṇaḥ khaṇḍasaṃjñakaḥ || 17 ||

tatrādhikamutonaṃ vā kramādṛṇadhanātmakam |
svasvamadhyamabhuktighnametad bhāskaraparyaye || 18 ||

hāraghnagirituṅgone dhanādisvaguṇāhate |
mandādisvahareṇāpte svarṇaṃ sāmyabhidāvaśāt || 19 ||

kramād dhanamṛṇaṃ kṛtvā tato hāreṇa saṃhṛtam |
kuryād graheṣu liptādyaṃ tadā te syurdhruvā iha || 20 ||

The result (quotient) obtained by dividing the number of civil days (*bhūdina*) in a *mahāyuga* by *ahargaṇa* is the *hāraka*. The number of revolutions of the planets divided by that [*hāraka*] would be the elapsed revolutions etc. of the planets. There, three *rāśis* should be added to the Moon's apogee.

In the case of *Rāhu*, the result has to be subtracted from twelve *rāśis* and then six *rāśis* have to be added. The group of civil days obtained by dividing the *bhūdina* by *hāraka* is a heap of days (*dyugaṇa*) called the *khaṇḍa*.

The remainder which is in excess or less would be negative or positive respectively. This is multiplied by the respective mean motions (*madhyamabhuktis*) of the planets [and] the result is applied [to the following]. The number 3623 (*girituṅga*) multiplied by the *hāraka* is subtracted from the number of revolutions of the Sun. [The remainder] is multiplied by *dhanādi-guṇakāras* and divided by the *mandādi-hārakas* of the respective planets.

Having added or subtracted [the result thus obtained] appropriately and dividing by the *hāraka*, [the result] in minutes should be applied to the mean planets obtained earlier. Then, the resulting quantities would be the *dhruvas*.

The guiding principle in any choice of the *khaṇḍa* is that the *dhruva* should be easily computable and the *khaṇḍaśeṣa* should be as small as possible. One method of obtaining the *khaṇḍa* and the *dhruva* was already discussed in verse 14. It may be noted that the value of the *khaṇḍa* K as obtained from that method would be necessarily a multiple of 3449, and hence the value of *khaṇḍaśeṣa* which is the difference between the *khaṇḍa* and the *ahargaṇa* can be as large as 3449. The purpose of the alternative method presented in this set of five verses (16-20) is to arrive at a value of *khaṇḍaśeṣa* that could be much smaller than the one obtained by the method outlined in verse 14.

We now proceed to explain this alternative procedure for finding the *khaṇḍa* as also the *dhruvas* of planets corresponding to the *khaṇḍa*. It is interesting to note that the *śakābdasaṃskāra* is also nicely incorporated in this procedure. First, a quantity known as *hāraka* (H) is defined as follows

$$H = \left[\frac{D_c}{A}\right]_{int}, \tag{1.46}$$

where A is the *ahargaṇa*, and '$[\,]_{int}$' denotes the integral part or quotient. Now, if R represents the revolutions made by a planet, then the uncorrected mean longitude for the *khaṇḍa* (which is defined below) is stated to be $(\theta)_0 = \frac{R}{H}$. As usual, 90° is to be added to Moon's apogee, and 180° to *Rāhu* whose motion is retrograde.

Having defined the *hāraka* and mean longitude, the text defines *khaṇḍa* K (see latter half of verse 17) as the ratio of the *yugasāvanadina* D_c and the *hāraka* H. That is,

$$K = \left[\frac{D_c}{H}\right]_{int}. \tag{1.47}$$

It is further stated that the *khaṇḍa* is to be taken as the quotient K itself when the remainder r, which is obtained when D_c is divided by H, is less than $\frac{H}{2}$; and when r is greater than $\frac{H}{2}$, the *khaṇḍa* is to be taken as $K + 1$. We now discuss these two cases.

Case i: $r < \frac{H}{2}$

In this case, it is said that the *khaṇḍa* is to be taken to be the quotient itself. That is,

$$K = \left[\frac{D_c}{H}\right]_{int}. \tag{1.48}$$

Hence (1.48) may be written as

$$D_c = (H \times K) + r, \tag{1.49}$$

where the remainder r is called *adhikaśeṣa*.[12]

Case ii: $r > \frac{H}{2}$

In this case it is said that the *khaṇḍa* is to be taken to be the quotient plus one. That is,

$$K = \left[\frac{D_c}{H}\right]_{int} + 1. \tag{1.50}$$

Hence,

$$\begin{aligned} D_c &= HK - (H - r) \\ &= HK - r'. \end{aligned} \tag{1.51}$$

Here the remainder $r' = H - r$, which is called '*ūnaśeṣa*' (*ūnātmakaḥ śeṣaḥ*) is subtractive in nature.

At this stage the text uses the mean daily motion of the planet which is given by $R \times \frac{21600}{D_c}$. The mean daily motion multiplied by r (*adhikaśeṣa*) or r' (*ūnaśeṣa*) is a quantity (which we denote by x) which will be considered below. To this is applied another quantity (which we denote by y) which is defined as follows:

$$y = [bh\bar{a}skaraparyaya - (h\bar{a}ra \times 3623)] \times \frac{g}{h}. \tag{1.52}$$

The term *bhāskaraparyaya* literally means the revolutions made by the Sun [in a *mahāyuga*], and hence refers to the number 4320000. Therefore, we have

$$y = (4320000 - 3623 \times H) \times \frac{g}{h}. \tag{1.53}$$

Now the corrected mean longitude at *khaṇḍāntya* K which is what is called '*Dhruva*' is stated to be:

[12] The remainder (*śeṣa*) is the excess (*adhika*) when the product $H \times K$ is removed from D_c, and hence called *adhikaśeṣa*.

$$D = (\theta)_0 + (y \pm x)\frac{1}{H}$$
$$= \frac{R}{H} + (y \pm x)\frac{1}{H}, \qquad (1.54)$$

where the '−' sign arises when $r < \frac{H}{2}$.

Rationale behind the expression for *dhruva*

We now present the rationale behind the expression (1.54) given above. It may be noted that the series of steps prescribed for finding the *dhruvas* commences with the definition of *hāraka* H, which is the integral part of the ratio of *yugasāvanadina* and *ahargaṇa*. This ratio itself can have a fractional part which we denote as f. That is,

$$\frac{D_c}{A} = H + f.$$

Rewriting the above $D_c = H \times A + f \times A$. Let $f \times A = qH + r$. Then,

$$D_c = H(A + q) + r$$
$$= Hq' + r, \qquad (1.55)$$

Depending upon the magnitude of H and r, we now consider two cases. If $r < \frac{H}{2}$, we write

$$D_c = K \times H + r, \qquad (1.56)$$

where $K = q' = A + q$ is the *khaṇḍa*, and r the remainder (*adhikaśeṣa*). If $r > \frac{H}{2}$, we write

$$D_c = K \times H - r', \qquad (1.57)$$

where $K = q' + 1 = A + q + 1$ is the *khaṇḍa*, and $r' = H - r$ is *ūnaśeṣa*.

We now return to the expression for *dhruva* D given in (1.54). This is valid for both the cases $r > \frac{H}{2}$ and $r < \frac{H}{2}$. In what follows will consider the case $r > \frac{H}{2}$. It is well known that the mean longitude of a planet is given by $R \times \frac{A}{D_c}$. The mean longitude at the *khaṇḍāntya* is what is known as *dhruva* (D) and may be expressed as

$$D = R \times \frac{K}{D_c} + \delta, \qquad (1.58)$$

where the additive δ represents the *śakābda* correction. Since the case under consideration is $r > \frac{H}{2}$, we use the expression for D_c given by (1.57) in (1.58). Doing so, we have

$$D = R \times \frac{K}{(K \times H - r')} + \delta$$
$$\approx \frac{R}{H}\left[1 + \frac{r'}{(K \times H)}\right] + \delta. \tag{1.59}$$

It may be noted that, in (1.59) we have used the approximation $\frac{1}{(1-z)} \approx 1 + z$, when $z \ll 1$, with $z = \frac{r'}{K \times H}$. Rewriting (1.59), we have

$$D = \frac{R}{H} + \frac{R}{(K \times H)} \times \frac{r'}{H} + \delta. \tag{1.60}$$

Considering (1.57), since $r' \ll K \times H$, to the "first order", $D_c \approx H \times K$. Therefore, $\frac{R}{(H \times K)} \approx \frac{R}{D_c}$ = number of revolutions/day. This when converted into minutes gives the mean daily motion g_m (*madhyamabhukti*). Hence the correction to the mean longitude in minutes due to the "remainder term" (second term in RHS of (1.60)) is given by

$$\frac{g_m \times r'}{H}, \tag{1.61}$$

which is what is represented as $\frac{x}{H}$ in (1.54). This is the case when the remainder is *ūnaśeṣa* $(r > \frac{H}{2})$. In the case of *adhikaśeṣa* $(r < \frac{H}{2})$ too, it can be seen that one is lead to a similar result.[13]

Now we shall show that the term $\frac{y}{H}$ occuring in (1.54) exactly corresponds to the *śakābda* correction δ. For this, we first find the number of years between the epoch *śakābda* 444 or *kalyabda* 3623 and our *khaṇḍa*, K. The number of years from the beginning of *kali* to the *khaṇḍa* K is

$$K \times \frac{4320000}{D_c}.$$

Therefore, the number of years between *śakābda* 444 and the *khaṇḍa* is

---

[13] In this case $D_c = KH + r$ and hence,

$$\frac{1}{D_c} = \frac{1}{KH}\left[\frac{1}{1 + \frac{r}{KH}}\right] \approx \frac{1}{KH}\left[1 - \frac{r}{KH}\right].$$

Hence, in place of (1.60), we have

$$D = \frac{R}{H} - \frac{R}{K \times H} \times \frac{r}{H} + \delta.$$

This leads to the '−' sign in (1.54).

$$\left(\frac{4320000}{D_c} \times K - 3623\right) \approx \frac{4320000}{H} - 3623$$
$$= \frac{1}{H} \times (4320000 - 3623 \times H). \qquad (1.62)$$

Hence the *śakābda* correction due to above number of years between *śakābda* 444 and the *khaṇḍa* K is given by

$$\frac{1}{H}(4320000 - 3623 \times H) \times \frac{g}{h}, \qquad (1.63)$$

which is the same as $\frac{y}{H}$ as given by (1.53). Hence the *dhruva* D is given by

$$D \approx \frac{R}{H} + (x + y)\frac{1}{H}.$$

This is the rationale behind (1.54), where the second term includes both the corrections due to the "remainder term" and due to the *śakābdasaṃskāra*, in minutes.

Illustrative example

We will now illustrate the above alternate method for finding the *khaṇḍa* and the *dhruva*. We shall in particular obtain the *dhruva* of Moon for the same *ahargaṇa* $A = 1754000$, considered earlier. The *hāraka* will be

$$H = \left[\frac{D_c}{A}\right]_{int} = \left[\frac{1577917500}{1754000}\right]_{int} = 899.$$

The uncorrected mean longitude is

$$(\theta)_0 = \frac{R}{H}$$
$$= \frac{57753336}{899} = 64241.75305895 \text{ revolutions} \qquad (1.64)$$
$$= 16266.073415', \qquad (1.65)$$

after subtracting the integral number of revolutions, 64241. In this case,

$$\frac{D_c}{H} = \frac{1577917500}{899}$$
$$\text{or} \qquad D_c = 899 \times 1755191 + 791. \qquad (1.66)$$

Since the number $r = 791$ is greater than half of the divisor H, the *khaṇḍa*, $K = 1755191 + 1 = 1755192$. The *ūnaśeṣa*

$$r' = 899 - r = 899 - 791 = 108.$$

The corrected mean longitude at the *khaṇḍa*, or the *dhruva* D, is given by

$$D = (\theta)_0 + \frac{r' \times g_m}{H} - \frac{1}{H}(4320000 - 3623 \times H) \times \frac{g}{h}. \qquad (1.67)$$

Here $(\theta)_0 = 16266.073415', r' = 108$, the *madhyamagati* of the Moon, g_m is $790.581'$, $H = 899$ and $\frac{g}{h} = -0.10588'$. Substituting these values, we have

$$\begin{aligned} D \text{ (Moon)} &= 16235.8597996 \\ &= 9^r 0^\circ 35' 51'' 35''' 16''''. \end{aligned} \qquad (1.68)$$

The *dhruvas* corresponding to the various planets are listed in Table 1.15.

Planet	*dhruva* (D)					
	sign (*rāśi*)	degrees (°)	minutes (′)	seconds (″)	thirds (‴)	fourths (⁗)
Sun	4	2	15	14	47	48
Moon	9	0	35	51	35	16
Moon's apogee	3	15	16	56	6	23
Mars	10	15	22	57	8	21
Mercury	3	14	35	12	24	17
Jupiter	1	17	43	34	29	32
Venus	3	17	5	45	7	4
Saturn	0	12	29	35	11	17
Moon's node	1	28	16	28	29	48

Table 1.15 *Dhruvas* of the planets corresponding to the *khaṇḍa* $K = 1755192$.

In this context the observation made in Malayalam Commentary II as a prelude to the set of five verses described above is worth noting:[14]

> *anantaraṃ "dhībhāvāṅga" ennatinekkāḷ aduppamuḷḷa iṣṭadinaṃ koṇṭu khaṇḍaṅaḷuṃ dhruvaṅaḷuṃ varuttuvān collunnū añjuśḷokaṅaḷe koṇḍu.*
>
> Hereafter, in the following five verses, [a procedure] is stated to obtain the *khaṇḍas* and *dhruvas* for any desired day which is more closer [to the day compared to that which is obtained from] *dhībhāvāṅga* [method].

What the commentator is implying is that the alternative method for finding the *khaṇḍas* given in verses 16-20 [Method II] leads to *khaṇḍas* which are closer to the given *ahargaṇa* than the *khaṇḍas* which are obtained by following the method outlined in verse 14 [Method I].

[14] {KP 1956}, p. 25.

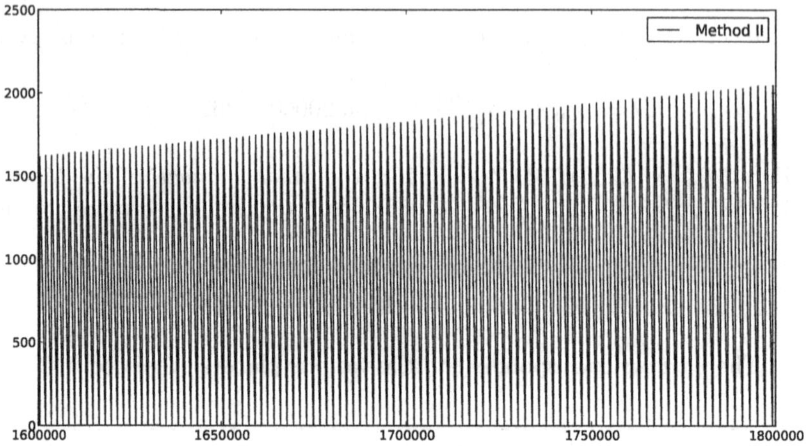

Fig. 1.1 Difference between the *ahargaṇa* and the *khaṇḍa* (obtained using Method II) for *ahargaṇas* between 1600000 to 1800000.

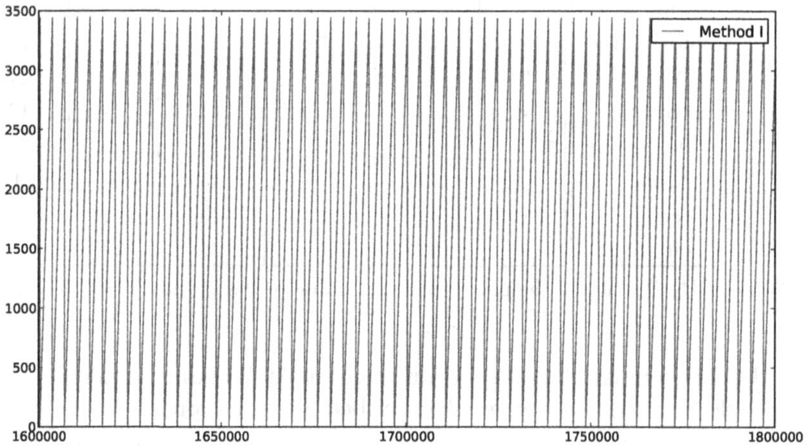

Fig. 1.2 Difference between the *ahargaṇa* and the *khaṇḍa* (obtained using Method I) for *ahargaṇas* between 1600000 to 1800000.

In order to check the veracity of the claim made by the commentator, we computed the *khaṇḍa* using both the methods for *ahargaṇas* ranging from 1600000 (15th century) to 1800000 (20th century). In Figure 1.1 we plot the difference between the *ahargaṇa* and the *khaṇḍa* computed using Method II against the *ahargaṇa* for this range at an interval of 10 days. We observe that the maximum difference is ≈ 2000. In Figure 1.2, we plot the difference between the *ahargaṇa* and the *khaṇḍa* obtained following Method I for the same range of *ahargaṇas* at an interval of 10 days. There we see the maximum difference is 3449.

It is clear that Method II is more advantageous for *ahargaṇas* $A \approx 10^6$. In Figure 1.3, we have plotted the difference between the *ahargaṇa* and the *khaṇḍa* following both the methods, for *ahargaṇas* ranging from 10^6 to 2.5×10^6 at an interval of 100 years. From the trend of the graph, it is seen that Method II may be less advantageous for $A > 2.1 \times 10^6$. However, this cannot be attributed as a drawback of Method II, since the *karaṇa* texts are not expected to be applicable for large range of values of A anyway.

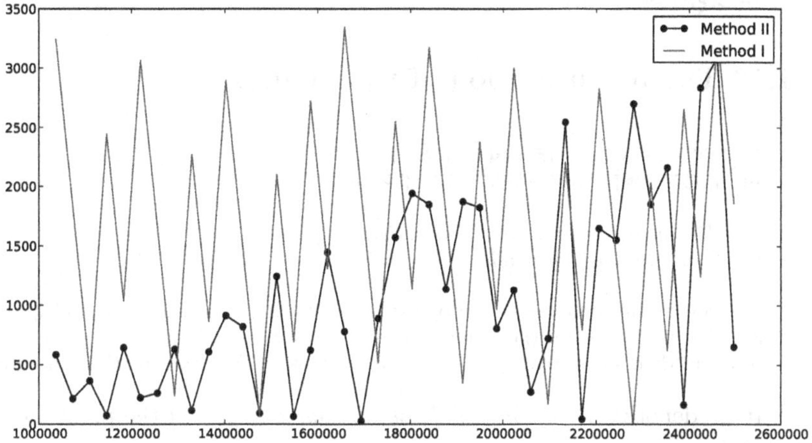

Fig. 1.3 Difference between the *ahargaṇa* and the *khaṇḍa* (using both Methods I, II) for *ahargaṇas* ranging from 10^6 to 2.5×10^6.

१.१५ राहोः शकाब्दसंस्कारे वैशिष्ट्यम्

1.15 Speciality in the *śakābdasaṃskāra* for *Rāhu*

शकाब्दसंस्कृतौ पाते शुद्ध एव भवेद्दणम् ।
ततोऽन्यत्र धनं विद्यात् तद्भुक्तिभगणादिके ॥ २१ ॥

śakābdasaṃskṛtau pāte śuddha eva bhavedṛṇam |
tato'nyatra dhanaṃ vidyāt tadbhuktibhagaṇādike || 21 ||

In the computed value of *Rāhu* [subtracted from 360°] the *śakābda* correction has to be subtracted from the [mean longitude]. In other places where the revolutions and *madhyamagati* are found, it has to be added.

In the case of *Rāhu*, after obtaining the mean longitude from its number of revolutions, as for other planets, it is subtracted from 6 *rāśis*, as its motion is retrograde and is located at 6 *rāśis* at *kalyādi*. The *śakābdasaṃskāra* for *Rāhu* is prescribed to be negative, considering the retrograde nature of its motion into account. This means that there is an increase in the revolution number and *madhyamagati* of *Rāhu* due to the *śakābdasaṃskāra*.

१.१६ ग्रहमध्यभुक्तिः

1.16 The mean motion of the planets

अर्कादेर्भगणाभ्यस्ता राशिचक्रस्य लिप्तिकाः ।
भूदिनैर्विहृतास्तेषां मध्यभुक्तिकलाः स्मृताः ॥ २२ ॥

arkāderbhagaṇābhyastā rāśicakrasya liptikāḥ |
bhūdinairvihṛtāsteṣāṃ madhyabhuktikalāḥ smṛtāḥ || 22 ||

The revolution numbers of the planets, starting from the Sun etc., multiplied by 21600 and divided by the number of civil days in a *mahāyuga* (*bhūdinas*) would result in their respective mean daily rate of motion (*madhyabhukti*) in minutes.

If g_m denotes the *madhyamabhukti*/*madhyamagati* of the planet, then it is given by

$$g_m = \frac{21600 \times R}{D_c}. \tag{1.69}$$

The integral part of the above result obtained would be in minutes. From the fractional part, the seconds etc. can be obtained. The *madhyamabhuktis* of different planets obtained using (1.69) are given in Table 1.16.

१.१७ मध्यभुक्तेः शकाब्दसंस्कारः

1.17 *Śakābdasaṃskāra* for the mean motions

नृपहतदिनकरभोगान्निजनिजगुणकैर्धनादिभिर्गुणितात् ।
मन्दादिस्वहराप्ताः चन्द्रादिगतौ प्रतत्पराः कार्याः ॥ २३ ॥

Planet	Mean motion (*madhyamagati*)			
	minutes (′)	seconds (″)	thirds (‴)	fourths (⁗)
Sun	59	8	10	13
Moon	790	34	52	39
Moon's apogee	6	40	59	30
Mars	31	26	27	48
Mercury	245	32	18	54
Jupiter	4	59	9	0
Venus	96	7	44	17
Saturn	2	0	22	41
Moon's node	3	10	44	7

Table 1.16 The mean motion of the planets.

nṛpahatadinakarabhogānnijanijaguṇakairdhanādibhirguṇitāt |
mandādisvaharāptāḥ candrādigatau pratatparāḥ kāryāḥ || 23 ||

The mean motion of the Sun (*dinakarabhoga*) multiplied by 10 (*nṛpa*) has to be [further] multiplied by the *guṇakāras, dhana* etc., and divided by the *hārakas, manda* etc. The result in fourths (*pratatparas*) should be applied to the [mean] motion (*madhyamagatis*) of the Moon etc.

If c_r represents the *śakābdasaṃskāra* to be applied to the *madhyamagati* of the planet, then it is stated that

$$\begin{aligned} c_r &= dinakarabhoga \times 10 \times \frac{g}{h} \\ &= 59'8''10'''13'''' \times 10 \times \frac{g}{h}. \end{aligned} \tag{1.70}$$

In (1.70), *dinakarabhoga* is the rate of motion of the Sun per day in minutes. It has been clearly mentioned in the verse that the magnitude of the correction given by (1.70) is in *pratatparas* (fourths).

The rationale behind (1.70) can be understood as follows. We know that the *śakābda* correction applied to the planets per year (in minutes) is $\frac{g}{h}$. Hence, the *saṃskāra* per day is

$$\frac{R_s}{D_c} \times \frac{g}{h} \text{ (in minutes)}, \tag{1.71}$$

where R_s represents the revolutions made by the Sun in a *mahāyuga* and D_c the total number of civil days in a *mahāyuga*. The ratio $\frac{R_s}{D_c}$ as such gives the rate of motion of the Sun in revolutions per day. Thus this ratio $\frac{R_s}{D_c}$ in (1.71) can be replaced by $\frac{dinakarabhoga}{21600}$. Doing so, the *saṃskāra* for one day is

$$\frac{\textit{dinakarabhoga}}{21600} \times \frac{g}{h} \text{ (in minutes).} \tag{1.72}$$

Now multipying this by 60^3, we would obtain the correction in fourths. Thus, we have

$$\begin{aligned} \textit{gati-saṃskāra}/\text{ day} &= \frac{\textit{dinakarabhoga}}{21600} \times \frac{g}{h} \times 60^3 \\ &= 59'8''10'''13'''' \times 10 \times \frac{g}{h}, \end{aligned} \tag{1.73}$$

which is the same as the expression (1.70) given in the text. The magnitude of the correction (c_r) to be applied to the mean motion of different planets in fourths (*tatparas*) are listed in Table 1.17.

Planet	*śakābdasaṃskāra* (c_r) in fourths ($''''$)
Sun	No *saṃskāra*
Moon	$62.6148 \approx 63$
Moon's apogee	$286.8546 \approx 287$
Mars	$113.2395 \approx 113$
Mercury	$1056.9018 \approx 1057$
Jupiter	$118.2723 \approx 118$
Venus	$385.0142 \approx 385$
Saturn	$50.3287 \approx 50$
Moon's node	$240.2407 \approx 240$

Table 1.17 *Saṃskāras* to be applied to the mean motion of planets.

The corrected mean motion of the planets (*saṃskṛta-madhyamagatis*) are given by

$$g'_m = g_m + c_r \quad \text{(for Mars, Mercury, Saturn and Moon's node),} \tag{1.74}$$

$$\text{and} \quad = g_m - c_r \quad \text{(for Moon, Jupiter, Venus and Moon's apogee).} \tag{1.75}$$

It may be mentioned that unlike in the computation of mean longitudes, c_r has to be added to the mean motion of Moon's node, *Rāhu.* The values of g'_m for different planets are listed in Table 1.18. These values have also been listed in the Commentary I in the form of *vākyas.*[15] They have been tabulated

[15] {KP 1956}, p. 30.

in Table 1.19, where we have also indicated the few places where the *vākyas* differ slightly from the computed values.

Planet	*saṃskṛta-madhyamagati* minutes (′)	seconds (″)	thirds (‴)	fourths (⁗)
Sun	59	08	10	13
Moon	790	34	51	36
Moon's apogee	6	40	54	43
Mars	31	26	29	41
Mercury	245	32	36	31
Jupiter	4	59	07	02
Venus	96	07	37	52
Saturn	2	00	23	31
Moon's node	3	10	48	7

Table 1.18 The corrected mean motion of planets.

Planet	*saṃskṛta madhyamagati* min. (′)	sec. (″)	thirds (‴)	fourths (⁗)	*vākya*
Sun	59	08	10	13	*gopājñayā dinadhāmāḥ*
Moon	790	34	51	36	*caṇḍikeśo bhargasnigdhosau*
Moon's apogee	6	40	54	43	*gūḍhavāśo'nubhāti*
Mars	31	26	29	(42)[16]	*prabhur dharācakrapālaḥ*
Mercury	245	32	36	(32)	*rāgī tumbururgaṇeśvaraḥ*
Jupiter	4	59	07	02	*prajñāsanno dharmavān*
Venus	96	07	37	(51)	*kāśīsāmbasannacoḍaḥ*
Saturn	2	00	23	(32)	*prabalaḥ prājño naraḥ*
Moon's node	3	10	(44)	(41)	*kavirvibhur nakulaḥ*

Table 1.19 The corrected mean motion of planets given in the Commentary I in the form of *vākyas*.

[16] The parentheses mark those places where the *vākya* values differ from those in Table 1.18.

१.१८ गुणकारस्य हारद्वितीयहारानयनम्

1.18 Obtaining the *hāra* and the *dvitīyahāra* of a *guṇakāra*

इष्टघ्नात् कुदिनात् स्वपर्ययहृतो हारस्त्वथेष्टो गुणः
तत्रोनाधिकमन्नतत्परहतं स्वर्णात्मकं पर्यये ।
सौरे हारहते धनादिगुणिते मन्दादिहारोद्धृते
कृत्वानेन हराहतक्षितिदिनाल्लब्धो द्वितीयो हरः ॥ २४ ॥

iṣṭaghnāt kudināt svaparyayahṛto hārastvatheṣṭo guṇaḥ
tatronādhikamannatatparahataṃ svarṇātmakaṃ paryaye |
saure hārahate dhanādiguṇite mandādihāroddhṛte
kṛtvānena harāhatakṣitidināllabdho dvitīyo haraḥ || 24 ||

The quotient obtained when any desired number is multiplied by the number of civil days in a *mahāyuga* (*bhūdina*) and divided by the number of revolutions of the planet would be the *hāraka*, and this desired number is indeed the multiplicand (*guṇa*). The remainder which is either less or in excess is multiplied by 21600 (*annatatpara*) and applied positively or negatively to the revolution number of the Sun multiplied by the *hāraka* and by the multipliers *dhana* etc., and divided by the divisors *manda* etc. By the [result thus obtained], the *hāraka* multiplied by the *bhūdina* has to be divided. The quotient would be the *dvitīyahāra*.

Let x be an arbitrary number which is considered as the multiplicand (*guṇa*) and R the number of revolutions of the planets. Now the product of x and D_c, divided by R may be expressed as

$$\frac{x \times D_c}{R} = Q + \frac{r}{R}, \tag{1.76}$$

where Q is the quotient and r is the remainder. Introducing the notation $\mathcal{H}_1$ for the *hāraka*, we consider the two possible cases that arise.

(i) If $r < \frac{R}{2}$, then

$$\mathcal{H}_1 = Q = \left[\frac{x \times D_c}{R}\right]_{int}, \tag{1.77}$$

and the remainder r is called *adhikaśeṣa* as $\frac{x \times D_c}{R}$ is in excess of the *hāraka*, $\mathcal{H}_1$.

(ii) If $r > \frac{R}{2}$, then

$$\mathcal{H}_1 = Q + 1 = \left[\frac{x \times D_c}{R}\right]_{int} + 1, \tag{1.78}$$

and the remainder $r' = R - r$ is called *ūnaśeṣa* as $\frac{x \times D_c}{R}$ is less than the *hāraka*.

Then, we compute one of the following quantities as the case may be

$$y = \mathcal{H}_1 \times 4320000 \times \frac{g}{h} + r' \times 21600 \qquad (\text{if } r > \frac{R}{2}),$$

$$\textit{and} \qquad y = \mathcal{H}_1 \times 4320000 \times \frac{g}{h} - r \times 21600 \qquad (\text{if } r < \frac{R}{2}).$$

Now, the *dvitīyahāra* $\mathcal{H}_2$ is defined in terms of $\mathcal{H}_1$ as:

$$\mathcal{H}_2 = \left[\frac{\mathcal{H}_1 \times D_c}{y}\right]_{int}. \tag{1.79}$$

Since $\mathcal{H}_1$ and D_c are always positive, $\mathcal{H}_2$ would be negative, if y is negative, and would be positive otherwise.

Explanation

This set of verses present an interesting method for simplifying the calculation of the *śakābda* corrected mean longitude of the planets.

Let θ_c be the *śakābda* corrected mean longitude for *ahargaṇa* A. That is,

$$\theta_c = \frac{A \times R_c}{D_c}, \tag{1.80}$$

where R_c is the corrected revolution number after incorporating the *śakābdasaṃskāra* (which is explained in the next verse). We can write θ_c as

$$\theta_c = \frac{A \times x}{\frac{x \times D_c}{R_c}}. \tag{1.81}$$

Now (as set forth in the next verse) $R_c = R + \frac{4320000}{21600} \times \frac{g}{h}$, as $\frac{g}{h}$ is the correction for the rate of motion per year in minutes and there are 4320000 years in a *mahāyuga*. We write R_c as

$$R_c = R\left[1 + \frac{4320000}{21600} \times \frac{g}{h \times R}\right], \tag{1.82}$$

and

$$\begin{aligned} \frac{x \times D_c}{R_c} &= \frac{x \times D_c}{R}\left[1 + \frac{4320000}{21600} \times \frac{g}{h \times R}\right]^{-1} \\ &\approx \frac{x \times D_c}{R} - \frac{x \times D_c}{R} \times \frac{4320000}{21600} \times \frac{g}{h \times R}. \end{aligned} \tag{1.83}$$

If $r < \frac{R}{2}$,

$$\frac{x \times D_c}{R} = \mathcal{H}_1 + \frac{r}{R}, \qquad \text{and}$$
$$\frac{x \times D_c}{R_c} \approx \mathcal{H}_1 + \frac{r}{R} - \mathcal{H}_1 \times \frac{4320000}{21600} \times \frac{g}{h \times R}, \tag{1.84}$$

where the term $\frac{x \times D_c}{R}$ is approximated by $\mathcal{H}_1$ in the *śakābdasaṃskāra* correction term. Hence,

$$\frac{x \times D_c}{R_c} \approx \mathcal{H}_1 - \frac{y}{R \times 21600}, \tag{1.85}$$

where

$$y = \mathcal{H}_1 \times 4320000 \times \frac{g}{h} - r \times 21600.$$

If $r > \frac{R}{2}$, then

$$\frac{x \times D_c}{R} = \mathcal{H}_1 - \frac{r'}{R}, \qquad \text{and}$$
$$\frac{x \times D_c}{R_c} \approx \mathcal{H}_1 - \frac{y}{R \times 21600}, \tag{1.86}$$

where $y = \mathcal{H}_1 \times 4320000 \times \frac{g}{h} + r' \times 21600$. Hence, the expression for θ_c will be

$$\begin{aligned}
\theta_c &= \frac{A \times x}{\frac{x \times D_c}{R_c}} \\
&\approx \frac{A \times x}{\mathcal{H}_1 - \frac{1}{R \times 21600} \times y} \\
&\approx \frac{A \times x}{\mathcal{H}_1 \left[1 - \frac{1}{R \times 21600} \times \frac{y}{\mathcal{H}_1}\right]} \\
&\approx \frac{A \times x}{\mathcal{H}_1} + \frac{A \times x}{\mathcal{H}_1}\left[\frac{1}{R \times 21600} \times \frac{y}{\mathcal{H}_1}\right].
\end{aligned} \tag{1.87}$$

This in terms of minutes reduces to

$$\theta_c \text{ (in min)} \approx \frac{A \times x}{\mathcal{H}_1} \times 21600 + \frac{A \times x}{\mathcal{H}_1 \times R} \times \frac{y}{\mathcal{H}_1}.$$

Now, $R\mathcal{H}_1 \approx x \times D_c$. Hence,

$$\theta_c \text{ (in min)} \approx \frac{A \times x}{\mathcal{H}_1} \times 21600 + \frac{A}{\left(\frac{\mathcal{H}_1 \times D_c}{y}\right)}. \tag{1.88}$$

The *dvitīyahāra* $\mathcal{H}_2$ is defined in the verse as $\left[\frac{\mathcal{H}_1 \times D_c}{y}\right]_{int}$. Therefore,

$$\theta_c \text{ (in min)} = \frac{A \times x}{\mathcal{H}_1} \times 21600 + \frac{A}{\mathcal{H}_2}. \tag{1.89}$$

The second term including $\mathcal{H}_2$ can be considered as a correction term as it is much smaller.

Illustrative example

For the purpose of convenience, we choose the multiplicand x to be 100. With this we will find out the *hāra* ($\mathcal{H}_1$) and the *dvitīyahāra* ($\mathcal{H}_2$) in the case of the Moon.

$$\frac{x \times D_c}{R_c\,(\text{Moon})} = 2732 + \frac{9636048}{57753336}.$$

Here, the remainder 9636048 is less than half of 57753336. So the remainder is *adhikaśeṣa* and the correction corresponding to that is negative. The *hāraka*, $\mathcal{H}_1 = 2732$. Now,

$$\begin{aligned} y &= 2732 \times 4320000 \times \frac{g}{h} - 9636048 \times 21600 \\ &= 2732 \times 4320000 \times \left(\frac{-9}{85}\right) - 9636048 \times 21600 \\ &= -209388285741.1765. \end{aligned}$$

Substituting the values in (1.79) the *dvitīyahāraka* $\mathcal{H}_2$ is found to be

$$\mathcal{H}_2 = \left[\frac{2732 \times 1577917500}{-209388285741.1765}\right]_{int} = -21.$$

Here $\mathcal{H}_2$ is negative. The value of $\mathcal{H}_1$ and $\mathcal{H}_2$ corresponding to $x = 100$ for the other planets are also found in a similar manner and these are listed in Table 1.20.

Planet	$\mathcal{H}_1$	$\mathcal{H}_2$
Sun	36526	4681
Moon	2732	-21
Moon's apogee	323199	-756
Mars	68700	1865
Mercury	8797	191
Jupiter	433227	-1817
Venus	22470	-1014
Saturn	1076606	4307
Moon's node	679475	-899

Table 1.20 The first and second *hārakas* of the planets for $x = 100$.

Similarly we have computed the first and second *hārakas* for $x = 1$ and $x = 50$ and the values are listed in Tables 1.21 and 1.22.

Planet	$\mathcal{H}_1$	$\mathcal{H}_2$
Sun	365	-23
Moon	27	0
Moon's apogee	88	11
Mars	225	7
Mercury	687	1865
Jupiter	4332	-1161
venus	10766	4525
Saturn	3232	-768
Moon's node	6795	-1004

Table 1.21 The first and second *hārakas* of the planets for $x = 1$.

Planet	$\mathcal{H}_1$	$\mathcal{H}_2$
Sun	18263	4681
Moon	1366	−2
Moon's apogee	4398	−44
Mars	11235	−1014
Mercury	3435	1865
Jupiter	216614	−18563
Venus	538303	4307
Saturn	161599	−744
Moon's node	339737	−895

Table 1.22 The first and second *hārakas* of the planets for $x = 50$.

१.१९ ग्रहपर्ययाणां शकाब्दसंस्कारः

1.19 *Śakābdasaṃskāra* for the number of revolutions of the planets

ज्ञानीन्द्रनिघ्ना गुणका धनाद्या मन्दादिहारैर्विहृता यथोक्तम् ।
चन्द्रादिकानां भगणेषु कार्याः तदा तु ते संस्कृतपर्ययाः स्युः ॥ २५ ॥

jñānīndranighnā guṇakā dhanādyā
mandādihārairvihṛtā yathoktam |
candrādikānāṃ bhagaṇeṣu kāryāḥ
tadā tu te saṃskṛtaparyayāḥ syuḥ || 25 ||

The number 200 (*jñānīndra*) is multiplied by the multipliers *dhana* etc., and divided by the divisors *manda* etc., of the respective planets. The results when applied to the revolution numbers (*bhagaṇas*) of the planets starting from Moon etc., would give the corrected revolutions (*saṃskṛtaparyayas*).

If R_c denotes the corrected *bhagaṇa* of the planets, referred to as *saṃskṛta-paryaya* in the above verse, then

$$R_c = R \pm \Delta. \qquad (1.90)$$

where, Δ is the *śakābda* correction and is given by

$$\Delta = \frac{j\tilde{n}\bar{a}n\bar{\imath}ndra \times gu\d{n}aka}{h\bar{a}ra} = \frac{200 \times g}{h}. \qquad (1.91)$$

It may be noted that $\frac{g}{h}$ is the correction to the rate of motion of the mean planet per year in minutes. Hence, $\frac{1}{21600} \times \frac{g}{h}$ is the correction per year in revolutions. Therefore, the correction to the number of revolutions in a *mahāyuga* is given by

$$\frac{4320000}{21600} \times \frac{g}{h} = \frac{200 \times g}{h},$$

as stated above.

It may also be noted that for the Sun, there is no *saṃskāra*. The corrected revolutions of all the planets are listed in Table 1.23.

Planet	*saṃskṛta-paryayas* (R_c)	*asaṃskṛta-paryayas* (R)
Sun	4320000	4320000
Moon	57753314. 823529	57753336
Moon's apogee	488121.985075	488219
Mars	2296862.297872	2296824
Mercury	17937377.446809	17937020
Jupiter	364184.000000	364224
Venus	7022257.787234	7022388
Saturn	146581.021277	146564
Moon's node	232307.250000	232226

Table 1.23 Corrected and uncorrected revolutions of planets in a *mahāyuga*.

१.२० ग्रहकक्ष्याणां परस्परसम्बन्धः

1.20 Relation between the *kakṣyās* of planets

अज्ञानान्तकरा हतेन्दुभगणस्त्वाकाशकक्ष्या ततः
खेटानां दिवसोक्तयोजनगतिर्धात्रीदिनैरुद्धृता ।
कक्ष्या या नभसः स्वपर्ययहृता कक्ष्या ग्रहाणां रवेः
कक्ष्या नीतिसमाहता निगदिता नक्षत्रकक्ष्या बुधैः ॥ २६ ॥

अथवा नयनाभ्यस्ता मध्यभुक्तिकला विधोः ।
स्वस्वकक्ष्यासु खेटानां दिनयोजनभुक्तयः ॥ २७ ॥

गुणो भवेद् योजनभुक्तिरेषा तदा स्वकक्ष्यैव हरो ग्रहाणाम् ।
कलागतिश्चेद् गुणकोऽत्र हारो भचक्रलिप्ता निजमध्यनीतौ ॥ २८ ॥

ajñānāntakarā hatendubhagaṇastvākāśakakṣyā tataḥ
kheṭānāṃ divasoktayojanagatirdhātrīdinairuddhṛtā |
kakṣyā yā nabhasaḥ svaparyayahṛtā kakṣyā grahāṇāṃ raveḥ
kakṣyā nītisamāhatā nigaditā nakṣatrakakṣyā budhaiḥ || 26 ||

athavā nayanābhyastā madhyabhuktikalā vidhoḥ |
svasvakakṣyāsu kheṭānāṃ dinayojanabhuktayaḥ || 27 ||

guṇo bhaved yojanabhuktireṣā tadā svakakṣyaiva haro grahāṇām |
kalāgatiśced guṇako'tra hāro bhacakraliptā nijamadhyanītau || 28 ||

[The number] 216000 (*ajñānāntakara*) multiplied by the revolutions of Moon is called the *ākāśakakṣyā*. [This *ākāśakakṣyā*] divided by the number of civil days gives the [common] mean daily motion in *yojanas* (*dinayojanagati*) of the planets. The *ākāśakakṣyā* divided by the revolutions of the planets give the orbits (*kakṣyās*) of the planets. It has been said by the learned that the orbit of the Sun multiplied by 60 (*nīti*) would be the *nakṣatra-kakṣyā*.

Or else, the mean motion (*madhyamabhukti*) of the Moon in minutes multiplied by 10 (*nayana*) would be the [common] mean daily motion in *yojanas* (*dinayojanagati*) of the planets in their respective orbits.

While finding the mean longitudes of the planets, if *dinayojanabhukti* is the *guṇakāra* then the respective *kakṣyās* would be the *hārakas*. If the daily motion of the planets in minutes would be *guṇaka* then 21600 (*bhacakraliptā*) would be the *hāra*.

The mean linear velocity of all the planets is assumed to be the same in the Indian astronomical texts. Hence the mean daily motion in *yojanas* (*dinayojanagati*) is the same for all the planets. The *ākāśakakṣyā* (S) or the total distance covered by each planet is stated to be

$$\begin{aligned} S &= 216000 \times \text{no. of Moon's revolutions} \\ &= 12474720576000 \text{ } yojanas, \end{aligned} \tag{1.92}$$

and *dinayojanagati* (V) is given by

$$V = \frac{S}{D_c} \simeq 7906. \tag{1.93}$$

Here, 216000 *yojanas* is the *kakṣyā* of the Moon itself. The *kakṣyā* of a planet (O), is defined by:

$$O = \frac{\bar{a}k\bar{a}\acute{s}akak\d{s}y\bar{a}}{svaparyaya} = \frac{S}{R_c}. \tag{1.94}$$

The value of the *kakṣyas* of different planets obtained by substituting their revolution number R_c in (1.94), are listed in Table 1.24.

Planet	*kakṣyā* (in *yojanas*)
Sun	2887666.800000
Moon	216000.000000
Moon's apogee	25551485.247400
Mars	5431291.459859
Mercury	695473.416208
Jupiter	34250133.368477
Venus	1776421.436127
Saturn	85114493.163396
Moon's node	53718018.550894

Table 1.24 The *kakṣyās* of planets.

In the last quarter of verse 26 it has been stated that the *kakṣyā* of the Sun multiplied by 60 gives the *nakṣatrakakṣyā*. That is,

$$naks\d{.}atrakaks\d{.}y\bar{a} = 2887666.799 \times 60 \approx 173260008.$$

Now the *madhyamabhukti* of the Moon is $790.581'$. This multiplied by 10 will also give the common *dinayojanabhukti* of the planets. That is,

$$V = 790.581 \times 10 \approx 7906, \tag{1.95}$$

which is the same as (1.93)

Let θ be the mean longitude at the end of the *ahargaṇa* A in revolutions, then

$$\begin{aligned}\theta &= \frac{A \times V}{O} \\ &= \frac{A \times g'_m \text{ (in min.)}}{21600}.\end{aligned} \tag{1.96}$$

These relations are a direct consequence of (1.93), as

$$\frac{V}{O} = \frac{S}{D_c O} = \frac{R_c}{D_c}, \tag{1.97}$$

and

$$g'_m \text{ (in min.)} = \frac{R_c \times 21600}{D_c}. \tag{1.98}$$

Chapter 2
अल्पगुणहारानयनम्
Obtaining smaller *guṇas* and *hāras*

२.१ ग्रहाणां महागुणकारहारनिर्वचनम्

2.1 Definition of the *mahāguṇakāras* and *mahāhāras* of the planets

मन्दादिहारगुणिता भगणा युतोना
ज्ञानीन्द्रसंगुणधनादिगुणैर्गुणाः स्युः ।
मन्दादिहारहतभूदिवसाश्च हाराः
प्रोक्ता महागुणहरास्त इमेऽपवर्त्याः ॥ १ ॥

mandādihāraguṇitā bhagaṇā yutonā
jñānīndrasaṃguṇadhanādiguṇairguṇāḥ syuḥ |
mandādihārahatabhūdivasāśca hārāḥ
proktā mahāguṇaharāsta ime'pavartyāḥ || 1 ||

The product of 200 (*jñānīndra*) and the [*śakābda*]-*guṇakāras*, beginning with *dhana* (9), has to be added to or subtracted from the product of the [*śakābda*]-*hāras* [of the planets], beginning with *manda* (85), and their respective revolution numbers. These are *guṇas* [of the planets]. The [*śakābda*]-*hāras*, beginning with *manda*, multiplied by the the number of civil days (*bhūdina*) in a *mahāyuga* are the *hāras*. These are the *mahāguṇas* and *mahāhāras*, whose *apavartana* is to be done (they have to be factored by their GCD).

In the previous chapter, the methods outlined for obtaining the mean planets essentially involved multiplication of the *ahargaṇa* A by the revolution numbers of the respective planets and division by the civil days in a *mahāyuga*. The mean longitudes of the planets thus obtained were further refined by applying the *śakābdasaṃskāra*. The same mean longitudes can also be obtained straightaway by multiplying and dividing the *ahargaṇa* by certain multipliers and divisors respectively called the *mahāguṇakāras* and *mahāhāras*.

Let G be the *mahāguṇakāra* of the planets. They are given by

V. Pai et al., *Karaṇapaddhati of Putumana Somayājī*, Sources and Studies in the History of Mathematics and Physical Sciences, https://doi.org/10.1007/978-981-10-6814-0_2

$$G = R \times h + 200 \times g \quad \text{(for Mars, Mer., Sat. \& Moon's nodes).} \qquad (2.1)$$
$$= R \times h - 200 \times g \quad \text{(for the other planets).} \qquad (2.2)$$

The *mahāhāra* (H) for all the planets is the product of the *bhūdina* and the respective *śakābdahāra*.

$$H = D_c \times h.$$

For the Sun, the *mahāguṇa* and *mahāhāra* are its own revolution number and the *bhūdina* respectively. That is,

$$G = 4320000,$$
$$\text{and} \qquad H = 1577917500. \qquad (2.3)$$

The values of *mahāguṇakāras* and *mahāhāras* of all the other planets are listed in Table 2.1, along with the *vākyas* given in Commentary I.[1]

Now the ratio of the *śakābda* corrected number of revolutions (see section 1. 13) to the number of civil days in a *mahāyuga* is

$$\frac{R_c}{D_c} = \frac{R \pm \dfrac{200 \times g}{h}}{D_c} = \frac{R \times h \pm 200 \times g}{D_c \times h}. \qquad (2.4)$$

It is the numerator and denominator of (2.4) that are referred to as the *mahāguṇakāras* and the *mahāhāras* respectively.

२.२ दृढगुणहारानयनाय अपवर्तनप्रक्रिया

2.2 The process of *apavartana* for obtaining the *dṛḍhaguṇahāras*

राश्योरन्योन्यहरणे शेषः स्यादपवर्तनम् ।
तेन तौ विहृतौ राशी दृढाख्यावपवर्तितौ ॥ २ ॥

rāśyoranyonyaharaṇe śeṣaḥ syādapavartanam |
tena tau vihṛtau rāśī dṛḍhākhyāvapavartitau || 2 ||

In the process of mutual division of the two *rāśis*, whatever remains is *apavartana*. [The results obtained by dividing] the two *rāśis* by that *apavartana* are called the *dṛḍhas*.

[1] {KP 1956}, pp. 61-62.

Planet	The *mahāguṇakāras* and *mahāhāras*			
		in *kaṭapayādi*		in numerals
Moon	*G*	नीतिसकलनिधिर्नळभृत्	*nītisakalanidhirnaḷabhṛt*	4909031760
	H	इनो मासादिधुरो राज्यवलये	*ino māsādidhuro rājyavalaye*	134122987500
Moon's apogee	*G*	तद्वर्गजो नाभिमतम्	*tadvargajo nābhimatam*	65408346
	H	अनेन शिवो धनी विभुः पापारिः	*anena śivo dhanī vibhuḥ pāpāriḥ*	211440945000
Mars	*G*	अवति प्रीतिं साधुगणः	*avati prītiṃ sādhugaṇaḥ*	539762640
	H	अग्निमरुत्पतिर्नृपजनस्थले	*agnimarutpatirnṛpajanasthale*	370810612500
Mercury	*G*	अनर्थाङ्गहारीशङ्करोभूत्	*anarthāṅgahārīśaṅkarobhūt*	4215283700
	H	अग्निमरुत्पतिर्नृपजनस्थले	*agnimarutpatirnṛpajanasthale*	370810612500
Jupiter	*G*	उर्वीन्द्रो गजमर्मजित्	*urvīndro gajamarmajit*	85583240
	H	अग्निमरुत्पतिर्नृपजनस्थले	*agnimarutpatirnṛpajanasthale*	370810612500
Venus	*G*	अहीशो नगरनाशतुष्टः	*ahīśo nagaranāśatuṣṭaḥ*	1650230580
	H	अग्निमरुत्पतिर्नृपजनस्थले	*agnimarutpatirnṛpajanasthale*	370810612500
Saturn	*G*	उर्वीमोक्षविभावगा	*urvīmokṣavibhāvagā*	34446540
	H	अग्निमरुत्पतिर्नृपजनस्थले	*agnimarutpatirnṛpajanasthale*	370810612500
Moon's node	*G*	रङ्गे जाबालिवासः	*raṅge jābālivāsaḥ*	7433832
	H	अनूननीतिर्बलधन्वी नाशे	*anūnanītirbaladhanvī nāśe*	50493360000

Table 2.1 The *mahāguṇakāras* and *mahāhāras* of planets given in Commentary I (see {KP 1956}, pp. 61-62).

It may be noted from Table 2.1, that the *mahāguṇakāras* and *mahāhāras* are very large numbers and performing computations with them as such would be a difficult task. The above verse prescribes a certain procedure to obtain what are known as *dṛḍhaguṇakāras* and *dṛḍhahārakas* by which the computations are made simpler.

The principle behind the desired simplification lies in finding the greatest common divisor (GCD), which is called *apavartana*.[2] The name *apavartana*, which literally means changing or transforming, is employed to refer to the GCD, using which the *mahāguṇakāras* and *mahāhāras* are transformed into *dṛḍhaguṇakāras* and *dṛḍhahāras*. The *dṛḍhaguṇakāras*, *dṛḍhahāras* and the *apavartana* corresponding to the *mahāguṇahāras* of all the planets are listed in Table 2.2.

By taking a specific example we shall now illustrate the *apavartana* process leading to the *dṛḍhaguṇakāras* and *dṛḍhahāras*. In the case of the Sun, the *mahāguṇakāra* and the *mahāhāra*, are: $G = 4320000$, $H = 1577917500$. Now dividing H by G,

$$\frac{1577917500}{4320000} = 365 + \frac{1117500}{4320000},$$

the remainder is 1117500. Again we divide G by this remainder, we have

$$\frac{4320000}{1117500} = 3 + \frac{967500}{1117500}.$$

The remainder now obtained is 967500. Again dividing 1117500 by 967500

$$\frac{1117500}{967500} = 1 + \frac{150000}{967500},$$

a remainder of 150000 is obtained. Continuing the process,

$$\frac{967500}{150000} = 6 + \cfrac{1}{2 + \cfrac{1}{4 + \cfrac{7500}{15000}}},$$

we obtain the remainder 7500. At this stage, when we divide 15000 by 7500, the remainder becomes zero. The number 7500, which is the last non-zero remainder, is called the *apavartana*. The quantities obtained by dividing 4320000 and 1577917500 by 7500 are called *dṛḍhaguṇakāra* (G_d) and *dṛḍhahāra* (H_d) respectively, and in the case of the Sun they are 576 and 210389 respectively. G_d and H_d found in a similar manner for all the other planets are listed in

[2] The word *apavartana* is employed in the above verse in two senses: (i) in the verbal form and (ii) in the noun form. In the noun form it refers to the GCD. The verbal form which appears towards the end of the verse refers to the process of factoring out the GCD from both the numerator and the denominator.

Table 2.2. They have also been presented in the form of *vākyas* in the Commentary I[3] and these are listed in Table 2.3.

Planet	*apavartana* (G.C.D.)	*dṛḍhaguṇakāra* G_d	*dṛḍhahāra* H_d
Sun	7500	576	210389
Moon	60	81817196	2235383125
Moon's apogee	6	10901391	35240157500
Mars	60	8996044	6180176875
Mercury	100	42152837	3708106125
Jupiter	940	91046	394479375
Venus	60	27503843	6180176875
Saturn	60	574109	6180176875
Moon's node	24	309743	2103890000

Table 2.2 The *dṛḍhaguṇakāras* and the *dṛḍhahāras* of planets.

Planet	*apavartana* (G.C.D.)	*dṛḍhaguṇakāra* G_d	*dṛḍhahāra* H_d
Sun	*ūnamāsam*	*tatsama*	*dhījagannūpuram*
Moon	*nītiḥ*	*co̱likathā kuñjapade*	*śatrukulaṃ halīmallārīndre*
Moon's apogee	*tajjñaḥ*	*yuddhagaḥ pīnadhānuṣkaḥ*	*jñānaśaṃsī medhyajño vidrumāṅgaḥ*
Mars	*nītiḥ*	*bhuvane stabdho dhījit*	*māsajātasampanno jayantaḥ*
Mercury	*anīka*	*saṅgaharo śaṅkaro'bhūt*	*mādrīpatirnayadānasaṅgaḥ*
Jupiter	*avadhī*	*tattvajñayoddhā*	*māṃsagandhisabhāviddholam*
Venus	*nṛttam*	*garbhajālaghno matsarī*	*māsajātasampanno jayantaḥ*
Saturn	*netā*	*dhaniṣko vasumān*	*māsajātasampanno jayantaḥ*
Moon's node	*vajram*	*gūḍhasiddhirnṛgaḥ*	*anūnanidhirjalānāṃ pūram*

Table 2.3 *Vākyas* for the *dṛḍhaguṇakāras* and the *dṛḍhahāras* (see Table 2.2) as given in the Commentary I.

[3] {KP 1956}, pp. 63-66.

२.३ महागुणकारमहाहाराभ्यां मध्यग्रहानयनम्

2.3 Mean longitudes of the planets from *mahā-guṇakāras* and *mahāhāras*

महागुणघ्नाद् द्युगणादभीष्टात् महाहराप्ता भगणादिखेटाः ।
कल्यादिजैः स्वध्रुवकैः समेताः फणी तु तत्र ध्रुवतो विशोध्यः॥ ३ ॥

mahāguṇaghnād dyugaṇādabhīṣṭāt
mahāharāptā bhagaṇādikheṭāḥ |
kalyādijaiḥ svadhruvakaiḥ sametāḥ
phaṇī tu tatra dhruvato viśodhyaḥ || 3 ||

The desired *ahargaṇa* when multiplied by the *mahāguṇakāras* and divided by the *mahāhāras* would give the elapsed revolutions etc. of the planets. The initial position of the planets at the beginning of *kali* (*kalyādidhruvas*) are to be added [to them]. In the case of *Rāhu*, the result has to be subtracted from the [*Kali*] *dhruva*.

The formula prescribed in the above verse for obtaining the mean longitude θ_c of a planet from the *mahāguṇakāras* (G) and the *mahāhāras* (H) may be represented as

$$\theta_c = \theta_{k0} \pm \frac{A \times G}{H}, \tag{2.5}$$

where A represents the *ahargaṇa* and θ_{k0} the *kalyādidhruva* of the planet (discussed in the next verse). While the sign '+' is to be chosen for all the planets including Moon's apogee; in the case of *Rāhu*, '−' is to be chosen since its motion is retrograde.

Illustrative example

We shall illustrate the procedure for finding the mean longitude of the Moon. Let the *ahargaṇa* chosen be $A = 1754000$.

From Table 2.1, the *mahāguṇakāra* (G) and *mahāhāra* (H) of the Moon are 4909031760 and 134122987500 respectively.[4] Hence,

$$\begin{aligned}\frac{A \times G}{H} &= \frac{1754000 \times 4909031760}{134122987500} \\ &= 64198.105541304023. \end{aligned} \tag{2.6}$$

That is, at the end of 1754000 civil days the Moon has completed 64198 revolutions, since the commencement of *kaliyuga*. The *rāśi*, degrees etc., traversed by the Moon in the 64199th revolution is to be found from the fractional

[4] We can use the *dṛḍhaguṇakāra* $G_d = 81817196$ and *dṛḍhahāra* $H_d = 2235383125$, instead of G and H in the computation of the mean longitude.

part. The mean longitude when expressed in *rāśis* etc., may be written as $1^r7°59'41''31'''48''''$.

Adding the *kalyādidhruva*[5] θ_{k0} to the value of the longitude obtained above, we get the actual value of the mean longitude of the Moon to be

$$\begin{aligned}\theta_c \ (\text{Moon}) &= 1^r7°59'41''31'''48'''' + 0^r6°23'36''42'''21'''' \\ &= 1^r14°23'18''14'''9''''.\end{aligned} \tag{2.7}$$

The mean longitudes of all the other planets obtained in a similar manner using the *mahāguṇakāras* and *mahāhāras* are listed in Table 2.4.

Planet	mean longitudes (θ_c) obtained from *mahāguṇakāras* and *mahāhāras* corresponding to $A = 1754000$					
	sign (*rāśi*)	degrees (°)	minutes (′)	seconds (″)	thirds (‴)	fourths (⁗)
Sun	0	27	24	55	48	22
Moon	1	14	23	18	14	9
Moon's apogee	11	2	32	8	58	42
Mars	1	20	44	35	3	3
Mercury	8	26	27	22	46	19
Jupiter	10	8	41	6	38	21
Venus	11	27	22	8	48	43
Saturn	11	2	37	47	39	8
Moon's node	4	1	26	18	55	30

Table 2.4 Mean longitudes of the planets corresponding to $A = 1754000$ computed using the *mahāguṇakāras* and *mahāhāras*.

२.४ ग्रहाणां कल्यादिध्रुवाः

2.4 *Kalyādidhruvas* of the planets

गोत्रोत्तुङ्गहताद् धनादिगुणतो मन्दादिहारोद्धृताः
कल्यादौ रजनीकरादिविहगा लिप्तादयः स्युः क्रमात् ।
तेषामिन्दुजभूमिजार्कतनयाः शोध्याः पुनर्मण्डलात्
तुङ्गो भत्रयसंयुतस्तुहिनगोः पातस्तु षड्भान्वितः ॥ ४ ॥

[5] The *kalyādidhruvas* are presented in Table 2.5.

gotrottuṅgahatād dhanādiguṇato mandādihāroddhṛtāḥ
kalyādau rajanīkarādivihagā liptādayaḥ syuḥ kramāt |
teṣāmindujabhūmijārkatanayāḥ śodhyāḥ punarmaṇḍalāt
tuṅgo bhatrayasaṃyutastuhinagoḥ pātastu ṣaḍbhānvitaḥ || 4 ||

The multipliers commencing with 9 (*dhana*) are to be multiplied by 3623 (*gotrottuṅga*) and divided by the divisors 85 (*manda*) etc. [The results obtained would be] the initial positions at the beginning of *kali* (*kalyādidhruva*), in minutes, of the planets starting with Moon (*rajanīkara*), in order. Further, among them [the *dhruvas* of] Mercury, Mars and Saturn have to be subtracted from 12 *rāśis*; and to [the *dhruva* of] Moon's apogee (*tuṅga*) three *rāśis* are to be added; and to that of *Rāhu* 6 *rāśis* should be added.

The number 3623 appearing in the above verse is the *kalyabda* corresponding to the *śakābda* 444. It may be recalled that in prescribing the *śakābdasaṃskāra* (see verse 27 of chapter 1) there is an implicit assumption that if we compute the mean longitudes of the planets based on the *Āryabhaṭan* parameters it should give accurate results at the end of *Kalyabda* 3623 (*śakābda* 444), as the *śakābda-saṃskāra* for that year is taken to be zero.

Now, the *śakābda* correction is applied starting from the end of the *kali* year 3623. However, if we want to make all our calculations by using the *kalyahargaṇa*, then we will have to modify the *kalyādidhruvas* prescribed by Āryabhaṭa. The *śakābdasaṃskāra* was such that the *madhyamagatis* of the Moon, its apogee, Jupiter and Venus got reduced. Therefore, the mean longitudes obtained by *mahāguṇakāra* would be less than those obtained by employing the values given in *Āryabhaṭīya*. The amounts by which they become deficient over a period of 3623 years have to be added to the *Āryabhaṭīya-dhruvas*. In the case of Mars, Mercury and Saturn, these amounts have to be subtracted from the *Āryabhaṭīya-dhruvas*. Since the *Āryabhaṭīya-dhruvas* for these planets are zero, they have to be subtracted from 12 *rāśis*.

We know that the *saṃskāra* to be applied per year is $\frac{g}{h}$. Therefore, for 3623 years it would be

$$3623 \times \frac{g}{h}.$$

This would be in minutes and has to be added to the *Āryabhaṭīya-dhruvas* in the case of the Moon, Jupiter and Venus. Since *Āryabhaṭīya-dhruvas* are zero for all the planets except for the apogee and the node of the Moon, the above values would themselves be the *kalyādidhruva*. That is,

$$\theta_{k0} = 3623 \times \frac{g}{h} \quad \text{(for Moon, Jupiter and Venus)}.$$

But in the case of the apogee of the Moon, three *rāśis* are to be added. Hence,

$$\theta_{k0} = 3623 \times \frac{g}{h} + 90^\circ \quad \text{(for the apogee of Moon)}.$$

In the case of the other three planets, namely Mars, Mercury and Saturn the above values have to be subtracted from 12 *rāśis*. That is,

$$\theta_{k0} = 360° - 3623 \times \frac{g}{h} \quad \text{(for Mars, Mercury and Saturn).}$$

In the case of *Rāhu*, the *madhyamagati* is increased because of *śakābdasaṃskāra*, but the motion is retrograde. The *Āryabhaṭīya-dhruva* for (*Rāhu*) is 180°. Hence its *kalyādidhruva* is

$$\theta_{k0} = 3623 \times \frac{g}{h} + 180° \quad \text{(for } \textit{Rāhu}\text{).}$$

Illustrative example

We now illustrate the prescription given in the above verse by finding out the *kalyādidhruvas* of two planets, namely Moon and Mercury. In the case of Moon, g and h are 9 and 85 respectively. Hence,

$$\theta_{k0} \ \text{(Moon)} = 3623 \times \frac{9}{85} = 383.6117647058',$$

since the *Āryabhaṭīya-dhruva* for Moon is zero. By converting this into degrees etc. we have

$$\theta_{k0} \ \text{(Moon)} = 6°23'36''42'''21''''.$$

The values of g and h for Mercury are 420 and 235 respectively. Hence,

$$|\theta_{k0} \ \text{(Mercury)}| = 3623 \times \frac{420}{235} = 6475.14893617'.$$

This has to be subtracted from 12 *rāśis*. Therefore, the *kalyādi-dhruva* for Mercury is

$$\theta_{k0} \ \text{(Mercury)} = 8^r 12°4'51''3'''49''''.$$

The *kalyādi-dhruvas* for the various planets are listed in Table 2.5. The Malayalam Commentaries I and II give the *kalyādidhruvas* in terms of *vākyas*.[6]

२.५ वल्ल्युपसंहारप्रक्रिया

2.5 The *Vallyupasaṃhāra* technique

Having delineated the procedure for obtaining the mean longitudes by making use of the *mahāguṇakāras*, *mahāhāras* and appropriate *dhruvas* of the planets, the text proceeds to explain two methods by which the ratio of two large integers can be systematically and successively approximated by ratios

[6] {KP 1956}, pp. 68-69.

Planet	*kalyādidhruvas* of the planets					
	sign (*rāśi*)	degrees (°)	minutes (′)	seconds (″)	thirds (‴)	fourths (⁗)
Sun	0	0	0	0	0	0
Moon	0	6	23	36	42	21
Moon's apogee	3	29	17	25	31	20
Mars	11	18	26	14	2	33
Mercury	8	12	4	51	3	49
Jupiter	0	12	4	35	59	59
Venus	1	9	18	48	15	19
Saturn	11	24	51	39	34	28
Moon's node	6	24	31	50	37	30

Table 2.5 *Kalyādidhruvas* of the planets.

of smaller integers. This technique, is referred to as *vallyupasaṃhāra* and the two methods by which the simplification is achieved are explained in the following sections. As we shall see, the *vallyupasaṃhāra* technique is essentially the same as the technique of continued fraction expansion.

२.५.१ वल्ल्युपसंहारे प्रथमप्रकारः

2.5.1 Vallyupasaṃhāra: ***Method I***

अन्योन्यं विभजन्महागुणहरौ यावद्विभक्तेऽल्पता
तावल्लब्धफलानि रूपमपि च न्यस्येदधोऽधः क्रमात् ।
प्रक्षिप्यान्त्यमुपान्तिमेन गुणिते स्वोर्ध्वे तदन्त्यं त्यजेत्
भूयोऽप्येष विधिर्भवेद् गुणहरौ स्यातां तदोर्ध्वस्थितौ ॥ ५ ॥

anyonyaṃ vibhajanmahāguṇaharau yāvadvibhakte'lpatā
tāvallabdhaphalāni rūpamapi ca nyasyedadho'dhaḥ kramāt |
prakṣipyāntyamupāntimena guṇite svordhve tadantyaṃ tyajet
bhūyo'pyeṣa vidhirbhaved guṇaharau syātāṃ tadordhvasthitau || 5 ||

Divide the *mahāguṇakāra* and *mahāhāra* mutually by each other till the remainder becomes small. Having placed the successive quotients (*labdaphalas*) obtained till then one below the other with 1 [at the bottom most place], the product of the penultimate (*upāntima*) number and the one preceding [to that] is added to the last number. [Then] the last number is dropped. Again this process is repeated. The two [numbers] that remain at the top would then be the [smaller] *guṇa* and *hāra*.

Let G and H denote the two large numbers corresponding to the *mahāguṇa* and *mahāhāra* respectively, usually $G < H$. The above verse presents a certain method by which we can obtain a set of optimal approximations to $\frac{G}{H}$. In other words, the method helps us to obtain smaller *guṇakāras* and *hārakas* whose ratios approximate the value of $\frac{G}{H}$. As we shall see, this method is nothing but finding approximation of a rational number by the convergents of the continued fraction expansion of that number. Assuming $H > G$, we divide H by G. Let

$$\frac{H}{G} = q_1 + \frac{r_1}{G}.$$

Here, q_1 is the first quotient and r_1 the first remainder. We now divide G by r_1 to obtain the second quotient (q_2) and the second remainder (r_2).

$$\frac{G}{r_1} = q_2 + \frac{r_2}{r_1}.$$

Repeating this process, we get the quotient q_3 and the remainder r_3.

$$\frac{r_1}{r_2} = q_3 + \frac{r_3}{r_2}.$$

In general,

$$\frac{r_{n-1}}{r_n} = q_{n+1} + \frac{r_{n+1}}{r_n}.$$

Hence, the ratio $\frac{H}{G}$ may be written as

$$\frac{H}{G} = q_1 + \cfrac{1}{q_2 + \cfrac{1}{q_3 + \ldots + \cfrac{1}{q_{n+1} + \ldots}}}.$$

This is the expression for $\frac{H}{G}$ in terms of a continued fraction.[7]

The successive quotients $q_1, q_2, q_3 \ldots\ldots q_{n+1}$ obtained are referred to as *vallīphalas.* These are placed one below the other with 1 at the bottom as shown in the Table 2.6. By doing certain operations known as *vallyupasaṃhāra* with these *vallīphalas*, one is led to the *alpaguṇakāra* and *alpahāraka* as explained below with the help of the Table 2.6.

Considering the first two columns in Table 2.6, it may be noted that the entries in the second column are essentially the same as those in the first but for two changes:

1. In the second column the last entry of the first column has been omitted.
2. The penultimate entry of the second column is equal to the sum of the last entry and the product of the penultimate entry of the first column and the one preceding it.

[7] Further details regarding the continued fraction expansion of a rational number are given in Appendix A.

q_1	q_1	q_1		H_n
q_2	q_2	q_2		G_n
q_3	q_3	q_3		—
q_4	q_4	q_4	... ...	—
$\vdots$	$\vdots$	$\vdots$		$\vdots$
$\vdots$	$\vdots$	$\vdots$		$\vdots$
q_{n-1}	q_{n-1}	$((q_{n-1} \times (q_n \times q_{n+1} + 1)) + q_{n+1})$	... ...	—
q_n	$(q_n \times q_{n+1} + 1)$	$(q_n \times q_{n+1} + 1)$		—
q_{n+1}	q_{n+1}	—	... ...	—
1	—	—		—

Table 2.6 *Vallyupasaṃhāra*: Method I.

By the phrase *"bhūyo'pyeṣa vidhirbhavet"* it is implied that the same procedure is to be repeated in constructing the successive columns from the preceding ones. This procedure has to be continued till we finally arrive at only two terms. These two are known as *alpaguṇakāra* (G_n) and *alpahāraka* (H_n).

Rationale behind *Vallyupasaṃhāra* Method I:

Consider the continued fraction expansion of $\frac{H}{G}$. When r_{n+1} is small, this may be approximated by

$$\frac{H}{G} \approx q_1 + \cfrac{1}{q_2 + \cfrac{1}{q_3 + \ldots + \cfrac{1}{q_{n-1} + \cfrac{1}{q_n + \cfrac{1}{q_{n+1}}}}}}$$

$$= q_1 + \cfrac{1}{q_2 + \cfrac{1}{q_3 + \ldots + \cfrac{1}{q_{n-1} + \cfrac{q_{n+1}}{q_n \times q_{n+1} + 1}}}}$$

$$= q_1 + \cfrac{1}{q_2 + \cfrac{1}{q_3 + \ldots + \cfrac{1}{q_{n-2} + \cfrac{q_n \times q_{n+1} + 1}{(q_{n-1} \times (q_n \times q_{n+1} + 1)) + q_{n+1}}}}}.$$

A comparison of the above expression with Table 2.6, clearly shows that the penultimate elements of the different columns in Table 2.6 are the same as the ones that are obtained in the continued fraction expansion. This explains the correspondence between the process of *vallyupasaṃhāra* described in verse 5 and the continued fraction expansion of a rational number.

In the first stage (first column), we can consider 1, q_{n+1} and q_n as the last, penultimate and preceding entries. In the next stage, from the above expression, we can see that q_{n+1}, $(q_n \times q_{n+1}+1)$ and q_{n-1} are the corresponding entries. Similarly, in the next stage, $(q_n \times q_{n+1} + 1)$, $(q_{n-1} \times (q_n \times q_{n+1} + 1) + q_{n+1})$ and q_{n-2} are the last, penultimate and the preceding entries. The process is repeated further. This is precisely the process of *vallyupasaṃhāra* described in verse 5.

We now illustrate this method of *vallyupasaṃhāra* by considering a specific numerical example.

Illustrative example

Consider $G = 576$ and $H = 210389$ which are the *dṛḍhaguṇakāra* and the *dṛḍhahāraka* of the Sun. Dividing H by G, we get

$$\frac{H}{G} = \frac{210389}{576} = 365 + \frac{149}{576}.$$

Here the quotient is 365 and the remainder is 149. We once again divide 576 by 149 to get 3 as the quotient and 129 as remainder. Repeating the process of division we obtain the sequence of quotients or the *vallīphalas*:

365, 3, 1, 6 and 2.

Here we have terminated the process of mutual divison with $q_{n+1} = q_5 = 2$, as $r_5 = 2$ may be considered small. Placing these *vallīphalas* one below the other with 1 at the bottom, we have

365
3
1
6
2
1

The bottom most entry 1 is called *antya* or *antima* and the preceding entry 2 is called *upāntima*. Multiplying 6 by 2 and adding 1, the result would be 13. According to the prescription given for constructing the next column, the penultimate entry 2 of the previous column has to be made the last entry of the present column. Then 6 has to be replaced by 13 and 1 has to be deleted. Thus, the new *vallī* becomes

365	365
3	3
1	1
6	13
2	2
1	

Repeating the process till only two terms remain in the last column, we obtain the following array of numbers.

365	365	365	365	21185
3	3	3	58	58
1	1	15	15	
6	13	13		
2	2			
1				

Here, 58 and 21185 are the *alpaguṇakāra* and *alpahāra* of the Sun respectively. Thus we have $\frac{58}{21185}$ as an approximation to $\frac{576}{210389}$ which is correct to more than four decimal places.

This procedure leads to only one pair of *alpaguṇakāra* and *alpahāraka*. The text now proceeds to describe another method for obtaining several such pairs.

२.५.२ वल्ल्युपसंहारे द्वितीयप्रकारः

2.5.2 Vallyupasaṃhāra: ***Method II***

अन्योन्याहृतभाज्यहारकफलं सर्वं त्वधोऽधो न्यसेत्
एकत्राद्यफलेन हीनमपरत्रैकं द्वयोश्चोपरि ।
कुर्याद् वल्ल्युपसंहृतिं ह्युपरितः पूर्वप्रणाशं विना
त्याज्यं तत्प्रथमोर्ध्वगं हरगुणाश्शिष्टाश्च वा स्वेच्छया ॥ ६॥

anyonyāhṛtabhājyahārakaphalaṃ sarvaṃ tvadho'dho nyaset
ekatrādyaphalena hīnamaparatraikaṃ dvayoścopari |
kuryād vallyupasaṃhṛtiṃ hyuparitaḥ pūrvapraṇāśaṃ vinā
tyājyaṃ tatprathamordhvagaṃ haraguṇāśśiṣṭāśca vā svecchayā || 6 ||

The quotients obtained by dividing the *guṇa* and *hāra*, mutually by each other, have to be kept one below the other at one place. Place the same quotients [one below the other] at another place after dropping the first result. Placing 1 above in both the places, may the *vallyupasaṃhāra* be done [as before] from the top without deleting the previous results. [However], 1 from the first column has to be dropped. One can freely choose any of the remaining pairs as *hāras* and *guṇas* respectively.

Let q_1, q_2, $q_3 \ldots$ be the successive quotients. It is said in the verse that, in the first place, all of them should be placed one below the other, whereas in the other place, the same results have to be placed after omitting the first quotient q_1.

I	II
1	
q_1	1
q_2	q_2
q_3	q_3
q_4	q_4
q_5	q_5
$\vdots$	$\vdots$
$\vdots$	$\vdots$

Table 2.7 Arrangement of successive quotients (*vallīphalas*) in the second method of *vallyupasaṃhāra*.

Now, placing 1 at the top in both the *vallīs*, we obtain columns I and II as shown in Table 2.7. The process of *vallyupasaṃhāra* has to be done from the top. The way the successive entries are generated is shown in Table 2.8. Here the subsequent columns generated from column I denoted by the index 'n' ($\mathrm{I}^{(n)}$) give the set of *hārakas* and those ($\mathrm{II}^{(n)}$) generated from column II give the corresponding set of *guṇakas*. Further it is said:

"*tyājyaṃ tatprathamordhvagam*".

That is, 1 at the top of the *hāraka* column $\mathrm{I}^{(n)}$ in Table 2.8 has to be dropped. Once we do this, the entries in column $\mathrm{I}^{(n)}$ and column $\mathrm{II}^{(n)}$ can be as such paired up to get various pairs of *alpahārakas* and *alpaguṇakāras*. For a ratio $\frac{H}{G}$, the different pairs (H_i, G_i) obtained this way are listed in Table 2.9.

Illustrative example

To illustrate this second method of *vallyupasaṃhāra*, we consider the *dṛḍhahāraka* (210389) and *dṛḍhaguṇakāra* (576) of the Sun, as in the previous example. Dividing the two numbers 210389 and 576 mutually, till the remainder becomes either 0 or 1, we get

$$365,\ 3,\ 1,\ 6,\ 2,\ 4 \text{ and } 2$$

I	$\text{I}^{(1)}$	$\text{I}^{(2)}$		$\text{I}^{(n)}$
1	1	1	... ...	$\not{1}$
q_1	q_1	q_1	... ...	$H_1 = q_1$
q_2	$q_1 \times q_2 + 1$	$q_1 \times q_2 + 1$	... ...	$H_2 = q_1 \times q_2 + 1$
q_3	q_3	$(q_1 \times q_2 + 1) \times q_3 + q_1$	... ...	$H_3 = (q_1 \times q_2 + 1) \times q_3 + q_1$
$\vdots$	$\vdots$	$\vdots$	$\vdots$ $\vdots$	$\vdots$
$\vdots$	$\vdots$	$\vdots$	$\vdots$ $\vdots$	$\vdots$
q_n	q_n	q_n	... ...	$H_n = H_{n-1} \times q_n + H_{n-2}$

II	$\text{II}^{(1)}$	$\text{II}^{(2)}$		$\text{II}^{(n)}$
1	1	1	... ...	$G_1 = 1$
q_2	q_2	q_2	... ...	$G_2 = q_2$
q_3	$q_2 \times q_3 + 1$	$q_2 \times q_3 + 1$	... ...	$G_3 = q_2 \times q_3 + 1$
q_4	q_4	$(q_2 \times q_3 + 1) \times q_4 + q_2$	... ...	$G_4 = (q_2 \times q_3 + 1) \times q_4 + q_2$
$\vdots$	$\vdots$	$\vdots$	$\vdots$ $\vdots$	$\vdots$
$\vdots$	$\vdots$	$\vdots$	$\vdots$ $\vdots$	$\vdots$
q_n	q_n	q_n	... ...	$G_n = G_{n-1} \times q_n + G_{n-2}$

Table 2.8 Generating the *alpaguṇakāras* and *alpahārakas* using the second method of *vallyupasaṃhāra*.

as successive quotients (*vallīphalas*). We arrange them in two places one below the other with 1 at the top.

Then we do *vallyupasaṃhāra*, as shown in Table 2.10. After deleting 1 in the $\text{I}^{(7)}$-th column of Tables 2.10 we get the successive *alpaguṇakāras* and the corresponding *alpahārakas* of the Sun which are listed in Table 2.11.

Tables C.1–C.8 in Appendix C list the *alpaguṇakāras* and *alpahārakas*, for all the other planets, obtained in a similar manner.

As explained in Appendix A, the above technique of obtaining *alpaguṇakāras* and the *alpahārakas* is essentially the same as the technique of computing the convergents of a continued fraction.

$$(H_1, G_1) = (q_1,\ \ 1)$$

$$(H_2, G_2) = (q_1 \times q_2 + 1,\ \ q_2)$$

$$(H_3, G_3) = ((q_1 \times q_2 + 1) \times q_3 + q_1,\ \ q_2 \times q_3 + 1)$$

$$(H_4, G_4) = (((q_1 \times q_2 + 1) \times q_3 + q_1) \times q_4 + (q_1 \times q_2 + 1),\ \ (q_2 \times q_3 + 1) \times q_4 + q_2)$$

$$\vdots \qquad \vdots$$

Table 2.9 Pairs of *alpaguṇakāras* and *alpahārakas* obtained using the second method of *vallyupasaṃhāra*.

I	$\text{I}^{(1)}$	$\text{I}^{(2)}$	$\text{I}^{(3)}$	...	$\text{I}^{(7)}$
1	1	1	1	...	1
365	365	365	365	...	$H_1 = 365$
3	1096	1096	1096	...	$H_2 = 1096$
1	1	1461	1461	...	$H_3 = 1461$
6	6	6	9862	...	$H_4 = 9862$
2	2	2	2	...	$H_5 = 21185$
4	4	4	4	...	$H_6 = 94602$
2	2	2	2	...	$H_7 = 210389$
II	$\text{II}^{(1)}$	$\text{II}^{(2)}$	$\text{II}^{(3)}$	...	$\text{II}^{(7)}$
1	1	1	1	...	$G_1 = 1$
3	3	3	3	...	$G_2 = 3$
1	4	4	4	...	$G_3 = 4$
6	6	27	27	...	$G_4 = 27$
2	2	2	58	...	$G_5 = 58$
4	4	4	4	...	$G_6 = 259$
2	2	2	2	...	$G_7 = 576$

Table 2.10 Generating the *alpaguṇakāras* and *alpahārakas* using the second method of *vallyupasaṃhāra* for the *hāraka* $H = 210389$ and *guṇakāra* $G = 576$ associated with Sun.

After presenting the method of *vallyupasaṃhāra*, Putumana Somayājī discribes a technique by which the deviations of these approximate ratios from the actual value can be minimized. This involves finding the so called *dvitīya* and *tṛtīyahārakās*.

alpaguṇakāras G_i	*alpahārakas* H_i
1	365
3	1096
4	1461
27	9862
58	21185
259	94602
576	210389

Table 2.11 The *alpaguṇakāras* and *alpahārakas* of Sun.

२.६ दिनगतिस्थौल्यपरिहाराय द्वितीयतृतीयहारकौ

2.6 Better approximations to the rate of motion: *Dvitīya* and *tṛtīya-hārakas*

The following verses describe how the approximate rates of motion, $\frac{G_i}{H_i}$, can be improved by means of correction terms involving what are termed *"dvitīya"* and *" trtīya-hārakas"*.

स्वल्पौ हारगुणौ महागुणहरक्षुण्णौ तयोरन्तरं
स्वर्णाख्यं क्रमशो महाहरहते स्वल्पे गुणेऽल्पेऽधिके ।
तेनानन्तपुराहतेन हरयोर्घाताद् द्वितीयो हरः
तच्छिष्टेन तथा हरत्रयवधाल्लब्धस्तृतीयो हरः ॥ ७ ॥

svalpau hāraguṇau mahāguṇaharakṣuṇṇau tayorantaram
svarṇākhyaṃ kramaśo mahāharahate svalpe guṇe'lpe'dhike |
tenānantapurāhatena harayorghātād dvitīyo haraḥ
tacchiṣṭena tathā haratrayavadhāllabdhastṛtīyo haraḥ || 7 ||

The difference between the products of the *alpahāra* and *mahā-guṇakāra*, and of the *mahāhāra* and *alpaguṇakāra* is considered to be positive or negative, depending on whether the product of the *mahāhāra* and *alpaguṇakāra* is respectively smaller or greater [than the other]. [The quotient obtained] from the product of the two *hāras* when divided by that [difference] multiplied by 21600 (*anantapura*) is [called] the *dvitīya-hāra* (second divisor). The quotient obtained by dividing the product of the three *hāras* by the remainder is [called] *tṛtīya-hāra* (third divisor).

Let G and H be the *mahāguṇakāra* and *mahāhāra*, and G_i and H_i be the *alpaguṇakāra* and *alpahāra*, for some i. The ratio $\frac{G}{H}$ represents the number of revolutions made by the planet per day. The ratio $\frac{G_i}{H_i}$ obtained by the process of *vallyupasaṃhāra* is an approximation to this. This being the case, $\frac{G_i}{H_i}$ can be greater or smaller than $\frac{G}{H}$. The difference between the two is the error in the rate of motion per day. Denoting this difference by Δ_i, we have

$$\Delta_i = \frac{G}{H} - \frac{G_i}{H_i} = \frac{GH_i - G_iH}{HH_i}. \tag{2.8}$$

Obviously, Δ_i is positive when $GH_i > G_iH$ and negative otherwise. As such, Δ_i represents the error in the rate of motion per day. The magnitude of this, $|\Delta_i|$ has to be added to $\frac{G_i}{H_i}$ if Δ_i is positive, and subtracted otherwise.

We define ϕ_i through the relation:

$$\phi_i = GH_i - G_iH.$$

(In the next verse, ϕ_i would be identified with $\pm r_i$, where r_i is the remainder in the process of mutual division of G and H, at the ith stage).

To obtain the difference between the actual rate of motion $\frac{G}{H}$ and its approximation $\frac{G_i}{H_i}$, in minutes per day, we have to multiply by 21600. Therefore,

$$\begin{aligned} \left(\frac{G}{H} - \frac{G_i}{H_i}\right) \times 21600 &= \left(\frac{GH_i - G_iH}{HH_i}\right) \times 21600 \\ &= \frac{\phi_i}{HH_i} \times 21600. \end{aligned} \tag{2.9}$$

Now,

$$\frac{H \times H_i}{\phi_i \times 21600} = \mathcal{H}_{2i} + \frac{\epsilon_{2i}}{\phi_i \times 21600},$$

where the *dvitīya-hāra*, $\mathcal{H}_{2i}$ is given by

$$\mathcal{H}_{2i} = \left[\frac{H \times H_i}{\phi_i \times 21600}\right]_{int}, \tag{2.10}$$

with '$[\;]_{int}$' denoting the integral part, and ϵ_{2i} is the remainder when $H \times H_i$ is divided by $\phi_i \times 21600$. We may rewrite equation (2.10) including the remainder term ϵ_{2i} as

$$H \times H_i = \phi_i \times 21600 \times \mathcal{H}_{2i} + \epsilon_{2i}. \tag{2.11}$$

Using (2.11) in (2.9),

$$\frac{\phi_i \times 21600}{H \times H_i} = \frac{1}{\mathcal{H}_{2i}} - \frac{\epsilon_{2i}}{\mathcal{H}_{2i} \times H \times H_i}. \tag{2.12}$$

Now the *tṛtīya-hāraka* $\mathcal{H}_{3i}$ is defined by

$$\mathcal{H}_{3i} = \left[\frac{\mathcal{H}_{2i} \times H \times H_i}{\epsilon_{2i}}\right]_{int}. \tag{2.13}$$

Using (2.13) in (2.12),

$$\left(\frac{G}{H} - \frac{G_i}{H_i}\right) \times 21600 = \frac{\phi_i \times 21600}{H \times H_i} \approx \frac{1}{\mathcal{H}_{2i}} - \frac{1}{\mathcal{H}_{3i}}. \tag{2.14}$$

Hence

$$\frac{G}{H} \times 21600 \approx \frac{G_i}{H_i} \times 21600 + \frac{1}{\mathcal{H}_{2i}} - \frac{1}{\mathcal{H}_{3i}}. \tag{2.15}$$

Thus it is seen that *dvitīya-hāraka* and *tṛtīya-hāraka* give the correction terms to be applied to the approximate rate of motion $\frac{G_i}{H_i}$, to make it closer to the actual rate of motion, $\frac{G}{H}$.

२.७ महागुणहारापवर्तनशेषैः द्वितीयहारानयनम्

2.7 The *dvitīya-hāraka* in terms of the remainders in the mutual division of *mahāguṇas* and *mahāhāras*

यद्वा मिथो विहृतहारगुणोत्थशेषैः नूत्नातपत्रगुणितैर्महतोऽत्र हारात् ।
तत्तद्धराभिनिहताद् विहृताद् द्वितीयहाराः भवन्त्यृणधनात्मकलिप्तिकानाम् ॥८॥

yadvā mitho vihṛtahāraguṇotthaśeṣaiḥ
nūtnātapatraguṇitairmahato'tra hārāt |
tattaddharābhinihatād vihṛtād dvitīya-
hārāḥ bhavantyṛṇadhanātmakaliptikānām || 8 ||

Or, [the quotients obtained] when the *mahā-hāra* multiplied by the [*alpa*] *hāra* at each stage is divided by the product of 21600 (*nūtnātapatra*) and the remainders arising from the mutual division of [*mahā*]*hāra* and [*mahā*]*guṇa*, are the *dvitīya-hāras* in negative or positive minutes.

Let r_i be the remainders in the process of mutual division of H and G. Then, this verse states that the *dvitīya-hāra* $\mathcal{H}_{2i}$ is given by

$$\mathcal{H}_{2i} = - \left[\frac{H \times H_i}{r_i \times 21600}\right]_{int} \quad (i = 1, 3, \ldots), \tag{2.16}$$

or

$$= + \left[\frac{H \times H_i}{r_i \times 21600}\right]_{int} \quad (i = 2, 4, \ldots). \tag{2.17}$$

Comparing these equations with (2.10) in the previous section, we need to show that

$$\phi_i \equiv GH_i - HG_i = (-1)^i r_i. \tag{2.18}$$

This result is well known in the theory of continued fractions and is explained in Appendix A. We demonstrate this explicitly for $i = 1, 2$. In the mutual division of G and H,

$$H = Gq_1 + r_1,$$

and the first approximation to $\frac{G}{H}$ is given by $\frac{G_1}{H_1}$, where $G_1 = 1$, $H_1 = q_1$. Then,

$$\begin{aligned} GH_1 - HG_1 &= Gq_1 - (Gq_1 + r_1) \\ &= -r_1. \end{aligned}$$

Now

$$G = r_1 q_2 + r_2,$$

and the second approximation to $\frac{G}{H}$ is given by $\frac{G_2}{H_2}$, where $G_2 = q_2$, $H_2 = q_1 q_2 + 1$. Then,

$$\begin{aligned} GH_2 - HG_2 &= G(q_1 q_2 + 1) - (Gq_1 + r_1)q_2 \\ &= G - r_1 q_2 \\ &= r_2. \end{aligned}$$

Now the *vallyupasaṃhāra* is based on the relations,

$$\begin{aligned} H_{i+1} &= H_i q_{i+1} + H_{i-1}, \\ G_{i+1} &= G_i q_{i+1} + G_{i-1}. \end{aligned}$$

Also

$$r_{i-1} = r_i q_{i+1} + r_{i+1},$$

from the process of mutual division. From these relations, we can prove the general result,

$$GH_i - HG_i = (-1)^i r_i, \qquad (2.19)$$

by induction as explained in Appendix A.

It is remarkable that Putumana Somayājī is clearly aware of this general result, and he makes repeated use of it in different contexts as we shall in the later chapters.

Illustrative examples

We present the *dvitīya* and *tritīya-hāras* ($\mathcal{H}_{2i}$, $\mathcal{H}_{3i}$), the corrected rates of motion and the associated errors for the case of the Moon and Venus, in Tables 2.12 and 2.13 below. In case of the Moon, for $i = 1$, $\mathcal{H}_{2i} = 0$, and the method cannot be applied.

From the tables, we notice that the inclusion of the *dvitīya* and *tṛtīya-hāras* lead to fairly good approximations to the rates of motion even for low values of i.

$G = 4909031760, H = 134122987500$ and $\frac{G}{H} \times 21600 \simeq 790.581003$

i	G_i	H_i	$\frac{G_i}{H_i} \times 21600$	$\mathcal{H}_{2i}$	$\mathcal{H}_{3i}$	$\frac{G_i}{H_i} \times 21600 + \frac{1}{\mathcal{H}_{2i}} - \frac{1}{\mathcal{H}_{3i}}$	Error
2	3	82	790.243902	2	6	790.577236	0.003767
3	28	765	790.588235	−139	26452	790.581007	6.06×10^{-11}
4	115	3142	790.579249	570	37347323	790.581003	$\approx 10^{-16}$

Table 2.12 The *dvitīya-* and *tṛtīya-hāras* for Moon, with the corresponding rates of motion and errors.

$G = 1650230580, H = 370810612500$ and $\frac{G}{H} \times 21600 \simeq 96.127185$

i	G_i	H_i	$\frac{G_i}{H_i} \times 21600$	$\mathcal{H}_{2i}$	$\mathcal{H}_{3i}$	$\frac{G_i}{H_i} \times 21600 + \frac{1}{\mathcal{H}_{2i}} - \frac{1}{\mathcal{H}_{3i}}$	Error
1	1	224	96.428571	−4	19	96.125939	0.001245
2	1	225	96.000000	7	63	96.126984	2.01×10^{-4}
3	3	674	96.142433	−66	10391	96.127185	3.95×10^{-9}
4	7	1573	96.122059	195	403639	96.127185	4.70×10^{-12}
5	10	2247	96.128170	−1015	5354575	96.127185	4.26×10^{-14}

Table 2.13 The *dvitīya-* and *tṛtīya-hāras* for Venus, with the corresponding rates of motion and errors.

२.८ भुक्त्यन्तरगुणहाराः

2.8 *Guṇakāras* and *hāras* for the difference in rates of motion

महागुणास्ते भगणाः प्रकल्प्याः महाहरा भूदिवसाश्च तद्वत् ।
सर्वत्र ते तुल्यहराश्च कार्याः परस्परं योगवियोगकाले ॥ ९ ॥
अन्योन्यहारगुणितौ गुणकौ हारौ च तुल्यहारौ स्तः ।
तत्रापवर्तिताभ्यां हाराभ्यां वा परस्परं गुणयेत् ॥१०॥

mahāguṇāste bhagaṇāḥ prakalpyāḥ mahāharā bhūdivasāśca tadvat |
sarvatra te tulyaharāśca kāryāḥ parasparaṃ yogaviyogakāle || 9 ||

anyonyahāraguṇitau guṇakau hārau ca tulyahārau staḥ |
tatrāpavartitābhyāṃ hārābhyāṃ vā parasparaṃ guṇayet || 10 ||

The *mahāguṇakāras* are to be considered as revolution numbers [of the planets in future operations], and similarly the civil days as the *mahāharas*. In all those instances where they are mutually added or subtracted, [one] has to find the common divisor.

The multipliers and the divisors mutually multiplied become the *guṇakāras* and [the product of] the divisors would be the common divisors. There, the divisors which have been already divided by the *apavartana* can be multiplied mutually.

These verses give a method for obtaining the effective *guṇas* and *hāras* which will appear while calculating the difference in the rates of motion of two planets. Let $G^{(1)}$ and $H^{(1)}$ be the *mahāguṇakāra* and *mahāhāra* of a particular planet. Let $G^{(2)}$ and $H^{(2)}$ be those of a second planet. Here $G^{(1)}$ and $G^{(2)}$ are to be considered as revolution numbers and $H^{(1)}$ and $H^{(2)}$ are to be considered as civil days. Now the ratio $\frac{G^{(1)}}{H^{(1)}}$ called *dinagati* represents the daily motion of the first planet. In order to find the difference in the daily motion of the two planets, we subtract one from the other, and obtain

$$\frac{G^{(1)}}{H^{(1)}} - \frac{G^{(2)}}{H^{(2)}} = \frac{\left(G^{(1)} \times H^{(2)} - G^{(2)} \times H^{(1)}\right)}{H^{(1)} \times H^{(2)}}.$$

Here the common divisor is the product of the two *hāras* $H^{(1)} \times H^{(2)}$ and the multiplier is $\left(G^{(1)} \times H^{(2)} - G^{(2)} \times H^{(1)}\right)$. It may be pointed out that the result would be the same even if we use *dṛḍhaguṇahāras* instead of *mahāguṇahāras*. Then the difference in the ratios of the *guṇakāras* and the *hāras* is

$$\frac{G_d^{(1)}}{H_d^{(1)}} - \frac{G_d^{(2)}}{H_d^{(2)}} = \frac{\left(G_d^{(1)} \times H_d^{(2)} - G_d^{(2)} \times H_d^{(1)}\right)}{H_d^{(1)} \times H_d^{(2)}}.$$

Illustrative example

We know that the *dṛḍhaguṇakāra* and *dṛḍhahāra* of the Sun are 576 and 210389, and those of the Moon are 81817196 and 2235383125 respectively.

Now

$$\textit{ravi-dinabhukti} = \frac{576}{210389} \tag{2.20}$$

and

$$\textit{candra-dinabhukti} = \frac{81817196}{2235383125}. \tag{2.21}$$

The difference in their daily motion known as *bhuktyantara* is given by

$$\begin{aligned} \textit{bhuktyantara} &= \left|\frac{576}{210389} - \frac{81817196}{2235383125}\right| \\ &= \left|\frac{(576 \times 2235383125 - 81817196 \times 210389)}{210389 \times 2235383125}\right| \\ &= \frac{15925857369244}{470300020285625}. \end{aligned} \tag{2.22}$$

The huge numbers appearing in the above equation signify that 15925857369244 is the difference in the number of revolutions (referred to as *paryayāntara* in chapter 4) of the planets in 470300020285625 number of days. We can use the *vallyupasaṃhāra* method for finding the smaller numerators and denominators which approximate the above *bhuktyantara*.

Chapter 3
शशिखण्डध्रुवाद्यानयनम्
Computation of the *khaṇḍa*, *dhruva*, etc. of the Moon

३.१ चन्द्रकेन्द्रस्य अल्पगुणकारहारकाः

3.1 The *alpa-guṇakāras* and *alpa-hārakas* of the *candra-kendra*

विधोस्तदुच्चस्य च पर्ययान्तरं धरादिनौघं च मिथोऽथ संहरेत् ।
फलैरमीभिर्गुणहारकान्नयेद् यथोदितं केन्द्रभवा भवन्ति ते ॥ १ ॥

vidhostaduccasya ca paryayāntaraṃ
dharādinaughaṃ ca mitho'tha saṃharet |
phalairamībhirguṇahārakānnayed
yathoditaṃ kendrabhavā bhavanti te || 1 ||

Let the heap of days (*dharādinaugha*), and the difference in the revolutions of the Moon and its apogee be mutually divided. Then, from these results, the *guṇakāras* and *hārakas* associated with the Moon's anomaly (*candra-kendra*) are to be obtained as stated earlier.

Let $G^{(1)}$, $G^{(2)}$ and $H^{(1)}$, $H^{(2)}$ be the *mahāguṇakāras* and *mahāhāras* of the Moon and that of its apogee respectively (see Table 2.1). Now the daily rate of motion (*dinagati*) of the Moon and its apogee are

$$\frac{G^{(1)}}{H^{(1)}} = \frac{4909031760}{134122987500}, \tag{3.1}$$

$$\frac{G^{(2)}}{H^{(2)}} = \frac{65408346}{211440945000}. \tag{3.2}$$

The difference between the two is the daily rate of motion of the anomaly (*kendra-dinabhukti*) and is given by

$$\frac{4909031760}{134122987500} - \frac{65408346}{211440945000}.$$

V. Pai et al., *Karaṇapaddhati of Putumana Somayājī*, Sources and Studies in the History of Mathematics and Physical Sciences, https://doi.org/10.1007/978-981-10-6814-0_3

Here the *samaccheda* or common-divisor is the heap of days (*dharādinaugha*)

$$134122987500 \times 211440945000 = 28359091223223187500000. \tag{3.3}$$

The number of revolutions made by the Moon (*candraparyaya*) in these many days is

$$4909031760 \times 211440945000 = 1037970314369413200000,$$

and that of its apogee (*tuṅgaparyayas*) is

$$65408346 \times 134122987500 = 8772762772953675000.$$

The difference in the *paryayas* (*paryayāntara*) given above is

$$1029197551596459525000. \tag{3.4}$$

The G.C.D. or the *apavartana* of the *paryayāntara* given by (3.4) and the *sama-ccheda* given by (3.3) is found to be 225000. Dividing these two quantities by their *apavartana*, we get the *dṛḍhaguṇahāras* of the *candra-kendra* (see the last row in Table 3.1), the ratio ($\frac{G}{H}$) of which gives the *dṛḍhakendragati* of the Moon which is given by

$$\frac{G}{H} = \frac{4574211340428709}{126040405436547500}. \tag{3.5}$$

The quotients obtained in the mutual division (*vallyupasaṃhṛta-phalas*) of the above ratio are 27, 1, 1, 4, 12, 4, 15, 7, 1, 13, 2, 9, 1, 2, 1, 1, 1, 1, 1, 1, 2, 1, 1 and 4. The *alpa-guṇakāras* and *alpa-hārakas* obtained by the process of *vallyupasaṃhāra* are listed in Table 3.1.

For a number of days equal to any of the *hārakas* in Table 3.1, the motion in anomaly will not be exactly given by a complete number of revolutions. The *kendrabhukti* of a *hāraka* is the change in Moon's anomaly after a number of days equal to the *hāraka*. We have also listed the *kendrabhuktis* corresponding to some of these *hārakas* in Table 3.2. The *kendrabhuktis* have been computed by choosing the value given by (3.5) as the daily motion of the anomaly.

३.२ चन्द्रखण्डध्रुवयोरानयनम्

3.2 Obtaining the *khaṇḍas* and *dhruvas* for the Moon

देवेन्द्रशर्मैक्यदिनेन्दुतुङ्गभेदार्धयुक्तेष्टदिनेन्दुकेन्द्रात् ।
लिप्तीकृतात् केन्द्रहरेष्वभीष्टेनाहत्य नानार्तिपरैरवाप्ताः ॥ २ ॥

i	*hārakas* (H_i)	*guṇakāras* (G_i)
1	27	1
2	28	1
3	55	2
4	248	9
5	3031	110
6	12372	449
7	188611	6845
8	1332649	48364
9	1521260	55209
10	21109029	766081
11	43739318	1587371
12	414762891	15052420
13	458502209	16639791
14	1331767309	48332002
15	1790269518	64971793
16	3122036827	113303795
17	4912306345	178275588
18	8034343172	291579383
19	12946649517	469854971
20	20980992689	761434354
21	54908634895	1992723679
22	75889627584	2754158033
23	130798262479	4746881712
24	599082677500	21741684881
25	126040405436547500	4574211340428709

Table 3.1 The *guṇakāras* and *hārakas* corresponding to the Moon's anomaly.

अभीष्टहारोर्ध्वहरेण हत्वा पूर्वोदिताभीष्टहरेण हृत्वा ।
तत्राधिकोनं द्युगणात् विशोध्यं हारौजयुग्मत्ववशात् क्रमेण ॥ ३ ॥
शिष्टं शशाङ्कोदितवाक्यखण्डः तस्य ध्रुवस्तद्दिवसस्फुटेन्दुः ।
तथा हराणां ध्रुवकाश्च तैस्तैः दिनैस्समानीतविधुस्फुटानि ॥ ४ ॥

devendraśarmaikyadinenduturiga-
bhedārdhayukteṣṭadinendukendrāt |
liptīkṛtāt kendrahareṣvabhīṣṭe-
nāhatya nānārtiparairavāptāḥ || 2 ||

hārakas (H_i)	*kendrabhukti* (in minutes)	deviation from 21600 (in minutes)
27	21165.27668	−434.72332
28	349.17581	349.17581
55	21514.45249	−85.54751
248	6.98579	6.98579
3031	21598.28195	−1.71805
12372	0.11358	0.11358
188611	21599.98568	−0.01432
1332649	0.01333	0.01333
1521260	21599.99901	−0.00099
21109029	0.00047	0.00047
43739318	21599.99995	-4.95911×10^{-5}
414762891	3.50095×10^{-5}	3.50095×10^{-5}
458502209	21599.99998	-1.33765×10^{-5}

Table 3.2 The *kendrabhuktis* corresponding to different *kendrahārakas* and their deviation from complete revolution.

abhīṣṭahārordhvahareṇa hatvā
pūrvoditābhīṣṭahareṇa hṛtvā |
tatrādhikonaṃ dyugaṇāt viśodhyaṃ
hāraujayugmatvavaśāt krameṇa || 3 ||

śiṣṭaṃ śaśāṅkoditavākyakhaṇḍaḥ
tasya dhruvastaddivasasphuṭenduḥ |
tathā harāṇāṃ dhruvakāśca taistaiḥ
dinaissamānītavidhusphuṭāni || 4 ||

Half of the difference between the longitudes (*bhuktis*) of the Moon and its apogee, corresponding to the sum of 248 (*devendra*) and 55 (*śarma*) days, has to be added to the longitude of Moon's anomaly in minutes for the desired day. This has to be multiplied by the desired divisor (*kendrahāraka*) and divided by the 21600 (*nānārtipara*).

The quotient of this division is multiplied by the previous divisor and divided by the desired divisor, and the resulting remainder is stored. In the case of odd divisors, this remainder (*adhikaśiṣṭa*) has to be subtracted from the *ahargaṇa*; in the case of even divisors the divisor minus the remainder (*ūnaśiṣṭa*) has to be subtracted from the *ahargaṇa*. The result of the subtraction would be the *khaṇḍa* spoken of [in *vākya* texts] (*vākyakhaṇḍa*) for the Moon.

The *dhruva* of that (*khaṇḍa*) is the true Moon (*sphuṭacandra*) at the end of that day (corresponding to the *khaṇḍa*). In the same manner, the *candrasphuṭas* obtained for the *kendrahārakas* are the *dhruvas* of the respective *hārakas*.

The above verses present an algorithm for obtaining a *khaṇḍa* corresponding to a given *hāraka*. A *khaṇḍa* (also referred to as *khaṇḍadina*) is a day close to given *ahargaṇa* when the anomaly is close to zero at the mean sunrise.

In order to determine the *khaṇḍa*, we need to know the *guṇakāras* and *hārakas* associated with the rate of motion of the anomaly (*kendragati*). In the case of Moon, these are listed in Table 3.1. Let us denote the odd *hārakas* by H_1, H_3, $H_5 \ldots$ and the odd *guṇas* by G_1, G_3, $G_5 \ldots$. Similarly, let H_2, H_4, $H_6 \ldots$ and G_2, G_4, $G_6 \ldots$ be the even *hārakas* and *guṇas* respectively. We now present the algorithm to find the *khaṇḍa* corresponding to an *ahargaṇa* given in the above verses.

३.२.१ खण्डानयनप्रकारः

3.2.1 Algorithm to determine the khaṇḍa

The algorithm essentially consists of the following steps:

1. Adding half the anomaly of the Moon in minutes corresponding to 303 days to the value of anomaly corresponding to the *ahargaṇa* A: That is we need to find
$$(\theta_0 - \theta_m)_A + \frac{(\theta_0 - \theta_m)_{303}}{2},$$
where θ_0 and θ_m are the longitudes of the mean Moon and its apogee respectively, and $(\theta_0 - \theta_m)_A$ and $(\theta_0 - \theta_m)_{303}$ are the longitudes of Moon's anomaly after A days and 303 days respectively.
2. Obtaining the quotient Q_i by finding the ratio of the product of the above quanity and the desired *hāraka* H_i (listed in Table 3.1) and dividing by 21600:
$$Q_i = \left[\frac{\left((\theta_0 - \theta_m)_A + \frac{(\theta_0 - \theta_m)_{303}}{2} \right) \times H_i}{21600} \right]_{int}. \tag{3.6}$$
3. Finding the remainder r_i by multiplying the quotient of the above division by the previous divisor (*ūrdhvahāraka*) and dividing it by the desired divisor (*iṣṭahāraka*): That is,
$$\begin{aligned} r_i &= \left[\frac{Q_i \times H_{i-1}}{H_i} \right]_{rem} \\ &= \left[\frac{\left[\frac{(\theta_0 - \theta_m)_A + \frac{(\theta_0 - \theta_m)_{303}}{2}}{21600} \times H_i \right]_{int} \times H_{i-1}}{H_i} \right]_{rem}. \end{aligned} \tag{3.7}$$
Here, '$[\ ]_{rem}$' denotes the remainder of the quantity obtained by finding the ratio.

4. Having obtained r_i, in order to obtain the *khaṇḍa*, the *adhikaśiṣṭa* (r_i) or *ūnaśiṣṭa* ($H_i - r_i$) has to be subtracted from A depending on whether H_i is odd or even respectively. That is, *khaṇḍa*

$$\begin{aligned} K_i &= A - r_i && \text{(for odd } H_i), \\ K_i &= A - (H_i - r_i) && \text{(for even } H_i). \end{aligned} \tag{3.8}$$

Except for the correction term $\frac{(\theta_0-\theta_m)303}{2}$ in (3.6), the rationale for which is not clear, the rest of the steps in the above algorithm can be explained as follows. Consider the anomaly $(\theta_0 - \theta_m)_A$ in minutes, which includes the zero-correction at *kali* beginning (*kalyādidhruva*). This is to be computed using the exact rate of motion of the anomaly $\frac{G}{H}$. This rate of motion can be reasonably approximated by $\frac{G_i}{H_i}$, for a reasonably large value of the *hāraka* H_i. $\frac{(\theta_0-\theta_m)_A}{21600}$ which is the anomaly in revolutions may be expressed as $\frac{n'}{H_i}$ (after subtracting the integral number of revolutions). Hence anomaly at *ahargaṇa* A in revolutions is can be expressed as

$$\frac{(\theta_0 - \theta_m)_A}{21600} = \frac{n'}{H_i}, \tag{3.9}$$

$$\text{or} \qquad \left[\frac{(\theta_0 - \theta_m)_A}{21600} \times H_i\right]_{int} = [n']_{int} = n, \tag{3.10}$$

Thus, we are essentially approximating the anomaly at A by $\frac{n}{H_i}$, where n is an integer. For large H_i (say 12372 or above, see Table 3.2), this is a good approximation. The *khaṇḍa*, K_i corresponds to the day when the anomaly is close to zero. Let $K_i = A - x$. So x is the number of days in which the *kendra* increases by $\frac{n}{H_i}$ revolutions. As the daily motion of the *kendra* is approximately given by $\frac{G_i}{H_i}$ revolutions, we have

$$\frac{G_i}{H_i} \times x - y = \frac{n}{H_i}, \tag{3.11}$$

where y is the number of completed revolutions of the *kendra*. The above equation can be rewritten as

$$G_i x - H_i y = n, \tag{3.12}$$

where we have to solve for x and y in integers. The solution of this is related to the solution of the equation

$$G_i x' - H_i y' = 1. \tag{3.13}$$

Here we may recall the relation between the successive "convergents" in the continued fraction expansion of $\frac{G}{H}$ as given by relation (A.13) of Appendix A,

$$G_i H_{i-1} - H_i G_{i-1} = (-1)^{i+1}. \tag{3.14}$$

Therefore, a solution of (3.13) is given by

$$\begin{aligned} x' &= H_{i-1} && \text{(odd } i\text{)}, \\ x' &= H_i - H_{i-1} && \text{(even } i\text{)}. \end{aligned} \tag{3.15}$$

Hence, a solution of (3.12) is given by

$$\begin{aligned} x &= nH_{i-1} && \text{(odd } i\text{)}, \\ x &= nH_i - nH_{i-1} && \text{(even } i\text{)}. \end{aligned} \tag{3.16}$$

For the linear indeterminate equation (3.12), a general solution for x may be obtained by adding an arbitrary integral multiple of H_i. Therefore, the smallest solution for x can be obtained by dividing the above solution by H_i and taking the remainder. Therefore, we have the solution

$$\begin{aligned} x &= \left[\frac{nH_{i-1}}{H_i}\right]_{rem} && \text{(odd } i\text{)}, \\ x &= H_i - \left[\frac{nH_{i-1}}{H_i}\right]_{rem} && \text{(even } i\text{)}. \end{aligned} \tag{3.17}$$

Thus from (3.10) and (3.17), it is clear that the *khaṇḍa* K_i is given by

$$\begin{aligned} K_i &= A - r_i && \text{(odd } i\text{)}, \\ K_i &= A - (H_i - r_i) && \text{(even } i\text{)}, \end{aligned} \tag{3.18}$$

where

$$r_i = \left[\frac{\left[\frac{(\theta_0-\theta_m)_A}{21600} \times H_i\right]_{int} \times H_{i-1}}{H_i}\right]_{rem},$$

as stated in the text except for the correction term $\dfrac{(\theta_0 - \theta_m)_{303}}{2}$. It may be noted that here it is not insisted upon that the *kendra* should exactly be zero at the *khaṇḍa*.

Illustrative example

We shall now illustrate the procedure for obtaining the *khaṇḍa* outlined above by means of an example. Let's consider the *ahargaṇa* A to be 1851974 corresponding to August 6, 1969 CE.

The *kalyādidhruva* of the *kendra* is $6°23'37'' - 129°17'26'' = -6773.836667'$. We take the daily motion of the *kendra* to be $\frac{766081}{21109029}$ revolutions. Hence the motion of the *kendra* in 1851974 days is

$$\left[\frac{766081}{21109029} \times 1851874\right]_{frc} = 0.14903 \text{ revolution}$$
$$= 3218.94200',$$

where we have left out the integral part or the total number of revolutions completed. Using the above rate of motion, we can also compute the quantity

$$\left|\frac{(\theta_0 - \theta_m)_{303}}{2}\right| = 39.28086' = 39'17''. \tag{3.19}$$

Hence

$$\begin{aligned}(\theta_0 - \theta_m)_A + \left|\frac{(\theta_0 - \theta_m)_{303}}{2}\right| &= -6773.816667' + 3218.94200' + 39.28086' \\ &= -3515.59381 \\ &= 18084.40619' \approx 18084'. \end{aligned}\tag{3.20}$$

Actually we need to find the number of civil days elapsed from the time when the *kendra* was zero to the point when the *kendra* becomes 18084′ close to sunrise. That many number of days when subtracted from the *ahargaṇa* 1851874 would give the *khaṇḍa.*

Now we take the *hāraka* $H_4 = 248$. Then, the daily motion of the *kendra* is $\frac{9}{248}$. That is, after each day (commencing from the day on which the *kendra* is zero), the increase of *kendra* would be in multiples of $\frac{9}{248}$. That is, at the end of the first day it would be $\frac{9}{248}$, at the end of the second day $\frac{18}{248}$ and so on. After the 28th day when the Moon's *kendra* has completed one revolution, it would be $\frac{4}{248}$, since $9\times 28 - 248 = 4$. Hence, at the sunrise of any arbitrary day, the *kendra* would be an integral multiple of $\frac{1}{248}$ (in *bhagaṇas*) or $\frac{1}{248} \times 21600$ (in minutes).

Let n' be an integer so that $\frac{n'}{248} \times 21600'$ is equal to 18084′. Then

$$\begin{aligned}\frac{n'}{248} &= \frac{18084}{21600}, \\ \text{or} \qquad n' &= \frac{18084 \times 248}{21600} \\ &= 207.631.\end{aligned}$$

Hence

$$n = [n']_{int} = 207,$$

As the *hāraka* previous to 248 is 55, we calculate

$$\left[\frac{207 \times 55}{248}\right]_{rem} = 225.$$

Now, the *hāraka* chosen is $H_4 = 248$, which is a *yugmahāraka*, corresponding to even i. Therefore, as per the prescription given in the verse

$$x = 248 - 225 = 23.$$

Hence, the *khaṇḍa* is given by

$$K_4 = 1851974 - 23 = 1851951. \tag{3.21}$$

The *kendrabhukti* corresponding to K_4 is obtained by multiplying it by the daily motion of the *kendra* which is taken to be $\frac{766081}{21109029}$ in this example. After subtracting the integral number of revolutions from this, we find it to be $6789.26186'$. Adding the *kalyādidhruva* -6773.81667, we find that *kendra* at *khaṇḍa* K_4 is $15.44519'$ which is small, as expected.

If we take the *hāraka* to be $H_6 = 12372$, then $n' = 10358.11333$, so that $n = 10358$. In this case, the *hāraka* previous to 12372 is 3031. Thus we have

$$\left[\frac{10358 \times 3031}{12372}\right]_{rem} = 7334.$$

As 12372 is also *yugmahāraka* ($i = 6$), $x = 12372 - 7334 = 5038$.

Hence, in this case, the *khaṇḍa* $K_6 = 1851974 - 5038 = 1846936$. The *kendra* at K_6 is then

$$\left[\frac{1846936 \times 766081}{21109029}\right]_{frc} \times 21600 - 6773.81667 = -38.72306',$$

where '$[\;]_{frc}$' denotes the fractional part. For this *ahargaṇa*, not only is the difference between the *ahargaṇa* and the *khaṇḍa* high (5038), the *kendra* is also slightly larger.

Now, we find the *khaṇḍa* K_4 and the *kendra* at the *khaṇḍa* without considering the correction term $\frac{(\theta_0-\theta_m)_{303}}{2} = 39.28086$. In this case,

$$(\theta_0 - \theta_m)_A = 18045.12533' \approx 18045',$$

for the same *ahargaṇa*, $A = 1851974$ days. If we take the *hāraka* as 248, then

$$\frac{n'}{248} = \frac{18045}{21600},$$
$$\text{or} \qquad n' = 207.183333,$$

and $n = 207$. This is the same as what we obtained earlier, and we obtain the same value of *khaṇḍa* $K_4 = 1851951$ and *kendra* at K_4 to be $15.44519'$.

If we take the *hāraka* to be $H_6 = 12372$, then

$$\frac{n'}{12372} = \frac{18045}{21600},$$

or, $n' = 10335.775$ or $n = 10335$. Then

$$\left[\frac{10335 \times 3031}{12372}\right]_{rem} = 11853,$$

and $12372 - 11853 = 519$. Therefore, *khaṇḍa*

$$K_6 = 1851974 - 519 = 1851455.$$

Now the *kendra* at K_6 is

$$\frac{1851455 \times 766081}{21109029} \times 21600 - 6773.81667$$
$$= 1.47',$$

where we have subtracted the integral multiple of 21600. It is to be noted that the magnitude of the *kendra* is small at the *khaṇḍa* for this *hāraka* when we do not use the correction term $\frac{(\theta_0 - \theta_m)_{303}}{2}$.

३.२.२ हारकखण्डान्ते ध्रुवाः

3.2.2 Dhruvas at the end of different *hārakas* and *khaṇḍas*

In verse 4 it is stated that the *dhruvas* corresponding to the *khaṇḍas* K_i would be the true longitudes of the Moon at the end of those *khaṇḍas*. Similarly, for a *hāraka* H_i, the *dhruva* is the increase in the true longitude during an interval equal to H_i days, at the beginning of which true anomaly is zero. It is well known that the true longitude (θ) in terms of the mean longitude (θ_0) and the *mandakendra* ($\theta_0 - \theta_m$) is given by

$$\theta = \theta_0 - \sin^{-1}\left(\frac{7}{80} \times \sin(\theta_0 - \theta_m)\right). \tag{3.22}$$

Using the above relation, we have computed the *dhruvas* for all the *hārakas* and tabulated them in Table 3.3.

For the *ahargaṇa* $A = 1851974$ considered above, the *khaṇḍa* was 1851951 corresponding to the *hāraka* 248, and the *mandakendra* was $15.45'$. The mean longitude of the Moon for this *ahargaṇa* including the *kalyādidhruva* is $\theta_0 = 74°40'$. Then the true longitude of the Moon is found to be

$$\theta = \theta_0 - \sin^{-1}\left(\frac{7}{80} \times \sin(15.45')\right)$$
$$= 74°38.6'.$$

Similarly for the *hāraka* $H_6 = 12372$, the *khaṇḍa* was $K_6 = 1846936$ with mean longitude $\theta_0 = 241°39.4'$ and $\theta_0 - \theta_m = -38.72306'$ as seen above. The true longitude of the Moon at the *khaṇḍa* would be

$$\begin{aligned}\theta &= \theta_0 - \sin^{-1}\left(\frac{7}{80} \times \sin(-38.72')\right) \\ &= 241°42.8'.\end{aligned}$$

३.३ केन्द्रफलानयनम्

3.3 Obtaining the *kendraphalas*

तत्तद्दिनोत्थेन्दुतदुच्चभेदलिप्ताहतात् केन्द्रहरादभीष्टात् ।
केन्द्राण्यनन्तोत्करसंहृतानि स्वर्णात्मकानीन्द्वधिकाल्पकत्वे ॥ ५ ॥

tattaddinotthendutaduccabhedaliptāhatāt kendraharādabhīṣṭāt |
kendrāṇyanantotkarasaṃhṛtāni svarṇātmakānīndvadhikālpakatve || 5 ||

The difference between the longitudes of the Moon and its apogee in minutes corresponding to the respective days,[1] multiplied by the desired *kendrahāra* and divided by 21600 (*anantotkara*) would give the *kendra*[*phalas*] of the respective days [corresponding to the chosen *kendra-hāra*]. [This] would be positive or negative depending on whether [the longitude of] the Moon is larger or smaller [than the longitude of the apogee].

If θ_0 and θ_m be the mean longitudes of the Moon and its apogee in minutes, then the *kendraphalas* k_{ij} of the different *hārakas* H_j (where $j = 1, 2, \ldots i-1$) corresponding to *iṣṭakendrahāraka* H_i may be expressed as

$$k_{ij} = +\left|\frac{(\theta_0 - \theta_m)_{H_j} \times H_i}{21600}\right| \quad (\text{if } \theta_0 > \theta_m), \tag{3.23}$$

$$k_{ij} = -\left|\frac{(\theta_0 - \theta_m)_{H_j} \times H_i}{21600}\right| \quad (\text{if } \theta_0 < \theta_m). \tag{3.24}$$

Similarly, given the *khaṇḍa* K_i and the *hāraka* H_i, the *kendraphala* k_i is given by

[1] Here, the word *tattaddina* (respective days) refers to both the *khaṇḍadinas* obtained and the number of days given by the *kendrahārakas*.

hārakas H_i	*dhruvas* (in minutes)
27	21383.62485
28	505.76731
55	289.43982
248	1663.47755
3031	20251.17118
12372	17868.16227
188611	7473.60524
1332649	5383.39894
1521260	12857.00418
21109029	21324.45330
43739318	12305.91079
414762891	2477.65039
458502209	14783.56116
1331767309	10444.77271
1790269518	3628.33398
3122036827	14073.10645
4912306345	17701.44043
8034343172	10174.54688
12946649517	6275.98828
20980992689	16450.53517
54908634895	17577.05469
75889627584	12427.59375
130798262479	8404.65625
599082677500	2846.18750

Table 3.3 The *dhruvas* corresponding to different *kendrahārakas.*

$$k_i = +\left|\frac{(\theta_0 - \theta_m)_{K_i} \times H_i}{21600}\right| \quad (\text{if } \theta_0 > \theta_m), \tag{3.25}$$

$$k_i = -\left|\frac{(\theta_0 - \theta_m)_{K_i} \times H_i}{21600}\right| \quad (\text{if } \theta_0 < \theta_m). \tag{3.26}$$

Here $(\theta_0 - \theta_m)_{H_j}$ or $(\theta_0 - \theta_m)_{K_i}$ represent the change in the longitude of the *kendra* or anomaly in a period of H_j or K_i days respectively, in minutes. Now, if the rate of motion of the *kendra* (anomaly) is taken to be $\frac{G_i}{H_i}$, then

$$(\theta_0 - \theta_m)_X = \left[X \times \frac{G_i}{H_i}\right]_{frc} \times 21600 \text{ (in minutes)},$$

where $[\ ]_{frc}$ in the RHS denotes the fractional part. Hence

$$\frac{(\theta_0 - \theta_m)_X}{21600} = \left[X \times \frac{G_i}{H_i}\right]_{frc} = \frac{\alpha_i}{H_i}, \tag{3.27}$$

where α_i is the remainder when XG_i is divided by H_i. This remainder α_i is the *kendraphala* denoted by k_{ij} when $X = H_j$, and by k_i when $X = K_i$.

३.४ केन्द्रफलानयने प्रकारान्तरम्

3.4 An alternative method for obtaining the *kendraphalas*

अभीष्टहारञ्च गुणं तदीयं मिथो हरेत्तद्गुणहारशेषाः ।
अभीष्टहारोर्ध्वगहारकाणां धनर्णकेन्द्राण्यथवा भवन्ति ॥ ६ ॥

abhīṣṭahārañca guṇaṃ tadīyaṃ mitho harettadguṇahāraśeṣāḥ |
abhīṣṭahārordhvagahārakāṇāṃ dhanarṇakendrāṇyathavā bhavanti || 6 ||

Or else, the remainders obtained in [the process] of mutual division (*vallyupasaṃhāra*) of the desired *hāraka* and its *guṇakāra* would be the positive and negative *kendra[phalas]* of the preceding *hārakas*.

Essentially the verse states that in the process of doing mutual division (*vallyupasaṃhāra*) between the desired *hāraka* H_i and its *guṇakāra* G_i, the remainders obtained $r_{ij}, j = 1, 2, \ldots, i-1$, are the *kendraphalas* of the *hārakas* $H_1, H_2, \ldots H_{i-1}$ respectively corresponding to H_i. This can be seen as follows. In equation (3.27), setting $X = H_j$, and correspondingly $\alpha_i = k_{ij}$, we have

$$\frac{(\theta_0 - \theta_m)_{H_j}}{21600} = \left[H_j \times \frac{G_i}{H_i}\right]_{frc} = \frac{k_{ij}}{H_i}. \tag{3.28}$$

But it has been shown in (A.29) of Appendix A that

$$G_i H_j = H_i G_j \pm r_{ij}, \qquad j = 1, 2, \ldots (i-1),$$

where we take '+' sign when j is even and '−' sign when j is odd. Therefore, we have

$$\frac{k_{ij}}{H_i} = \left[G_j \pm \frac{r_{ij}}{H_i}\right]_{frc}. \tag{3.29}$$

The G_j term does not contribute to the fractional part. Hence $k_{ij} = \pm r_{ij}$. Here r_{ij} are the remainders obtained during the process of mutual division of the given *hāraka* and its *guṇakāra*. The *kendraphalas* are just these remainders apart from a sign.

Now we shall illustrate the procedure to obtain the *kendraphalas* by choosing $H_i = 248$ as an example. If we divide 248 by 9, the quotient would be 27 and the remainder is 5. Now dividing 9 by 5, we get the quotient as 1 and the remainder is 4. In the next step we get the remainder as 1. Thus the ratio $\frac{9}{248}$ can be written in the form of a continued fraction as

$$\frac{9}{248} = \cfrac{1}{27 + \cfrac{1}{1 + \cfrac{1}{1 + \cfrac{1}{4}}}} . \tag{3.30}$$

The remainders are 5, 4 and 1. Hence, $-5, +4$ and -1 are the *kendraphalas* of the *ūrdhvahārakas* 27, 28 and 55, respectively, corresponding to the *hāraka* 248.

३.५ ध्रुवसंस्कारहारकानयनम्

3.5 Obtaining the *dhruva-saṃskārahārakas*

अभीष्टहारस्य गुणोऽयमुक्तः संस्कारहारानयने तु भाज्यः ।
केन्द्राण्यमूनि स्वहरोद्धृतानां क्रमात् फलानां गुणका भवेयुः ॥ ७ ॥

abhīṣṭahārasya guṇo'yamuktaḥ saṃskārahārānayane tu bhājyaḥ |
kendrāṇyamūni svaharoddhṛtānāṃ kramāt phalānāṃ guṇakā bhaveyuḥ || 7 ||

While obtaining the divisor which is used for correcting the *dhruva* (*dhruva-saṃskārahāra*), the multiplier corresponding to the desired divisor would become the dividend. The *kendra*-[*phalas*] would be the multipliers of the quotients obtained when dividing [*khaṇḍaśeṣas*] by their own *hāras*.

The expression for *dhruva-saṃskārahāraka* (the divisor which appears in the correction for *dhruva*) given in the above verse is the following:

$$\textit{dhruva-saṃskārahāraka} \;=\; \frac{G_i}{\sum_{j=l}^{i-1} k_{ij} q_j}, \tag{3.31}$$

where q_j, $j = l, l+1, \ldots i$, are defined below. In (3.31), l is a suitable small number, not specified. In practice $l = 4$, corresponding to the fourth *hāraka*, 248 (see Table 3.1).

Let K_i be the *khaṇḍa*, as given by (3.8), corresponding to the *hāraka* H_i. The *khaṇḍaśeṣa* $A - K_i$ is to be divided by H_{i-1}. The quotient obtained is

q_{i-1}; and the remainder obtained ν_{i-1} is divided by the H_{i-2} and so on. In this way the *khaṇḍaśeṣa* $A - K_i$ may be expressed as:

$$\begin{aligned} A - K_i &= q_{i-1}H_{i-1} + q_{i-2}H_{i-2} + \ldots q_l H_l + \nu_l \\ &= S_l + \nu_l. \end{aligned} \tag{3.32}$$

The longitude of the *kendra* at S_l in revolutions is given by

$$(\theta_0 - \theta_m)_{S_l} = q_{i-1}(\theta_0 - \theta_m)_{H_{i-1}} + q_{i-2}(\theta_0 - \theta_m)_{H_{i-2}} + \ldots + q_l(\theta_0 - \theta_m)_{H_l}.$$

Now

$$(\theta_0 - \theta_m)_{H_j} = \frac{k_{ij}}{H_i}.$$

Hence

$$(\theta_0 - \theta_m)_{S_l} = \frac{\sum_{j=l}^{i-1} k_{ij} q_j}{H_i}. \tag{3.33}$$

Verses 15 – 18 of the *Vākyakaraṇa* of Parameśvara[2] define *dhruva-saṃskārahāraka* (H_D) to be

$$\begin{aligned} \textit{kendra} \text{ at } \textit{vākyārambhakāla} &= \frac{\textit{kendra-madhyama-gati}}{\textit{dhruva-saṃskārahāra}}, \\ \text{or} \qquad (\theta_0 - \theta_m)_{S_l} &= \frac{\textit{kendra-madhyama-gati}}{H_D} \\ &= \frac{G_i}{H_i \times H_D}. \end{aligned} \tag{3.34}$$

Equating (3.33) and (3.34), we get

$$H_D = \frac{G_i}{\sum_{j=l}^{i-1} k_{ij} q_j}, \tag{3.35}$$

which is the same as the expression given in the text.

The sum of the increases in the mean longitude of the Moon corresponding to $H_{i-1}, \ldots, H_l$ days multiplied by the quotients $q_{i-1}, \ldots, q_l$ respectively added to the mean longitude at the *khaṇḍa* will give the mean longitude at S_l. From these the true longitude of the Moon at S_l can be obtained. If $(\theta)_{S_l}$ is the true longitude of Moon at the *vākyārambhakāla* S_l, then

$$(\theta)_{S_l} = q_{i-1}d_{i-1} + q_{i-2}d_{i-2} + \ldots + q_l d_l. \tag{3.36}$$

where d_i's are the *dhruvas* giving the true longitudes of the Moon after H_i days. For the remaining days ν_l, the corresponding *candravākya* can be used

[2] *Vākyakaraṇam* by Parameśvara, Mss. no. KVS 242, K V Sarma Library, Adyar.

to find the change in longitude. In this way, the true longitude of the Moon can be found for the desired *ahargaṇa* A.

३.६ हारकेण खण्डानयनम्

3.6 Obtaining the *khaṇḍa* using a specific *hāraka*

शीतांशोर्विकलादितस्सकळगैः युक्तात् स्वतुङ्गोनितात्
लिप्तीकृत्य कपोतदुर्जयहतात् नूत्नातपत्रोद्धृतम् ।
श्रीसङ्गप्रियताडितं विभजतात् कापोतदेहायनैः
तच्छिष्टं द्युगणात् त्यजेत् तुहिनगोः वाक्योक्तखण्डाप्तये ॥ ८ ॥

śītāṃśorvikalāditassakaḷagaiḥ yuktāt svatuṅgonitāt
liptīkṛtya kapotadurjayahatāt nūtnātapatroddhṛtam |
śrīsaṅgapriyatāḍitaṃ vibhajatāt kāpotadehāyanaiḥ
tacchiṣṭaṃ dyugaṇāt tyajet tuhinagoḥ vākyoktakhaṇḍāptaye || 8 ||

Having added the [mean longitude of the] Moon [accurate upto] the seconds to 39′17″ (*sakaḷaga*) subtract the mean longitude of its apogee and [the result] has to be converted into minutes. [Then] multiply by 188611 (*kapotadurjaya*) and divide by 21600 (*nūtnātapatra*). [The result thus obtained] has to be multiplied (*tāḍitam*) by 12372 (*śrīsaṅgapriya*) and divided by 188611 (*kāpota-dehāyana*). The remainder has to be subtracted from the *ahargaṇa* in order to obtain the *khaṇḍa* of the Moon (*tuhinagu*) that is mentioned in *vākya* [texts].

The expression given by the above verse, using which the *khaṇḍa* can be obtained, is

$$r_7 = \left[\frac{\left[\dfrac{((\theta_0 - \theta_m)_A + 39'17'') \times 188611}{21600} \right]_{int} \times 12372}{188611} \right]_{rem} . \tag{3.37}$$

The extra term 39′17″ which appears in the above equation is the same as the term $\frac{(\theta_0 - \theta_m)_{303}}{2}$ which appears in the definition (3.7) of the r_i. This is because the motion in anomaly in 303 days is very nearly equal to $2 \times 39'17'' = 78'34''$. This remainder when subtracted from A would give the necessary *khaṇḍa*.

This prescription (3.37) is a particular case of general procedure described in verses 2 and 3 (see (3.7)) with $H_i = 188611$ and $H_{i-1} = 12372$ ($i = 7$). As this H_7 is a *ojahāraka* (odd divisor) the remainder itself is subtracted from the *ahargaṇa* to obtain the *khaṇḍa*.

Illustrative example

We choose the same *ahargaṇa* $A = 1851974$ as in the previous example. The longitude of the *kendra* at the end of this *ahargaṇa* added to $39'17''$ is $18084'$. Using this value in (3.37), we get the reminder as 17410. Therefore, the *khaṇḍa*

$$K_7 = 1851974 - 17410 = 1834564.$$

३.७ वाक्योक्तखण्डाप्तिप्रकारः

3.7 A method for finding different *khaṇḍas*

लब्धं तत्र तु कालनागगुणितं श्रीरङ्गरम्यैर्भजेत्
तच्छिष्टोनहरोनितो दिनगणो वा वाक्यखण्डो भवेत् ।
तत्राप्ते किल देवरैर्विनिहते कालानलैः संहृते
शेषं वा द्युगणात् त्यजेत् तुहिनगोर्वाक्योक्तखण्डाप्तये ॥ ९ ॥

labdhaṃ tatra tu kālanāgaguṇitaṃ śrīraṅgaramyairbhajet
tacchiṣṭonaharonito dinagaṇo vā vākyakhaṇḍo bhavet |
tatrāpte kila devarairvinihate kālānalaiḥ saṃhṛte
śeṣaṃ vā dyugaṇāt tyajet tuhinagorvākyoktakhaṇḍāptaye || 9 ||

The quotient obtained there (i.e., in the previous calculation) has to be multiplied by 3031 (*kālanāga*) and divided by 12372. The reminder deducted from the *hāraka* (as it is an *ūnaśiṣṭa*) when subtracted from the *ahargaṇa* will also result in a *vākyakhaṇḍa*. The quotient obtained in this case may be multiplied by 248 (*devara*) and divided by 3031 (*kālānala*). The remainder [thus obtained] when subtracted from the *ahargaṇa* will also lead to a *khaṇḍa* that is mentioned in the *vākya* [texts].

The above verse presents an algorithm to find approximately the *khaṇḍas* $K_{i-1}, K_{i-2}, \ldots$ given the *khaṇḍa* K_i. The rationale for this method is briefly as follows.

Let us recall the definition of *khaṇḍa* K_i

$$\begin{aligned} K_i &= A - r_i \qquad &\text{(for odd } i\text{)},\\ K_i &= A - (H_i - r_i) \qquad &\text{(for even } i\text{)}, \end{aligned}$$

where

$$r_i = \left[\frac{\left[\dfrac{(\theta_0 - \theta_m)_A + \frac{(\theta_0 - \theta_m)_{303}}{2}}{21600} \times H_i \right]_{int} \times H_{i-1}}{H_i} \right]_{rem}.$$

Now let us define *phala* P_i by

$$P_i = \left[\frac{\left[\frac{(\theta_0 - \theta_m)_A + \frac{(\theta_0-\theta_m)_{303}}{2}}{21600} \times H_i \right]_{int} \times H_{i-1}}{H_i} \right]_{int} . \tag{3.38}$$

When H_i is fairly large, (see Table 3.2) we can approximate

$$\left[\frac{(\theta_0 - \theta_m)_A + \frac{(\theta_0-\theta_m)_{303}}{2}}{21600} \times H_i \right]_{int} \simeq \frac{(\theta_0 - \theta_m)_A + \frac{(\theta_0-\theta_m)_{303}}{2}}{21600} \times H_i \tag{3.39}$$

Hence

$$P_i \simeq \left[\frac{(\theta_0 - \theta_m)_A + \frac{(\theta_0-\theta_m)_{303}}{2}}{21600} \times H_{i-1} \right]_{int} . \tag{3.40}$$

Therefore we obtain

$$r_{i-1} \simeq \left[\frac{P_i \times H_{i-2}}{H_{i-1}} \right]_{rem} . \tag{3.41}$$

Further

$$P_{i-1} \simeq \left[\frac{P_i \times H_{i-2}}{H_{i-1}} \right]_{int} , \tag{3.42}$$

and so on.

The prescription given in the above verse, and the numerical values presented therein amount to finding r_6 and r_5 starting from r_7. To start with

$$r_7 = \left[\frac{\left[\frac{((\theta_0 - \theta_m)_A + 39'17'') \times 188611}{21600} \right]_{int} \times 12372}{188611} \right]_{rem} . \tag{3.43}$$

Then the *phala* P_7 is defined by

$$P_7 = \left[\frac{\left[\frac{((\theta_0 - \theta_m)_A + 39'17'') \times 188611}{21600} \right]_{int} \times 12372}{188611} \right]_{int} . \tag{3.44}$$

It is then said that

$$r_6 = \left[\frac{P_7 \times 3031}{12372} \right]_{rem} .$$

Now P_6 is given by

$$P_6 = \left[\frac{P_7 \times 3031}{12372}\right]_{int}. \qquad (3.45)$$

Then it is said that r_5 is given by

$$r_5 = \left[\frac{P_6 \times 248}{3031}\right]_{rem}.$$

The *khaṇḍas* K_6 and K_5 are given by

$$K_6 = A - (12372 - r_6),$$

$$\text{and} \qquad K_5 = A - r_5. \qquad (3.46)$$

Clearly, K_6 is closer to A than K_7 and K_5 is closer to A than K_6.

Illustrative example

Let us consider the *ahargaṇa*, $A = 1851974$ as in the previous example. It was found (see 3.20) that the *kendra* $(\theta_0 - \theta_m)_A + 39'17''$ is $18084'$. Using this in (3.43),

$$r_7 = \left[\frac{\left[\frac{18084' \times 188611}{21600}\right]_{int} \times 12372}{188611}\right]_{rem} = 17410.$$

Hence the *khaṇḍa* given by (3.8) is

$$K_7 = A - 17410 = 1851974 - 17410 = 1834564. \qquad (3.47)$$

Now, from (3.44) P_7 is given by

$$P_7 = \left[\frac{\left[\frac{18084' \times 188611}{21600}\right]_{int} \times 12372}{188611}\right]_{int}$$

$$= 10358. \qquad (3.48)$$

We see that we obtain the same result 10358, if we use the approximation

$$P_7 \simeq \left[\frac{18084 \times 12372}{21600}\right]_{int}. \qquad (3.49)$$

Proceeding this way, the remainder

$$r_6 = \left[\frac{10358 \times 3031}{12372}\right]_{rem} = 7334,$$

and the *khaṇḍa*, in this case the *ūnaśiṣṭa*, is given by

$$K_6 = A - (12372 - 7334) = 1851974 - 5038$$
$$= 1846936. \tag{3.50}$$

Now P_6 and r_5 are given by

$$P_6 = \left[\frac{10358 \times 3031}{12372}\right]_{int} = 2537,$$
$$r_5 = \left[\frac{2537 \times 248}{3031}\right]_{rem} = 1759. \tag{3.51}$$

The next *khaṇḍa* K_5 is

$$K_5 = 1851974 - 1759 = 1850215. \tag{3.52}$$

Again, P_5 and r_4 are given by

$$P_5 = \left[\frac{2537 \times 248}{3031}\right]_{int} = 207,$$

$$r_4 = \left[\frac{207 \times 55}{248}\right]_{rem} = 225.$$

Therefore, the *ūnaśiṣṭa* is $248 - 225 = 23$, and the *khaṇḍa* K_4 is given by

$$K_4 = 1851974 - 23 = 1851951. \tag{3.53}$$

which is the same as was obtained in (3.21).

३.८ हारकविशेषेण केन्द्रफलानयनम्

3.8 Obtaining *kendraphalas* using a special *hāraka*

तद्दिनेन्दूच्चयोर्भेदात् पर्याप्तहृदयाहतात् ।
केन्द्राख्यं चक्रलिप्ताप्तं स्फुटेन्दुस्तत्र हि ध्रुवः ॥ १० ॥

taddinendūccayorbhedāt paryāptahṛdayāhatāt |
kendrākhyaṃ cakraliptāptaṃ sphuṭendustatra hi dhruvaḥ || 10 ||

The difference between the longitude of the Moon and its apogee for that day[3] has to be multiplied by 188611 (*paryāptahṛdaya*) and divided by 21600. [The quotient] would be the corresponding *kendraphala*. The *dhruva* obtained there would be the true longitude of the Moon for that day.

This is a particular case of (3.28) for the *hāraka* $H_7 = 188611$.

[3] The term *taddina* refers to the day associated with a given *khaṇḍa* and a *hāraka*.

३.९ गुणकानयनप्रकारः

3.9 The process of obtaining the multipliers

शिवोदितं कम्पितदुर्जयं च छिन्द्यान्मिथो व्युत्क्रमतोऽत्र शेषाः ।
श्रीसङ्गरम्यादिहराहृतानां धनर्णरूपा गुणका भवन्ति ॥ ११ ॥

śivoditaṃ kampitadurjayaṃ ca chindyānmitho vyutkramato'tra śeṣāḥ |
śrīsaṅgaramyādiharāhṛtānāṃ dhanarṇarūpā guṇakā bhavanti || 11 ||

The remainders obtained by the mutual division of 6845 (*śivodita*) and 188611 (*kampitadurjaya*) in reverse order would become the positive and negative multipliers of the quotients obtained while dividing [*khaṇḍaśeṣa*] by the *hāras* 12372 etc. (*śrīsaṅgaramyādi*).

The method prescribed in this verse has already been discussed in connection with verses 6, 7. The process of mutual division prescribed here especially gives the *kendraphalas*. These positive and negative *kendraphalas* would be the multipliers (*guṇakas*).

An important point needs to be emphasized here. We have seen that the anomaly or the *kendra* is not zero at the end of the *khaṇḍa*, which is obtained using the successive *hārakas* $H_i, \ldots, 248$. In fact we have seen in (3.33) that, after finding the *khaṇḍa* closest to A using the *hāraka* 248 at the last stage, the *kendra* at the *khaṇḍa* would be given by

$$\sum_{j=4}^{i-1} \frac{q_j k_{ij}}{H_i} \tag{3.54}$$

revolutions. In the text *Vākyakaraṇa*, while prescribing the calculation of the true longitude of the Moon using *dhruvas* and the *vākyas*, a correction term has been specified which takes into account the fact that the *kendra* does not complete exactly integral number of revolutions in 248 or 3031 days, as the anomaly increases by the amounts $7'$ and $-1.75'$ respectively. This correction is explained in Appendix D.

३.१० चन्द्रवाक्यानयनप्रकारः

3.10 The procedure for obtaining the *candravākyas*

देवेन्द्रसङ्ख्यावधिकैः एकाद्येकोत्तरैर्दिनैः ।
पृथक् पृथक् स्फुटीकुर्यादिन्दुं तद्वाक्यसिद्धये ॥ १२ ॥

devendrasaṅkhyāvadhikaiḥ ekādyekottarairdinaiḥ |
pṛthak pṛthak sphuṭīkuryādinduṃ tadvākyasiddhaye || 12 ||

The true longitudes of the Moon may be computed for each day separately starting from day number 1 till day number 248 (*devendra*) in order to obtain the *candravākyas*.

This verse gives a general prescription as to how the *candravākyas* have to be obtained. The *candravākyas* accurate to a minute were enunciated by Vararuci and these are the famous *gīrnaśreyādi-vākyas.* The great astronomer-mathematician Mādhava enunciated the *vākyas, śīlaṃ rājñaḥ śriye* etc., which are accurate to a second. The true longitudes of the Moon for 248 days, starting from a day when the longitude of the Moon and its apogee was zero, have been computed and listed along with the values encoded in the *vākyas* in Appendix E.

३.११ शशितुङ्गयोगस्य सूर्योदयसन्निकर्षः

3.11 The relationship between the magnitude of *hāraka* and the proximity to sunrise

तुङ्गोनितादेव विधोर्यथोक्तं नीते च खण्डे शशितुङ्गयोगः ।
तत्रेष्टहाराधिकतावशात् तद्योगस्य सूर्योदयसन्निकर्षः ॥ १३ ॥

tuṅgonitādeva vidhoryathoktaṃ nīte ca khaṇḍe śaśituṅgayogaḥ |
tatreṣṭahārādhikatāvaśāt tadyogasya sūryodayasannikarṣaḥ || 13 ||

[For any given day], having obtained the longitude of the Moon and subtracting just the longitude of its apogee (*tuṅgonitādeva*[4]), if the *khaṇḍa* is [obtained], then it ensures that the conjunction of the Moon and its apogee [is close to the sunrise]. There, larger the magnitude of the *hāraka* chosen, the closer is the conjunction (of the Moon and its apogee) to the sunrise.

In the previous sections, we have seen how the *khaṇḍa* for the Moon can be obtained corresponding to a given *ahargaṇa.* The purpose of obtaining the *khaṇḍa* is to find a heap of days since the beginning of the *kali* in such a way that, at the *khaṇḍāntya*, the longitude of the Moon should be equal to the longitude of apogee. Here, the word *khaṇḍāntya* refers to the number of days elapsed after the beginning of *kaliyuga.* That is, if K is the *khaṇḍa* obtained, then the *khaṇḍāntya* corresponds to the beginning of the $(K+1)^{th}$ day which is the sunrise on that day. But, since the *khaṇḍa* is obtained from the approximate divisors, the conjunction between the Moon and its apogee will not exactly coincide with the sunrise. There will be a small time difference

[4] In the Commentary II, it is noted that the word *eva* associated with the word *tuṅgonita* implies that only *tuṅga* has to be subtracted from the mean longitude and the *ad hoc* correction term given by half the motion of the anomaly in 303 days or $39'17''$ need not be added ({KP 1956}, p. 109). That is, the word '*eva*' is used as *anyayoga-vyavacchedaka.*

(*antara*) between the instant at which conjunction (*yoga*) happens and the time of sunrise (*udaya*). This time difference in terms of the time unit *prāṇas*[5] is called *yogodayāntaraprāṇas*. The method to obtain the *yogodayāntaraprāṇas* is explained in the next verse.

३.१२ योगोदयान्तरप्राणयोगध्रुवयोरानयनम्

3.12 Obtaining the *yogodayāntaraprāṇas* and *yogadhruvas*

तद्वासरोत्थतुहिनांशुतदुच्चभेदलिप्ता हताभिमतकेन्द्रहराद् गुणाप्ताः।
योगोदयान्तरभवाः ह्यसवः क्रमेण स्वर्णात्मकाः तुहिनगोरधिकाल्पकत्वे ॥ १४ ॥

प्राणैस्तैः स्वगुणाभ्यस्तैः स्वहाराप्तैश्च संस्कृतौ ।
चन्द्रतुङ्गाविमौ स्यातां तुल्यौ योगध्रुवाह्वयौ ॥ १५ ॥

tadvāsarotthatuhināṃśutaduccabhedaliptā
hatābhimatakendraharād guṇāptāḥ |
yogodayāntarabhavāḥ hyasavaḥ krameṇa
svarṇātmakāḥ tuhinagoradhikālpakatve || 14 ||

prāṇaistaiḥ svaguṇābhyastaiḥ svahārāptaiśca saṃskṛtau |
candratuṅgāvimau syātāṃ tulyau yogadhruvāhvayau || 15 ||

The difference in the longitudes of the Moon and its apogee, in minutes, corresponding to that day (*tadvāsarottha*) has to be multiplied by the desired *kendrahāra* and divided by the *guṇakāra*. The [result thus obtained] would be positive or negative *yogodayāntara-prāṇas* depending on whether [the longitude of] the Moon is larger or smaller [than that of the apogee].

The [longitudes of the] Moon and its apogee when corrected by [the result obtained by] multiplying these [*yogodayāntara*]*prāṇas* by their *guṇas*, and dividing by their own (*sva*)[6] *hāras* respectively, would become equal. [These correction terms] are known as *yogadhruvas*.

If $(\theta_0 - \theta_m)_K$ represents the longitude of the *kendra* at the *khaṇḍāntya* (at the sunrise), then the expression for the time difference between the sunrise and the time of conjunction of Moon and its *mandocca* (*yogodayāntaraprāṇas*), (Δt) given in the above verse can be written as

$$\Delta t = +\left|\frac{(\theta_0 - \theta_m)_K \times H_i}{G_i}\right| \qquad (\text{if } \theta_0 > \theta_m), \tag{3.55}$$

$$\Delta t = -\left|\frac{(\theta_0 - \theta_m)_K \times H_i}{G_i}\right| \qquad (\text{if } \theta_0 < \theta_m). \tag{3.56}$$

[5] The word *prāṇa* is generally used to refer to life. However, in the context of astronomy it has to be understood as a time unit which is 4 sidereal seconds.

[6] The word *sva* indicates that appropriate values corresponding to the Moon and its apogee are to be used.

Here $(\theta_0 - \theta_m)_K$ is the value of the anomaly in minutes, at sunrise after K days have elapsed. The rate of motion of the anomaly is $\frac{G_i}{H_i}$ revolutions per day, or $\frac{G_i}{H_i} \times 21600$ minutes of arc per day or $\frac{G_i}{H_i}$ minutes per *prāṇa*, as there are 21600 *prāṇas* in a day. Hence the time interval between sunrise and the instant at which the anomaly is zero (that is, instant of conjunction of the Moon and its apogee) is given by

$$\left|\frac{(\theta_0 - \theta_m)_K \times H_i}{G_i}\right|,$$

which is the magnitude of the *yogodayāntaraprāṇas*. Here if $\theta_0 > \theta_m$ at sunrise, then the *yoga* (conjunction) has already occured and the *yogodayāntaraprāṇa* is positive. It is negative if $\theta_0 < \theta_m$. In that case, the mean Moon will be in conjunction with the apogee, after the sunrise. Thus, from Δt one can obtain the instant at which the conjuction between the Moon and its apogee occurs.

It may be recalled that from the *khaṇḍadina*, we can only obtain the longitudes of the Moon and its apogee at the time of sunrise. But by making use of Δt, we can find the longitude at the instant when the conjuction of the Moon and its apogee happens. This can be achieved by applying the correction terms to the longitudes at the sunrise. These correction terms are known as *yogadhruvas*. Verse 15 explains the method to obtain the *yogadhruvas* as a function of Δt.

If $G_i^{(1)}$ and $H_i^{(1)}$ are the *guṇakāras* and *hāras* of the Moon, and $G_i^{(2)}$ and $H_i^{(2)}$ are those of the apogee repectively, then the *yogadhruvas* for Moon and its apogee are given by

$$\textit{yogadhruva} \text{ of the Moon} = \frac{\Delta t \times G_i^{(1)}}{H_i^{(1)}}, \tag{3.57}$$

$$\textit{yogadhruva} \text{ of the apogee} = \frac{\Delta t \times G_i^{(2)}}{H_i^{(2)}}. \tag{3.58}$$

Illustrative example

We know that the *khaṇḍa* corresponding to the *ahargaṇa* $A = 1851974$ is 1851951 (see (3.21)). Now, the longitudes of the Moon and its apogee at the *khaṇḍāntya* can be written as

$$\theta_0 = 4857.14276',$$

$$\theta_m = 4845.36580'.$$

$$\text{Hence} \quad |\theta_0 - \theta_m| = 11.77696. \tag{3.59}$$

Since $\theta_0 > \theta_m$, we use (3.55) to find Δt. With the choice of *hāraka* $H_i = 188611$, corresponding to the *khaṇḍa* $K = 1851951$, we obtain[7]

$$\begin{aligned}\Delta t &= +\left[\frac{11.77696 \times 188611}{6845}\right] \\ &= +324.50910. \end{aligned} \tag{3.60}$$

The *yogodayāntaraprāṇas* corresponding to different *hārakas* were computed in the similar manner and the values obtained are listed in Table 3.4.

hāraka (H_i)	*guṇakāra* (G_i)	*yogodayāntara* (in *prāṇas*)
2	55	+ 323.86649
9	248	+ 324.52076
110	3031	+ 324.50887
449	12372	+ 324.50911
6845	188611	+ 324.50910

Table 3.4 The *yogodayāntara-prāṇas* corresponding to different *hārakas* for the *khaṇḍa* $K = 1851951$.

By substituting the value of Δt given by (3.60) and the respective *guṇahāras* in (3.57) and (3.58), we get

$$\begin{aligned}\textit{yogadhruva}\text{ of the Moon} &= \frac{324.5091 \times G_5^{(1)}}{H_5^{(1)}} \\ &= \frac{324.5091 \times 143}{3907} \\ &= 11.87735'. \end{aligned} \tag{3.61}$$

$$\begin{aligned}\textit{yogadhruva}\text{ of the apogee} &= \frac{324.5091 \times G_3^{(2)}}{H_3^{(2)}} \\ &= \frac{324.5091 \times 2}{6465} \\ &= 0.10039'. \end{aligned} \tag{3.62}$$

Applying[8] these values to the respective mean longitudes of the Moon and its apogee at the sunrise, we get

[7] Since the accuracy of the Δt increases with the increase in the magnitude of the *hārakas*, we have used the values corresponding to the *hāraka* 188611.

[8] As per the Commentary II, the magnitude of the *yogadhruvas* are to be added to, or subtracted from the mean longitudes depending on whether the *yogodayāntaraprāṇas* are negative or positive respectively ({KP 1956}, p. 111).

$$\begin{aligned}\text{Longitude of the Moon at } yoga &= 4857.14276' - 11.87735' \\ &= 4845.26541', \qquad (3.63)\\ \text{Longitude of the apogee at } yoga &= 4845.36580' - 0.10039' \\ &= 4845.26541'. \qquad (3.64)\end{aligned}$$

३.१३ ग्रहयुतिः

3.13 General rule for conjunction of planets

एवमेव पुनरिष्टखेटयोः योगमिष्टसमये समानयेत् ।
इष्टकालखगमध्यमान्तरक्ष्मादिनेष्टखगपर्ययान्तरैः ॥ १६ ॥

evameva punariṣṭakheṭayoḥ yogamiṣṭasamaye samānayet |
iṣṭakālakhagamadhyamāntarakṣmādineṣṭakhagaparyayāntaraiḥ || 16 ||

The conjunction of any two desired planets for any desired day can be obtained by making use of the difference between the mean longitudes of the planets (*khagamadhyamāntara*), the number of civil days (*kṣmādina*) and the difference in revolutions (*khagaparyayāntara*), in a similar manner.

This has already been explained in section 3.1.

Chapter 4
ग्रहहारकानयनम्
Obtaining the *hārakas* for the planets

४.१ ग्रहमन्दकेन्द्रहारकाः

4.1 The *mandakendrahārakas* of the planets

चन्द्रादन्यविहङ्गानां मध्यानयनहारकाः ।
मन्दकेन्द्रहरा ज्ञेयाः सौरा एव ज्ञशुक्रयोः ॥ १ ॥

candrādanyavihaṅgānāṃ madhyānayanahārakāḥ |
mandakendraharā jñeyāḥ saurā eva jñaśukrayoḥ || 1 ||

It is to be understood that the *madhyānayana-hārakas* of the planets other than the Moon are also their *mandakendrahārakas*. For Mercury and Venus [*madhyānayanahārakas*] of the Sun would be the *mandakendrahārakas*.

The above verse defines the *mandakendrahārakas* of the planets. It may be recalled that the *mandakendra* is the difference between the planet and its *mandocca* (apogee). The *mandakendra-guṇakāras* and the *mandakendra-hārakas* determine the successive approximations to the rate of motion of the *mandakendra*.

The term *madhyānayanahāraka* appearing in the earlier half of the verse has to be understood as '*madhyamānayane hārakāḥ*', i.e., the *hārakas* employed in obtaining the mean longitudes of the planets. The statement that 'the *madhyānayana-hārakas* are to be understood as the *mandakendra-hārakas*', implies that the *mandocca* of the planets have a negligible motion. Therefore, the *mandakendra-hārakas* and the *mandakendra-guṇakāras* of the planets are the same as the *dṛḍhahārakas* and *dṛḍhaguṇakāras* listed in Table 2.2.

It may also be mentioned here that the text *Karaṇapaddhati* in certain respects follows the traditional planetary model used by the Indian astronomers at least since the time of Āryabhaṭa. In this model while computing the *mandasaṃskāra* or the equation of centre, the mean Sun is taken as the mean planet in the case of Mercury and Venus. *Karaṇapddhati* in fact makes no

V. Pai et al., *Karaṇapaddhati of Putumana Somayājī*, Sources and Studies in the History of Mathematics and Physical Sciences, https://doi.org/10.1007/978-981-10-6814-0_4

reference to the revised planetary model introduced by Nīlakaṇṭha Somayāji in his *Tantrasaṅgraha* (c. 1500), which proposed that what were considered as the *śīghroccas* of Mercury and Venus should be taken to be the mean planets and that the mean Sun is to be taken as the *śīghrocca* for all the planets including Mercury and Venus.[1]

४.२ ग्रहशीघ्रकेन्द्रहारकानयनम्

4.2 Obtaining the *śīghrakendra-hārakas* of the planets

भास्करेष्टखगपर्ययान्तरं भूदिनं च विभजेत् परस्परम् ।
हारकानिह फलैः समानयेत् ते भवन्ति चलकेन्द्रहारकाः ॥ २ ॥

bhāskareṣṭakhagaparyayāntaraṃ bhūdinaṃ ca vibhajet parasparam |
hārakāniha phalaiḥ samānayet te bhavanti calakendrahārakāḥ || 2 ||

Divide the difference in the revolutions (*paryayāntara*) of the Sun and the desired planet and the number of civil days (*bhūdina*) mutually. The *hārakas* are to be obtained from the quotients (*phalas*) [by the process of *vallyupasaṃhāra*]. These [*hārakas*] would be the *śīghrakendra-hārakas*.

The method for obtaining the *śīghrakendra-hārakas* of the planet is explained in the above verse. If G_p and H_p are the *dṛḍhaguṇakāras* and the *dṛḍhahārakas* of the planet respectively, and G_s and H_s are those of the Sun, then the "difference in revolutions" (*paryayāntara*) mentioned in the verse is

$$\pm(G_s \times H_p - G_p \times H_s),$$

where the '+' sign is to be used in the case of the exterior planets Mars, Jupiter and Saturn, and '−' sign for the interior planets Mercury and Venus. The "number of civil days" (*bhūdina*) is $H_s \times H_p$. In other words, $(G_s \times H_p - G_p \times H_s)$ is the difference in the number of revolutions made by the planet and the Sun in $H_s \times H_p$ civil days. Doing *vallyupasaṃhāra* by mutually dividing them, we obtain the *śīghrakendra-hārakas*.

[1] Later, in verse 25 of Chapter 7, there is a statement that the *manda* correction for Mercury and Venus should be applied to their "*nijamadhyama*" and this is interpreted, in the Commentary II, as a reference to their *śīghroccas*. If we accept this interpretation, then it would imply that *Karaṇapaddhati* is adopting the revised planetary model of Nīlakaṇṭha, at least as regards the application of the *manda* correction in the case of the interior planets. For details of Nīlakaṇṭha's revised planetary model, see {TS 2011}, Appendix F, pp. 487-535.

Illustrative example

We will now find the *śīghrakendra-hārakas* for Saturn. This can be done by choosing the *dṛḍhahāraka* (6180176875) and *dṛḍhaguṇakāra* (574109) of the Saturn. The *dṛḍhahāraka* and *dṛḍhaguṇakāra* of the Sun are 210389 and 576 respectively. Now the "difference in revolutions" between the Sun and the Saturn is given by

$$576 \times 6180176875 - 574109 \times 210389 = 3438995661599.$$

The corresponding "number of civil days" is given by

$$210389 \times 6180176875 = 1300241232554375. \qquad (4.1)$$

Dividing the above quantities mutually till the remainder becomes zero and doing *vallyupasaṃhāra*, we get *śīghrakendra-guṇakāras* and *śīghrakendra-hārakas* related to Saturn which are listed along with the quotients and remainders obtained in the process of *vallyupasaṃhāra* in Table 4.1. Tables 4.2 – 4.5 list the *śīghrakendra-guṇakāras* and *śīghrakendra-hārakas* for all other planets, which are obtained in a similar manner.

quotient	remainder	*śīghrakendra-guṇakāra*	*śīghrakendra-hāraka*
378	300872469953	1	378
11	129398492116	11	4159
2	42075485721	23	8696
3	3172034953	80	30247
13	839031332	1063	401907
3	654940957	3269	1235968
1	184090375	4332	1637875
3	102669832	16265	6149593
1	81420543	20597	7787468
1	21249289	36862	13937061
3	17672676	131183	49598651
1	3576613	168045	63535712
4	3366224	803363	303741499
1	210389	971408	367277211
16	0	16345891	6180176875

Table 4.1 The *śīghrakendra-guṇakāras* and *śīghrakendra-hārakas* of Saturn.

quotient	remainder	*śīghrakendra-guṇakāra*	*śīghrakendra-hāraka*
779	1560066203739	1	779
1	107046975145	1	780
14	61408551709	15	11699
1	45638423436	16	12479
1	15770128273	31	24178
2	14098166890	78	60835
1	1671961383	109	85013
8	722475826	950	740939
2	227009731	2009	1566891
3	41446633	6977	5441612
5	19776566	36894	28774951
2	1893501	80765	62991514
10	841556	844544	658690091
2	210389	1769853	1380371696
4	0	7923956	6180176875

Table 4.2 The *śīghrakendraguṇkāras* and *śīghrakendrahārakas* of Mars.

quotient	remainder	*śīghrakendra-guṇakāra*	*śīghrakendra-hāraka*
115	5892968539430	1	115
1	839655556163	1	116
7	15379646289	8	927
54	9154656557	433	50174
1	6224989732	441	51101
1	2929666825	874	101275
2	365656082	2189	253651
8	4418169	18386	2130483
82	3366224	1509841	174953257
1	1051945	1528227	177083740
3	210389	6094522	706204477
5	0	32000837	3708106125

Table 4.3 The *śīghrakendra-guṇakāras* and *śīghrakendra-hārakas* of Mercury.

४.३ ग्रहखण्डानयनम्

4.3 Obtaining the *khaṇḍas* of the planets

मध्यात् ग्रहाणां स्वमृदूच्चहीनः शीघ्रोच्चतो मध्यविवर्जिताच्च ।
खण्डं नयेत् केन्द्रहरैस्तदीयैः इन्दूक्तवन्निर्गणितोदितास्ते ॥ ३ ॥

quotient	remainder	*śīghrakendra-guṇakāra*	*śīghrakendra-hāraka*
398	184234070687	1	398
1	23830972419	1	399
7	17417263754	8	3191
1	6413708665	9	3590
2	4589846424	26	10371
1	1823862241	35	13961
2	942121942	96	38293
1	881740299	131	52254
1	60381643	227	90547
14	36397297	3309	1319912
1	23984346	3536	1410459
1	12412951	6845	2730371
1	11571395	10381	4140830
1	841556	17226	6871201
13	631167	234319	93466443
1	210389	251545	100337644
3	0	988954	394479375

Table 4.4 The *śīghrakendra-guṇakāras* and *śīghrakendra-hārakas* of Jupiter.

quotient	remainder	*śīghrakendra-guṇakāra*	*śīghrakendra-hāraka*
583	2061056061934	1	583
1	165668082993	1	584
12	73039066018	13	7591
2	19589950957	27	15766
3	14269213147	94	54889
1	5320737810	121	70655
2	3627737527	336	196199
1	1693000283	457	266854
2	241736961	1250	729907
7	841556	9207	5376203
287	210389	2643659	1543700168
4	0	10583843	6180176875

Table 4.5 The *śīghrakendra-guṇakāras* and *śīghrakendra-hārakas* of Venus.

madhyāt grahāṇāṃ svamṛdūccahīnaḥ śīghroccato madhyavivarjitācca |
khaṇḍaṃ nayet kendraharaistadīyaiḥ indūktavannirgaṇitoditāste || 3 ||

From [the *mandakendras* that are obtained by] subtracting the *mandoccas* from the mean [longitudes] of the planets, as well as from [the *śīghrakendras* that are obtained by] subtracting the mean [longitudes of the planets] from the *śīghroccas*, the respective *kendrahārakas* are obtained. From them the *khaṇḍas* may be obtained in the same way as was done in the case of Moon. These are the *khaṇḍas* mentioned in the *Nirgaṇita* system.

Here the author discusses the procedure to obtain the *khaṇḍas* as enunciated in the *Nirgaṇita* system.[2] If θ_0, θ_m and θ_s represent the longitudes of the *madhyamagraha*, the *mandocca* and the *śīghrocca* of the planets respectively, then the *mandakendra* (θ_{mk}) and the *śīghrakendra* (θ_{sk}) are given by

$$\theta_{mk} = \theta_0 - \theta_m, \qquad \text{and} \qquad \theta_{sk} = \theta_s - \theta_0.$$

In the case of planets, the *khaṇḍa* corresponding to a specific *ahargaṇa* A, can be of two kinds:

1. The *śaighrakhaṇḍa*, is obtained from the *śīghrakendra* $(\theta_{sk})_A$ (expressed in minutes), corresponding to the *ahargaṇa* A, by making use of the *śīghrakendra-hārakas* $H_i^{(s)}$. As in the case of the Moon these *khaṇḍas* are given by the following expressions.

$$A - \left(H_i^{(s)} - \left[\left[\frac{((\theta_{sk})_A)}{21600} \times H_i^{(s)} \right]_{int} \times \frac{H_{i-1}^{(s)}}{H_i^{(s)}} \right]_{rem} \right), \tag{4.2}$$

$$A - \left[\left[\frac{((\theta_{sk})_A)}{21600} \times H_i^{(s)} \right]_{int} \times \frac{H_{i-1}^{(s)}}{H_i^{(s)}} \right]_{rem}. \tag{4.3}$$

   Expressions (4.2) and (4.3) correspond to even and odd i values respectively.

2. The *mandakhaṇḍa*, is obtained from the *mandakendra* $(\theta_{mk})_A$ (expressed in minutes), corresponding to the *ahargaṇa* A, by making use of the *mandakendra-hārakas* $H_i^{(m)}$. As in the case of the Moon these *khaṇḍas* are given by the following expressions.

$$A - \left(H_i^{(m)} - \left[\left[\frac{((\theta_{mk})_A)}{21600} \times H_i^{(m)} \right]_{int} \times \frac{H_{i-1}^{(m)}}{H_i^{(m)}} \right]_{rem} \right), \tag{4.4}$$

$$A - \left[\left[\frac{((\theta_{mk})_A)}{21600} \times H_i^{(m)} \right]_{int} \times \frac{H_{i-1}^{(m)}}{H_i^{(m)}} \right]_{rem}. \tag{4.5}$$

[2] The *Nirgaṇita* system is perhaps the same as the *Agaṇita* system discussed by Mādhava.

Here again (4.4) and (4.5) correspond to even and odd i values respectively.

The above expressions for the *śaighrakhaṇḍa* and *mandakhaṇḍa* give rise to *ahargaṇas* close to the chosen *ahargaṇa* A, such that the *śīghrakendra* or *mandakendra* is nearly zero to a level of accuracy dependent on the *hāraka* chosen. The rationale for these expressions have already been explained earlier in section 3.2.

४.४ हारस्वीकृतौ सामान्यनियमः

4.4 Convention in the choice of *hāraka*

गुणहाराविशेषोक्तौ [3] सर्वत्रापि हरो महान् ।
हाराधिके गुणे हारान् अनाद्यैश्च फलैर्नयेत् ॥ ४ ॥

guṇahārāviśeṣoktau sarvatrāpi haro mahān |
hārādhike guṇe hārān anādyaiśca phalairnayet || 4 ||

Whenever the *guṇakāras* and the *hārakas* are not mentioned explicitly, the *hāraka* is always chosen to be larger [than the *guṇakāra*]. When the *guṇakāra* [stated] is larger than the *hāraka*, then the *hārakas* may be obtained by using the quotients (in *vallyupasaṃhāra*) after excluding the first one.

This definition is helpful in formulating a common strategy for computing the *dhruvas* of the mean planet/*śīghrocca* at the end of *maṇḍalas* as will be explained later.

४.५ वाक्यकरणोक्तमण्डलध्रुवयोः आनयनम्

4.5 Procedure for obtaining *maṇḍalas* and *dhruvas* given in the *Vākyakaraṇa*

हारः सूर्यविहङ्गयोर्भगणयोर्भेदो गुणोऽल्पस्तयोः
ताभ्यामत्र परस्परासफलजाः हाराः धराहाहताः ।
भक्तास्तेनहरेण वाक्यकरणोक्ता मण्डलाः स्युस्तथै-
वान्योन्याहृतशिष्टचक्रकलिकाभ्यासात् धनर्णध्रुवाः ॥ ५ ॥

[3] Here we need to first form a compound of the words *viśeṣa* and *ukti* which is then compunded with *nañ*. That is, विशेषत: उक्ति: = विशेषोक्ति: । न विशेषोक्ति: = अविशेषोक्ति: । (*viśeṣataḥ uktiḥ = viśeṣoktiḥ. na viśeṣoktiḥ = aviśeṣoktiḥ*).

hāraḥ sūryavihaṅgayorbhagaṇayorbhedo guṇo'lpastayoḥ
tābhyāmatra parasparāptaphalajāḥ hārāḥ dharāhāhatāḥ |
bhaktāstenahareṇa vākyakaraṇoktā maṇḍalāḥ syustathai-
vānyonyāhṛtaśiṣṭacakrakalikābhyāsāt dhanarṇadhruvāḥ || 5 ||

The "difference in revolutions" between the Sun and the planet is the divisor (*hāra*). Of those two [quantities], the smaller would be the multiplier (*guṇakāra*). Here, the *hārakas* derived from the quotients obtained in the mutual division of these two [by doing *vallyupasaṃhāra*], are multiplied by the "number of civil days" (*dharāha*). [The results] when divided by the divisor (*hāra*) are the *maṇḍalas* mentioned in *Vākyakaraṇa*. In the same way, the remainders obtained in the mutual division [of *guṇakāra* and *hāra*] when multiplied by 21600 (*cakrakalikābhyāsa*) [and divided by the *hāra*] give the positive or negative *dhruvas*.

Before getting into the details of the procedure outlined in the above verse for finding *maṇḍala* and *dhruva*, we need to be familiar with some technical terms. Hence we first define them below.

४.५.१ मण्डलध्रुवानयनसूत्रम्

4.5.1 Expression for the maṇḍalas *and* dhruvas

If G_p and H_p are the *dṛḍhaguṇakāras* and the *dṛḍhahārakas* of the planets respectively, and G_s and H_s are those of the Sun, then the "difference in revolutions" mentioned in the above verse is given by

$$|G_s \times H_p - G_p \times H_s|.$$

This is the *hāra*, and the smaller of the two terms above is the *guṇakāra*. Let G_i and H_i be the successive *guṇakāras* and *hārakas* obtained respectively by doing the process of *vallyupasaṃhāra* with the above *guṇakāra* and *hāra*. Another term mentioned in the above verse, "number of civil days" (*dharāha*) is given by $H_s \times H_p$ (as indicated in verse 2). Now, the *maṇḍalas* are given by

$$\begin{aligned} maṇḍala \;\; (M_i) &= \frac{dharāha \times H_i}{hāra} \\ &= \frac{(H_s \times H_p) \times H_i}{|G_s \times H_p - G_p \times H_s|}. \end{aligned} \tag{4.6}$$

The *dhruvas* can be obtained from the remainders r_i at various stages in the above process of *vallyupasaṃhāra*. Depending upon whether i is odd or even, the *dhruva* is given by

$$dhruva \;\; (D_i) = \pm \left| \frac{r_i \times 21600}{G_s \times H_p - G_p \times H_s} \right|. \tag{4.7}$$

We shall now explain the rationale behind the above expressions for the *maṇḍala* and *dhruva*.

४.५.२ मण्डलानयने युक्तिः

4.5.2 Rationale behind the expression for maṇḍala

For this purpose, let us consider the case of an exterior planet (Mars, Jupiter and Saturn). In this case the rate of revolution of the planet $\frac{G_p}{H_p}$, is smaller than that of the *śīghrocca* (Sun) $\frac{G_s}{H_s}$. Hence, the rate of revolution of the *śīghrakendra* (number of revolutions per day) is

$$\frac{G_s}{H_s} - \frac{G_p}{H_p} = \frac{G_s H_p - G_p H_s}{H_s H_p}. \tag{4.8}$$

In this case, clearly $G_p H_s$ being smaller than $H_p G_s$, will be the *guṇakāra*. And $G_s H_p - G_p H_s$ will be the *hāra*. In other words,

$$G = G_p H_s,$$
$$\text{and} \qquad H = G_s H_p - G_p H_s. \tag{4.9}$$

To start with let us consider the case when $G < H$. In the case of Saturn, we have already evaluated H (see (4.1)) to be 1300241232554375. From the same calculation, we can see that $G = 120786218401$, and clearly $G < H$. From the above discussion, it follows that the synodic period, or the number of civil days for one revolution of the *śīghrakendra*, is given by

$$\frac{H_s H_p}{G_s H_p - G_p H_s}.$$

Now H revolutions of the *śīghrakendra* corresponds to

$$\left(\frac{H_s H_p}{G_s H_p - G_p H_s} \right) \times H \text{ days}.$$

As the denominator is actually H, this corresponds to an integral number of days $H_s H_p$. Moreover, the rate of revolution of the planet is $\frac{G_p}{H_p}$. Hence, the number of revolutions of the planet after H revolutions of the *śīghrakendra* is

$$H_s H_p \times \frac{G_p}{H_p},$$

which is again an integer $H_s G_p$. Hence the planet and the *śīghrocca* (Sun for an exterior planet) both complete a full number of revolutions after H revolutions of the *śīghrakendra*, or in H_sH_p days. This means that if the planet and the *śīghra* are at the *mandocca* of the planet at some instant, they would come back to the same *mandocca* after H revolutions of the *śīghrakendra*. As we saw above, H is a very large number, and it is impractical to work with this large time interval between two instants at which both the *mandakendra* and *śīghrakendra* are zero.

Now, consider H_i revolutions of the *śīghrakendra*, which correspond to

$$\frac{H_sH_p}{G_sH_p - G_pH_s} \times H_i = \frac{H_sH_p}{H} \times H_i \qquad \text{(in days)}. \tag{4.10}$$

The number of revolutions of the planet in this interval is

$$\frac{H_s \times H_p}{H} \times H_i \frac{G_p}{H_p} = H_sG_p \times \frac{G_i}{H} \times \frac{H_i}{G_i}. \tag{4.11}$$

Since, H_i, G_i arise from the *vallyupasaṃhāra* of H and G, $\frac{H_i}{G_i}$ would be a good approximation to $\frac{H}{G}$. Also $G = H_sG_p$. Hence the number of revolutions of the planet in H_i revolutions of the *śīghrakendra* is equal to

$$G \times \frac{G_i}{H} \times \frac{H_i}{G_i} \approx G \times \frac{G_i}{H} \times \frac{H}{G} = G_i. \tag{4.12}$$

In other words, the planet also makes very nearly a complete integral number of revolutions during the period where the *śīghrakendra* makes H_i revolutions, which corresponds to

$$\frac{H_s \times H_p}{G_sH_p - G_pH_s} \times H_i,$$

days. This is the reason for considering *maṇḍalas* defined in (4.6), as the planet and the *śīghrocca* are both expected to come back close to the *mandocca*, after a *maṇḍala* if their initial positions coincided with it. We now proceed to explain the *dhruvas*.

४.५.३ ध्रुवानयने युक्तिः

4.5.3 Rationale behind the expression for dhruva

After a *maṇḍala*, the change in the longitude of the planet (or the *śīghrocca*) in revolutions, or the *dhruva* in revolutions is given by

$$\frac{H_sH_p}{(G_sH_p - G_pH_s)} \times H_i \times \frac{G_p}{H_p} = \frac{H_sG_pH_i}{H} = \frac{GH_i}{H}, \tag{4.13}$$

where we have used (4.9). This may be written as

$$\frac{GH_i}{H} = \frac{GH_i - HG_i}{H} + \frac{HG_i}{H}. \tag{4.14}$$

The second term above can be dropped as it is an integer.

Now, H_i and G_i are the *hāraka* and the *guṇakāra* obtained doing the process of mutual division of H and G, and it has been shown in equation (A.24) of Appendix A that

$$GH_i - HG_i = \pm r_i, \tag{4.15}$$

where r_i is the remainder at the ith stage, and we have to choose '−' sign when i is odd and '+' when i is even. The remainder r_i is the *guṇaśiṣṭa* when i is even and the *hāraśiṣṭa* when i is odd. Hence, (4.14) in minutes is given by

$$\frac{GH_i}{H} = \pm\frac{r_i}{H} \times 21600, \tag{4.16}$$

where the RHS is the same as in (4.7).

We may now consider the case when $G > H$. Let

$$G = Ha_1 + r_1, \tag{4.17}$$

where a_1 is an integer, and $r_1 < H$ is the remainder when G is divided by H. Then the *dhruva* (in revolutions) is given by

$$\frac{GH_i}{H} = \frac{r_1 H_i}{H} + a_1 H_i. \tag{4.18}$$

As the second term is an integer, it would not contribute to the *dhruva*. Now we can consider the *vallyupasaṃhāra* of H and r_1 for computing the *dhruva*. This is the reason for the prescription in verse 4, that the *hārakas* H_i may be obtained after excluding the first quotient (a_1) when $G > H$.

It is easy to see that similar results can be obtained for the interior planets (Mercury and Venus) also where the mean planet is the Sun and the actual heliocentric mean planet is the *śīghra*.

The *maṇḍalas* and *dhruvas* of all the planets, corresponding to different *hārakas* are calculated following the above process, and listed in Tables G.6 – G.10 in Appendix G.[4] There the *maṇḍalas* and *dhruvas* listed in the *Vākyakaraṇa* are also tabulated and it is seen that they coincide fairly accurately with the computed values.

[4] Some of these *maṇḍalas* and *dhruvas* have also been presented in the form of *vākyas* in the Commentaries I, II ({KP 1956}, pp. 119-129).

४.६ सामान्यतः शोध्यानयनप्रकारः

4.6 A general prescription for obtaining *śodhyas*

मण्डलानयने नीताः हाराः मण्डलहारकाः ।
तैः शोध्यमानयेत् यद्वा सध्रुवैः स्वल्पमण्डलैः ॥ ६ ॥

maṇḍalānayane nītāḥ hārāḥ maṇḍalahārakāḥ |
taiḥ śodhyamānayet yadvā sadhruvaiḥ svalpamaṇḍalaiḥ || 6 ||

The *hāras* employed in the process of obtaining *maṇḍalas* are the corresponding *maṇḍala-hārakas*. The *śodhyas* are to be obtained from these [*hārakas*]. Or else, [the *śodhyas* can also be obtained by] using smaller *maṇḍalas* (*svalpamaṇḍalas*) along with their *dhruvas*.

The *śodhya* corresponds to a time interval (not necessarily an integer) close to the desired *ahargaṇa*, at which the *śīghrakendra* is zero, and the longitudes of the mean planet and the *śīghrocca* (which are equal) are close to the longitude of *mandocca*. The method of calculating *śodhyas* is discussed later in this chapter (verses 9-12).

४.७ शीघ्रोच्चग्रहयोगकालानयनम्

4.7 Obtaining the instant at which the *śīghrocca-grahayoga* occurs

अभिमतदिनविहगोनात् शीघ्रोच्चात् भुक्तिविवरलब्धोनः ।
अभिमतदिवसगणोऽयं शीघ्रोच्चविहङ्गयोगसमयः स्यात् ॥ ७ ॥

abhimatadinavihagonāt śīghroccāt bhuktivivaralabdhonaḥ |
abhimatadivasagaṇo'yaṃ śīghroccavihaṅgayogasamayaḥ syāt || 7 ||

The mean longitude of the planet corresponding to the desired *ahargaṇa* is subtracted from its *śīghrocca* and divided by the difference between their daily motions (*dinabhuktis*). The result thus obtained when subtracted from the desired *ahargaṇa* gives the instant of conjunction (*yoga*) of the *śīghrocca* with the planet.

The above verse presents an expression for finding the instant, close to the desired *ahargaṇa*, when the *śīghrakendra* is zero. This phenomenon is referred to as *graha-śīghrocca-yoga* or *śīghrocca-grahayoga*. If θ_0 and θ_s are the mean longitudes of the planet and the *śīghrocca* respectively, then the instant at which the *śīghrakendra* becomes zero can be obtained from the relation

$$t_c = A - \frac{(\theta_s - \theta_0)_A}{(\dot{\theta_s} - \dot{\theta_0})}, \tag{4.19}$$

where $\dot{\theta}_s$ and $\dot{\theta}_0$ represent the daily motion of the *śīghrocca* and the planet respectively.

४.८ शीघ्रोच्चग्रहयोगकालानयने द्वितीयः प्रकारः

4.8 Another method to obtain the instant of the *śīghrocca-grahayoga*

यद्वेष्टमध्यविहगोनचलोच्चलिप्ता-
संवर्धितक्षितिदिनात् भगणान्तराप्तम् ।
अज्ञातपारहृतमिष्टदिनात् विशोध्यं
शिष्टं चलोच्चखगमध्यमयोगकालः ॥ ८ ॥

yadveṣṭamadhyavihagonacaloccaliptā-
saṃvardhitakṣitidināt bhagaṇāntarāptam |
ajñātapārahṛtamiṣṭadināt viśodhyaṃ
śiṣṭaṃ caloccakhagamadhyamayogakālaḥ || 8 ||

Or else, the mean longitude of the planet is subtracted from its *śīghrocca*, expressed in minutes, and multiplied (*saṃvardhita*) by the number of civil days (*kṣitidina*), and divided by the difference in revolutions (*bhagaṇāntara*). [Again, the result thus obtained] is divided by 21600 (*ajñātapāra*) and is subtracted from the *ahargaṇa*. The [end] result gives the instant at which the conjunction of the mean planet and its *śīghrocca* occurs.

The above verse gives an alternative expression for obtaining the instant at which the *śīghrakendra* is zero. If $(\theta_s - \theta_0)_A$ (expressed in minutes) is the *śīghrakendra* corresponding to the *ahargaṇa* A, then the instant at which the conjunction of the planet and its *śīghrocca* occurs is now expressed as

$$t_c = A - \left| \frac{(\theta_s - \theta_0)_A \times (H_s H_p)}{((G_s H_p - G_p H_s) \times 21600)} \right| . \tag{4.20}$$

This is in fact the same as (4.19) since the denominator in the latter can be expressed as

$$\begin{aligned} (\dot{\theta}_s - \dot{\theta}_0) \;\; (\text{in min.}) &= \left(\frac{G_s}{H_s} - \frac{G_p}{H_p} \right) \times 21600 \\ &= \frac{(G_s H_p - G_p H_s)}{H_s H_p} \times 21600, \end{aligned} \tag{4.21}$$

since $\frac{G_s}{H_s}$ and $\frac{G_p}{H_p}$ are the rates motion in revolutions of the *śīghrocca* (Sun) and the mean planet, for the exterior planets. For the interior planets, $\frac{G_s}{H_s}$ is the rate of motion of the mean Sun which is also the rate of motion of the mean planet. This is smaller than $\frac{G_p}{H_p}$ which is the rate of motion of the *śīghrocca* of the planet. Hence for the interior planets

$$
\begin{aligned}
(\dot{\theta}_s - \dot{\theta}_0) \text{ (in min.)} &= \left(\frac{G_p}{H_p} - \frac{G_s}{H_s}\right) \times 21600 \\
&= \frac{(G_p H_s - G_s H_p)}{H_s H_p} \times 21600. \qquad (4.22)
\end{aligned}
$$

४.९ शोध्यदिनानयनप्रकारः

4.9 Procedure for obtaining the *śodhyadina*

तत्कालमध्यविहगं स्वमृदूच्चहीनं लिप्तीकृतं तु निजमण्डलहारकेषु ।
इष्टेन संगुणमनन्तपुरेण भक्तम् इष्टोर्ध्वहारहतमिष्टहरेण हृत्वा ॥ ९ ॥

ऊनाधिकं तदिह हारसमासमत्वे धात्रीदिनघ्नमुभयोर्भगणान्तराप्तम् ।
शीघ्रोच्चमध्यमविहङ्गमयोगकालात् शोध्यं तदा भवति शोध्यदिनं ग्रहाणाम् ॥ १० ॥

tatkālamadhyavihagaṃ svamṛdūccahīnaṃ
liptīkṛtaṃ tu nijamaṇḍalahārakeṣu |
iṣṭena saṃguṇamanantapureṇa bhaktam
iṣṭordhvahārahatamiṣṭahareṇa hṛtvā || 9 ||

ūnādhikaṃ tadiha hārasamāsamatve
dhātrīdinaghnamubhayorbhagaṇāntarāptam |
śīghroccamadhyamavihaṅgamayogakālāt
śodhyaṃ tadā bhavati śodhyadinaṃ grahāṇām || 10 ||

After subtracting its own *mandocca* from the mean planet at that moment (when *śīghrocca-grahayoga* occurs), convert [the result] into minutes. Multiply [the result] by a desired *hāraka* among the *maṇḍala-hārakas* of the [planet], and divide by 21600 (*anantapura*). [The quotient thus obtained] is multiplied by the penultimate *hāraka* and divided by the desired *hāraka*.

Depending on whether the *hāraka* is odd or even, the remainder [of the above division] (*adhikaśiṣṭa*), or the remainder subtracted from the *hāraka* (*ūnaśiṣṭa*) is multiplied by the number of civil days (*dhātrīdina*) and divided by the difference in revolutions (*bhagaṇāntara*) of the two (planet and its *śīghrocca*). The result thus obtained has to be subtracted from the time of conjunction of the planet and its *śīghrocca*. Then we obtain the *śodhyadina* of the planets.

The instant of time at which the *mandocca* will be close to the *śīghrocca-grahayoga* is called the *śodhyadina*. The verses above present an algorithm to find the *śodhyadina* from the instant t_c at which the *śīghrocca-grahayoga* occurs.

If H_i and H_{i-1} are the desired *maṇḍalahāraka* and the one immediately preceding it, and if $(\theta_{mk})_{t_c} = (\theta_0 - \theta_m)_{t_c}$ is the *mandakendra* at time t_c, then the remainder r_i (*adhikaśiṣṭa* or *ūnaśiṣṭa*) mentioned in the above verse may be written as

$$r_i = \left[\left[\frac{((\theta_{mk})_{t_c})}{21600} \times H_i\right]_{int} \times \frac{H_{i-1}}{H_i}\right]_{rem} \qquad \text{(for odd } H_i\text{)},$$
$$r_i = H_i - \left[\left[\frac{((\theta_{mk})_{t_c})}{21600} \times H_i\right]_{int} \times \frac{H_{i-1}}{H_i}\right]_{rem} \qquad \text{(for even } H_i\text{)}. \tag{4.23}$$

The *śodhyadina* t_s, can then be obtained as follows.

$$t_s = t_c - \frac{r_i \times (H_p \times H_s)}{(G_s H_p - G_p H_s)}. \tag{4.24}$$

We shall explain the rationale behind the above expression by considering the case of an exterior planet. The period of the *śīghrakendra* (synodic period) of the planet T_s is given by

$$T_s = \frac{H_s H_p}{(G_s H_p - G_p H_s)}. \tag{4.25}$$

Hence, the motion of the mean planet (and the *śīghrocca*) after one synodic period, in revolutions, is

$$T_s \times \frac{G_p}{H_p} = \frac{H_s H_p}{(G_s H_p - G_p H_s)} \times \frac{G_p}{H_p} = \frac{G}{H}, \tag{4.26}$$

where, following verse 6, we have used the definition of G and H given in Section 4.5 (see the discussion after equation (4.8)). Therefore,

$$\frac{G_p}{H_p} = \frac{1}{T_s} \times \frac{G}{H}. \tag{4.27}$$

Now $\frac{G_i}{H_i}$ are the successive approximants of $\frac{G}{H}$ in the process of *vallyupasaṃhāra*. Hence, the rate of motion of the mean planet may be approximated as

$$\frac{G_p}{H_p} \approx \frac{1}{T_s} \times \frac{G_i}{H_i}. \tag{4.28}$$

Now let $(\theta_{mk})_{t_c}$ in minutes be the longitude of the *mandakendra* at the instant of conjunction of the mean planet and the *śīghrocca*, t_c. In terms of revolutions it is $\frac{(\theta_{mk})_{t_c}}{21600}$. This can be written in the form

$$\frac{(\theta_{mk})_{t_c}}{21600} = \frac{n'}{H_i}, \tag{4.29}$$

Further, n' can be approximated by

$$n = \left[\frac{(\theta_{mk})_{t_c}}{21600} \times H_i\right]_{int}. \tag{4.30}$$

Let

$$\overline{x} = x \times T_s, \tag{4.31}$$

be the number of days in which the mean planet (or *mandakendra*, since the *mandocca* is assumed to be stationary) increases by $\frac{(\theta_{mk})_{t_c}}{21600}$. Here, we assume x to be an integer, as we want the *śīghrakendra* also to complete an integral number of revolutions so that the mean planet and the *śīghrocca* are again in conjunction. If we subtract $\overline{x}$ from t_c, we would obtain the *śodhyadina*, at which the *śīghrocca* and the mean planet are in conjunction and also close to the *mandocca*.

As the rate of motion of the mean planet is $\frac{1}{T_s} \times \frac{G_i}{H_i}$, we should have

$$\frac{1}{T_s} \times \frac{G_i}{H_i} \times \overline{x} - y = \frac{n}{H_i}, \tag{4.32}$$

where y is the integral number of completed revolutions. Using (4.31) in the above equation we have,

$$G_i x - H_i y = n.$$

Following our discussion in Section 3.2 of Chapter 3, the solution of the above equation can be expressed in the form

$$x = \left[\frac{n \times H_{i-1}}{H_i}\right]_{rem} = r_i \quad \text{(for odd } H_i\text{)},$$

$$\text{and} \qquad x = H_i - \left[\frac{n \times H_{i-1}}{H_i}\right]_{rem} = r_i \quad \text{(for even } H_i\text{)}, \tag{4.33}$$

where n is given by (4.30). Now the number of days to be subtracted from t_c to find the *śodhyadina* is given by

$$\overline{x} = T_s x = \frac{x(H_p \times H_s)}{(G_s H_p - G_p H_s)} = \frac{r_i(H_p \times H_s)}{(G_s H_p - G_p H_s)}. \tag{4.34}$$

Hence, we obtain the expression for the *śodhyadina* t_s as given by (4.24).

४.१० शोध्यध्रुवानयनम्

4.10 Finding the *śodhyadhruva*

तत्कालग्रहमध्यस्य मन्दतुङ्गस्य चान्तरम् ।
शोध्यध्रुवं धनर्णाख्यं उच्चान्मध्याधिकेऽल्पके ॥ ११ ॥

tatkālagrahamadhyasya mandatuṅgasya cāntaram |
śodhyadhruvaṃ dhanarṇākhyaṃ uccānmadhyādhike'lpake || 11 ||

The difference between the mean longitude of the planet and its apogee (*mandatuñga*) at that instant (t_s) would be the *dhruva* at the end of the *śodhyadina* (*śodhyadhruva*). It is positive or negative depending on whether the mean planet is larger or smaller than the apogee.

The *śodhyadhruva* at the end of the *śodhyadina* is stated to be

$$\acute{s}odhyadhruva = (\theta_0 - \theta_m)_{t_s}.$$

From the procedure for finding the *śodhyadina*, it is clear that $\theta_0 - \theta_m$ (*mandakendra*) would be close to zero but not exactly zero at the instant t_s (*śodhyadina* which is not necessarily an integer).

४.११ शोध्यदिनाप्तौ प्रकारान्तरम्

4.11 Another method to obtain the *śodhyadina*

शीघ्रोच्चग्रहमध्ययोः सदृशयोः तन्मध्यतुङ्गान्तरं
लिप्तीकृत्य हरेत् ध्रुवैः ऋणधनैःमध्यग्रहेऽल्पेऽधिके ।
शिष्टं शोध्यदिनध्रुवं ध्रुवफलक्षुण्णाश्च तन्मण्डलात्
शीघ्रोच्चग्रहमध्यसाम्यसमयात् शोध्याः स्वशोध्याप्तये ॥ १२ ॥

śīghroccagrahamadhyayoḥ sadṛśayoḥ tanmadhyatuṅgāntaraṃ
liptīkṛtya haret dhruvaiḥ ṛṇadhanaiḥ madhyagrahe'lpe'dhike |
śiṣṭaṃ śodhyadinadhruvaṃ dhruvaphalakṣuṇṇāśca tanmaṇḍalāt
śīghroccagrahamadhyasāmyasamayāt śodhyāḥ svaśodhyāptaye || 12 ||

The difference between the mean longitude of the planet, when it is the same as that of the *śīghrocca*, and the *mandocca*, expressed in minutes, is divided by [any suitable] negative or positive *maṇḍaladhruva* depending on whether the mean longitude is smaller or larger [than the *mandocca*]. The remainder [of this division] would be the *śodhyadhruva*. The quotients [of the division] (*dhruvaphalas*) are multiplied by the corresponding *maṇḍalas* and are to be subtracted from the time of conjunction of the planets and their *śīghrocca*, in order to obtain their own *śodhyadinas*.

If D_i represents a *maṇḍaladhruva*, then the *śodhya-dhruva* or *śodhyadinadhruva* given in the above verse is

$$\acute{s}odhyadinadhruva = \left[\frac{(\theta_0 - \theta_m)_{t_c}}{D_i}\right]_{rem}, \tag{4.35}$$

where we need to choose i to be odd ($D_i < 0$) if $\theta_0 < \theta_m$, and i to be even ($D_i > 0$) if $\theta_0 > \theta_m$. In (4.35), t_c is the instant of conjunction of *śīghrocca* and the planet (*śīghrocca-grahayoga*) close to the desired *ahargaṇa* A.

The expression for the *śodhyadina*, given in the verse, is

$$t_s = t_c - M_i \times \left[\frac{(\theta_0 - \theta_m)_{t_c}}{D_i}\right]_{int}, \tag{4.36}$$

where M_i is the *maṇḍala* with which the *dhruva* D_i is associated.

We can easily see the rationale for the above expressions for the *śodhyadina* and the *śodhyadinadhruva*. As we have already seen, the *maṇḍalas* M_i are chosen in such a way that in M_i days the planet and the *śīghrocca* will be in conjunction and close to the *mandocca*. Also, in M_i days, the *mandakendra* of the planet changes by an amount equal to the *dhruva* D_i. At the time t_c (*śīghrocca-grahayoga*), the *mandakendra* is $(\theta_0 - \theta_m)_{t_c}$. Hence, the *mandakendra* will be zero at the time

$$t_c - M_i \times \left(\frac{(\theta_0 - \theta_m)_{t_c}}{D_i}\right).$$

Therefore, the *mandakendra* will be nearly zero at the time

$$t_s = t_c - M_i \times \left[\frac{(\theta_0 - \theta_m)_{t_c}}{D_i}\right]_{int}. \tag{4.37}$$

Since t_s as given by the above equation, differs from t_c by an integral multiple of the *maṇḍala* M_i, the *śīghrakendra* will also be zero at the instant t_s. This is the rationale for the above equation (4.36) for *śodhyadina*.

Now we shall consider the rationale for the expression for *śodhyadhruva*. Let the *mandakendra* at time t_c be expressed in terms of D_i as

$$(\theta_0 - \theta_m)_{t_c} = q \times D_i + r,$$

where

$$q = \left[\frac{(\theta_0 - \theta_m)_{t_c}}{D_i}\right]_{int} \quad \text{and} \quad r = \left[\frac{(\theta_0 - \theta_m)_{t_c}}{D_i}\right]_{rem}. \tag{4.38}$$

Then the longitude of the *mandakendra* on the *śodhyadina* will be given by

$$\begin{aligned}(\theta_0 - \theta_m)_{t_c} - qD_i &= r \\ &= \left[\frac{(\theta_0 - \theta_m)_{t_c}}{D_i}\right]_{rem}.\end{aligned} \tag{4.39}$$

which coincides with the (4.35) as prescribed in the verse. Here r is negative if D_i is negative.

४.१२ मौढ्यावसानखण्डाः

4.12 Obtaining the *mauḍhyāvasāna-khaṇḍas*

शीघ्रोच्चमध्यग्रहयोगकालः मौढ्योत्थकालार्धदिनैः समेतः ।
मौढ्यावसानद्युगणः सखण्डः मौढ्योऽत्र हाराश्चलकेन्द्रहाराः ॥ १३ ॥

śīghroccamadhyagrahayogakālaḥ
mauḍhyotthakālārdhadinaiḥ sametaḥ|
mauḍhyāvasānadyugaṇaḥ sakhaṇḍaḥ
mauḍhyo'tra hārāścalakendrahārāḥ || 13 ||

Half the duration of the period of invisibility (*mauḍhyakāla*) added to the instant of conjunction of the *śīghrocca* and the mean planet, would give the *mauḍhyāvasāna-khaṇḍa* days. Here, the divisors are the *śīghrakendrahārakas*.

During the course of revolution of a planet, it will not be visible for a certain period of time when its longitude is close to that of the Sun. This period of invisibility of the planet is called the *mauḍhyakāla* and the starting of this *mauḍhya* (*mauḍhyārambha*) is known as the setting (*astamana*) of the planet. Similarly, the time of ending of this *mauḍhya* (*mauḍhyāvasāna*) is known as the rising (*udaya*) of the planet.

The term *mauḍhyāvasāna-khaṇḍa* refers to the number of days since the beginning of the *kali* to the recent *mauḍhyāvasāna* (ending of *mauḍhya*). The procedure to obtain it is explained in the verse above.

Let the time interval between the setting and rising of the planets called the *mauḍhyakāla* be denoted by Δt_m. Here, it is important to note that the *śīghrocca-grahayoga* occurs exactly at the middle of the *mauḍhyakāla*. If t_c represents the instant at which the conjunction of the mean planet and its *śīghrocca* takes place, then the *mauḍhyāvasāna-khaṇḍa* given by the above verse may be written as

$$\textit{mauḍhyāvasāna-khaṇḍa} \;= t_c + \frac{\Delta t_m}{2}. \tag{4.40}$$

It may be noted that, in the above formula (4.40), the variations in the rate at which the planet and the *śīghrocca* themselves move have not been considered.

४.१३ अगणितोक्तहारकानयनम्

4.13 Obtaining the *hārakas* mentioned in the *Agaṇita*

गुणहारौ ग्रहादित्यभगणौ तौ मिथो हरेत् ।
हारकास्तत्फलैर्नीता भवन्त्यगणितोदिताः ॥ १४ ॥

guṇahārau grahādityabhagaṇau tau mitho haret |
hārakāstatphalairnītā bhavantyagaṇitoditāḥ || 14 ||

The *guṇakāras* and *hārakas* which give the number of revolutions of the planet and the Sun are divided mutually. The *hārakas* obtained by the quotients of the above division are the ones that are mentioned in the *Agaṇita.*

As stated earlier, the ratios $\frac{G_p}{H_p}$ and $\frac{G_s}{H_s}$ are the rates of revolutions of the planet and the Sun per day, where G_p and G_s are the *dṛḍhaguṇakāras* and H_p and H_s are the corresponding *dṛḍhahārakas.* Their inverses are proportional to the number of revolutions in a *mahāyuga.* The ratio of the number of revolutions of the planet to that of the Sun is therefore given by

$$\frac{G_p \times H_s}{H_p \times G_s}.$$

This in fact gives the number of revolutions made by a planet in one sidereal year (one revolution of the Sun). Using the values of *dṛḍhaguṇakāras* and *dṛḍhahārakas* listed in Table 2.2 of the second chapter of the text, we can compute the above numerator and the denominator for each planet. They are listed in Table 4.6.

Name of the planet	*guṇakāra*	*hāraka*
Mars	8996044	16920000
Mercury	42152837	10152000
Jupiter	91046	1080000
Venus	27503843	16920000
Saturn	574109	16920000

Table 4.6 The *guṇakāras* and *hārakas* for the revolutions made by the planets in a solar year.

By doing the *vallyupasaṃhāra* of the *guṇakāras* and the *hārakas* listed in Table 4.6, we get a set of *hārakas.* We have computed these *hārakas* for different planets and listed them in Appendix H, Tables H.6 – H.10. The text notes that, these are the *hārakas* given in the *Agaṇita* system of Mādhava. The Commentary I also gives some of these *hārakas* in the form of *vākyas.* These *vākyas* and *hārakas* are listed in Appendix G, Table H.1– H.5.

४.१४ ग्रहशोध्याब्दानयनम्

4.14 Obtaining the *śodhyābdas* of the planets

मध्याब्दान्तसमानीतमध्यग्रहदिनेशयोः ।
अल्पभोगं महाभोगात् त्यक्त्वा शिष्टं कलीकृतम् ॥ १५ ॥

हारेष्वगणितप्रोक्तेष्वभीष्टेन समाहतम् ।
चक्रलिप्ताप्तमिष्टोर्ध्वहारकेण हतं पुनः ॥ १६ ॥

इष्टहारेण संहृत्य तत्रोनमधिकं तु वा ।
इष्टहारयुगोजत्ववशात् त्याज्यं शकाब्दतः ॥ १७ ॥

शिष्टाब्दान्ते भवेद् योगः इष्टग्रहदिनेशयोः ।
तस्मादगणितप्रोक्तः शोध्याब्दः सोऽयमीरितः ॥ १८

madhyābdāntasamānītamadhyagrahadineśayoḥ |
alpabhogaṃ mahābhogāt tyaktvā śiṣṭaṃ kalīkṛtam || 15 ||

hāreṣvagaṇitaprokteṣvabhīṣṭena samāhatam |
cakraliptāptamiṣṭordhvahārakeṇa hataṃ punaḥ || 16 ||

iṣṭahāreṇa saṃhṛtya tatronamadhikaṃ tu vā |
iṣṭahārayugojatvavaśāt tyājyaṃ śakābdataḥ || 17 ||

śiṣṭābdānte bhaved yogaḥ iṣṭagrahadineśayoḥ |
tasmādagaṇitaproktaḥ śodhyābdaḥ so'yamīritaḥ || 18 ||

Having obtained the mean longitude of the planet and the Sun at the end of the mean solar year (*madhyābdānta*),[5] subtract [the mean longitude of] that which has smaller daily motion from the other which has larger daily motion and convert the remainder into minutes.

Multiply [the result] by any desired divisor mentioned in the *Agaṇita* and after dividing it by 21600 (*cakraliptā*), the quotient obtained is multiplied by the immediately preceding divisor (*iṣṭordhvahāraka*).

The result is divided by the desired divisor. Then, depending on whether the desired *hāraka* is odd or even, the remainder in the above division or the *hāraka* minus the remainder, respectively, is subtracted from the *śakābda*.

The conjunction of the mean planet and the mean Sun occurs at the end of these remaining years (*śiṣṭābda*). Therefore, this gives that *śodhyābda* as mentioned in the *Agaṇita*.

The *śodhyābda* is the number of years elapsed since the beginning of the *śaka* era to the beginning of that year when the mean planet and the *śīghrocca* are in conjunction at the *meṣādi*. This notion seems to have been introduced in the *Agaṇita* system. The four verses given above explain how to compute the *śodhyābda* of a planet which is close to the time when y_s *śaka* years have elapsed.

[5] The term *madhyābdānta* refers to the instant at which the transit of the mean Sun occurs from *Mīna* to *Meṣa-rāśi*. In other words, it refers to the instant at which the mean longitude of the Sun is zero.

The algorithm presented here for finding the *śodhyābda* from a given *śakābda* (y_s) may be written as

- Let $(\theta_0)_{y_s}$ represent the mean longitude of the planet in minutes at the end of the mean *śakābda* year y_s. Since at that time the mean longitude of the Sun is zero, the difference between the mean longitude of the planet and that of the Sun would be $(\theta_0)_{y_s}$ itself. This in minutes has to be first multiplied by any desired *agaṇita-hāraka*, say H_i (see Tables H.6 – H.10) and divided by 21600. The integral part of this is to be obtained. That is, we need to find
$$\left[\frac{((\theta_0)_{y_s}) \times H_i}{21600}\right]_{int}.$$
- This result should be further multiplied by previous *hāraka* (H_{i-1}) and divided by H_i. The remainder (r_i) of the last division is found.
$$r_i = \left[\frac{\left[\frac{((\theta_0)_{y_s}) \times H_i}{21600}\right]_{int} \times H_{i-1}}{H_i}\right]_{rem}. \tag{4.41}$$
- Now, r_i or $H_i - r_i$ is to be subtracted from the *śakābda* (y_s) depending on whether i is odd or even respectively and the *śodhyābda* is given by
$$\begin{aligned} \textit{śodhyābda} &= y_s - r_i \qquad \text{(for odd } i\text{)},\\ &= y_s - (H_i - r_i) \quad \text{(for even } i\text{)}. \end{aligned} \tag{4.42}$$

Now we explain the rationale behind the procedure outlined above. At the end of the mean *śakābda* year, y_s, the mean longitude of the Sun is zero by definition. The mean longitude of the planet is $(\theta_0)_{y_s}$ in minutes, that is $\frac{(\theta_0)_{y_s}}{21600}$ in revolutions. Let

$$\frac{(\theta_0)_{y_s}}{21600} = \frac{n'}{H_i}.$$

If
$$\left[\frac{(\theta_0)_{y_s}}{21600} \times H_i\right]_{int} = [n']_{int} = n, \tag{4.43}$$

we may approximate $\frac{(\theta_0)_{y_s}}{21600}$ by $\frac{n}{H_i}$.

Let x be the number of years in which the longitude of the planet increases by $\frac{n}{H_i}$. Then, the *śodhyābda*, that is, the *śaka* year at the end of which the planet is at *meṣādi* is given by $y_s - x$. The rate of motion of the planet per solar year is $\frac{G}{H}$. This can be approximated by $\frac{G_i}{H_i}$. Therefore,

$$\frac{G_i}{H_i} \times x = \frac{(\theta_0)_{y_s}}{21600} + z \approx \frac{n}{H_i} + z, \tag{4.44}$$

where z is an integer giving the completed number of revolutions. That is, x is a solution of the indeterminate equation

$$G_i x - H_i z = n. \tag{4.45}$$

Following the discussion in Section 3.2, the solution of the above equation can be expressed in the form

$$x = r_i = \left[\frac{n \times H_{i-1}}{H_i}\right]_{rem} \qquad \text{(for odd } i\text{)},$$

and

$$x = H_i - r_i = H_i - \left[\frac{n \times H_{i-1}}{H_i}\right]_{rem} \qquad \text{(for even } i\text{)}. \tag{4.46}$$

Subtracting this from the *śaka* year y_s, we obtain

$$\acute{s}odhyabda = y_s - x,$$

which is what is stated in the verses.

४.१५ अधिमासखण्डानयनम्

4.15 Obtaining the *adhimāsakhaṇḍa*

कल्यब्दघ्नाधिमासाः दिनकरभगणैः संहृता भूदिनघ्नाः
कल्यादीन्दुध्रुवांशक्षितिदिनवधतो निश्चलाप्तैर्विहीनाः ।
भक्तास्तत्राधिमासैर्भवति दिनगणः सोऽधिमासोक्तखण्डोऽ-
थान्योन्याप्ताधिमासक्षितिदिनरचिता हारकास्तत्र हाराः ॥ १९ ॥

kalyabdaghnādhimāsāḥ dinakarabhagaṇaiḥ saṃhṛtā bhūdinaghnāḥ
kalyādīndudhruvāṃśakṣitidinavadhato niścalāptairvihīnāḥ |
bhaktāstatrādhimāsairbhavati dinagaṇaḥ so'dhimāsoktakhaṇḍo-
'thānyonyāptādhimāsakṣitidinaracitā hārakāstatra hārāḥ || 19 ||

The number of *adhikamāsas* in a *mahāyuga* (*yugādhmāsas*), multiplied by the number of elapsed years since the beginning of *kali*, is divided by the number of solar revolutions [in a *mahāyuga*]. [The result] is multiplied by the number of civil days (*bhūdina*) [in a *mahāyuga*]. [Separately], the *dhruva* of Moon, in degrees, at the beginning of *kali* is multiplied by the number of civil days in a *mahāyuga* and divided by 360 (*niścala*). The quotient of this division is subtracted [from the first result]. The result [thus obtained] when divided by the number of *adhikamāsas* [in a *mahāyuga*] gives the number of civil days which is stated as the *adhimāsakhaṇḍa*. The divisors obtained by the mutual division of the *yugādhimāsas* and the *bhūdina* are to be used as the *hārakas* here.

The *adhimāsakhaṇḍa* is the number of civil days elapsed since the beginning of the *kaliyuga* to the beginning of an *adhimāsa* which is close to the

Meṣādi when y_k *kali* years have elapsed. If R_s and D_c represent the number of revolutions of the Sun and the number of civil days in a *mahāyuga*, then the expression for *adhimāsakhaṇḍa* given in the above verse may be written as

$$\frac{\left[\frac{y_k \times yug\bar{a}dhim\bar{a}sa}{R_s} - \frac{kaly\bar{a}didhruva \text{ of Moon (in deg)}}{360}\right]_{int} \times D_c}{yug\bar{a}dhim\bar{a}sa}.$$

The rationale behind the above expression can be understood as follows. If the longitude of the Moon were to be zero at *kalyādi*, the number of *adhimāsas* elapsed (n_a) when y_k *kali* years have elapsed is given by the rule of three to be

$$n_a = \frac{y_k \times yug\bar{a}dhim\bar{a}sas}{R_s}. \tag{4.47}$$

Because of the *kalyādidhruva* (initial position being non-zero), the Moon would have to cover that much less before the new Moon, compared to the zero-*dhruva* case. So $\frac{kaly\bar{a}didhruva \text{ (in degrees)}}{360}$ should be subtracted from n_a. Hence

$$adhim\bar{a}sakha\d{n}\d{d}a = \left[n_a - \frac{kaly\bar{a}didhruva \text{ of Moon (in deg)}}{360}\right]_{int}.$$

The above result is in *adhimāsās*. It may be noted that only the integral part is taken, because in the calculation of the *adhimāsakhaṇḍa*, only the integral number of *adhimāsas* before the *kali* year y_k should be considered. Now the duration corresponding to one *adhimāsa* is $\frac{D_c}{yug\bar{a}dhim\bar{a}sa}$. Hence the *adhimāsakhaṇḍa* is

$$\frac{\left[n_a - \frac{kaly\bar{a}didhruva \text{ of Moon (in deg)}}{360}\right]_{int} \times D_c}{yug\bar{a}dhim\bar{a}sa} \text{ civil days.}$$

We need to find $\frac{D_c}{yug\bar{a}dhim\bar{a}sa}$ for computing the above. The last quarter of the verse prescribes that this ratio can be approximated by the standard *vallyupasaṃhāra* method. The commentators note that here we should actually use the *śakābda* corrected number of *adhimāsas* and the *śakābda* corrected number of civil days in a *mahāyuga*.[6] These are given by 135431760 and 134122987500 respectively. Using the above values, we have computed the *guṇakāras* and the corresponding *hārakas* which have been listed in Table 4.7.

[6] {KP 1956}, pp. 150-152.

guṇakāra	*hāraka*
1	990
2	1981
3	2971
110	108937
113	111908
675	668477
1463	1448862
2138	2117339
10015	9918218
12153	12035557
46474	46024889
337471	334209780
383945	380234669
2257196	2235383125
135431760	134122987500

Table 4.7 The *guṇakāras* and the *hārakas* for the *adhimāsakhaṇḍa*.

४.१६ ग्रहणहारानयनप्रकारः

4.16 Procedure for obtaining the *grahaṇahārakas*

हृत्वा परस्परमथो युगचान्द्रमासं
द्विघ्नार्कपातभगणैक्यमपीह लब्धैः ।
हारान्नयेत् पुनरमी धरणीदिनघ्नाः
स्युश्चान्द्रमासविहृता ग्रहणोक्तहाराः ॥ २० ॥

hṛtvā parasparamatho yugacāndramāsaṃ
dvighnārkapātabhagaṇaikyamapīha labdhaiḥ |
hārānnayet punaramī dharaṇīdinaghnāḥ
syuścāndramāsavihṛtā grahaṇoktahārāḥ || 20 ||

The number of lunar months in a *yuga* and twice the sum of the revolutions of the Sun and the node of the Moon (*dvighnārkapāta-bhagaṇaikya*) have to be divided mutually and from the quotients obtained here, let the *hārakas* be obtained. These [*hārakas*] when multiplied by the number of civil days (*dharaṇīdina*) and divided by the number of lunar months give the *grahaṇahārakas*.

The above verse describes the procedure to obtain the *grahaṇahārakas* which are used in finding the *grahaṇakhaṇḍas* (explained in the following verses). Both the Commentaries I and II mention that the revolution numbers to be used while obtaining these *hārakas* are the *Dṛggaṇita* revolution numbers that are listed in Table 4.8.[7]

Planet	number of revolutions	
	in *kaṭapayādi*	in numerals
Sun	*jñānājñānapralobham*	4320000
Moon	*nikhilaguṇasusīmā*	57753320
Rāhu	*anaṅgo rāgakhinnaḥ*	232300

Table 4.8 The *Dṛggaṇita* parameters used in the computation of *grahaṇahārakas*.

From Table 4.8, we find that twice the sum of revolutions of the Sun and the node of the Moon amounts to

$$G = 2 \times (4320000 + 232300) = 9104600.$$

The number of lunar months is given by $H = 53433320$. Using them, we have computed the *guṇakāras* G_i and *hārakas* H_i obtained by the process of *vallyupasaṃhāra* and these are listed in Table 4.9.

The *grahaṇahārakas* (H_{gi}), which are used for obtaining the *grahaṇakhaṇḍa*, are given by

$$\frac{H_i \times D_c}{\text{lunar months in a } mah\bar{a}yuga}. \tag{4.48}$$

These have also been listed in the third column of Table 4.9.

४.१७ ग्रहणखण्डानयनप्रकारः

4.17 Procedure for obtaining the *grahaṇakhaṇḍas*

नीत्वा मध्यार्कचन्द्रौ फणिनमपि दृशा मध्यपर्वान्तकाले
पातोनार्केन्दुलिप्ता लुनदगगुणिताश्चक्रलिप्ता विभक्ताः ।
तापस्थानेन हत्वा लुनदगविहृते शिष्टतो भूदिनघ्नात्
चान्द्रैर्मासैरवाप्तं त्यजतु दिनगणात् सोपरागोक्तखण्डः ॥ २१ ॥

nītvā madhyārkacandrau phaṇinamapi dṛśā madhyaparvāntakāle
pātonārkenduliptā lunadagaguṇitāścakraliptā vibhaktāḥ |

---

[7] {KP 1956}, pp. 152-154. The revolution numbers are perhaps inferred from the parameters given in *Dṛggaṇita*. The published edition of the *Dṛggaṇita* of Parameśvara ({DG 1963}) gives only *guṇakāras* and *hārakas* for planets.

guṇakāra (G_i)	*hāraka* (H_i)	*grahaṇahāraka* (H_{gi})
1	5	148
1	6	177
7	41	1211
8	47	1388
15	88	2599
23	135	3987
38	223	6585
61	358	10572
587	3445	101733
648	3803	112305
3179	18657	550952
3827	22460	663257
22314	130957	3867238
26141	153417	4530495
100737	591208	17458721
227615	1335833	39447938
9104600	53433320	1577917500

Table 4.9 The *grahaṇahārakas.*

> *tāpasthānena hatvā lunadagavihṛte śiṣṭato bhūdinaghnāt*
> *cāndrairmāsairavāptaṃ tyajatu dinagaṇāt soparāgoktakhaṇḍaḥ* || 21 ||

> Having obtained the mean longitudes of the Sun, Moon and its node (*phaṇi*) at the instant of mean conjunction/opposition (*madhyaparvānta*)[8] using [the parameters given in] *Dṛg*[*gaṇita*], subtract [the mean longitude of] the node (*pāta*) from [that of] the Sun and the Moon and convert [the results] into minutes. They are multiplied by 3803 (*lunadaga*) and divided by 21600 (*cakraliptā*). [The quotients thus obtained] are multiplied by 716 (*tāpasthāna*) and divided by 3803 (*lunadaga*). The remainders thus obtained are multiplied by the number of civil days in a *mahāyuga* and divided by the number of lunar months in a *mahāyuga.* The results thus obtained, when subtracted from the *ahargaṇa* [at the end of the conjunction/opposition], are said to be the *grahaṇa-khaṇḍas.*

Here, the verse gives the prescription to obtain the day on which the conjunction of the Sun or Moon and the Moon's node (*Rāhu*) occurs before any given new Moon day or full Moon day (*parvāntakāla*). Let's assume that the Sun is close to one of the nodes of the Moon and that it is a *parvāntakāla.* On this day the longitude of the Moon will be same as that of the node in

[8] The instant at which the difference in longitudes of the mean Sun and the Moon is either 0° or 180° is called the *madhyaparvāntakāla.*

the case of solar eclipse or at the other node which is six *rāśis* away in the case of lunar eclipse. The procedure outlined in the above verse for obtaining *grahaṇa-khaṇḍa* (for a solar/lunar eclipse) corresponding to a new Moon/full Moon day with *ahargaṇa* A_n/A_f is as follows.

- Let θ_{s0}, θ_{m0} and θ_{n0} represent the mean longitudes in minutes of the Sun, the Moon and its node respectively. We evaluate them corresponding to the *aharagaṇa* A_n when the Sun and the Moon are in conjunction, and then find
$$\left[\frac{(\theta_{s0}-\theta_{n0})_{A_n}\times 3803}{21600}\right]_{int}.$$
Similarly, we find these longitudes when the Sun and the Moon are in opposition, i.e., when *ahargaṇa* $A_f \approx A_n + 15$. Then we determine the following quantity
$$\left[\frac{(\theta_{m0}-\theta_{n0})_{A_f}\times 3803}{21600}\right]_{int}.$$
- These two results have to be further multiplied by 716 and divided by 3803. The remainders (r_s and r_m) of the resulting quantities are then found.
$$r_s = \left[\frac{\left[\frac{(\theta_{s0}-\theta_{n0})_{A_n}\times 3803}{21600}\right]_{int}\times 716}{3803}\right]_{rem},$$
and
$$r_m = \left[\frac{\left[\frac{(\theta_{m0}-\theta_{n0})_{A_f}\times 3803}{21600}\right]_{int}\times 716}{3803}\right]_{rem}. \tag{4.49}$$
- These remainders are now multiplied by the number of civil days (D_c) in a *mahāyuga* and divided by the number of lunar months (M_m). Then the *grahaṇa-khaṇḍas* are given by
$$\textit{grahaṇa-khaṇḍa} = A_n - \frac{r_s\times D_c}{M_m} \quad \text{(solar eclipse)}, \tag{4.50}$$
$$= A_f - \frac{r_m\times D_c}{M_m} \quad \text{(lunar eclipse)}. \tag{4.51}$$

We shall now provide the rationale behind the above procedure, by first considering the case when the Sun and the Moon are in conjunction corresponding to an *ahargaṇa* A_n. Let the mean longitude of the Sun and the Moon be given by $\theta_{s0}(=\theta_{m0})$ in minutes, and that of the ascending node be θ_{n0} in minutes. Let

$$\frac{\theta_{s0} - \theta_{n0}}{21600} = \frac{n'}{H_i} \approx \frac{n}{H_i}, \tag{4.52}$$

where n is the integral part and H_i is a suitable *grahaṇa-hāraka*.

Now, we have to find the instant prior to the specified *ahargaṇa*, when the Sun, Moon and one of its two nodes are at conjunction. This gives the *khaṇḍa* for solar eclipse. Let the difference between the two instants be x lunar months. The motion of the nodes of the Moon (*Rāhu* and *Ketu*) is retrograde. The nodes individually separate from the Sun at the rate $\frac{G}{2H}$ revolution per lunar month, where

$$\frac{G}{H} = \frac{2 \times (\text{revolutions of the Sun} + \text{revolutions of nodes})}{\text{number of lunar months}}.$$

We shall approximate the above rate by $\frac{G_i}{H_i}$ by making use of the *grahaṇa-guṇakāras* and *grahaṇahārakas* discussed in the previous section (see Table 4.9). At the *sūryagrahaṇakhaṇḍa*, which is x lunar months before the given *ahargaṇa*, either *Rāhu* or *Ketu* should be in conjunction with the mean Sun. Hence

$$\frac{G_i}{2H_i} \times x - \frac{y}{2} = \frac{n}{H_i}, \tag{4.53}$$

where y is an integer. Thus, $G_i x - H_i y = 2n$. Following the discussion in Section 3.2 of Chapter 3, the solution of the above equation can be expressed in the form

$$x = H_i - \left[\frac{2nH_{i-1}}{H_i}\right]_{rem} \qquad \text{(for even } i\text{)},$$

$$\text{and} \qquad x = \left[\frac{2nH_{i-1}}{H_i}\right]_{rem} \qquad \text{(for odd } i\text{)}. \tag{4.54}$$

Since x is in the units of lunar months, the number of civil days between the new Moon (A_n) (*parvānta*) and the *sūryagrahaṇa-khaṇḍa* is given by

$$\left[\frac{2nH_{i-1}}{H_i}\right]_{rem} \times \frac{D_c}{M_m} \qquad \text{(for odd } i\text{)}, \tag{4.55}$$

and

$$\left(H_i - \left[\frac{2nH_{i-1}}{H_i}\right]_{rem}\right) \times \frac{D_c}{M_m} \qquad \text{(for even } i\text{)}. \tag{4.56}$$

We choose the even (10th) *hāraka* $H_i = 3803$, so that $H_{i-1} = 3445$. Using these values in (4.54), we have,

$$x = 3803 - \left[\frac{2n \times 3445}{3803}\right]_{rem}. \tag{4.57}$$

Now $2n \times 3445 = 2n(3803 - 358)$. Let

$$2n \times 358 = n \times 716 = l \times 3803 + r, \tag{4.58}$$

where l is an integer. That is

$$r = \left[\frac{n}{3803} \times 716\right]_{rem}. \tag{4.59}$$

Then

$$2n \times 3445 = (2n - l - 1) \times 3803 + (3803 - r).$$

The first term does not contribute to the remainder, and hence

$$\left[\frac{2n \times 3445}{3803}\right]_{rem} = 3803 - r. \tag{4.60}$$

Therefore,

$$\begin{aligned} r_s = x = 3803 &- \left[\frac{2n \times 3445}{3803}\right]_{rem} \\ &= 3803 - (3803 - r) \\ &= r = \left[\frac{n}{3803} \times 716\right]_{rem}, \end{aligned} \tag{4.61}$$

where

$$n = \left[\frac{(\theta_{s0} - \theta_{n0})_{A_n} \times 3803}{21600}\right]_{int}.$$

This is what is prescribed in the text. Now r_s lunar months correspond to $r_s \times \frac{D_c}{M_m}$ days. Hence the *sūryagrahaṇa-khaṇḍa* corresponding to the *parvānta*, which is a New Moon, is

$$A_n - r_s \times \frac{D_c}{M_m}. \tag{4.62}$$

Similarly, the *candragrahaṇa-khaṇḍa* corresponding to the same *parvānta*, is $A_f - r_m \times \frac{D_c}{M_m}$, where A_f is half a lunar month more than A_n, and r_m is given by

$$r_m = \left[\frac{n}{3803} \times 716\right]_{rem}, \tag{4.63}$$

where

$$n = \left[\frac{(\theta_{m0} - \theta_{n0})_{A_f} \times 3803}{21600}\right].$$

Illustrative example:

We take the *Dṛggaṇita* values for the revolution numbers. The longitude of the Sun and Moon at *kalyādi* are zero, whereas that of the *Rāhu* is 180°. Consider the *ahargaṇa* $A = 1851981$. The mean longitudes of the Sun and the Moon are not exactly equal at A. It can be easily checked that

$$A_n = 1851981.462035,$$

corresponds to the instant of conjunction with $\theta_{s0} = \theta_{m0} = 118.2247°$. At A_n, the longitude of *Rāhu* is found to be $\theta_{n0} = -53.1073°$. Hence, at the *parvānta* (here the new Moon day),

$$\begin{aligned} \theta_{s0} - \theta_{n0} &= 171.3320° \\ &= 10279.92', \end{aligned}$$

and

$$\begin{aligned} \frac{\theta_{s0} - \theta_{n0} \text{ (minutes)}}{21600} &= 0.47592 \\ &= \frac{1809.9323}{3803} \approx \frac{1810}{3803}. \end{aligned}$$

With this we find the remainder r_s to be

$$\begin{aligned} r_s &= \left[\frac{n \times 716}{3803}\right]_{rem} \\ &= \left[\frac{1810 \times 716}{3803}\right]_{rem} \\ &= 2940 \text{ lunar months}, \end{aligned}$$

and

$$\frac{r_s \times D_c}{M_m} = 86819.93651 \text{ days}.$$

Hence the *sūryagrahaṇa-khaṇḍa* is given by

$$\begin{aligned} \textit{sūryagrahaṇa-khaṇḍa} &= A_n - \frac{r_s \times D_c}{M_m} \\ &= 1765161.526 \text{ days}. \end{aligned}$$

One can check that at this *khaṇḍa*,

$$\begin{aligned} \theta_{s0} &= 228.2474, \\ \theta_{m0} &= 228.2472, \\ \theta_{n0} &= 228.2597. \end{aligned}$$

It is also worth noting that θ_{n0} and θ_{s0} (or θ_{m0}) differ only by $0.0125° \approx 0.6'$.

Along the same lines, we shall now proceed to obtain the *candragrahaṇa-khaṇḍa*. If $A_n = 1851981.462035517$ days corresponds to the time of new moon, then

$$A_f = A_n + \frac{1}{2} \text{ lunar month} = 1851996.227 \text{ days},$$

corresponds to the time of full moon. At this *ahargaṇa* we find that

$$\begin{aligned}\theta_{m0} &= 312.7773756^\circ, \\ \theta_{s0} &= 132.7773756^\circ \qquad (\theta_{m0} - 180^\circ), \\ \theta_{n0} &= -53.88983997^\circ.\end{aligned}$$

At the *parvānta* (here, the time of full moon)

$$\begin{aligned}\theta_{m0} - \theta_{n0} &= 6.66721556^\circ \\ &= 400.032934', \\ \text{and} \qquad \frac{\theta_{m0} - \theta_{n0} \text{ (min)}}{21600} &= 0.01852004324 \\ &= \frac{70.4317444}{3803} \approx \frac{70}{3803}.\end{aligned}$$

Now

$$\begin{aligned}\left[\frac{n \times 716}{3803}\right]_{rem} &= 681 \text{ lunar months} \\ &= 20110.33223 \text{ days}.\end{aligned}$$

Hence, the *candragrahaṇa-khaṇḍa* is

$$\begin{aligned}A_f - 20110.33223 &= 1851996.227 - 20110.33223 \\ &= 1831885.895 \text{ days}.\end{aligned}$$

At this *khaṇḍa*,

$$\begin{aligned}\theta_{s0} &= 111.9763^\circ, \\ \theta_{m0} &= 291.9721^\circ, \\ \theta_{n0} &= 291.9370^\circ.\end{aligned}$$

It is seen that the longitude of Sun differs from (longitude of Moon -180°) only by $0.0042^\circ \approx 0.24'$. Similarly, the longitude of *Rāhu* and mean Moon differ by $0.0351^\circ \approx 2.1'$. This again shows the accuracy of the procedure outlined in the text.

Chapter 5
भगणादिपरीक्षा
Examination of the revolution numbers etc.

५.१ कल्पग्रहभगणस्फुटीकरणप्रकारः

5.1 Procedure for correcting the revolutions of the planets in a *kalpa*

ग्रहणग्रहयोगाद्यैः ये ग्रहाः सुपरीक्षिताः ।
दृक्समास्तत्समाः कल्पे कल्प्या वा भगणादयः ॥ १ ॥
परीक्षितस्य खेटस्य तन्त्रानीतस्य चान्तरम् ।
लिप्तीकृतार्कभगणैः कल्पोक्तैश्च समाहतम् ॥ २ ॥
तन्त्रनिर्माणकालस्य परीक्षासमयस्य च ।
अन्तरालगतैरब्दै राशिचक्रकलाहतैः ॥ ३ ॥
हृत्वाप्तं तन्त्रनीतस्य ग्रहस्याल्पाधिकत्वतः ।
स्वर्णं तत् कल्पभगणे कुर्यान्नैषविधी रवेः ॥ ४ ॥

grahaṇagrahayogādyaiḥ ye grahāḥ suparīkṣitāḥ |
dṛksamāstatsamāḥ kalpe kalpyā vā bhagaṇādayaḥ || 1 ||
parīkṣitasya kheṭasya tantrānītasya cāntaram |
liptīkṛtārkabhagaṇaiḥ kalpoktaiśca samāhatam || 2 ||
tantranirmāṇakālasya parīkṣāsamayasya ca |
antarālagatairabdai rāśicakrakalāhataiḥ || 3 ||
hṛtvāptaṃ tantranītasya grahasyālpādhikatvataḥ |
svarṇaṃ tat kalpabhagaṇe kuryānnaiṣavidhī raveḥ || 4 ||

Based on a thorough examination of the position of the planets at the time of eclipses, planetary conjunctions, and so on, the revolution numbers etc. [of the planets] in a *kalpa* have to be conceived of for achieving concordance with observations.

Multiply the magnitude of the difference between the textually computed (*tantrānīta*) and the observed (*parīkṣita*) values of the [longitude of] a planet, expressed in minutes, by the revolutions of the Sun in a *kalpa* and divide by the product of the

V. Pai et al., *Karaṇapaddhati of Putumana Somayājī*, Sources and Studies in the History of Mathematics and Physical Sciences, https://doi.org/10.1007/978-981-10-6814-0_5

number of years elapsed between the time of observation and that of the composition of the text (employed for making the computation) and 21600 (*cakrakalā*). The quotient thus obtained should be added to or subtracted from the number of revolutions in a *kalpa* (*kalpabhagaṇa*) of the planet, depending on whether the textually computed values are smaller or larger [than the observed values], respectively. This procedure is not to be employed in the case of the Sun.

Let the magnitude of the difference between the observed and the computed mean longitudes of the planet be $\Delta\theta$ (in minutes). Assuming that at the time of composition of the text there was no difference between the two, the entire difference $\Delta\theta$ can be considered to be the error accumulated during the period starting from the date of composition till the date of observation. If y be the number of years between the date of composition of the text and that of observation, then the discrepancy per year is

$$\frac{\Delta\theta \text{ (in min.)}}{y} = \frac{\Delta\theta}{21600 \times y} \text{ (\textit{bhagaṇas})}. \tag{5.1}$$

Multiplying (5.1) by the number of years in a *kalpa* would give the error accumulated in the *kalpa*. The number of years in a *kalpa* is by definition equal to the number of solar revolutions in a *kalpa*, R_s. The result obtained can be applied to the *bhagaṇas* in a *kalpa* to obtain the corrected revolutions of the planets in a *kalpa*. Thus, the correction ΔR to be applied to the revolutions of the planet is given by

$$\Delta R = \frac{\Delta\theta}{21600 \times y} \times R_s. \tag{5.2}$$

The above correction ΔR is to be added or subtracted from the revolution number of the planet depending on whether the observed values are smaller or larger, respectively, than the computed values.

५.२ ग्रहाणां कल्पादिध्रुवानयनम्

5.2 Obtaining the *kalpādidhruvas* of the planets

तत्पर्ययाभिहतकल्पगताब्दवृन्दात्
कल्पार्कपर्ययहृतं भगणादिमध्यम् ।
त्यक्त्वा परीक्षितसमान्तविहङ्गमध्यात्
शिष्टं वदन्ति किल कल्पमुखध्रुवाख्यम् ॥ ५ ॥

tatparyayābhihatakalpagatābdavṛndāt
kalpārkaparyayahṛtaṃ bhagaṇādimadhyam |
tyaktvā parīkṣitasamāntavihaṅgamadhyāt
śiṣṭaṃ vadanti kila kalpamukhadhruvākhyam || 5 ||

The product of those [corrected] revolutions of the planets and the number of years elapsed since the beginning of the *kalpa* when divided by the number of revolutions of the Sun in a *kalpa*, gives the mean longitude of the planet. [This is] subtracted from the observed mean longitude of the planet at the end of the elapsed year. The remainder indeed is stated to be the *kalpādidhruva* of the planet.

This verse lays down the procedure for obtaining the initial position of the planets at the beginning of *kalpa* (known as *kalpādidhruvas*). It is stated that difference between the computed mean longitude (θ_c) and the observed mean longitude[1] (θ_o) gives the *kalpādidhruvas* or the *grahadhruvas.*

If Y be the number of elapsed years since the beginning of *kalpa*, then the computed mean longitudes of the planets are given by

$$\theta_c = \frac{(R \pm \Delta R) \times Y}{R_s}, \tag{5.3}$$

where R_s represents the number of revolutions of the Sun in *kalpa* and $R \pm \Delta R$ represents the corrected *bhagaṇas* of the planets. The integral part of the above equation gives the elapsed number of revolutions. From the fractional part one can obtain the longitudes in terms of sign, degrees, minutes etc., as done earlier. Subtracting θ_c from θ_o we obtain the *grahadhruva.* That is,

$$grahadhruva = \theta_o - \theta_c. \tag{5.4}$$

५.३ कल्पादौ सङ्क्रमणध्रुवः

5.3 *Saṅkramaṇadhruva* at the beginning of the *kalpa*

कल्पोक्तभूदिवसकल्पगताब्दघातात्
कल्पोक्तभानुभगणाहृतवासरादेः ।
सप्ताप्तशिष्टरहितेष्टसमान्तकालः
कल्पादिजो भवति सङ्क्रमणध्रुवोऽयम् ॥ ६ ॥

kalpoktabhūdivasakalpagatābdaghātāt
kalpoktabhānubhagaṇāhṛtavāsarādeḥ |
saptāptaśiṣṭarahiteṣṭasamāntakālaḥ
kalpādijo bhavati saṅkramaṇadhruvo'yam || 6 ||

Multiplying the number of elapsed years in the [present] *kalpa* by the number of civil days in a *kalpa*, and dividing the result by the number of solar revolutions in a *kalpa*, whatever is obtained has to be divided by 7. The remainder of this, when subtracted from [the number corresponding] to the observed weekday at the

[1] The term 'observed' mean longitude may sound strange as it is only the true longitude that can be observed. However, at those instances when the true coincides with the mean, the observed value straightaway gives the mean longitude.

end of the elapsed year (*iṣṭasamāntakāla*), would be the *saṅkramaṇadhruva* at the beginning of the *kalpa*.

The term *saṅkramaṇadhruva* represents the weekday at the beginning of the *kalpa*. The procedure outlined above is based on the assumption that the mean Sun is at *meṣādi* (with zero degree longitude) at the beginning of *kalpa*. If D_c is the number of civil days in a *kalpa*, and Y is the elapsed number of years since the beginning of the *kalpa*, then the elapsed number of civil days d_c elapsed is given by

$$d_c = \frac{Y \times D_c}{R_s}. \tag{5.5}$$

The weekday w_c at the end of the Yth year is given by

$$w_c = \left[\frac{d_c}{7}\right]_{rem}. \tag{5.6}$$

Here, $w_c = 0, 1, 2 \ldots$ and 6 correspond to Sunday, Monday ... and Saturday respectively.

If w_o be the observed weekday at the end of the elapsed year, then the *saṅkramaṇa-dhruva* or the weekday at the beginning of the *kalpa* would be

$$\begin{aligned} Saṅkramaṇadhruva &= w_o - w_c \quad [\text{for } w_o > w_c], \\ &= (7 + w_o) - w_c \quad [\text{for } w_o < w_c]. \end{aligned} \tag{5.7}$$

५.४ सङ्क्रमणग्रहध्रुवयोः स्फुटीकरणम्

5.4 Corrections to *saṅkramaṇa* and *graha-dhruvas*

तन्त्रानीतपरीक्षितारुणभिदालिप्ताः पृथग् भूदिनैः
कल्पोक्तैर्ग्रहपर्ययैश्च गुणिताः कल्पार्कवर्षाहृताः ।
प्राणाः सङ्क्रमणग्रहेषु कलिकाः कल्पादिगेषु क्रमात्
स्वर्णं तत्र परीक्षिते दिनकरे स्वल्पेऽधिके ते स्फुटाः ॥ ७ ॥

tantrānītaparīkṣitāruṇabhidāliptāḥ pṛthag bhūdinaiḥ
kalpoktairgrahaparyayaiśca guṇitāḥ kalpārkavarṣāhṛtāḥ |
prāṇāḥ saṅkramaṇagraheṣu kalikāḥ kalpādigeṣu kramāt
svarṇaṃ tatra parīkṣite dinakare svalpe'dhike te sphuṭāḥ || 7 ||

The difference in [the longitudes] of the textually computed (*tantrānīta*) and the observed values (*parīkṣita*) of the Sun in minutes are separately multiplied by the number of civil days in a *kalpa* and by the number of revolutions of the planets in a *kalpa*, and divided by the number of solar years in a *kalpa*. The results thus obtained should be applied to the *saṅkramaṇadhruva* in *prāṇas* and to the *grahadhruvas* of the respective planets, respectively. This [correction] has to be applied positively or negatively depending on whether the observed [longitude of the] Sun is smaller or larger [than the computed value]. These give the true values [of the *saṅkramaṇa* and *grahadhruvas*].

The above verse gives the corrections to the *saṅkramaṇadhruva* and the *grahadhruvas* of the planets at the beginning of the *kalpa* which arise due to a shift in the *kalpādi* itself as inferred from the difference between the observed and computed longitudes of the Sun. These corrections are obtained by first finding the difference between the observed and the computed longitudes of the Sun (θ_{so} and θ_{sc} respectively) $\delta\theta_s$ (in revolutions) given by

$$\delta\theta_s = \theta_{so} - \theta_{sc}. \tag{5.8}$$

This multiplied by the number of civil days in a solar year (ratio of the number of civil days to the number of solar revolutions in a *kalpa*) would give the fraction of a day corresponding to $\delta\theta_s$. Thus, the correction added to the *saṅkramaṇadhruva* is

$$\Delta_s \text{ (in days)} = \frac{\delta\theta_s \times D_c}{R_s}. \tag{5.9}$$

As one day is equal to 21600 *prāṇas*, multiplying both sides of the above equation by 21600, we get the correction in *prāṇas*.

$$\begin{aligned} \Delta_s \text{ (in } \textit{prāṇas}) &= \frac{\delta\theta_s \times D_c \times 21600}{R_s} \\ &= \frac{\delta\theta_s \text{ (in min.)} \times D_c}{R_s}. \end{aligned} \tag{5.10}$$

If $\theta_{so} > \theta_{sc}$, it means that the *ahargaṇa* for the day of observation is actually more, that is, the *kalpādi* is actually earlier and hence Δ_s (in *prāṇas*) must be subtracted from the *saṅkramaṇadhruva*. If $\theta_{so} < \theta_{sc}$, Δ_s has to be added to it.

The shift in the *kalpādi* would alter the *grahadhruvas* also. The change in *grahadhruva* in minutes would be

$$\begin{aligned} \Delta \text{ (in minutes)} &= \frac{\Delta_s \text{ (in } \textit{prāṇas}) \times (R \pm \Delta R)}{D_c} \\ &= \frac{\delta\theta_s \text{ (in minutes)} \times (R \pm \Delta R)}{R_s}, \end{aligned} \tag{5.11}$$

as the rate of motion of the planet is

$$\frac{(R \pm \Delta R)}{D_c} \text{ revolutions} = \frac{(R \pm \Delta R)}{D_c} \times 21600 \text{ minutes per day.}$$

If the *kalpādi* is earlier (if $\theta_{so} > \theta_{sc}$), Δ has to be subtracted from the *grahadhruva*. If $\theta_{so} < \theta_{sc}$, Δ has to be added to the *grahadhruva* of the planet.

५.५ कल्पादिध्रुवसद्भावपरिहरणम्

5.5 Eliminating the *kalpādidhruvas*

एवं तु ध्रुवसद्भावः कल्पादौ नैव युज्यते ।
इति तत्परिहारार्थं संस्कारान्तरमिष्यते ॥ ८ ॥

हृत्वा मिथः कल्पगताब्दवृन्दं कल्पोदितं भास्करपर्ययञ्च ।
लब्धैः समानीतहरेष्वभीष्टेनाहत्य कल्पादिखगध्रुवांशान् ॥ ९॥

हृत्वातुलैराप्तमभीष्टहारस्योर्ध्वस्थहारेण निहत्य हृत्वा ।
पूर्वोदिताभीष्टहरेण शिष्टं हारौजयुग्मत्ववशाद् धनर्णम् ॥ १० ॥

कल्पोदिते संस्कृतपर्ययौघे ग्रहस्य कुर्यात् स तदा स्फुटः स्यात् ।
इष्टघ्नहारोनयुतः स दृष्टः क्वचिद् ग्रहो दृष्टिसमो यतः स्यात् ॥ ११ ॥

ताभ्यां हराभ्यां तु तथा मृदूच्चात् पाताच्च नीतो भगणस्तदीयः ।
हाराद्विशोध्योऽयमृणात्मकश्चेत् पातस्य सर्वं विपरीतमेव ॥ १२ ॥

evaṃ tu dhruvasadbhāvaḥ kalpādau naiva yujyate |
iti tatparihārārthaṃ saṃskārāntaramiṣyate || 8 ||

hṛtvā mithaḥ kalpagatābdavṛndaṃ
kalpoditaṃ bhāskaraparyayañca |
labdhaiḥ samānītahareṣvabhīṣṭe-
nāhatya kalpādikhagadhruvāṃśān || 9 ||

hṛtvātulairāptamabhīṣṭahāra-
syordhvasthahāreṇa nihatya hṛtvā |
pūrvoditābhīṣṭahareṇa śiṣṭaṃ
hāraujayugmatvavaśād dhanarṇam || 10 ||

kalpodite saṃskṛtaparyayaughe
grahasya kuryāt sa tadā sphuṭaḥ syāt |
iṣṭaghnahāronayutaḥ sa dṛṣṭaḥ
kvacid graho dṛṣṭisamo yataḥ syāt || 11 ||

tābhyāṃ harābhyāṃ tu tathā mṛdūccāt
pātācca nīto bhagaṇastadīyaḥ |
hārādviśodhyo'yamṛṇātmakaścet
pātasya sarvaṃ viparītameva || 12 ||

This way of having *dhruvas* at the beginning of *kalpa* does not seem to be appropriate. Hence, in order to avoid that, an alternative correction procedure is sought.

[For this purpose], the elapsed number of years since the beginning of *kalpa* and the number of years or solar revolutions (*bhāskaraparyaya*) in a *kalpa* are mutually divided. Of the [successive multipliers and] divisors thus obtained, choose any divisor and multiply it by the *grahadhruva* of the planet in degrees (*kalpādikhaga-dhruvāṃśa*) and divide by 360 (*atula*).

Multiply the quotient thus obtained by the preceding divisor and divide by the previously stated desired divisor. The remainder thus obtained is applied to the corrected revolution (*saṃskṛtaparyaya*) of the planet positively or negatively depending on whether the divisor is odd or even respectively. Then, we get the true revolution (*sphuṭa*). An integral multiple of the *hāraka* may be added to or subtracted from this (*sphuṭa*) so as to ensure that the planetary longitude coincides with what is observed.

With the same *hārakas*, obtain the revolutions of the apogees and the nodes. If it (the *grahadhruva*) turns out to be negative then it has to be subtracted from the divisor. All the operations have to be reversed in the case of the node.

The verses 9-11 prescribe a correction term which is to be applied to the *saṃskṛtaparyayas* R' of the planets ($R \pm \Delta R$ discussed in Section 5.1), in order to eliminate the *kalpādidhruvas*. The prescription is as follows:

- Find the successive approximations for the ratio of the number of years elapsed since the beginning of the *kalpa* (Y) to the number of years in the *kalpa* (R_s) by the method of mutual division. The successive *guṇakāras* and *hārakas* that arise in this mutual division are to be noted down. Let these be denoted by G_i's and H_i's ($i = 1, 2, \ldots$) respectively.
- Multiply the *grahadhruva* in degrees ($\theta_o - \theta_c$) by any desired *hāraka* (H_i), divide by 360 and find the quotient. That is, find

$$\left[\frac{(\theta_o - \theta_c) \times H_i}{360}\right]_{int}.$$

- Multiply the above result by the penultimate *hāraka* (H_{i-1}) and divide by the desired *hāraka*. The remainder obtained has to be applied to R'. The result would be the corrected revolution number of the planet, *sphuṭaparyaya* R''. That is

$$R'' = R' \pm \left[\frac{\left[\frac{(\theta_o - \theta_c) \times H_i}{360}\right]_{int} \times H_{i-1}}{H_i}\right]_{rem}.$$

- Here we need to choose '+' for odd and '−' for even *hāraka*. An integral multiple of *hāra* can be added to or subtracted from this (*sphuṭaparyaya*) to make the revolution number tally with observations.

Now we proceed to explain the rationale behind the above correction term. The mean longitude of a planet corresponding to elapsed years Y since the beginning of the *kalpa* is given by

$$R' \times \frac{Y}{R_s} \pm \frac{(\theta_o - \theta_c)}{360}, \tag{5.12}$$

where $(\theta_o - \theta_c)$ is the magnitude of *grahadhruva* in degrees (*dhruvāṃśa*). In the above equation, the mean longitude computed corresponds to the elapsed number of years Y. It may also be noted that the *dhruva* has been applied to the result. Now instead of applying *dhruva*, we could as well modify the *saṃskṛta-paryayas* or the corrected number of revolutions in a *kalpa*, R', itself by adding a quantity x to it and still obtain the same result. Here, the quantity

x has the same dimension as that of the *bhagaṇa*. Thus new expression for mean longitude is

$$(R' + x) \times \frac{Y}{R_s}. \tag{5.13}$$

We will consider the case where the quantity $\theta_o - \theta_c$ is positive. Equating (5.12) and (5.13), we have

$$\left[x \times \frac{Y}{R_s}\right]_{frc} = \frac{(\theta_o - \theta_c)}{360},$$

$$\text{or} \qquad x \times \frac{Y}{R_s} - y = \frac{(\theta_o - \theta_c)}{360}, \tag{5.14}$$

where y is an integer. Rewriting the above equation, we have

$$x \times \frac{Y}{R_s} - \frac{(\theta_o - \theta_c)}{360} = y. \tag{5.15}$$

Now we have an indeterminate equation of first order, wherein we need to find integer solutions for x and y.

It is noted that by solving (5.15) and applying the value of x obtained to R', we can correct the *bhagaṇas* of the planets which takes into account the zero correction due to *dhruva*. To find x, first find successive multipliers and divisors of the ratio $\frac{Y}{R_s}$ by doing *vallyupasaṃhāra*. Let G_i's and H_i's be the successive multipliers and divisors. Now we approximate $\frac{Y}{R_s}$ by $\frac{G_i}{H_i}$ for a suitable *hāraka* H_i. Also by writing

$$\frac{(\theta_o - \theta_c)}{360} = \frac{z'}{H_i}, \tag{5.16}$$

we have $z' = \frac{(\theta_o - \theta_c)}{360} \times H_i$. We approximate z' by its integral part z. Hence

$$\frac{(\theta_o - \theta_c)}{360} \approx \frac{z}{H_i} = \frac{1}{H_i}\left[\frac{(\theta_o - \theta_c)}{360} \times H_i\right]_{int}, \tag{5.17}$$

Then x satisfies the equation

$$G_i x - H_i y = z. \tag{5.18}$$

Following our discussion in Section 3.2, the solution of the above equation can be written as

$$x = \left[\frac{z}{H_i} \times H_{i-1}\right]_{rem} \qquad \text{(when } i \text{ is odd)},$$

$$\text{and} \qquad x = H_i - \left[\frac{z}{H_i} \times H_{i-1}\right]_{rem} \qquad \text{(when } i \text{ is even)}. \tag{5.19}$$

Now the corrected revolution number is $R' + x$. If (x, y) is a solution of the indeterminate equation, then $(x + mH_i, y + mG_i)$ is also a solution for any integer m. So, a suitable multiple of H_i can be added to the above to make the corrected revolution number tally with observations. For the same reason, the term H_i in the above solution for even i can be dropped.

Now when the *dhruvāṃśa* $(\theta_o - \theta_c)$ is negative, we would have the equation

$$x \times \frac{G_i}{H_i} - y = -\frac{(\theta_o - \theta_c)}{360} \approx \frac{-z}{H_i} \approx \frac{H_i - z}{H_i} - 1,$$

$$\text{or} \qquad x \times \frac{G_i}{H_i} - (y - 1) = \frac{H_i - z}{H_i}. \tag{5.20}$$

Here, the same procedure is to be followed, as in the case of positive *dhruvāṃśa*, except that $H_i - z$ is used instead of z, where $\frac{z}{H_i}$ is the magnitude of $\frac{dhruv\bar{a}\dot{m}\acute{s}a}{360}$.

५.६ कल्पादौ सङ्क्रमणध्रुवस्य शून्यतासंपादनम्

5.6 Obtaining zero *saṅkramaṇadhruva* at the beginning of the *kalpa*

कल्पगतं वर्षगणं सप्ताहतकल्पभानुभगणमपि ।
हृत्वा पुनरन्योन्यं तत्राप्तैर्हारकाः कार्याः ॥ १३ ॥

हारेषु तेष्वभिमतेन पुनर्विनिघ्नात्
कल्पादिसङ्क्रमदिनात् खलु सप्तभक्तम् ।
इष्टोर्ध्वहारहतमिष्टहरेण तष्टं
प्राग्वच्च कल्पकुदिने स्वमृणं प्रकुर्यात् ॥ १४ ॥

kalpagataṃ varṣagaṇaṃ saptāhatakalpabhānubhagaṇamapi |
hṛtvā punaranyonyaṃ tatrāptairhārakāḥ kāryāḥ || 13 ||

hāreṣu teṣvabhimatena punarvinighnāt
kalpādisaṅkramadināt khalu saptabhaktam |
iṣṭordhvahārahatamiṣṭahareṇa taṣṭaṃ
prāgvacca kalpakudine svamṛṇaṃ prakuryāt || 14 ||

Obtain the *hārakas* by mutually dividing the elapsed number of years since the beginning of the *kalpa* and seven times the number of years in a *kalpa*.

Multiply the *saṅkramaṇadina* corresponding to the beginning of *kalpa* by the desired *hāraka* and divide by seven. [The quotient] thus obtained has to be multiplied by the previous divisor and divided by the desired divisor. As earlier, [the remainder] should be added to or subtracted from the number of civil days in a *kalpa*.

The correction term which is to be added to the number of civil days (D_c) in order to make the *saṅkramaṇadhruva* zero as prescribed in the above verse is of the form

$$\left[\frac{\left[\frac{sa\dot{n}kramana\d{n}adhruva\times H_i}{7}\right]_{int}\times H_{i-1}}{H_i}\right]_{rem}, \tag{5.21}$$

where H_i and H_{i-1} are the *iṣṭahāraka* and *iṣṭordhvahārakas* which are obtained by doing *vallyupasaṃhāra* of the ratio $\frac{Y}{7R_s}$. The above correction has to be added to or subtracted from the number of civil days (D_c) depending on whether the *hāraka* H_i is odd or even respectively. We now proceed to explain the rationale behind the correction term (5.21).

Let R_s be the number of revolutions of the Sun in a *kalpa* and D_c the tabulated number of civil days in a *kalpa*. Hence the number of days, d_c corresponding to number of elapsed years Y starting from *kalpādi* can be represented as

$$d_c = \frac{Y \times D_c}{R_s} = 7l + w_c, \tag{5.22}$$

where l is an integer, w_c is the computed weekday which could be different from the actual count of the weekday w_o. Note that w_o and w_c can have integral and fractional parts. Earlier this quantity $(w_o - w_c)$ was considered as the *saṅkramaṇadhruva* for the beginning of the *kalpa*. However, this could also be attributed to the number of civil days in a *kalpa* being equal to a number D_c' different from D_c. Then

$$d_o = \frac{Y \times D_c'}{R_s} = 7k + w_o, \tag{5.23}$$

where k is an integer. Now from (5.22) and (5.23),

$$d_o - d_c = \frac{Y}{R_s} \times (D_c' - D_c) = 7(k - l) + (w_o - w_c). \tag{5.24}$$

Dividing this by 7, we have

$$\frac{Y}{7 \times R_s} \times (D_c' - D_c) = (k - l) + \frac{(w_o - w_c)}{7}. \tag{5.25}$$

Let G_i, H_i be the successive *guṇakāras* and *hāras* in the *vallyupasaṃhāra* of $\frac{Y}{7R_s}$. Then for a sufficiently large H_i, $\frac{Y}{7R_s}$ can be approximated by $\frac{G_i}{H_i}$. Also let

$$\frac{(w_o - w_c)}{7} = \frac{n'}{H_i} \approx \frac{n}{H_i},$$

where

$$n = \left[\frac{(w_o - w_c)}{7} \times H_i\right]_{int}. \tag{5.26}$$

Let $k - l = y$, and $D'_c - D_c = x$. We then have the indeterminate equation,

$$G_i x - H_i y = n.$$

The solution of this is given by

$$x = + \left[\frac{nH_{i-1}}{H_i} \right]_{rem} \qquad \text{(when } i \text{ is odd)},$$
$$x = - \left[\frac{nH_{i-1}}{H_i} \right]_{rem} \qquad \text{(when } i \text{ is even)}. \tag{5.27}$$

In the case of "even" H_i, use is made of the fact that if (x, y) is a solution of the indeterminate equation, then $(x - H_i, y - G_i)$ is also a solution. Hence the corrected number of civil days D'_c is given by

$$D'_c = D_c \pm \left[\frac{nH_{i-1}}{H_i} \right]_{rem}, \tag{5.28}$$

where we have to choose '+' when i is odd and '−' when i is even. Clearly the correction term (5.28) is the same as that prescribed by the verse in the form (5.21).

५.७ विभिन्नकल्पपरिमाणानि

5.7 Different measures of *kalpa*

कल्पादीनां प्रमाणं तु बहुधा कल्प्यते बुधैः ।
उपेयस्यैव नियमो नोपायस्येति यत् ततः ॥ १५ ॥

kalpādīnāṃ pramāṇaṃ tu bahudhā kalpyate budhaiḥ |
upeyasyaiva niyamo nopāyasyeti yat tataḥ || 15 ||

The duration of a *kalpa* has been conceived differently by different scholars. This is due to the fact that there are no constraints on the means (*upāya*), but only on the end result (*upeya*).

Quantities such as the number of years in a *kalpa* are only the means (*upāya*) for obtaining the end results (*upeya*), such as the longitudes of the planets. In the following verses, the author presents views of two different schools (*pakṣas*), which take the duration of the *kalpa* to be 1000 years instead of 1008 years considered in Section 1.5.

५.८ कल्पस्य विभिन्नविभागाः

5.8 Different divisions of a *kalpa*

कल्पे युगानि तु सहस्रमुशन्ति केचित् तत्रैकसप्ततियुगानि पृथङ् मनूनाम् ।
आद्यन्तयोश्च विवरे च तथैव तेषां स्युः सन्धयो युगदशांशचतुष्कतुल्याः ॥ १६ ॥

मनवोऽथ चतुर्दशैव कल्पे पृथुतुल्यानि युगानि चैव तेषाम् ।
त्रियुगानि गतानि सृष्टितः प्राक् परतः स्युः प्रलयात् तथाहुरन्ये ॥ १७ ॥

kalpe yugāni tu sahasramuśanti kecit
tatraikasaptatiyugāni pṛthaṅ manūnām |
ādyantayośca vivare ca tathaiva teṣāṃ
syuḥ sandhayo yugadaśāṃśacatuṣkatulyāḥ || 16 ||

manavo'tha caturdaśaiva kalpe
pṛthutulyāni yugāni caiva teṣām |
triyugāni gatāni sṛṣṭitaḥ prāk
parataḥ syuḥ pralayāt tathāhuranye || 17 ||

Some [teachers] say that there are 1000 *caturyugas* in a *kalpa*. In it, there are 71 *caturyugas* corresponding to 14 *Manus*. At the beginning and towards the end [of the *kalpa*], as well as in the interval between (different *Manu*'s) periods, there will be *sandhis* whose durations are equal to $\left(\frac{4}{10}\right)^{th}$ of a *caturyuga*.

Some other [teachers] say that the number of *Manus* in a *kalpa* is 14 only and that the *yugas* associated with each of them is equal to 71 (*pṛthu*). [Also, according to them] a period of three *yugas* is said to have elapsed before the creation (*sṛṣṭi*) and there will be [three *yugas*] after the dissolution (*pralaya*).

According to some teachers, the number of *caturyugas* in a *Manvantara* is stated to be 71 and a *kalpa* corresponds to a period which is equal to the sum of 14 *Manus* and 15 *sandhis*, where each *sandhi* corresponds to $\left(\frac{4}{10}\right)^{th}$ of a *caturyuga*. Thus the total number of *caturyugas* in a *kalpa* is equal to

$$71 \times 14 + 15 \times \frac{4}{10} = 1000.$$

Having explained what constitutes a *kalpa* of one thousand[2] *mahāyugas* as per one school, the author presents the view held by another school in this regard.

Even according to them the total number of *caturyugas* in a *kalpa* is 1000 only. However, the internal composition of the number of *caturyugas* in a *kalpa* is different. In this school, instead of conceiving of 15 *sandhi* periods of shorter duration in between every *manvantara*, they conceive of two long periods each

[2] It may be recalled that in *Bhagavadgītā* too the number of *caturyugas* in a *kalpa* is stated to be 1000, which is said to be the same as the duration of the day of Brahmā.

sahasrayugaparyantaṃ aharyad brahmaṇo viduḥ |
rātriṃ yugasahasrāntāṃ te'horātravido janāḥ ||

equal to 3 *mahāyugas*, for creation and dissolution. Thus according to them also the total number of *mahāyugas* in a *kalpa* will be

$$71 \times 14 + 3 + 3 = 1000.$$

५.९ कृतादियुगानां परिमाणम्

5.9 The duration of *kṛta* and other *yugas*

युगस्य दशमो भागो भोगप्रियहतः क्रमात् ।
कृतादीनां प्रमाणं स्यात् पक्षयोरनयोर्द्वयोः ॥ १८ ॥

yugasya daśamo bhāgo bhogapriyahataḥ kramāt |
kṛtādīnāṃ pramāṇaṃ syāt pakṣayoranayordvayoḥ || 18 ||

According to both these schools (*pakṣas*) the period of the [*yugapādas*] *kṛta* etc. are one-tenth of a [*mahā*]*yuga* multiplied by 4, 3, 2 and 1 (*bhogapriya*) respectively .

A *mahāyuga* consists of 43, 20, 000 years. One-tenth of it is 4, 32, 000 years. The number of years in the four *pādas* according to both the schools referred to in the previous section are the same, and are presented in Table 5.1. It may be recalled that this is different from the conception of equal *yugapādas* presented in Section 1.6.

yugapāda		no. of years
कृत	*kṛta*	1728000
त्रेता	*tretā*	1296000
द्वापर	*dvāpara*	864000
कलि	*kali*	432000

Table 5.1 The number of years in the four *yugas* constituting a *mahāyuga*.

५.१० वर्तमानकल्पे गतकालः

5.10 Time elapsed in the present *kalpa*

कल्पेऽस्मिन् सप्तमस्यास्य वैवस्वतमनोर्युगे ।
अष्टाविंशे कलिः सर्वैर्वर्तमान इह स्मृतः ॥ १९ ॥

kalpe'smin saptamasyāsya vaivasvatamanoryuge |
aṣṭāviṃśe kaliḥ sarvairvartamāna iha smṛtaḥ || 19 ||

As per the civilizational memory (*smṛti*), all [the schools] recall that, we are in the *Vaivasvata-manvantara* associated with the seventh *Manu* of the present *kalpa* [and in this *manvantara*], we are in the 28th *kali* [*yuga*].

Chapter 6
परिधिव्याससंबन्धः ज्यानयनञ्च
Relation between the circumference and the diameter and computation of Rsines

६.१ माधवीयश्रेढी

6.1 The Mādhava series

व्यासाच्चतुर्घ्नात् बहुशः पृथक्स्थात् त्रिपञ्चसप्ताद्ययुगाहृतानि ।
व्यासे चतुर्घ्ने क्रमशस्त्वृणं स्वं कुर्यात्तदा स्यात् परिधिः सुसूक्ष्मः ॥ १ ॥

vyāsāccaturghnāt bahuśaḥ pṛthaksthāt
tripañcasaptādyayugāhṛtāni |
vyāse caturghne kramaśastvṛṇaṃ svaṃ
kuryāttadā syāt paridhiḥ susūkṣmaḥ || 1 ||

May the diameter be multiplied by four, kept separately at several places, and divided by the odd numbers 3, 5, 7 etc. [The results] may be sequentially applied negatively and positively to the diameter multiplied by four. Then we obtain a very accurate [value of the] circumference.

If C be the circumference of a circle whose diameter is D, then the above verse enunciates the following infinite series:

$$C = 4D - \frac{4D}{3} + \frac{4D}{5} - \frac{4D}{7} + \frac{4D}{9} - \dots \, . \tag{6.1a}$$

The above series is the well known series— first enunciated by Mādhava (14th century)—relating the circumference and the diameter of a circle. It is generally referred to as the Gregory-Leibniz series (who rediscovered it in the latter half of the 17th century) and is expressed in the form

$$\frac{C}{4D} = \frac{\pi}{4} = 1 - \frac{1}{3} + \frac{1}{5} - \frac{1}{7} + \dots \, . \tag{6.1b}$$

The series in (6.1a) may also be expressed in the form

V. Pai et al., *Karaṇapaddhati of Putumana Somayājī*, Sources and Studies in the History of Mathematics and Physical Sciences, https://doi.org/10.1007/978-981-10-6814-0_6

$$C = 4D \sum_{n=1}^{\infty} (-1)^{n+1} \frac{1}{(2n-1)} \ . \tag{6.1c}$$

As the series (6.1) converges terribly slowly, Mādhava considered a finite number of terms in the series, to which an end-correction (*antyasaṃskāra*) denoted by $\frac{1}{a_p}$, is applied as follows:

$$C \approx 4D \left(1 - \frac{1}{3} + \frac{1}{5} - \ldots + (-1)^{\frac{p-1}{2}} \frac{1}{p} + (-1)^{\frac{p+1}{2}} \frac{1}{a_p}\right). \tag{6.2}$$

Here p represents the last odd number appearing as the divisor, at which the series is terminated, and $(-1)^{p+\frac{1}{2}} \left(\frac{1}{a_p}\right)$ is an approximation to the remaining terms in the series. The three successive approximations to the end-correction $\left(\frac{1}{a_p}\right)$ that have been given by Mādhava, are the following:[1]

$$\frac{1}{a_p(1)} = \frac{1}{2(p+1)}, \tag{6.3}$$

$$\frac{1}{a_p(2)} = \frac{\left(\frac{p+1}{2}\right)}{(p+1)^2+1}, \tag{6.4}$$

$$\frac{1}{a_p(3)} = \frac{\left(\frac{p+1}{2}\right)^2 + 1}{\left[\left(\left(\frac{p+1}{2}\right)^2 + 1\right) 4 + 1\right]\left(\frac{p+1}{2}\right)}. \tag{6.5}$$

Now, by applying the end-correction given by (6.3), the series (6.2) becomes,

$$C \approx 4D \left(1 - \frac{1}{3} + \frac{1}{5} - \ldots + (-1)^{\frac{p-1}{2}} \frac{1}{p} + (-1)^{\frac{p+1}{2}} \frac{1}{2(p+1)}\right). \tag{6.6}$$

६.१.१ परिणतमाधवीयश्रेढी

6.1.1 Transformed Mādhava series

व्यासाद् वनसङ्गुणितात् पृथगाप्तं त्र्याद्ययुग्विमूलघनैः ।
त्रिगुणव्यासे स्वमृणं क्रमशः कृत्वापि परिधिरानेयः ॥ २ ॥

vyāsād vanasaṅguṇitāt pṛthagāptaṃ tryādyayugvimūlaghanaiḥ |
triguṇavyāse svamṛṇaṃ kramaśaḥ kṛtvāpi paridhirāneyaḥ || 2 ||

[1] {GYB 2008}, Section 6.8, pp. 201–205.

> The diameter multiplied by four (*vana*) is divided separately by the cubes of the odd numbers, starting with three, diminished by their bases. By applying this positively and negatively, in order, to the diameter multiplied by three, the circumference may be obtained.

The above verse gives the following infinite series for the circumference of a circle in terms of its diameter.

$$\begin{aligned} C &= 3D + \frac{4D}{(3^3-3)} - \frac{4D}{(5^3-5)} + \frac{4D}{(7^3-7)} - \cdots \\ &= 4D\left[\frac{3}{4} + \frac{1}{(3^3-3)} - \frac{1}{(5^3-5)} + \frac{1}{(7^3-7)} - \cdots\right], \end{aligned} \tag{6.7}$$

The rationale behind the above expression can be understood as follows. We can rewrite the equation (6.1) in terms of the so called *sthaulyas* as follows:[2]

$$\begin{aligned} C &= 4D\left[\left(1-\frac{1}{a_1}\right) + \left(\frac{1}{a_1}+\frac{1}{a_3}-\frac{1}{3}\right) - \left(\frac{1}{a_3}+\frac{1}{a_5}-\frac{1}{5}\right) + \cdots\right] \\ &= 4D\left[\left(1-\frac{1}{a_1}\right) + E(3) - E(5) + E(7) - \cdots\right], \end{aligned} \tag{6.8}$$

where the *sthaulya* $E(p)$ is given by

$$E(p) = \left[\frac{1}{a_{p-2}} + \frac{1}{a_p}\right] - \frac{1}{p}.$$

By choosing $a_p = 2p+2$ and $a_{p-2} = 2p-2$ and substituting them in (6.8), we get the transformed series

$$\begin{aligned} C &= 3D + \frac{4D}{(3^3-3)} - \frac{4D}{(5^3-5)} + \frac{4D}{(7^3-7)} - \cdots \\ &= 4D\left[\frac{3}{4} + \frac{1}{(3^3-3)} - \frac{1}{(5^3-5)} + \frac{1}{(7^3-7)} - \cdots\right]. \end{aligned} \tag{6.9}$$

It can be easily seen that the above transformed Mādhava series (6.9) would have faster convergence since the cubes of odd numbers appear in the denominator unlike the original Mādhava series (6.1*c*) wherein only the first power of odd numbers appear in the denominator. The series given in (6.9) can be rewritten as

$$C = 3D + \sum_{n=1}^{\infty}(-1)^{n+1}\frac{4D}{(2n+1)^3-(2n+1)}. \tag{6.10}$$

[2] {GYB 2008}, Section 6.8, pp. 201–205.

६.१.२ किञ्चिदव्यक्तसूत्रम्

6.1.2 An algebraic identity

हार्यं हारैक्यभेदघ्नं हारघातेन वा हरेत् ।
हार्याद्धारयुगावाप्तफलयोगान्तराप्तये ॥ ३ ॥

hāryaṃ hāraikyabhedaghnaṃ hāraghātena vā haret |
hāryāddhārayugāvāptaphalayogāntarāptaye || 3 ||

The dividend multiplied by the sum or the difference of the divisors should be divided by product of the divisors in order to obtain the sum or the difference of the results obtained by dividing the dividend by these divisors.

If d be the dividend and h_1 and h_2 be the *hāras* (divisors), then the above verse essentially gives the following algebraic identity

$$\frac{d}{h_1} \pm \frac{d}{h_2} = \frac{d \times (h_2 \pm h_1)}{h_1 \times h_2}. \tag{6.11}$$

६.१.३ पुतुमनसोमयाजीश्रेढी

6.1.3 The Putumana Somayājī series

वर्गैर्युजां वा द्विगुणैर्निरेकैः वर्गीकृतैर्वर्जितयुग्मवर्गैः ।
व्यासं च षड्घ्नं विभजेत् फलं स्वं व्यासे त्रिनिघ्ने परिधिस्तदा स्यात् ॥ ४ ॥

vargairyujāṃ vā dviguṇairnirekaiḥ
vargīkṛtairvarjitayugmavargaiḥ |
vyāsaṃ ca ṣaḍghnaṃ vibhajet phalaṃ svaṃ
vyāse trinighne paridhistadā syāt || 4 ||

Or, from the square of even numbers multiplied by two, subtract one, and from the square [of that] subtract the square of the same (even number). Divide the diameter multiplied by six by the above [quantities]. When [the sum of] these is added to three times the diameter, the result will be the circumference.

The series presented by the above verse is,

$$C = 3D + \frac{6D}{(2.2^2-1)^2-2^2} + \frac{6D}{(2.4^2-1)^2-4^2} + \frac{6D}{(2.6^2-1)^2-6^2} + \dots ,$$

which may be written in the form

$$C = 3D + \sum_{n=1}^{\infty} \frac{6D}{(2.(2n)^2-1)^2-(2n)^2}. \tag{6.12}$$

The above series (6.12) is a new series not found either in the *Gaṇitayuktibhāṣā* of Jyeṣṭhadeva or in the *Yuktidīpikā*, commentary of Śaṅkara Vāriyar on the *Tantrasaṅgraha*, and is perhaps due to Putumana Somayājī himself.[3]

We may note that the series (6.12) converges faster than (6.10) or (6.9) since the fourth powers of odd numbers appear as the denominators in (6.12). It can be shown that this series can easily be obtained from the series (6.10) given above. The series (6.10) can be re-written as:

$$C = 3D + \sum_{n=1}^{\infty}\left(\frac{4D}{(4n-1)^3-(4n-1)}\right) - \left(\frac{4D}{(4n+1)^3-(4n+1)}\right). \tag{6.13}$$

After some algebraic manipulations, the argument of the summation in the second term in RHS of the above can be shown to be

$$\begin{aligned}
&\frac{4D \times (((4n+1)^3-(4n+1)) - ((4n-1)^3-(4n-1)))}{((4n-1)^3-(4n-1)) \times ((4n+1)^3-(4n+1))} \\
&= \frac{4D \times 96n^2}{((4n-1)((4n-1)^2-1)) \times ((4n+1)((4n+1)^2-1))} \\
&= \frac{4D \times 96n^2}{((4n-1)(16n^2-8n)) \times ((4n+1)(16n^2+8n))} \\
&= \frac{4D \times 96n^2}{64n^2\,(64n^4-20n^2+1)} \\
&= \frac{6D}{((8n^2-1)^2-4n^2)} \\
&= \frac{6D}{((2.(2n)^2-1)^2-(2n)^2)},
\end{aligned} \tag{6.14}$$

which is the same as (6.12).

Figure 6.1 graphically depicts the rates of convergence of the three series given by (6.1c), (6.10) and (6.12).

[3] This verse of *Karaṇapaddhati* has been cited in the article of Whish (Whish 1834). There, Whish also gives the following transformed version of the series

$$\begin{aligned}
C &= 6 \times \left(\frac{1}{1.3.3.5} + \frac{1}{3.5.7.9} + \frac{1}{5.7.11.13} + \frac{1}{7.9.15.17.} + \ldots\right) \\
&= 3 + 6 \times \sum_{n=1}^{\infty}\left(\frac{1}{(2n-1)(2n+1)(4n-1)(4n+1)}\right).
\end{aligned}$$

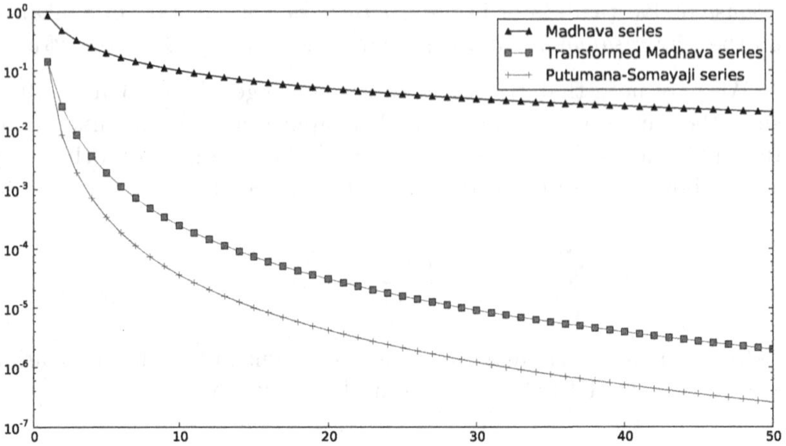

Fig. 6.1 Graph depicting the convergence of the Mādhava series and its transformed versions.

६.२ परिधिव्यासानुपातः

6.2 Ratio of the circumference to the diameter

व्यासादभीष्टान्महतोऽस्य वृत्तं नीत्वैवमाभ्यां तु मिथो विभक्तैः ।
कार्या यथोक्तं गुणकारहाराः ते व्यासवृत्तानि तदा भवन्ति ॥ ५ ॥

गुणहारकभूतैस्तैः व्यासवृत्तैर्यथोदितम् ।
इष्टवृत्तान्नयेद् व्यासं व्यासाद् वृत्तं विपर्ययात् ॥ ६ ॥

अनूननूत्नाननुनुन्ननित्यैः समाहताश्चक्रकलाविभक्ताः ।
चण्डांशुचन्द्राधमकुंभिपालैः व्यासस्तदर्धं त्रिभमौर्विका स्यात् ॥ ७ ॥

vyāsādabhīṣṭānmahato'sya vṛttaṃ
nītvaivamābhyāṃ tu mitho vibhaktaiḥ |
kāryā yathoktaṃ guṇakārahārāḥ
te vyāsavṛttāni tadā bhavanti || 5 ||

guṇahārakabhūtaistaiḥ vyāsavṛttairyathoditam |
iṣṭavṛttānnayed vyāsaṃ vyāsād vṛttaṃ viparyayāt || 6 ||

anūnanūtnānanununnanityaiḥ
samāhatāścakrakalāvibhaktāḥ |
caṇḍāṃśucandrādhamakuṃbhipālaiḥ
vyāsastadardhaṃ tribhamaurvikā syāt || 7 ||

Having obtained the circumference corresponding to a chosen large value of the diameter, and by doing the mutual division of the two (considering the diameter as the multiplier and the circumference as the divisor), various [pairs of] multipliers

> and divisors are obtained as stated earlier. These [pairs] then form the [successive approximations to] diameters and circumferences.
>
> With [any of] these [pairs of] diameters and circumferences serving as multipliers and divisors, by the process described earlier, from the desired [value of the] circumference, the diameter can be obtained. [Similarly] the circumference can be obtained from the diameter through the reverse process.
>
> The product of 10000000000 (*anūnanūtnānanunnanitya*) and the number of minutes in a circle (*cakrakalā*) when divided by 31415926536 (*caṇḍāṃśucandrādhamakuṃbhipāla*) would give the diameter. Half of that would be the Rsine (*maurvikā*) corresponding to three signs (*tribha*).

The first of the three verses above states that from an accurate value of the ratio of the circumference to the diameter, approximate values of the ratio can be obtained by the method of *vallyupasaṃhāra* described in chapter 2.

The next verse then mentions that from any of these ratios of $\frac{C}{D}$, the circumference can be calculated if the diameter is known and vice versa.

Diameter (D)	Circumference (C)	Order of $\left\lvert \frac{C}{D} - \pi \right\rvert$
1	3	1.41592654×10^{-1}
7	22	1.26448927×10^{-3}
106	333	8.32196275×10^{-5}
113	355	2.66764189×10^{-7}
33102	103993	5.77890624×10^{-10}
33215	104348	3.31628058×10^{-10}
66317	208341	1.22356347×10^{-10}
99532	312689	2.91433544×10^{-11}
464445	1459097	7.51132490×10^{-12}
563977	1771786	1.13287157×10^{-11}
1028422	3230883	9.60476143×10^{-12}
1592399	5002669	1.02153841×10^{-11}
72686377	228350988	1.02069464×10^{-11}
74278776	233353657	1.02069464×10^{-11}
146965153	461704645	1.02069464×10^{-11}
1250000000	3926990817	1.02069464×10^{-11}
10000000000	31415926536	1.02069464×10^{-11}

Table 6.1 Successive approximations for the ratio $\frac{31415926536}{10000000000}$.

Verse 7 gives a fairly accurate value of the radius of a circle when its circumference is given by 21600 (minutes). That is,

$$R \approx \frac{1}{2} \times \left[\frac{10000000000 \times 21600}{31415926536}\right]$$
$$\approx 3437.7467707737701'$$
$$\approx 3437'44''48'''22'''', \qquad (6.15)$$

which is indeed accurate upto the fourths ($''''$) of a degree.

From the ratio of circumference ($C = 31415926536$) to the diameter ($D = 10000000000$) given in the last verse

$$\frac{C}{D} \approx \frac{31415926536}{10000000000},$$

by doing *vallyupasaṃhāra*, with D as the multiplier and C as the divisor, we obtain different approximations to the above ratio that are given in the Table 6.1. As we can see, this table contains a few of the well known rational approximations to π such as $\frac{3}{1}$, $\frac{22}{7}$ and $\frac{355}{113}$.

६.३ एकराश्यध्यर्धराशिज्याभ्यां पठितज्यानयनम्

6.3 Generation of the tabular Rsines from Rsin 30 and Rsin 45

त्रिज्यार्द्धमेकराशिज्या त्रिज्यावर्गार्द्धतः पदम् ।
भवेदध्यर्धराशिज्या ताभ्यामन्यगुणान् नयेत् ॥ ८ ॥

trijyārddhamekarāśijyā trijyāvargārddhataḥ padam |
bhavedadhyardharāśijyā tābhyāmanyaguṇān nayet || 8 ||

Half of *trijyā* would be the *jyā* of one *rāśi* (*ekarāśijyā*) and the square root of half of the square of *trijyā* would be the *jyā* of one and a half *rāśis* (*adhyardharāśijyā*). From these two, the other sine values (*guṇas*) may be obtained.

Having given an accurate value of *trijyā* (in the previous verse) this verse states how to obtain the values of $R\sin 30°$ (*ekarāśijyā*) and $R\sin 45°$ (*adhyardharāśijyā*) from the value of *trijyā* (R).

$$R\sin 30 = \frac{R}{2},$$
$$R\sin 45 = \frac{R}{\sqrt{2}}.$$

Further it is said that from these two values all other Rsine values may be obtained.

By *'anyaguṇān'* ('all other *jyās*') what is meant is the rest of the twenty-one Rsine values, leaving $R\sin 90 = R$, $R\sin 30°$ and $R\sin 45°$, since most texts on

Indian astronomy divide the quadrant of a circle into twenty-four equal parts, each division corresponding to $\frac{5400'}{24} = 225'$. The scheme by which the rest (21) of the tabular Rsines is obtained, is outlined in Table 6.2. The rationale behind this scheme is explained in the next section.

६.४ इष्टचापार्धभुजकोटिकानयनम्

6.4 Procedure for obtaining the Rsine and Rcosine of the half of any desired arc

त्रिज्येष्टज्यावधाढ्योनत्रिज्यावर्गोत्थमूलयोः ।
भेदयोगदलेत्विष्टचापार्धभुजकोटिके ॥ ९ ॥

trijyeṣṭajyāvadhāḍhyonatrijyāvargotthamūlayoḥ |
bhedayogadaletviṣṭacāpārdhabhujakoṭike || 9 ||

Having multiplied *trijyā* with any desired *jyā*, let it be added to as well as subtracted from the square of the *trijyā*. Half of the difference and sum of the square roots of the results thus obtained would yield the Rsine (*bhujajyā*) and Rcosine (*koṭijyā*) respectively of half of the desired arc.

In the previous verse, having given the values of $R\sin 30°$ and $R\sin 45°$ it was simply mentioned that with these two Rsine values the remaining tabular Rsines can be obtained. Two expressions that could be used for this purpose are presented in this verse.

$$R\sin\left(\frac{\alpha}{2}\right) = \frac{\sqrt{R^2 + R^2\sin\alpha} - \sqrt{R^2 - R^2\sin\alpha}}{2}, \tag{6.16}$$

$$\text{and} \qquad R\cos\left(\frac{\alpha}{2}\right) = \frac{\sqrt{R^2 + R^2\sin\alpha} + \sqrt{R^2 - R^2\sin\alpha}}{2}. \tag{6.17}$$

The rationale behind these expressions can be understood with the help of Figure 6.2. Here, O is the center of a circle whose diameter is AE. CD and OD are Rsine and Rcosine respectively corresponding to the arc $\overset{\frown}{AC}$. The mid-point of the $\overset{\frown}{AC}$ is denoted by M. Then AB is the Rsine of the $\overset{\frown}{AM}$, which is half of $\overset{\frown}{AC}$. Similarly OB is Rcosine of the $\overset{\frown}{AM}$. That is, if $\overset{\frown}{AC} = \alpha$, then

$$CD = R\sin\alpha, \qquad OD = R\cos\alpha,$$

$$\text{and,} \qquad AB = R\sin\left(\frac{\alpha}{2}\right), \qquad OB = R\cos\left(\frac{\alpha}{2}\right).$$

Now the area of the $\triangle OAC$ can be expressed in two ways:

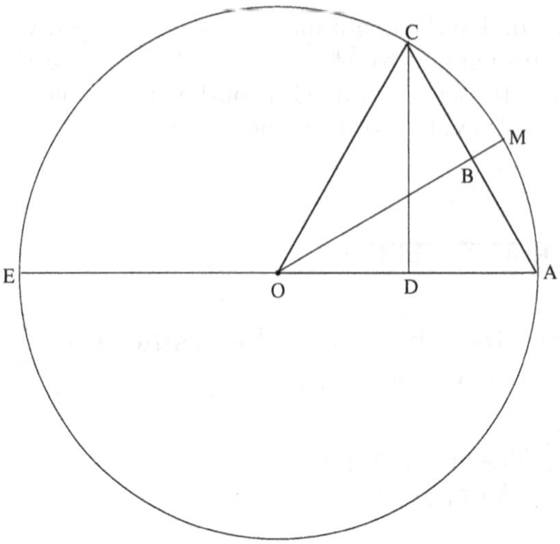

Fig. 6.2 The expression for the Rsine and Rcosine of half the desired arc.

$$\begin{aligned}\text{Area} &= \frac{1}{2} AC \times OB, \\ &= \frac{1}{2} OA \times CD. \qquad (6.18)\end{aligned}$$

As $AC = 2AB$, we have

$$\begin{aligned}AC \times OB &= 2AB \times OB \\ &= OA \times CD \\ &= R.R \sin \alpha. \qquad (6.19)\end{aligned}$$

The square of the sum/difference of OB and AB is given by

$$\begin{aligned}(OB \pm AB)^2 &= OB^2 + AB^2 \pm 2(OB \times AB) \\ &= R^2 \pm R.R \sin \alpha. \qquad (6.20)\end{aligned}$$

Hence,

$$OB + AB = \sqrt{R^2 + R.R \sin \alpha}, \qquad (6.21)$$

$$\text{and} \qquad OB - AB = \sqrt{R^2 - R.R \sin \alpha}. \qquad (6.22)$$

Subtracting (6.22) from (6.21), we get

$$AB = R \sin\left(\frac{\alpha}{2}\right) = \frac{\sqrt{R^2 + R.R \sin \alpha} - \sqrt{R^2 - R.R \sin \alpha}}{2}.$$

Adding (6.21) and (6.22), we get

$$OB = R\cos\left(\frac{\alpha}{2}\right) = \frac{\sqrt{R^2 + R.R\sin\alpha} + \sqrt{R^2 - R.R\sin\alpha}}{2}.$$

The above two equations are the same as (6.16) and (6.17). Using these two equations and the values of $R\sin(30°)$ (i.e., J_8, the 8th *jyā*) and $R\sin(45°)$ (i.e., J_{12}, the 12th *jyā*) one can find all the 24 tabular sines known as *paṭhita-jyās* and denoted by J_1, J_2 ...J_{24}.

Table 6.2 lists all the 24 Rsines and also indicates how each of them can be found from the two base values J_8 and J_{12} once *trijyā* J_{24} is known. For instance in the second row against J_2 we find $J_8 \rightharpoonup J_4 \rightharpoonup J_2$. Here the symbol $\rightharpoonup$ conveys that (6.16) is employed to first obtain J_4 from J_8, and further J_2 from J_4. Similarly, the symbol $\rightharpoondown$ in the 20th row $J_8 \rightharpoondown J_{20}$ conveys that (6.17) is employed to obtain J_{20} from J_8, as

$$J_{20} = R\sin(20 \times 225') = R\cos(4 \times 225') = R\cos\left(\frac{8 \times 225'}{2}\right).$$

६.५ इष्टचापार्धज्यानयने प्रकारान्तरम्

6.5 An alternative expression for the Rsine of half the desired arc

यद्वेष्टचापगुणतच्छरवर्गयोग-
मूलार्धमिष्टधनुरर्धगुणः प्रदिष्टः ।
ज्यानां निजत्रिगुणवर्गविशेषमूलं
कोटिस्तदूनसहितौ त्रिगुणौ स्वबाणौ ॥ १० ॥

yadveṣṭacāpaguṇataccharavargayoga-
mūlārdhamiṣṭadhanurardhaguṇaḥ pradiṣṭaḥ |
jyānāṃ nijatriguṇavargaviśeṣamūlaṃ
koṭistadūnasahitau triguṇau svabāṇau || 10 ||

Alternatively, it has been stated that half of the square root of the sum of squares of the *jyā* and the *śara* of a desired arc (*iṣṭacāpa*)[4] would be the Rsine of half of the desired arc (*dhanurardhaguṇa*). The square root of the square of the *jyā* subtracted from the square of *trijyā*, gives the *koṭijyā*. This, when subtracted from or added to *trijyā*, would give the pair of versines (*bāṇas*) corresponding to that *jyā*.

---

[4] In the compound word *yadveṣṭacāpaguṇataccharavargayogaḥ*, the author has employed the *nyāya*, *'dvandvānte śrūyamāṇaṃ padaṃ pratyekamabhisaṃpadyate'* Hence, the word *varga* is to be associated with both *guṇa* (Rsine) and *tacchara* (Rversine).

The 24 *jyās*	The scheme of obtaining them
J_1	$J_8 \rightharpoonup J_4 \rightharpoonup J_2 \rightharpoonup J_1$
J_2	$J_8 \rightharpoonup J_4 \rightharpoonup J_2$
J_3	$J_{12} \rightharpoonup J_6 \rightharpoonup J_3$
J_4	$J_8 \rightharpoonup J_4$
J_5	$J_8 \rightharpoondown J_{20} \rightharpoonup J_{10} \rightharpoonup J_5$
J_6	$J_{12} \rightharpoonup J_6$
J_7	$J_8 \rightharpoondown J_{20} \rightharpoondown J_{14} \rightharpoonup J_7$
J_8	$\frac{R}{2}$
J_9	$J_{12} \rightharpoondown J_{18} \rightharpoonup J_9$
J_{10}	$J_8 \rightharpoondown J_{20} \rightharpoonup J_{10}$
J_{11}	$J_8 \rightharpoonup J_4 \rightharpoondown J_{22} \rightharpoonup J_{11}$
J_{12}	$\frac{R}{\sqrt{2}}$
J_{13}	$J_8 \rightharpoonup J_4 \rightharpoondown J_{22} \rightharpoondown J_{13}$
J_{14}	$J_8 \rightharpoondown J_{20} \rightharpoondown J_{14}$
J_{15}	$J_{12} \rightharpoondown J_{18} \rightharpoondown J_{15}$
J_{16}	$\sqrt{(R^2 - J_8^2)}$
J_{17}	$J_8 \rightharpoondown J_{20} \rightharpoondown J_{14} \rightharpoondown J_{17}$
J_{18}	$J_{12} \rightharpoondown J_{18}$
J_{19}	$J_8 \rightharpoondown J_{20} \rightharpoonup J_{10} \rightharpoondown J_{19}$
J_{20}	$J_8 \rightharpoondown J_{20}$
J_{21}	$J_{12} \rightharpoonup J_6 \rightharpoondown J_{21}$
J_{22}	$J_8 \rightharpoonup J_4 \rightharpoondown J_{22}$
J_{23}	$J_8 \rightharpoonup J_4 \rightharpoonup J_2 \rightharpoondown J_{23}$
J_{24}	R

Table 6.2 The scheme for obtaining the twenty-four tabular Rsines from a set of two values (J_8 and J_{12}) along with *trijyā*.

In the first half of this verse, another formula has been presented (which is different from (6.16)) for obtaining $R\sin(\frac{\alpha}{2})$ from $R\sin\alpha$. In the later half of the verse, the author defines the *koṭijyā* and the two *śarajyās* (Rversines). The formulae given in the verse may be written as:

$$\begin{aligned} \textit{iṣṭadhanurardhaguṇa} &= \frac{\sqrt{(\textit{iṣṭacāpaguṇa})^2 + (\textit{iṣṭaśara})^2}}{2}, \\ R\sin\left(\frac{\alpha}{2}\right) &= \frac{\sqrt{R^2\sin^2\alpha + (R - R\cos\alpha)^2}}{2}. \end{aligned} \tag{6.23}$$

Also,

$$R\cos\alpha = \sqrt{R^2 - R^2\sin^2\alpha}. \tag{6.24}$$

The sum or difference of R and $R\cos\alpha$ is referred to as *bāṇa*. Hence the dual usage *bāṇau* here refers to both $R+R\cos\alpha$ and $R-R\cos\alpha$. For instance, in Figure 6.2 if $\overset{\frown}{AC}=\alpha$, then ED and DA refer to the two *bāṇas*, which add up to the diameter of the circle.

We shall now show how to obtain (6.23) from the expression (6.16) discussed in the earlier verses. Squaring (6.16), we get

$$\begin{aligned} R^2\sin^2\left(\frac{\alpha}{2}\right) &= \frac{2R^2 - 2\sqrt{R^2 + R.R\sin\alpha}.\sqrt{R^2 - R.R\sin\alpha}}{4}, \\ \text{or,}\quad 4R^2\sin^2\left(\frac{\alpha}{2}\right) &= 2R^2 - 2\sqrt{R^4 - R^4\sin^2\alpha} \\ &= 2R^2 - 2R\sqrt{R^2 - R^2\sin^2\alpha} \\ &= R^2 + R^2 - 2R^2\cos\alpha. \end{aligned} \tag{6.25}$$

By replacing one R^2 by $R^2\sin^2\alpha + R^2\cos^2\alpha$ and doing some algebraic manipulations, and finally taking the square root on both sides, (6.25) becomes

$$\begin{aligned} R\sin\left(\frac{\alpha}{2}\right) &= \frac{\sqrt{R^2 + R^2\sin^2\alpha + R^2\cos^2\alpha - 2R^2\cos\alpha}}{2} \\ &= \frac{\sqrt{R^2\sin^2\alpha + (R - R\cos\alpha)^2}}{2}, \end{aligned} \tag{6.26}$$

which is the same as the expression (6.23) indicated in the text.

६.६ शरात् इष्टचापार्धदोःकोटिज्यानयनम्

6.6 Finding the Rsine and Rcosine of half of the arc from the *bāṇa*

यद्वेष्टकोट्याहतविस्तरार्धेनोनान्वितौ व्यासदलस्य वर्गौ ।
अर्धीकृतौ तौ पदितावभीष्टचापार्धदोःकोटिगुणौ भवेताम् ॥ ११ ॥

yadveṣṭakoṭyāhatavistarārdhenonānvitau vyāsadalasya vargau |
ardhīkṛtau tau paditāvabhīṣṭacāpārdhadoḥkoṭiguṇau bhavetām || 11 ||

Or else, the product of the *trijyā* and the *koṭijyā* of a desired arc, added to and subtracted from the square of half the diameter, are halved and the square root is taken. The results will be the *koṭijyā* and *dorjyā* respectively of half of the desired arc.

The formulae given in the above verse may be expressed as

$$\sqrt{\frac{R^2 - R^2\cos\alpha}{2}} = R\sin\left(\frac{\alpha}{2}\right), \tag{6.27}$$

$$\sqrt{\frac{R^2 + R^2\cos\alpha}{2}} = R\cos\left(\frac{\alpha}{2}\right). \tag{6.28}$$

It may be noted that the LHS of the above equations may be expressed as

$$\sqrt{\frac{R(R \pm R\cos\alpha)}{2}} = \sqrt{\frac{trijy\bar{a} \times b\bar{a}\d{n}a}{2}}.$$

It can be easily seen that equations (6.27) and (6.28) can be obtained by taking the square root of (6.25). Then, we have

$$\begin{aligned} R\sin\left(\frac{\alpha}{2}\right) &= \sqrt{\frac{2R^2 - 2R^2\cos\alpha}{4}} \\ &= \sqrt{\frac{R^2 - R^2\cos\alpha}{2}}. \end{aligned} \tag{6.29}$$

We know that $R\cos\left(\frac{\alpha}{2}\right) = \sqrt{R^2 - R^2\sin^2\left(\frac{\alpha}{2}\right)}$. Now squaring (6.29) and subtracting it from R^2, we obtain

$$\begin{aligned} R^2 - R^2\sin^2\left(\frac{\alpha}{2}\right) &= R^2 - \left(\sqrt{\frac{R^2 - R^2\cos\alpha}{2}}\right)^2 \\ &= \frac{2R^2 - R^2 + R^2\cos\alpha}{2}. \end{aligned} \tag{6.30}$$

Taking the square root on both sides of (6.30), we get

$$R\cos\left(\frac{\alpha}{2}\right) = \sqrt{\frac{R^2 + R^2\cos\alpha}{2}},$$

which is the same as the expression (6.28).

६.७ ज्याकोटिभ्यां श्रेणिविस्तारः

6.7 Series expansion for the Rsine and Rcosine

चापाच्च तत्तत्फलतोऽपि तद्वत् चापाहताद् द्व्यादिहतत्रिमौर्व्या ।
लब्धानि युग्मानि फलान्यधोऽधः चापादयुग्मानि च विस्तरार्धात् ॥ १२ ॥

विन्यस्य चोपर्युपरि त्यजेत् तच्छेषौ भुजाकोटिगुणौ भवेताम् ।
एकादिसङ्ख्याहतभाष्टमांशात् एवं चतुर्विंशतिमौर्विकाः स्युः ॥ १३ ॥[5]

cāpācca tattatphalato'pi tadvat
cāpāhatād dvyādihatatrimaurvyā |
labdhāni yugmāni phalānyadho'dhaḥ
cāpādayugmāni ca vistarārdhāt || 12 ||

vinyasya coparyupari tyajet tac-
cheṣau bhujākoṭiguṇau bhavetām |
ekādisaṅkhyāhatabhāṣṭamāṃśāt evaṃ
caturviṃśatimaurvikāḥ syuḥ || 13 ||

The [desired] arc and the results obtained are successively multiplied by the arc divided by the product of the radius with two, three, etc. Having placed [all] the even terms [sequentially] below the arc, and the odd terms below the radius, subtract [all] the succeeding terms from the immediately preceding ones. The [resultant] remainders are the Rsine (*bhujājyā*) and the Rcosine (*koṭijyā*) respectively. Thus [all] the twenty-four Rsines are obtained by using the integral multiples of the arc equal to one-eighth of 30 degrees.

The above verses present the well known series for the sine and cosine functions. Having presented both the series in one and a half verses, in the latter half of verse 13, it is stated that the 24 tabular sines can also be computed using the series presented here. If $R\alpha$ represents the desired arc (*cāpa*), then the successive terms to be generated in order to obtain the series are:

$$\frac{(R\alpha)\times(R\alpha)}{2R},\ \frac{(R\alpha)\times(R\alpha)^2}{2R\times 3R},\ \frac{(R\alpha)\times(R\alpha)^3}{2R\times 3R\times 4R},\ \frac{(R\alpha)\times(R\alpha)^4}{2R\times 3R\times 4R\times 5R}\cdots\cdots$$

Now the odd and even terms obtained have to be arranged as shown in Table 6.3.

Then it is said that all the successive terms are to be subtracted from the immediately preceding term. Thus we have

[5] For convenience, we present the prose order of the verse:

चापात् चापाहतात् द्विहतत्रिमौर्व्या [विभज्य लब्धं फलं], तद्वत् तत्फलतोऽपि च चापाहतात् त्रिहतत्रिमौर्व्या [विभज्य लब्धं फलं], तद्वत् तत्फलतोऽपि च चापाहतात् चतुर्हतत्रिमौर्व्या [विभज्य लब्धं फलं], [इत्येवं] लब्धानि युग्मानि फलानि चापादधोऽधः विन्यस्य, अयुग्मानि च फलानि विस्तरार्धादधोऽधः विन्यस्य उपर्युपरि त्यजेत्। तत्र लब्धौ शेषौ (तच्छेषौ) भुजाकोटिगुणौ [क्रमशः] भवेताम्।

(*cāpāt cāpāhatāt dvihatatrimaurvyā [vibhajya labdhaṃ phalaṃ], tadvat tatphalato'pi ca cāpāhatāt trihatatrimaurvyā [vibhajya labdhaṃ phalaṃ], tadvat tatphalato'pi ca cāpāhatāt caturhatatrimaurvyā [vibhajya labdhaṃ phalaṃ], [ityevaṃ] labdhāni yugmāni phalāni cāpādadho'dhaḥ vinyasya, ayugmāni ca phalāni vistarārdhādadho'dhaḥ vinyasya uparyupari tyajet / tatra labdhau śeṣau (taccheṣau) bhujākoṭiguṇau [kramaśaḥ] bhavetām.*)

odd terms	even terms
$R,$	$R\alpha,$
$\dfrac{(R\alpha)\times(R\alpha)}{2R},$	$\dfrac{(R\alpha)\times(R\alpha)^2}{2R\times 3R},$
$\dfrac{(R\alpha)\times(R\alpha)^3}{2R\times 3R\times 4R},$	$\dfrac{(R\alpha)\times(R\alpha)^4}{2R\times 3R\times 4R\times 5R},$
$\dfrac{(R\alpha)\times(R\alpha)^5}{2R\times 3R\times 4R\times 5R\times 6R},$	$\dfrac{(R\alpha)\times(R\alpha)^6}{2R\times 3R\times 4R\times 5R\times 6R\times 7R},$
$\vdots$	$\vdots$

Table 6.3 Odd and even terms for generating the cosine and sine series.

$$\begin{aligned}
R\sin\alpha &= (R\alpha) - \left(\frac{(R\alpha)^3}{3!R^2} - \left(\frac{(R\alpha)^5}{5!R^4} - \left(\frac{(R\alpha)^7}{7!R^6} - \ldots\ldots\right)\right)\right) \\
&= (R\alpha) - \frac{(R\alpha)^3}{3!R^2} + \frac{(R\alpha)^5}{5!R^4} - \frac{(R\alpha)^7}{7!R^6} + \ldots\ldots \\
&= (R\alpha) + \sum_{i=1}^{\infty}(-1)^i \frac{(R\alpha)^{(2i+1)}}{(2i+1)!R^{(2i)}}, \qquad (6.31) \\
R\cos\alpha &= (R) - \left(\frac{(R\alpha)^2}{2!R} - \left(\frac{(R\alpha)^4}{4!R^3} - \left(\frac{(R\alpha)^6}{6!R^5} - \ldots\ldots\right)\right)\right) \\
&= (R) - \frac{(R\alpha)^2}{2!R} + \frac{(R\alpha)^4}{4!R^3} - \frac{(R\alpha)^6}{6!R^5} + \ldots\ldots \\
&= R + \sum_{i=1}^{\infty}(-1)^i \frac{(R\alpha)^{2i}}{(2i)!R^{(2i-1)}}, \qquad (6.32)
\end{aligned}$$

which are the well known series for sine and cosine functions.

६.८ विद्वानित्यादिना ज्याकोट्यानयनम्

6.8 Computation of the Rsine and Rcosine values using the *vākyas vidvān* etc.

विद्वान् तुन्नबलः कवीशनिचयः[6] सर्वार्थशीलस्थिरः
निर्विद्धाङ्गनरेन्द्ररुङ् निगदितेष्वेषु क्रमात् पञ्चसु ।

[6] {KP 1937} and {KP 1953} have the incorrect reading *kapīśanicayaḥ*. {KP 1956} has the correct reading *kavīśanicayaḥ* which also tallies with the computed value.

आधस्थ्यात् गुणितादभीष्टधनुषः कृत्या विहृत्यान्तिम-
स्याप्तं शोध्यमुपर्युपर्यथ घनेनैवं धनुष्यन्ततः ॥ १४ ॥[7]

vidvān tunnabalaḥ kavīśanicayaḥ sarvārthaśīlasthiraḥ
nirviddhāṅganarendraruṅ nigaditeṣveṣu kramāt pañcasu |
ādhasthyāt guṇitādabhīṣṭadhanuṣaḥ kṛtyā vihṛtyāntima-
syāptaṃ śodhyamuparyuparyatha ghanenaivaṃ dhanuṣyantataḥ || 14 ||

Among the five values stated in order [as] *vidvān, tunnabalaḥ, kavīśanicayaḥ, sarvārthaśīlasthiraḥ, nirviddhāṅganarendraruk*, the last value is to be multiplied by the square of the desired arc [and] divided by the [square] of the ultimate [value of the arc, 5400]. The result obtained should be subtracted from the [immediately] preceding one above. Similarly the process is repeated with revised last value. Then [the result is to be multiplied] by the cube [of the arc divided by the last arc]. Finally [the result is to be subtracted] from the [desired] arc.

In the first half of the above verse, five numbers ($\beta_1, \beta_2, \beta_3, \beta_4$ and β_5) are presented using the *vākyas*: *vidvān*, ..., *nirviddhāṅganarendraruk*, which are listed in Table 6.4. The operations to be carried out with these numbers in order to obtain the Rsine values are outlined in the latter part of the verse. We shall present the sequence of these operations in the form of an algorithm.

β_i	*vākyas*		computed values
	in *kaṭapayādi*	in numerals	
β_5	*vidvān*	$44'''$	$44'''32.33''''$
β_4	*tunnabalaḥ*	$33''06'''$	$33''05'''36.13''''$
β_3	*kavīśanicayaḥ*	$16'05''41'''$	$16'05''40'''51.99''''$
β_2	*sarvārthaśīlasthiraḥ*	$273'57''47'''$	$273'57''47'''02.80''''$
β_1	*nirviddhāṅganarendraruk*	$2220'39''40'''$	$2220'39''39'''33.89''''$

Table 6.4 Comparing the values encoded by the *vākyas*, *vidvān* etc., with the computed values.

[7] For convenience, we present the prose order of the verse:

विद्वान् तुन्नबलः कवीशनिचयः सर्वार्थशीलस्थिरः निर्विद्धाङ्गनरेन्द्ररुक् [इति] एषु पञ्चसु क्रमात् निगदितेषु आधस्थ्यात् अभीष्टधनुषः कृत्या गुणितात् अन्तिमस्य [धनुषः कृत्या] विहृत्य, आप्तम् उपरि [विद्यमानात्] शोध्यम्। .....एवम् उपरि उपरि [कार्यम्]। अथ [अभीष्टधनुषः] घनेन [गुणितात्] [अन्तिमस्य धनुषः] घनेन [विहृत्य] आप्तम् अन्ततः धनुषि [शोध्यम्]।
(*vidvān tunnabalaḥ kaviśanicayaḥ sarvārthaśīlasthiraḥ nirviddhāṅganarendraruk [iti] eṣu pañcasu kramāt nigaditeṣu ādhastyāt abhīṣṭadhanuṣaḥ kṛtyā guṇitāt antimasya [dhanuṣaḥ kṛtyā] vihṛtya, āptam upari [vidyamānāt] śodhyam |...evam upari upari [kāryam]. atha [abhīṣṭadhanuṣaḥ] ghanena [guṇitāt] [antimasya dhanuṣaḥ] ghanena [vihṛtya] āptam antataḥ dhanuṣi [śodhyam].*)

- β_5 is multiplied by the square of the desired arc ($R\alpha$) and divide by $(5400)^2$ (*antimasya kṛtiḥ*) and the result thus obtained is subtracted from β_4. That is,

$$\beta_4 - \left(\frac{\beta_5 \times (R\alpha)^2}{(5400)^2} \right).$$

- The above result, is further multiplied by $(R\alpha)^2$ and divided by $(5400)^2$. The result is subtracted from β_3,

$$\beta_3 - \left(\beta_4 - \left(\frac{\beta_5 \times (R\alpha)^2}{(5400)^2} \right) \right) \times \frac{(R\alpha)^2}{(5400)^2}.$$

- This process has to be repeated till we reach β_1. At this stage we have

$$\beta_1 - \left(\beta_2 - \left(\beta_3 - \left(\beta_4 - \left(\frac{\beta_5 \times (R\alpha)^2}{(5400)^2} \right) \right) \times \frac{(R\alpha)^2}{(5400)^2} \right) \times \frac{(R\alpha)^2}{(5400)^2} \right) \times \frac{(R\alpha)^2}{(5400)^2}.$$

- The above result is then multiplied by $(R\alpha)^3$ (*atha ghanenaivam*) and divided by $(5400)^3$. The result thus obtained is subtracted from $R\alpha$ (*dhanuṣyantataḥ*). Thus, we get

$$R\alpha - \frac{\beta_1 \times (R\alpha)^3}{(5400)^3} + \frac{\beta_2 \times (R\alpha)^5}{(5400)^5} - \frac{\beta_3 \times (R\alpha)^7}{(5400)^7} + \frac{\beta_4 \times (R\alpha)^9}{(5400)^9} - \frac{\beta_5 \times (R\alpha)^{11}}{(5400)^{11}}. \quad (6.33)$$

The rationale behind the expression (6.33), as well as the origin of the numbers listed in Table 6.4 can be understood with the help of the Mādhava series (6.31) for $R\sin\alpha$. Multiplying and dividing each term in the sum in the RHS of (6.31) by $(5400)^{(2i+1)}$, we obtain

$$\begin{aligned} R\sin\alpha &\approx (R\alpha) + \sum_{i=1}^{5} (-1)^i \frac{(R\alpha)^{(2i+1)}}{(2i+1)!R^{(2i)}} \times \frac{(5400)^{(2i+1)}}{(5400)^{(2i+1)}} \\ &= (R\alpha) + \sum_{i=1}^{5} (-1)^i \frac{\beta_i \times (R\alpha)^{(2i+1)}}{(5400)^{(2i+1)}}, \end{aligned} \quad (6.34)$$

where $\beta_i = \dfrac{(5400)^{(2i+1)}}{(2i+1)!R^{(2i)}}$.

The above expression is the same as (6.33). It can easily be verified that the values of β_i's ($i = 1, \ldots, 5$) given in Table 6.4 are obtained by choosing the value of R to be $3437'44''48'''$ in the expression for β_i. In this table, we also present, for the sake of comparison, the computed values (accurate upto fourths) by taking the value of R to be $\frac{21600}{2\pi}$. As we can see from the table, the computed values tally with the values given by the *vākyas* except for β_5 which when rounded off to the nearest thirds should be $45'''$.

The following verse presents the truncated series for the Rcosine function. The numerical values represented by the *vākyas*, *stenaḥ*, ... *ūnadhanakṛdbhūreva*, are listed in Table 6.5.

स्तेनः स्त्रीपिशुनः सुगन्धिनगनुद् भद्राङ्गभव्यासनो
मीनाङ्गो नरसिंह ऊनधनकृद्भूरेव षट्स्वेषु तु ।
आधस्थ्यात् गुणितादभीष्टधनुषः कृत्या विहृत्यान्तिम-
स्याप्तं शोध्यमुपर्युपर्यथ फलं स्यादुत्क्रमस्यान्त्यजम् ॥ १५ ॥

stenaḥ strīpiśunaḥ sugandhinaganud bhadrāṅgabhavyāsano
mīnāṅgo narasiṃha ūnadhanakṛdbhūreva ṣaṭsveṣu tu |
ādhasthyāt guṇitādabhīṣṭadhanuṣaḥ kṛtyā vihṛtyāntima-
syāptaṃ śodhyamuparyuparyatha phalaṃ syādutkramasyāntyajam || 15 ||

Among the six values [listed in order as] *stenaḥ, strīpiśunaḥ, sugandhinaganud, bhadrāṅgabhavyāsanaḥ, mīnāṅgo narasiṃhaḥ, ūnadhanakṛdbhūreva*, the last value is to be multiplied by the square of the desired arc [and] divided by the [square] of the ultimate [value of the arc, 5400]. The result obtained should be subtracted from the [immediately] preceding one above. Similarly the process is repeated with revised last value. The result is to be [multiplied by the square of the desired arc divided by the square of the last arc]. The final result obtained (*antyajam phalam*) will give the value of the Rversine (*utkramasya*) [of the desired arc].

This verse presents the *vākyas*, *stena* etc., for the coefficients $\phi_1, \phi_2 \dots, \phi_6$ appearing in Rversine (*utkramajyā*) series

$$R - R\cos\alpha \approx \sum_{i=1}^{6} (-1)^i \frac{\phi_i \times (R\alpha)^{2i}}{(5400)^{2i}},$$

which is the same as presenting the Rcosine series

$$R\cos\alpha \approx R + \sum_{i=1}^{6} (-1)^i \frac{\phi_i \times (R\alpha)^{2i}}{(5400)^{2i}}.$$

As with the series for Rsine, here too the rationale behind the given expression can be understood by multiplying and dividing each term in the sum in the RHS of (6.32) by $(5400)^{(2i)}$. Doing so, we get

$$\begin{aligned} R\cos\alpha &\approx R + \sum_{i=1}^{6} (-1)^i \frac{(R\alpha)^{2i}}{(2i)!R^{(2i-1)}} \times \frac{(5400)^{2i}}{(5400)^{2i}} \\ &= R + \sum_{i=1}^{6} (-1)^i \frac{\phi_i \times (R\alpha)^{2i}}{(5400)^{2i}}, \end{aligned} \tag{6.35}$$

where $\phi_i = \dfrac{(5400)^{2i}}{(2i)!R^{(2i-1)}}$.

It can easily be verified that the values of ϕ_i's ($i = 1, \dots, 6$) given in Table 6.5 are obtained by choosing the value of R to be $3437'44''48'''$ in the expression for $\phi_i = \frac{(5400)^{2i}}{(2i)!R^{(2i-1)}}$. In this table, we also present, for the sake of comparision, the computed values (accurate upto fourths) by taking the value of R to be $\frac{21600}{2\pi}$. As we can see from the table, the computed values tally with the values given by the *vākyas* except for ϕ_2 which when rounded off to the nearest thirds should be $872'03''06'''$.

ϕ_i	*vākyas*		computed values
	in *kaṭapayādi*	in numerals	
ϕ_6	*stenaḥ*	$06'''$	$05'''49.81''''$
ϕ_5	*strīpiśunaḥ*	$05''12'''$	$05''11'''53.86''''$
ϕ_4	*sugandhinaganud*	$03'09''37'''$	$03'09''36'''39.75''''$
ϕ_3	*bhadrāṅgabhavyāsano*	$071'43''24'''$	$71'43''24'''06.54''''$
ϕ_2	*mīnāṅgo narasiṃha*	$872'03''05'''$	$872'03''05'''30.83''''$
ϕ_1	*ūnadhanakṛdbhūreva*	$4241'09''00'''$	$4241'09''00'''17.79''''$

Table 6.5 Comparing the values encoded by the *vākyas*, *stena* etc., with the computed values.

६.९ अन्त्योपान्त्यज्याभ्यां चतुर्विंशतिज्यानयनम्

6.9 Obtaining the tabular Rsines from the last and the penultimate Rsine

अन्त्योपान्त्यगुणान्तरेण विहृता त्रिज्यात्र हारो भवेत्
आद्यज्या स्वहरांशकेन रहिता द्विघ्ना द्वितीया भवेत् ।
जीवा सा द्विगुणा स्वतो हरहृतेनाद्यज्यया चोनिता
जीवा सैव तृतीयकैवमवराः कार्यास्तुरीयादयः ॥ १६ ॥

antyopāntyaguṇāntareṇa vihṛtā trijyātra hāro bhavet
ādyajyā svaharāṃśakena rahitā dvighnā dvitīyā bhavet |
jīvā sā dviguṇā svato harahṛtenādyajyayā conitā
jīvā saiva tṛtīyakaivamavarāḥ kāryāsturīyādayaḥ || 16 ||

The result obtained by dividing the radius by the difference between the last and the penultimate *jyās* is the divisor (*hāra*). Dividing the first *jyā* by the *hāra* and subtracting the result from the first *jyā*, and further multiplying [the result] by two would yield the second *jyā*. The second *jyā* is multiplied by two and from that when the same [quantity] divided by *hāra*, as well as the previous *jyā*, are subtracted, that indeed will give the third *jyā*. In a similar manner, the later *jyās* commencing with the fourth can be obtained.

The text first defines a quantity called the *hāra* (H) as follows:

$$H = \frac{\textit{trijyā}}{\textit{antyajyā} - \textit{upāntyajyā}} = \frac{\text{radius}}{\text{diff. of last and last but one Rsines}}. \quad (6.36a)$$

Using the *hāra*, explicit expressions are given for the second and the third Rsines.

$$\textit{dvitīyajyā} = 2 \times \left(\textit{ādyajyā} - \frac{\textit{ādyajyā}}{H} \right), \quad (6.36b)$$

$$\textit{tritīyajyā} = 2 \times \left(\textit{dvitīyajyā} - \frac{\textit{dvitīyajyā}}{H} \right) - \textit{ādyajyā}. \quad (6.36c)$$

Then it is said that all other *jyās* can be obtained in a similar manner.

Essentially what is presented here is a recursive relation using which all the tabular Rsines can be obtained. We shall now explain this recursive formula. First we note the expression for *hāra* (H) given in the verse is

$$H = \frac{R}{R\sin 24\alpha - R\sin 23\alpha} = \frac{R}{R - R\cos\alpha}. \quad (6.37)$$

Using this *hāra* and $R\sin\alpha$ or the first Rsine, all the tabular Rsines can be obtained.

To start with, it is well known that

$$R\sin 2\alpha = 2R\sin\alpha\cos\alpha. \quad (6.38)$$

Adding and subtracting $2R\sin\alpha$ to the RHS of the above equation, we get

$$\begin{aligned} R\sin 2\alpha &= 2R\sin\alpha - 2R\sin\alpha + 2R\sin\alpha\cos\alpha \\ &= 2\left(R\sin\alpha - R\sin\alpha\left(1-\cos\alpha\right)\right) \\ &= 2\left(R\sin\alpha - \frac{R\sin\alpha\left(R - R\cos\alpha\right)}{R}\right). \end{aligned} \quad (6.39)$$

Using (6.37) we may rewrite the above equation as

$$R\sin 2\alpha = 2\left(R\sin\alpha - \frac{R\sin\alpha}{H}\right), \quad (6.40)$$

which is the same as (6.36b) given in the text. Now, $R\sin 3\alpha$ can be expressed as

$$R\sin(2\alpha + \alpha) = R\sin 2\alpha\cos\alpha + R\cos 2\alpha\sin\alpha. \quad (6.41)$$

Applying the relation $\cos 2\alpha = (2\cos^2\alpha - 1)$, the second term in the RHS of (6.41) becomes

$$\begin{aligned} R\cos 2\alpha\sin\alpha &= 2R\cos^2\alpha\sin\alpha - R\sin\alpha \\ &= R\sin 2\alpha\cos\alpha - R\sin\alpha. \end{aligned} \quad (6.42)$$

Now substituting (6.42) in (6.41) and adding and subtracting $2R\sin 2\alpha$, we get

$$\begin{aligned} R\sin 3\alpha &= 2R\sin 2\alpha - 2R\sin 2\alpha + 2R\sin 2\alpha\cos\alpha - R\sin\alpha \\ &= 2R\sin 2\alpha - 2R\sin 2\alpha(1-\cos\alpha) - R\sin\alpha \\ &= 2\left(R\sin 2\alpha - \frac{R\sin 2\alpha}{H}\right) - R\sin\alpha, \end{aligned} \tag{6.43}$$

which is the same as (6.36*c*). In the same way, $R\sin((i+1)\alpha)$ can be expressed as

$$R\sin((i+1)\alpha) = 2\left(R\sin(i\alpha) - \frac{R\sin(i\alpha)}{H}\right) - R\sin((i-1)\alpha). \tag{6.44}$$

This is the recursive relation that is implied in the text, when it mentions that in a similar manner the remaining *jyās* can be obtained.

६.१० ज्यानयने प्रकारान्तरम्

6.10 Another recursive relation for obtaining the Rsines

आद्येष्टजीवाकृतिभेदवर्गात् इष्टाद्यजीवाकृतिसंहृता तु ।
इष्टोर्ध्वजीवाकृतिरेतया तत्रेष्टाद्यजीवाकृतिरेव लब्धा ॥ १७ ॥

ādyeṣṭajīvākṛtibhedavargāt iṣṭādyajīvākṛtisaṃhṛtā tu |
iṣṭordhvajīvākṛtiretayā tatreṣṭādyajīvākṛtireva labdhā || 17 ||

When the square of the difference between the squares of the desired Rsine (*iṣṭa-jīvā*) and the first Rsine (*ādyajīvā*) is divided by the square of the Rsine that is immediately preceding to the desired Rsine (*iṣṭādyajīvā*),[8] it results in the square of the Rsine that is the successor of the desired Rsine (*iṣṭordhvajīvā*). [If the previous result is divided] by this (*iṣṭordhvajīvā*) then the resultant would be the square of the preceding Rsine.

The formulae presented in the above verse may be expressed as:

$$\begin{aligned} (\textit{iṣṭordhvajīvā})^2 &= \frac{\left[(\textit{iṣṭajīvā})^2 - (\textit{ādyajīvā})^2\right]^2}{(\textit{iṣṭādyajīvā})^2} \\ \text{and} \qquad (\textit{iṣṭādyajīvā})^2 &= \frac{\left[(\textit{iṣṭajīvā})^2 - (\textit{ādyajīvā})^2\right]^2}{(\textit{iṣṭordhvajīvā})^2} \end{aligned} \tag{6.45}$$

[8] Here the term *iṣṭādyajīvā* should be considered as a *tatpuruṣa* compound (and not *dvandva* compound), and hence has to be derived as *iṣṭajīvāyāḥ ādyajīvā*, meaning the Rsine prior to desired Rsine.

Let the desired Rsine *iṣṭajīvā* be denoted by $R\sin(i\alpha)$ (where $\alpha = 225'$ and $i = 1, 2 \ldots 24$). Then, the above expressions translate to

$$R^2\sin^2((i+1)\alpha) = \frac{\left[R^2\sin^2(i\alpha) - R^2\sin^2\alpha\right]^2}{R^2\sin^2((i-1)\alpha)},$$
$$\text{and} \qquad R^2\sin^2((i-1)\alpha) = \frac{\left[R^2\sin^2(i\alpha) - R^2\sin^2\alpha\right]^2}{R^2\sin^2((i+1)\alpha)}. \qquad (6.46)$$

In fact, these relations are a consequence of the following identity.

$$R\sin((i-1)\alpha) \times R\sin((i+1)\alpha) = R^2\sin^2(i\alpha) - R^2\sin^2\alpha. \qquad (6.47)$$

६.११ अभीष्टजीवायाः चापानयनम्

6.11 Computation of the arc from the corresponding chord

व्यासार्धेन हतादभीष्टगुणतः कोट्याप्तमाद्यं फलं
ज्यावर्गेण विनिघ्नमादिमफलं तत्तत्फलं चाहरेत् ।
कृत्या कोटिगुणस्य तत्र तु फलेष्वेकत्रिपञ्चादिभिः
भक्तेष्वोजयुतैस्त्यजेत् समयुतिं जीवाधनुश्शिष्यते ॥ १८ ॥

vyāsārdhena hatādabhīṣṭaguṇataḥ koṭyāptamādyaṃ phalaṃ
jyāvargeṇa vinighnamādimaphalaṃ tattatphalaṃ cāharet |
kṛtyā koṭiguṇasya tatra tu phaleṣvekatripañcādibhiḥ
bhakteṣvojayutaistyajet samayutiṃ jīvādhanuśśiṣyate || 18 ||

The first *phala* is the Rsine of the desired arc mutliplied by the radius and divided by the Rcosine of the arc. The first *phala* multiplied by the square of the Rsine and divided by the square of the Rcosine (*koṭiguṇa*) [gives the second *phala*] and [in a similar manner] the successive *phalas* are obtained. Here these *phalas* are divided by one, three, five etc., [successively]. [Then] the sum of the even terms is to be subtracted from that of the odd ones. What remains is the arc of the Rsine.

If s be the arc-length corresponding to a given *jyā*, j_s, and *koṭi*, k_s, then the above verse presents the following series:

$$s = R\left(\frac{j_s}{k_s}\right) - \frac{R}{3}\left(\frac{j_s}{k_s}\right)^3 + \frac{R}{5}\left(\frac{j_s}{k_s}\right)^5 - \cdots. \qquad (6.48)$$

The above expression, with $s = Rx$, is equivalent to the series for the inverse tan function (the so called Gregory series).

$$x = \tan x - \frac{1}{3}\tan^3 x + \frac{1}{5}\tan^5 x - \ldots,$$

or

$$\tan^{-1} x = x - \frac{x^3}{3} + \frac{x^5}{5} - \dots.$$

The Mādhava series for $\frac{\pi}{4}$ given in the first verse of this chapter can be readily obtained from this by substituting $x = 1$.

६.१२ अल्पचापस्य ज्यानयनम्

6.12 Computation of the Rsine value of a small arc

स्वल्पचापघनषष्ठभागतो विस्तरार्धकृतिभक्तवर्जितम् ।
शिष्टचापमिह शिञ्जिनी भवेत् तद्युतोऽल्पकगुणोऽसकृद् धनुः ॥ १९ ॥

svalpacāpaghanaṣaṣṭhabhāgato vistarārdhakṛtibhaktavarjitam |
śiṣṭacāpamiha śiñjinī bhavet tadyuto'lpakaguṇo'sakṛd dhanuḥ || 19 ||

The cube of a small arc is divided by six and the result is [further] divided by the square of the radius. The result obtained by subtracting this from the arc will be the Rsine (*śiñjinī*) [of that arc]. The Rsine of the small arc added to that (cube of the arc divided by the square of the radius multiplied by six) would be the arc when the process is iterated.

In the above verse Putumana Somayājī gives an approximation for the Rsine of an arc when it is small, that is, an arc that is much smaller than 225′ which is taken to be the unit in tabulating Rsine values. If $R\delta\theta$ be the length of a small arc along the circle, corresponding to an angle $\delta\theta$, then the expression for its Rsine (*śiñjinī*)[9] given in the above verse is :

$$R \sin \delta\theta \approx R\,\delta\theta - \frac{(R\,\delta\theta)^3}{6\,R^2},$$

$$\text{or,} \qquad \sin \delta\theta \approx \delta\theta - \frac{(\delta\theta)^3}{6}. \tag{6.49}$$

The above equation is a well known approximation for the sine function and gives a fairly accurate value when the angle is very small.

In the last quarter of the verse, the text indicates an iterative process for evaluating the arc $R\delta\theta$ from the equation

$$R\,\delta\theta = R \sin \delta\theta + \frac{(R\,\delta\theta)^3}{6\,R^2}. \tag{6.50}$$

The last part of the verse is "*tadyuto'lpakaguṇaḥ asakṛddhanuḥ*". The word '*tad*' being a pronoun, always points to something that was referred to ear-

[9] The term *śiñjinī* is synonymous with *jyā* and stands for Rsine.

lier.[10] Here it refers to $\frac{(R\delta\theta)^3}{6R^2}$ as this was the quantity that was prescribed in the first half of the verse for further operations. Hence it is stated that this quantity added to the Rsine corresponding to a small arc (*alpakaguṇaḥ*) gives the value of the arc (*dhanuḥ*) itself when an iterative process is employed.

To start with, we have as the first approximation

$$R\delta\theta \approx R\sin\delta\theta.$$

Substituting this in the RHS of (6.50), we get

$$R\delta\theta \approx R\sin\delta\theta + \frac{(R\ \sin\delta\theta)^3}{6\ R^2}. \tag{6.51}$$

The iterative procedure prescribed above may be captured by the following recursive relation:

$$R\delta\theta_{i+1} = R\sin(\delta\theta) + \frac{(R\delta\theta_i)^3}{6}. \tag{6.52}$$

The successive approximations obtained in this way are listed in Table 6.6. It may be noted that the successive terms do differ from the well known series for the arc-sine in higher orders.

	Successive approximations to $\delta\theta$	Corresponding terms in arc-sine series
$\delta\theta_1$	$\sin\delta\theta$	$\sin\delta\theta$
$\delta\theta_2$	$\sin\delta\theta + \frac{(\sin\delta\theta)^3}{6}$	$\sin\delta\theta + \frac{(\sin\delta\theta)^3}{6}$
$\delta\theta_3$	$\sin\delta\theta + \frac{(\sin\delta\theta)^3}{6} + \frac{(\sin\delta\theta)^5}{12} + \frac{(\sin\delta\theta)^7}{72} + \frac{(\sin\delta\theta)^9}{1296}$	$\sin\delta\theta + \frac{(\sin\delta\theta)^3}{6} + \frac{3(\sin\delta\theta)^5}{40} + \frac{15(\sin\delta\theta)^7}{336} + \dots$

Table 6.6 The successive approximations for $\delta\theta$ in terms of $\sin\delta\theta$.

[10] The rule is: *sarvanāmnāṃ pūrvaprakrāntaparāmarśitvam.*

६.१३ अल्पचापज्याखण्डेभ्यः ज्यानयनम्

6.13 Finding the Rsines from a table of arc-sine differences for small arcs

एकद्वित्र्यादिसंख्याघ्नत्रिज्यावर्गनयांशतः ।
घनमूलं हि चापज्या स्वसंख्योनविलिप्तिकम् ॥ २० ॥

ekadvitryādisaṃkhyāghnatrijyāvarganayāṃśataḥ |
ghanamūlaṃ hi cāpajyā svasaṃkhyonaviliptikam || 20 ||

After multiplying the square of the radius by the numbers 1, 2, 3 etc. and dividing by 10 (*naya*), [let] the cube root of the resulting quantity be obtained. The integral numbers 1, 2, 3, ... in seconds when subtracted from the above [*cāpa*] would be the corresponding *jyās* (Rsines).

This verse gives a method for calculating the value of the arc (c_n) and its Rsine (j_n) such that their difference (*jyācāpāntara*) $c_n - j_n$ is equal to n seconds where $n = 1, 2, 3, \ldots$.

It commences with the prescription of an approximate value of c_n

$$c_n \approx \sqrt[3]{\frac{n \times R^2}{10}}. \tag{6.53}$$

Then, it is said that the Rsine is to be obtained using

$$j_n = c_n - n. \tag{6.54}$$

Following is the rationale for the above approximation. From (6.49) we know that, when the *cāpa* is small, the difference between the *cāpa* (arc) and its *jyā* (Rsine) called *jyācāpāntara*, to a good approximation, may be expressed as

$$R\,\delta\theta - R\sin\delta\theta \approx \frac{(R\,\delta\theta)^3}{6\,R^2}, \tag{6.55}$$

Here, all the quantities are expressed in minutes. If the *jyācāpāntara* $= n$, where $n = 1, 2, 3, \ldots$, expressed in seconds, then using (6.55) the corresponding *cāpa* ($R\delta\theta = c_n$) can be calculated using the relation

$$\frac{(c_n)^3}{6R^2} \approx \frac{n}{60}. \tag{6.56}$$

This implies that when *jyācāpāntara* $= n$, the corresponding *cāpa* (arc) is given by

$$c_n \approx \sqrt[3]{\frac{n \times 6R^2}{60}}$$
$$= \sqrt[3]{\frac{n \times R^2}{10}}. \tag{6.57}$$

Now the Rsine, j_n, corresponding to the arc, c_n, can be obtained using (6.55) as

$$j_n = c_n - n. \tag{6.58}$$

In Table 6.7, we compile the values of c_n and j_n given by the formulae (6.57) and (6.58). In fact, Commentary II lists (even more accurate) values of the j_n in terms of the *vākyas*, *gūḍhāmenaka* etc.[11]

A similar method for computing the arc and the Rsine when the difference between them is equal to n seconds, where n is a positive integer, has been given in *Tantrasaṅgraha* of Nīlakaṇṭha Somayājī. However, in the explanation of this verse, Śaṅkara Vāriyar seems to imply that the expression $\sqrt[3]{\frac{n \times R^2}{10}}$ should be considered as the *jyā* value to which n seconds have to be added to obtain the corresponding *cāpa*.[12] This interpretation is clearly flawed as it can be shown that $\sqrt[3]{\frac{n \times R^2}{10}}$ more closely approximates the value of the arc for which *jyācāpāntara* is equal to n seconds.

In Table 6.8, we have presented the accurate values of *jyā* and *cāpa* for which the *jyācāpāntara* is equal to n seconds. These may be compared with the computed values of $\sqrt[3]{\frac{n \times R^2}{10}}$ presented in Table 6.7. From that table, it is clear that $\sqrt[3]{\frac{n \times R^2}{10}}$ is very good approximation for the *cāpa* such that the *jyācāpāntara* is n seconds. Hence, n seconds have to be subtracted from $\sqrt[3]{\frac{n \times R^2}{10}}$, as in equation (6.58), in order to obtain the *jyā* values for which the *jyācāpāntara* is n seconds.

Interestingly, Śaṅkara Vāriyar also gives the *vākyas*, *lavaṇaṃ nindyaṃ* etc.,[13] for the *jyā* values such that the *jyācāpāntara* is n seconds. They seem to coincide mostly with the *jyā* values given by the *gūḍhāmenakādi-vākyas* mentioned in Commentary II of *Karaṇapaddhati*. Numerical values corresponding to both these set of *vākyas* have been included in Table 6.8 along with the accurate values of *jyās* and *cāpas*.

---

[11] {KP 1956} p. 200A.

[12] {TS 1958}, pp. 29-31.

[13] {TS 1958}, pp. 29-31.

The difference (*jyācāpāntara*) (n)	$\sqrt[3]{\frac{n\times R^2}{10}}$ (c_n)		$c_n - n$ (j_n)	
	min.	sec.	min.	sec.
1	105	43.57149473	105	42.57149473
2	133	12.39925772	133	10.39925772
3	152	29.01326248	152	26.01326248
4	167	49.79206396	167	45.79206396
5	180	47.35467201	180	42.35467201
6	192	07.03439517	192	01.03439517
7	202	14.81572241	202	07.81572241
8	211	27.14298945	211	19.14298945
9	219	55.16044655	219	46.16044655
10	227	46.81048694	227	36.81048694
11	235	07.97670738	234	56.97670738
12	242	03.15327734	241	51.15327734
13	248	35.85969959	248	22.85969959
14	254	48.90976526	254	34.90976526
15	260	44.59261467	260	29.59261467
16	266	24.79851544	266	08.79851544
17	271	51.10860342	271	34.10860342
18	277	04.86040335	276	46.86040335
19	282	07.19663531	281	48.19663531
20	286	59.10221742	286	39.10221742
21	291	41.43276140	291	20.43276140
22	296	14.93682505	295	52.93682505
23	300	40.27350818	300	17.27350818
24	304	58.02652497	304	34.02652497

Table 6.7 The values of c_n and j_n given by (6.57) and (6.58).

n	The values of *jyā* given in *vākyas*				Accurate values of *jyā* and *cāpa* such that *cāpa* − *jyā* = n seconds			
(sec.)	*Gūḍhāmenakādi*		*Lavaṇaṃ-nindyādi*		*jyā*		*cāpa*	
	min.	sec.	min.	sec.	min.	sec.	min.	sec.
1	105	43	105	43	105	42.67150	105	43.67150
2	133	11	133	11	133	10.59927	133	12.59927
3	152	26	152	26	152	26.31329	152	29.31329
4	167	46	167	46	167	46.19210	167	50.19210
5	180	43	180	43	180	42.85473	180	47.85473
6	192	02	192	02	192	01.63447	192	07.63447
7	202	09	202	08	202	08.51582	202	15.51582
8	211	20	211	20	211	19.94312	211	27.94312
9	219	47	219	47	219	47.06060	219	56.06060
10	227	34	227	38	227	37.81067	227	47.81067
11	234	58	234	58	234	58.07693	235	9.07693
12	241	52	241	52	241	52.35353	242	4.35353
13	248	24	248	24	248	24.15999	248	37.15999
14	254	36	254	36	254	36.31009	254	50.31009
15	260	31	260	31	260	31.09298	260	46.09298
16	266	10	266	10	266	10.39892	266	26.39892
17	271	36	271	36	271	35.80906	271	52.80906
18	276	49	276	48	276	48.66090	277	6.66090
19	281	51	281	50	281	50.09718	282	9.09718
20	286	40	286	40	286	41.10281	287	1.10281
21	291	22	291	22	291	22.53341	291	43.53341
22	295	55	295	55	295	55.13752	296	17.13752
23	300	19	300	18	300	19.57426	300	42.57426
24	304	36	304	36	304	36.42733	305	0.42733

Table 6.8 *Vākyas* for *jyās* along with the accurate values of *jyā* and *cāpa* such that *jyācāpāntara* is exactly n seconds. See {KP 1956}, p. 200, and {TS 1958}, pp. 30–31.

Chapter 7
ग्रहस्फुटानयनम्
Obtaining the planetary longitudes

७.१ ग्रहाणां मन्दशीघ्रपरिधिमानम्

7.1 The circumferences of the *manda* and *śīghra* epicycles of the planets

भानोर्गानं विधोः स्थानं स्फुटवृत्तकलाः सदा ।
वन्द्यो दिव्यो गुणी कृष्णः सूनुर्मानी कलाधरः ॥ १ ॥

स्थानं दानं तपो मान्यं भानुः प्राज्ञो धमः समः ।
धनी लोको धनं दानं भौमात् परिधिलिप्तिकाः ॥ २ ॥

मन्दशीघ्रक्रमात् केन्द्रस्यौजयुग्मपदादिगाः ।
पृथगेकैकपादोक्ता असुरैरपवर्तिताः ॥ ३ ॥

bhānorgānaṃ vidhoḥ sthānaṃ sphuṭavṛttakalāḥ sadā |
vandyo divyo guṇī kṛṣṇaḥ sūnurmānī kalādharaḥ || 1 ||

sthānaṃ dānaṃ tapo mānyaṃ bhānuḥ prājño dhamaḥ samaḥ |
dhanī loko dhanaṃ dānaṃ bhaumāt paridhiliptikāḥ || 2 ||

mandaśīghrakramāt kendrasyaujayugmapadādigāḥ |
pṛthagekaikapādoktā asurairapavartitāḥ || 3 ||

The true epicycle circumference in minutes (*sphuṭavṛttakalās*) of the Sun is 03 (*gānam*) and that of the Moon is 07 (*sthānam*) always (in all the quadrants). The [dimensions of the] circumferences of the *manda* and the *śīghra* [epicycles] of the planets starting from Mars in minutes, at the beginning of the odd and the even quadrants, in order, are 14 (*vandya*), 18 (*divya*), 53 (*guṇī*) and 51 (*kṛṣṇa*) [for Mars]; 07 (*sūnu*), 05 (*mānī*), 31 (*kalā*) and 29 (*dhara*) [for Mercury]; 07 (*sthānaṃ*), 04 (*dānaṃ*), 16 (*tapo*) and 15 (*mānyaṃ*) [for Jupiter]; 04 (*bhānuḥ*), 02 (*prājño*), 59 (*dhamaḥ*) and 57 (*samaḥ*) [for Venus]; 09 (*dhanī*), 13 (*loko*), 09 (*dhanaṃ*) and 08 (*dānaṃ*) [for Saturn]. These are stated separately for different quadrants after being divided (*apavartita*) by 270.

V. Pai et al., *Karaṇapaddhati of Putumana Somayājī*, Sources and Studies in the History of Mathematics and Physical Sciences, https://doi.org/10.1007/978-981-10-6814-0_7

The above set of verses essentially present the dimensions of the *manda* and *śīghra* epicycles.[1] From the latter half of the first verse beginning with *vandyaḥ*, for the next five quarters, only numbers are stated—each quarter presenting a list of four numbers. Of the four, the first two correspond to *manda* epicycle and the latter two to the *śīghra* epicycle (*mandaśīghrakramāt*). Of these pairs, the first gives the dimension of the epicycle at the beginning of the odd quadrants and the second at the beginning of the even quadrants respectively (*ojayugmapadādigāḥ*). The values of the *manda* and *śīghra-paridhis* at the beginning of the odd and even quadrants of all the planets are listed in Table 7.1. It may be noted that these values are the same as those prescribed in *Āryabhaṭīya*.

The numbers given in Table 7.1 multiplied by 270 give the *manda* and *śīghra-paridhis* (circumference of the epicycles) in minutes, when the circumference of the *kakṣyāmaṇḍala* (deferent) is taken to be 21600 minutes. The given number divided by 80 would be the ratio of the radius of the epicycle and the deferent. For example, if C'_{mo} and C'_{so} denote the circumferences, and r_{mo} and r_{so} denote the radii, of the *manda* and *śīghra* epicycles of Mars at the beginning of the odd quadrants in minutes, then

$$C'_{mo} = 14 \times 270, \text{ and } C'_{so} = 53 \times 270 \qquad \text{(for Mars)}. \tag{7.1}$$

Similarly

$$\frac{r_{mo}}{R} = \frac{14}{80}, \text{ and } \frac{r_{so}}{R} = \frac{53}{80} \qquad \text{(for Mars)}. \tag{7.2}$$

Planet	dimensions of the *manda* epicycle at the beginning of		dimensions of the *śīghra* epicycle at the beginning of	
	odd quadrant	even quadrant	odd quadrant	even quadrant
Sun	03	03	–	–
Moon	07	07	–	–
Mars	14	18	53	51
Mercury	07	05	31	29
Jupiter	07	08	16	15
Venus	04	02	59	57
Saturn	09	13	09	08

Table 7.1 The dimensions of the *manda* and *śīghra* epicycles of the planets.

[1] The epicycle and eccentric circle models which form the basis for the computations in this chapter are explained in Appendix B.

The *manda* and *śīghra-paridhis* at the beginning of the odd quadrants refer to their values when the *mandakendra/śīghrakendra* is equal to 0° or 180°. Similarly *manda* and *śīghra-paridhis* at the beginning of the even quadrants refer to their values when the *mandakendra/śīghrakendra* is equal to 90° or 270°. Having defined the values at the beginning of the quadrants the text proceeds to describe the procedure to find *manda* and *śīghra-paridhis* for other values of the *kendra*.

७.२ मन्दशीघ्रपरिधीनां स्फुटीकरणम्

7.2 Finding the true circumference of the epicycle

वृत्तान्तरेणाहतबाहुजीवां हृत्वा[2] त्रिमौर्व्या फलमोजवृत्ते ।
क्रमाद् धनर्णं विदुरोजवृत्तस्याल्पाधिकत्वे स्फुटवृत्तसिद्ध्यै ॥ ४ ॥

vṛttāntareṇāhatabāhujīvāṃ hṛtvā trimaurvyā phalamojavṛtte |
kramād dhanarṇaṃ vidurojavṛttasyālpādhikatve sphuṭavṛttasiddhyai || 4 ||

The difference between [the dimensions of] the epicycles at the beginning of the odd and even quadrants is to be multiplied by the Rsine [of the *kendra*] and divided by the *trijyā*. The result has to be added to or subtracted from the dimension of the epicycle at the beginning of an odd quadrant, depending on whether the epicycle at the beginning of odd quadrant is smaller or bigger [than the epicycle at the beginning of even quadrant] respectively, to obtain the true epicycle (*sphuṭavṛtta*).

Let C_o and C_e represent the dimensions of the *mandaparidhis* or *śīghra-paridhis* at the beginning of the odd and even quadrants respectively. Then the dimension at any other place (*sphuṭaparidhi*, C) is given by

$$C = C_o \pm \frac{\Delta \times |R\sin(\theta_k)|}{R}, \tag{7.3}$$

where $\Delta = |C_o - C_e|$ and θ_k is the anomaly or *kendra*. Here '+' and '−' signs are chosen depending on whether $C_o < C_e$ and $C_o > C_e$ respectively.

The above equation provides the general expression for finding the circumference of the *manda* or the *śīghra* epicycle for a given *kendra*. That is, if θ_{mk}, C_{mo} and C_{me} are the *mandakendra*, *ojādiparidhi* and *yugmādiparidhi* of *manda* epicycle respectively, then *manda-sphuṭaparidhi* (C_m) for an arbitrary value of θ_{mk} is given by

$$C_m = C_{mo} \pm \frac{\Delta_m \times |Rsin(\theta_{mk})|}{R}, \tag{7.4}$$

[2] In all the earlier editions {KP 1937}, {KP 1953} and {KP 1956}, we find the reading *hatvā*. However, it is clear from the context that it should be *hṛtvā*.

where $\Delta_m = |C_{mo} - C_{me}|$. When the dimension of the epicycle at the beginning of the odd quadrant ($\theta_{mk} = 0°$ or $180°$) is less than that of the even quadrant ($C_{mo} < C_{me}$), then, as the *kendra* increases, the dimension of the epicycle also increases till it reaches a maximum at the beginning of the even quadrant. Then it will start decreasing. This explains the prescription of '+' and '−' sign in the text.

Similarly, if θ_{sk}, C_{so} and C_{se} are the *śīghrakendra*, *ojādiparidhi* and *yugmādiparidhi* of *śīghra* epicycle respectively, then the *śīghra-sphuṭaparidhi* (C_s) for an arbitrary value of θ_{sk} is given by

$$C_s = C_{so} \pm \frac{\Delta_s \times |Rsin\theta_{sk}|}{R}, \tag{7.5}$$

where $\Delta_s = |C_{so} - C_{se}|$. As earlier, '+' and '−' signs correspond to the cases $C_{so} < C_{se}$ and $C_{so} > C_{se}$ respectively.

Note that for any value of the *manda* or *śīghrakendra* the radii of the *manda* and *śīghra* epicycle are given by

$$r_m = \frac{C_m}{80} \times R, \qquad \text{and} \qquad r_s = \frac{C_s}{80} \times R. \tag{7.6}$$

७.३ ग्रहमन्दफलानयनम्

7.3 Obtaining the *mandaphalas* of the planets

मान्देन स्फुटवृत्तेन निहतादिष्टदोर्गुणात् ।
नन्दाप्तं चापितं मान्दमर्कादीनां भुजाफलम् ॥ ५ ॥

māndena sphuṭavṛttena nihatādiṣṭadorguṇāt |
nandāptaṃ cāpitaṃ māndamarkādīnāṃ bhujāphalam || 5 ||

The Rsine [of the *mandakendra*][3] has to be multiplied by the true epicycle (*sphuṭavṛtta*) and divided by 80. When converted into arc (taking Rsine-inverse) it would give the *mandaphala* of the Sun, etc.

We refer the reader to the discussion in Appendix B, where (B.9) gives the *mandaphala*. From (7.6) we have

---

[3] We may recall the third verse of the fourth chapter of the text which defines the *manda* and the *śīghrakendras*.

madhyāt grahāṇāṃ svamṛdūccahīnāt śīghroccato madhyavivarjitācca...

[The *mandakendra* can be obtained by] subtracting their *mandoccas* from the mean [longitudes] and [the *śīghrakendra* can be obtained] by subtracting the mean planets from their *śīghroccas*.

$$\frac{r_m}{R} = \frac{C_m}{80}.$$

Therefore, (B.9) reduces to:

$$\sin(\Delta\theta) = \frac{C_m}{80}|\sin(\theta_{mk})|,$$
$$\text{or} \qquad \Delta\theta = \sin^{-1}\left(\frac{C_m}{80}|\sin(\theta_{mk})|\right). \qquad (7.7)$$

The quantity $R\Delta\theta$ which is the arc corresponding to the *mandaphala* is referred to as *māndaṃ* in the verse.

७.४ ग्रहशीघ्रफलानयनम्

7.4 Obtaining the *śīghraphalas* of the planets

शैघ्रे दोःकोटिजीवे स्फुटपरिधिहते नन्दभक्ते फले ते
व्यासार्धे कोटिजं तद् धनमिह मकरादावृणं कर्कटादौ ।
कृत्वा तद्वर्गयुक्ताद् भुजफलकृतितः स्यात् पदं शीघ्रकर्णः
त्रिज्याघ्नाद् दोःफलात्तद्विहृतफलधनुः कर्किनक्रादिजीवाः ॥ ६ ॥

śaighre doḥkoṭijīve sphuṭaparidhihate nandabhakte phale te
vyāsārdhe koṭijaṃ tad dhanamiha makarādāvṛṇaṃ karkaṭādau |
kṛtvā tadvargayuktād bhujaphalakṛtitaḥ syāt padaṃ śīghrakarṇaḥ
trijyāghnād doḥphalāttadvihṛtaphaladhanuḥ karkinakrādijīvāḥ ||6||

Multiply the Rsine (*bhujājyā*) and Rcosine (*koṭijyā*) of the *śīghrakendra* by [*śīghra*]-*sphuṭaparidhi* (C_s), and divide by 80 (*nanda*). [Among] the results, the one which is obtained from *koṭi* has to be added to or subtracted from the radius [depending upon whether the *śīghrakendra* is within the six signs] beginning with the *makara* or *karkaṭa*. The square root of the sum of the squares of this result and the result obtained from *bhujājyā* or the *doḥphala* is called *śīghrakarṇa*. The *doḥphala* multiplied by the radius [*trijyā*] and divided by that (*śīghrakarṇa*) gives the Rsine [of *śīghraphala*] that is either *makarādi* or *karkyādi*.

The *doḥphala* (*bhujāphala*), B_p, and the *koṭiphala*, K_p, in the *śīghra* process are given by

$$B_p = |r_s \sin\theta_{sk}| = \frac{C_s}{80}|R\sin\theta_{sk}|,$$
$$\text{and} \qquad K_p = |r_s \cos\theta_{sk}| = \frac{C_s}{80}|R\cos\theta_{sk}|. \qquad (7.8)$$

Then, from (B.13) and (B.14) in Appendix B, the *śīghrakarṇa* is given by

$$K_s = \left[(R \pm K_p)^2) + (B_p)^2\right]^{\frac{1}{2}}$$
$$= \left[\left(R \pm \frac{C_s}{80}|R\cos\theta_{sk}|\right)^2 + \left(\frac{C_s}{80}|R\sin\theta_{sk}|\right)^2\right]^{\frac{1}{2}}, \qquad (7.9)$$

where the '+' and '−' signs correspond to the six signs beginning with *makarādi* and *karkyādi* respectively. Also the Rsine of the *śīghraphala* is given by

$$R\sin\Delta\theta_s = \frac{R}{K_s}|r_s \sin\theta_{sk}| = \frac{R}{K_s} \times \frac{C_s}{80}|R\sin\theta_{sk}|. \qquad (7.10)$$

७.५ प्रकारान्तरेण स्फुटभुजाकोटिफलानयनम्

7.5 Obtaining the true *bhujā* and *koṭiphalas*

अभिमतदोःकोटिगुणावोजपरिध्याहतौ च नन्दाप्तौ ।
दोःकोटिफले स्यातामनयोर्वा वृत्तसंस्कृतिं कुर्यात् ॥ ७ ॥

द्विघ्नात् तद्बाहुचापाद् भुजगुणमपि तत्कोटिजीवां च नीत्वा
कर्क्येणादौ तु कोटीगुणयुतरहिता त्रिज्यका तत्र बाणः ।
बाणार्धात् दोर्गुणार्धादपि परिधिभिदा संगुणान्नन्दभक्तं
जह्याद् दोःकोटिजाभ्यां क्षिपतु च समवृत्तेऽधिके ते स्फुटे स्तः ॥ ८ ॥

abhimatadoḥkoṭiguṇāvojaparidhyāhatau ca nandāptau |
doḥkoṭiphale syātāmanayorvā vṛttasaṃskṛtiṃ kuryāt || 7 ||

dvighnāt tadbāhucāpād bhujaguṇamapi tatkoṭijīvāṃ ca nītvā
karkyeṇādau tu koṭīguṇayutarahitā trijyakā tatra bāṇaḥ |
bāṇārdhāt dorguṇārdhādapi paridhibhidā saṃguṇānnandabhaktaṃ
jahyād doḥkoṭijābhyāṃ kṣipatu ca samavṛtte'dhike te sphuṭe staḥ || 8 ||

The *bhujāphala* and *koṭiphala* are obtained by multiplying the desired Rsine [of the *śīghrakendra*] and the corresponding Rcosine by the circumference of the *śīghra* epicycle at [the beginning of] the odd quadrant and divided by 80 (*nanda*). The corrections (*paridhi-saṃskāra*) can also be applied to these [*phalas*].

Having obtained the Rsine and Rcosine of the double of the arc (i.e., twice the *śīghrakendra* ($2\theta_{sk}$)) corresponding to the *iṣṭajyā*, the *koṭijyā* has to be added to or subtracted from the radius when the double of the arc is *karkyādi* or is *makarādi* respectively. Half of the [result obtained which is called the] *bāṇa* and the half of the Rsine (*dorguṇārdha*) have to be multiplied by the difference between the *paridhis* and divided by 80 (*nanda*). [The results are] applied positively to the *bhujāphala* and *koṭiphala*, when the epicycle at the beginning of even quadrants (*samavṛtta*) is greater than that at the beginning of odd quandrants [or negatively otherwise], to obtain the *sphuṭas*.

The *bhujāphala* and the *koṭiphala* at the beginning of the odd quadrant are given by $\frac{C_{so}}{80}|R\sin\theta_{sk}|$ and $\frac{C_{so}}{80}|R\cos\theta_{sk}|$ respectively. The *bhujāphala* and *koṭiphala* for an arbitrary value of the *śīghrakendra* θ_{sk} will have in addition the correction terms which are to be added. Recalling the expression for *śīghra-sphuṭaparidhi* (C_s) given by (7.5), the expressions for the *bhujāphala* B_p and *koṭiphala* K_p discussed in the previous section (7.8) become,

$$B_p = \left(\frac{C_{so} \pm \dfrac{\Delta_s \times |R \sin \theta_{sk}|}{R}}{80} \right) |R \sin \theta_{sk}|$$

$$= B_{po} \pm \left(\frac{\Delta_s \times R \sin^2 \theta_{sk}}{80} \right), \tag{7.11}$$

where $B_{po} = \frac{C_{so}}{80} R \sin \theta_{sk}$ is the *bhujāphala* which is defined in the seventh verse. As per the relation (6.27) given in the previous chapter we have,

$$\sin^2 \theta_{sk} = \frac{1 - \cos(2\theta_{sk})}{2} = \frac{1 \pm |\cos(2\theta_{sk})|}{2}. \tag{7.12}$$

Here '+' and '−' signs are applicable when $2\theta_{sk}$ is *karkyādi* (that is, $90° \leq 2\theta_{sk} \leq 270°$), and *makarādi* (that is, $270° \leq 2\theta_{sk} \leq 360°$ or $0° \leq 2\theta_{sk} \leq 90°$), respectively. Using (7.12) in (7.11),

$$B_p = B_{po} \pm \left(\frac{\Delta_s}{80} \times \frac{(R \pm |R \cos(2\theta_{sk})|)}{2} \right). \tag{7.13}$$

The term $\frac{(R \pm |R \cos(2\theta_{sk})|)}{2}$ is referred to as *bāṇārdha* in the verse, since *bāṇā* (BD in Figure 7.1) is given by

$$\begin{aligned} BD &= R - |R \cos 2\theta_{sk}| \qquad \text{(when } 2\theta_{sk} \text{ is } \textit{makarādi}\text{)}, \\ &= R + |R \cos 2\theta_{sk}| \qquad \text{(when } 2\theta_{sk} \text{ is } \textit{karkyādi}\text{)}. \end{aligned} \tag{7.14}$$

Similarly, the expression for the *koṭiphala* may be written as

$$K_p = \frac{C_{so}}{80} |R \cos \theta_{sk}| \pm \left(\frac{\Delta_s \times |\sin \theta_{sk}|}{80} \right) |R \cos \theta_{sk}|$$

$$= K_{po} \pm \left(\frac{\Delta_s \times |R \sin \theta_{sk} \cos \theta_{sk}|}{80} \right)$$

$$= K_{po} \pm \left(\frac{\Delta_s \times \dfrac{|R \sin(2\theta_{sk})|}{2}}{80} \right). \tag{7.15}$$

In the verse, the term $\dfrac{|R \sin(2\theta_{sk})|}{2}$ is referred to as *dorguṇārdha*.

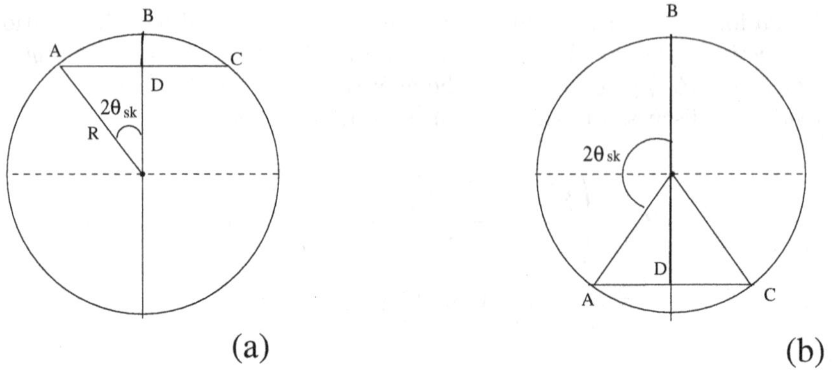

Fig. 7.1 *Bāṇā*, when the *kendra* is *makarādi* (a), and *karkāyadi* (b).

७.६ मन्दशीघ्रफलयोः संस्कारप्रकारः

7.6 Application of the *manda* and the *śīghraphalas*

मान्दे त्वेवं समानीतं दोःफलं चापितं स्फुटम् ।
शैघ्रे त्रिज्याहतं कर्णभक्तं चापीकृतं तथा ॥ ९ ॥

mānde tvevaṃ samānītaṃ doḥphalaṃ cāpitaṃ sphuṭam |
śaighre trijyāhataṃ karṇabhaktaṃ cāpīkṛtaṃ tathā || 9 ||

In the case of *manda*[*saṃskāra*], the arc of the *doḥphala* thus obtained itself would be the *sphuṭa*. [While] in the case of *śīghra* [*saṃskāra*], it (the *sphuṭa*) is the arc of the [*doḥphala*] which is multiplied by the radius and divided by the *karṇa*.

It is stated that in the case of *mandasaṃskāra*, the arc of the *doḥphala* given by[4]

$$R\Delta\theta = R\sin^{-1}\left(|r_m \sin(\theta_{mk})|\right). \tag{7.16}$$

is to be applied as it is to obtain the *manda-sphuṭa*.

This is because, as discussed in Appendix B, in the case of *mandasaṃskāra*, the radius of the epicycle is propotional to the *manda-karṇa* K and this would imply that K does not appear in the expression for *mandaphala*.

However, in *śīghrasaṃskāra*, the radius of the epicycle is the tabulated value itself, and the *śīghraphala* would involve the *śīghrakarṇa* K_s. Hence the magnitude of correction is to be obtained from

$$R\Delta\theta_s = R\sin^{-1}\left(|r_s \sin\theta_{sk}|.\frac{R}{K_s}\right). \tag{7.17}$$

[4] We shall be using the symbol $R\sin^{-1}(\alpha)$ or *cāpa* (α) to denote the arc or *cāpa*, in a circle of radius R, associated with an Rsine or *jyā* of magnitude α.

For further details the reader is referred to section B.2 of Appendix B.

७.७ अन्त्यफलानयनप्रकारः

7.7 Procedure for obtaining the *antyaphala*

स्फुटवृत्तमिनादीनां कामानलसमाहतम् ।
कंसेन विभजेल्लब्धं भवेदन्त्यफलाह्वयम् ॥ १० ॥

sphuṭavṛttaminādīnāṃ kāmānalasamāhatam |
kaṃsena vibhajellabdhaṃ bhavedantyaphalāhvayam || 10 ||

The true epicycle circumference (*sphuṭavṛtta*) of the Sun etc., have to be multiplied by 3051 (*kāmānala*) and divided by 71 (*kaṃsa*). The result thus obtained is called the *antyaphala*.

The term *antyaphala*, has different meanings depending upon the context. In this verse, it denotes the epicycle radius r_s, when the radius of the deferent circle is the *trijyā* R. Since the ratio $\frac{C_s}{80}$ is same as the ratio $\frac{r_s}{R}$, we have

$$r_s = C_s \times \frac{R}{80}. \tag{7.18}$$

In the sixth chapter (verse 7), we have seen that the radius R corresponding to the circumference of 21600 is

$$R \approx \frac{10000000000 \times 21600}{2 \times 31415926536}. \tag{7.19}$$

Using (7.19) in (7.18), we have

$$r_s \approx C_s \times \frac{2700000000000}{62831853072}. \tag{7.20}$$

The successive rational approximations of the $\frac{2700000000000}{62831853072}$, using a continued fraction expansion, are listed in Table 7.2.

The numbers 3051 and 71 in the fifth row of this table are referred to in the verse as *kāmānala* and *kaṃsa* respectively. Therefore, we have the simpler approximation

$$r_s \approx C_s \times \frac{3051}{71}, \tag{7.21}$$

which is what is stated in the verse.

multiplier	divisor
42	1
43	1
1504	35
1547	36
3051	**71**
163250	3799
166301	3870
1161056	27019
7132637	165984
29691604	690955
333740281	7766489
363431885	8457444
3604627246	83883485
7572686377	176224414
18750000000	436332313
2700000000000	62831853072

Table 7.2 Successive approximations for $\frac{270 \times R}{C} = \frac{270}{2\pi}$.

७.८ अन्त्यफलान्तर्भावितशीघ्रफलानयनम्

7.8 Obtaining the expression for the *śīghraphala* in terms of the *antyaphala*

कोटिज्यान्त्यफलेन हीनसहिता या कर्किनक्रादितः
तद्दोर्ज्याकृतियोगमूलमुदितः कर्णोऽमुना संहरेत् ।
दोर्ज्यां संस्कृतकोटिकामपि तथा व्यासार्धसंवर्धितां
तच्चापं भुजकोटिचापरहितं ज्याकर्किनक्रादिका ॥ ११ ॥

कोटिज्याल्पा चरमफलतः कर्कटादौ यदि स्यात्
त्रिज्याभ्यस्तादिह भुजगुणात् कर्णभक्तस्य चापम् ।
दोश्चापाढ्यं भगणदलतस्त्यज्यतां संस्कृतायाः
कोट्यास्तद्वच्[5] छ्रुतिहृतधनुः क्षिप्यतां कोटिचापे ॥ १२ ॥

[5] The term तद्वत् (*tadvat*) means 'similarly'. Which mathematical operation is to be carried out similarly has not be stated explicitly. From the context we understand it to be "multiplied by the radius". Thus, संस्कृतायाः कोट्याः तद्वत् ... = संस्कृतायाः कोट्याः तद्वत् [त्रिज्याभ्यस्तात्] ... (*saṃskṛtāyāḥ koṭyāḥ tadvat ... = saṃskṛtāyāḥ koṭyāḥ tadvat [trijyābhyastāt] ...*).

koṭijyāntyaphalena hīnasahitā yā karkinakrāditaḥ
taddorjyākṛtiyogamūlamuditaḥ karṇo'munā saṃharet |
dorjyāṃ saṃskṛtakoṭikāmapi tathā vyāsārdhasaṃvardhitāṃ
taccāpaṃ bhujakoṭicāparahitaṃ jyākarkinakrādikā || 11 ||

koṭijyālpā caramaphalataḥ karkaṭādau yadi syāt
trijyābhyastādiha bhujaguṇāt karṇabhaktasya cāpam |
doścāpāḍhyaṃ bhagaṇadalatastyajyatāṃ saṃskṛtāyāḥ
koṭyāstadvacchrutihṛtadhanuḥ kṣipyatāṃ koṭicāpe || 12 ||

The *antyaphala* is subtracted from or added to the Rcosine [of the *śīghrakendra*] depending on whether [the *kendra*] is *karkyādi* or *makarādi* respectively. The square root of the sum of the squares of that and the Rsine [of the *kendra*] would be the *karṇa*. The Rsine and the [*antyaphala*] corrected Rcosine have to be multiplied by the radius and divided by this [*karṇa*]. When the arc of these results is subtracted from the arc of the *kendrabhujā* and *kendrakoṭi* respectively, the results obtained are *karkyādi* and *makarādijyās* [or the *śīghraphalas*].

In the case [where *śīghrakendra* is] *karkyādi*, if the Rcosine [of the *kendra*] happens to be smaller than the *antyaphala*, then the *karṇa* should be obtained by subtracting the *koṭijyā* from the *antyaphala*. Multiply that by the *kendrabhujā* and by the radius and divide by the [*karṇa*]. The arc of this is to be added to the arc of the *kendrabhujā*, and the result obtained is subtracted from six signs *bhagaṇadala* [to obtain the *śīghraphala*]. [Or,] the *saṃskṛtakoṭijyā* having been multiplied by the radius and divided by the *karṇa*, the arc of the result can be added to the arc (*dhanus*) of the *kendrakoṭijyā* [to obtain the *śīghraphala*].

In these two verses Putumana Somayājī presents four different expressions for computing the *śīghraphala* of a planet. Since the expressions are a bit involved and lengthy, he has aptly chosen *Śārdūlavikrīḍita* metre[6] while composing the verses. The expressions for *śīghraphala* ($R\Delta\theta_s$) given are:

$$R\Delta\theta_s = c\bar{a}pa\ (bhuj\bar{a}) - c\bar{a}pa\left(\frac{dorjy\bar{a} \times vy\bar{a}s\bar{a}rdha}{kar\d{n}a}\right), \tag{7.22}$$

$$R\Delta\theta_s = c\bar{a}pa\ (ko\d{t}i) - c\bar{a}pa\left(\frac{sa\d{m}sk\d{r}tako\d{t}i \times vy\bar{a}s\bar{a}rdha}{kar\d{n}a}\right), \tag{7.23}$$

$$R\Delta\theta_s = bhaga\d{n}adala - c\bar{a}pa\left(\frac{bhujagu\d{n}a \times trijy\bar{a}}{kar\d{n}a}\right) - c\bar{a}pa\ (do\d{h}), \tag{7.24}$$

$$R\Delta\theta_s = c\bar{a}pa\ (ko\d{t}i) + c\bar{a}pa\left(\frac{sa\d{m}sk\d{r}tako\d{t}i \times trijy\bar{a}}{\acute{s}ruti}\right). \tag{7.25}$$

Of the four formulae, the first two are presented in verse 11, and the latter two in verse 12. While presenting the rationale behind these formulae, we too maintain this division, by considering two separate cases based on the value of *koṭijyā* and *antyaphala*.

Case I : When the *koṭijyā* is greater than *antyaphala* ($|R\cos\theta_{sk}| > r_s$)
Case II : When the *koṭijyā* is less than *antyaphala* ($|R\cos\theta_{sk}| < r_s$)

[6] This is one of the long metres having 19 syllables per quarter.

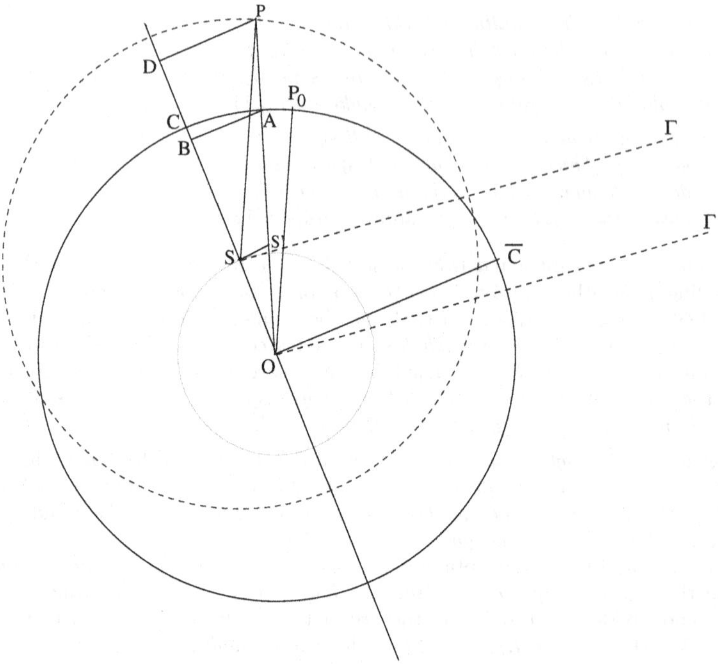

Fig. 7.2 Obtaining the *śīghraphala* when the *śīghrakendra* is *makarādi* with $|R\cos\theta_{sk}| > r_s$.

We first derive the expression for the *śīghrakarṇa* which is given in the first half of verse 11.

७.८.१ शीघ्रकर्णानयनसूत्रम्

7.8.1 Expression for the śīghrakarṇa

In Figure 7.2, O is the center for both the epicycle and the deferent circle. The dashed circle which is centered at *śīghrocca* S is the eccentric circle. P_0 and P are the *manda-sphuṭa* and the *śīghra*-corrected (true) planets respectively. The longitude of the *manda-sphuṭa* and the *śīghrocca* are

$$\begin{aligned} \theta_{ms} &= \Gamma\hat{O}P_0 = \Gamma\hat{S}P, \\ \text{and} \quad \theta_s &= \Gamma\hat{O}S = \Gamma\hat{S}D. \end{aligned} \tag{7.26}$$

The *śīghrakendra* is obtained by subtracting the *śīghrocca* from the *manda-sphuṭa*. That is,

$$\theta_{sk} = D\hat{O}P_0 = D\hat{S}P.$$

Since in Figure 7.2 the value of θ_{sk} is less than 90°, it is *meṣādi* as well as *makarādi*. The *antyaphala* or the radius of the *śīghravṛtta* (r_s) is OS. Further,

$$OC = OP_0 = SP = R,$$

and $PD = R\sin\theta_{sk}$ is the *kendrabhujājyā* while $SD = R\cos\theta_{sk}$ is the *kendrakoṭijyā*.

Let the line OP intersect the deferent at A. The arc $\overset{\frown}{AP_0}$ is $(R\Delta\theta_s)$, which is the arc of the *śīghraphala*. This can be obtained by subtracting $\overset{\frown}{AC}$ from the arc $\overset{\frown}{P_0C}$ of the *śīghrakendra*. That is,

$$R\Delta\theta_s = \overset{\frown}{AP_0} = \overset{\frown}{P_0C} - \overset{\frown}{AC} = R\theta_{sk} - \overset{\frown}{AC}\,. \tag{7.27}$$

It may be noted that $\overset{\frown}{AC}$ is the arc of the half-chord AB. Since the triangles OPD and OAB are similar, we have

$$AB = \frac{OA \times PD}{OP} = \frac{R \times R\sin\theta_{sk}}{K_s}, \tag{7.28}$$

$$\text{and} \qquad OB = \frac{OA \times OD}{OP} = \frac{R \times (R\cos\theta_{sk} + r_s)}{K_s}, \tag{7.29}$$

where $K_s = OP$ is the *śīghrakarṇa* and is given by

$$\begin{aligned} K_s &= \sqrt{OD^2 + PD^2} \\ &= \sqrt{(SD + OS)^2 + PD^2} \\ &= \sqrt{(R\cos\theta_{sk} + r_s)^2 + R^2\sin^2\theta_{sk}}. \end{aligned} \tag{7.30}$$

This is precisely the expression for *karṇa* that is given in the first half of the verse 11.

The expression for *śīghraphala* in each of the two cases mentioned above, depends on whether the value of the *śīghrakendra* is *makarādi* or *karkyādi*.

७.८.२ प्रथमकल्पे केन्द्रस्य मकारादौ शीघ्रफलम्

7.8.2 Expression for the *śīghraphala* in case I when the *śīghrakendra* is *makarādi*

In Figure 7.2, the position of the *manda-sphuṭa* P_0, is depicted in such a way that the *śīghrakendra* is *makarādi* ($270° < \theta_{sk} < 90°$). In order to obtain the *śīghraphala* given by (7.27), we need to know $\overset{\frown}{AC}$. It may be noted $\overset{\frown}{AC} = R\sin^{-1}(AB)$. Hence the expression for *śīghraphala* (7.27) becomes,

$$R\Delta\theta_s = R\theta_{sk} - R\sin^{-1}(AB), \tag{7.31}$$

$$\text{or,} \qquad R\Delta\theta_s = R\theta_{sk} - R\sin^{-1}\left(\frac{R \times R\sin\theta_{sk}}{K_s}\right). \tag{7.32}$$

In (7.17), the first term in RHS, may be expressed as $R\theta_{sk} = R\sin^{-1}(R\sin\theta_{sk})$. Hence it may be written as

$$R\Delta\theta_s = R\sin^{-1}(R\sin\theta_{sk}) - R\sin^{-1}\left(\frac{R \times R\sin\theta_{sk}}{K_s}\right). \tag{7.33}$$

It is clear that the expression for $R\Delta\theta_s$ in (7.33) is the same as (7.22). We now consider the case of *śīghrakendra* being *karkyādi.*

७.८.३ प्रथमकल्पे केन्द्रस्य कर्क्यादौ शीघ्रफलम्

7.8.3 Expression for *śīghraphala* in case I when the *śīghrakendra* is *karkyādi*

In Figure 7.3, the position of the *manda-sphuṭa* P_0' on the deferent is such that the *śīghrakendra* is *karkyādi* ($90° < \theta_{sk} < 270°$). The true planet which is along the direction of the *śīghrocca* from P_0' is shown at P' on the eccentric circle. Then, the *śīghrakarṇa*, $K_s = OP'$ is given by

$$K_s = \sqrt{(|R\cos\theta_{sk}| - r_s)^2 + R^2\sin^2\theta_{sk}}. \tag{7.34}$$

Let $O\bar{C}$ be the radius of the deferent perpendicular to OS.

Now the *śīghraphala*, $R\Delta\theta_s$ can be written as

$$\begin{aligned} R\Delta\theta_s &= \overset{\frown}{A'P_0'} = \overset{\frown}{P_0'\bar{C}} - \overset{\frown}{A'\bar{C}} \\ &= R\sin^{-1}(OF') - R\sin^{-1}(OB'). \end{aligned} \tag{7.35}$$

Since the triangles $OD'P'$ and $OB'A'$ an similar, we have

$$\begin{aligned} OB' &= OD'.\frac{OA'}{OP'} \\ &= (|R\cos\theta_{sk}| - r_s).\frac{R}{K_s}. \end{aligned} \tag{7.36}$$

Also it is clear from Figure 7.3, that $OF' = |R\cos\theta_{sk}|$. Using this, and (7.36) in (7.35) we have

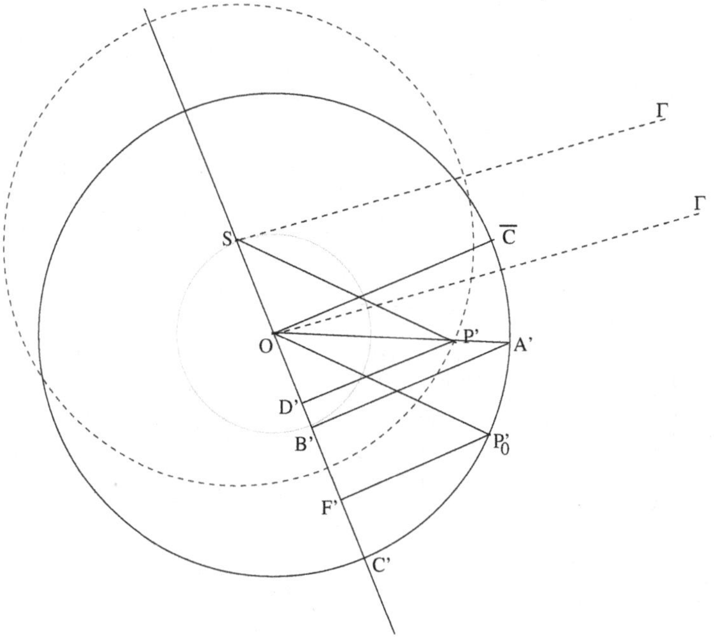

Fig. 7.3 Obtaining the *śīghraphala* when the *śīghrakendra* is *karkyādi* with $|R\cos\theta_{sk}| > r_s$.

$$R\Delta\theta_s = R\sin^{-1}(|R\cos\theta_{sk}|) - R\sin^{-1}\left(\frac{R}{K_s} \times (|R\cos\theta_{sk}| - r_s)\right), \quad (7.37)$$

It is straightforward to see that (7.37) is the same as (7.23). We now consider case II ($|R\cos\theta_{sk}| < r_s$) as discussed in verse 12. Here again there are two possible cases depending on whether the *śīghrakendra* θ_{sk} is *makarādi* or *karkyādi*. When θ_{sk} is *makarādi*, even when $|R\cos\theta_{sk}| < r_s$, the *śīghraphala* is the same as given in equation (7.32). Hence, both the formulae given in verse 12 correspond to θ_{sk} being *karkyādi*.

७.८.४ द्वितीयकल्पे केन्द्रस्य कर्क्यादौ शीघ्रफलम्

7.8.4 *Expression for the śīghraphala in case II when the śīghrakendra is karkyādi*

In Figure 7.4, the *manda-sphuṭa* P_0' is such that the *śīghrakendra* SOP_0' is *karkyādi*. Now the *śīghraphala* is given by

$$R\Delta\theta_s = \overset{\frown}{A'P_0'} = \overset{\frown}{A'C'} - \overset{\frown}{P_0'C'}\ . \quad (7.38)$$

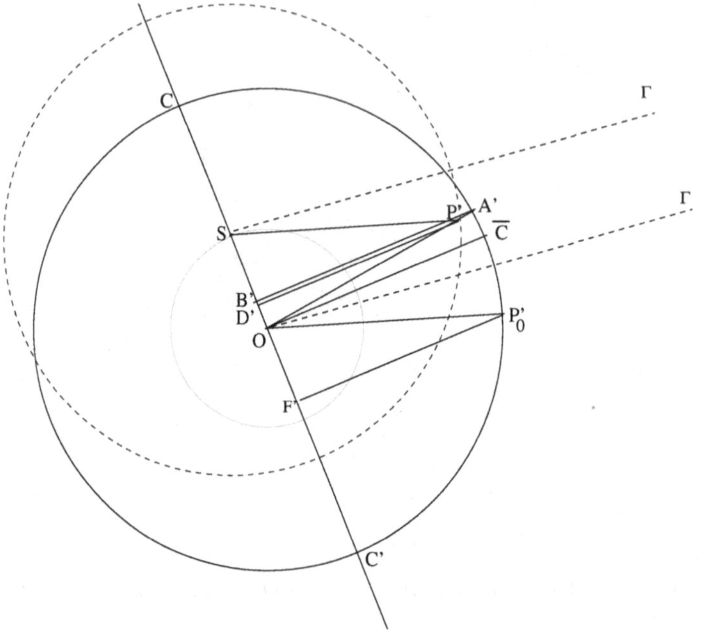

Fig. 7.4 Obtaining the *śīghraphala* when the *śīghrakendra* is *karkyādi* with $|R\cos\theta_{sk}| < r_s$.

It is evident from the figure that

$$\widehat{A'C'} = R.180° - \widehat{A'C},$$
$$\text{and} \qquad \widehat{A'C} = R\sin^{-1}(A'B'). \tag{7.39}$$

As the triangles $OA'B'$ and $OP'D'$ are similar, we have

$$A'B' = \frac{P'D' \times OA'}{OP'} = \frac{R\sin\theta_{sk} \times R}{K_s}, \tag{7.40}$$

since $P'D' = P'_0F' = R\sin\theta_{sk}$. Using (7.40) in (7.39) we have

$$\widehat{A'C'} = R.180° - R\sin^{-1}\left(\frac{R\sin\theta_{sk} \times R}{K_s}\right). \tag{7.41}$$

Now,

$$\overset{\frown}{P'_0C'} = R\sin^{-1}(P'_0F') = R\sin^{-1}(R\sin\theta_{sk}). \tag{7.42}$$

Using (7.42) and (7.41) in (7.38) we have,

$$\begin{aligned} R\Delta\theta_s &= R.180^\circ - R\sin^{-1}(R\sin\theta_{sk}) - R\sin^{-1}\left(R\sin\theta_{sk}.\frac{R}{K_s}\right) \\ &= \frac{C}{2} - \left(R\theta_{sk} + R\sin^{-1}\left(R\sin\theta_{sk}.\frac{R}{K_s}\right)\right), \end{aligned} \tag{7.43}$$

which is the same as (7.24) given in the verse, once we identify *cāpa* (*doḥ*) = $R\theta_{sk}$, and *bhagaṇadala* = $\frac{C}{2}$.

Alternatively, the expression for *śīghraphala* may be written as

$$R\Delta\theta_s = \overset{\frown}{A'P'_0} = \overset{\frown}{A'\bar{C}} + \overset{\frown}{\bar{C}P'_0}\,. \tag{7.44}$$

Now,

$$\begin{aligned} \overset{\frown}{A'\bar{C}} = R\sin^{-1}(OB') &= R\sin^{-1}\left(OD'.\frac{OA'}{OP'}\right) \\ &= R\sin^{-1}\left[(OS - SD').\frac{R}{K_s}\right] \\ &= R\sin^{-1}\left[(r_s - |R\cos\theta_{sk}|)\frac{R}{K_s}\right], \end{aligned} \tag{7.45}$$

as $SD' = SP'\cos(P'\hat{S}D') = R\cos(180^\circ - \theta_{sk}) = |R\cos\theta_{sk}|$. Also

$$\overset{\frown}{\bar{C}P'_0} = R\sin^{-1}(OF') = R\sin^{-1}(|R\cos\theta_{sk}|). \tag{7.46}$$

Using (7.45) and (7.46) in (7.44) we have,

$$R\Delta\theta_s = R\sin^{-1}\left[(r_s - |R\cos\theta_{sk}|).\frac{R}{K_s}\right] + R\sin^{-1}|R\cos\theta_{sk}|. \tag{7.47}$$

This is the same as (7.25) given in the verse once we identify *cāpa* (*koṭi*) with the second term in the *RHS* of the above equation (7.47).

७.९ शीघ्रफलानयने प्रकारान्तरम्

7.9 Another method to obtain the *śīghraphala*

अथवान्त्यफलाभ्यस्ताद् दोर्गुणात् कर्णसंहृतम् ।
चापीकृतं भवेदत्र कर्किनक्रादिमौर्विकाः ॥ १३ ॥

athavāntyaphalābhyastād dorguṇāt karṇasaṃhṛtam |
cāpīkṛtaṃ bhavedatra karkinakrādimaurvikāḥ || 13 ||

Otherwise, the arc of the result obtained by multiplying the Rsine of the *śīghrakendra* by the *antyaphala* and dividing by the *karṇa* would give the *karkimakarādi-jyās* [or the *śīghraphalas*].

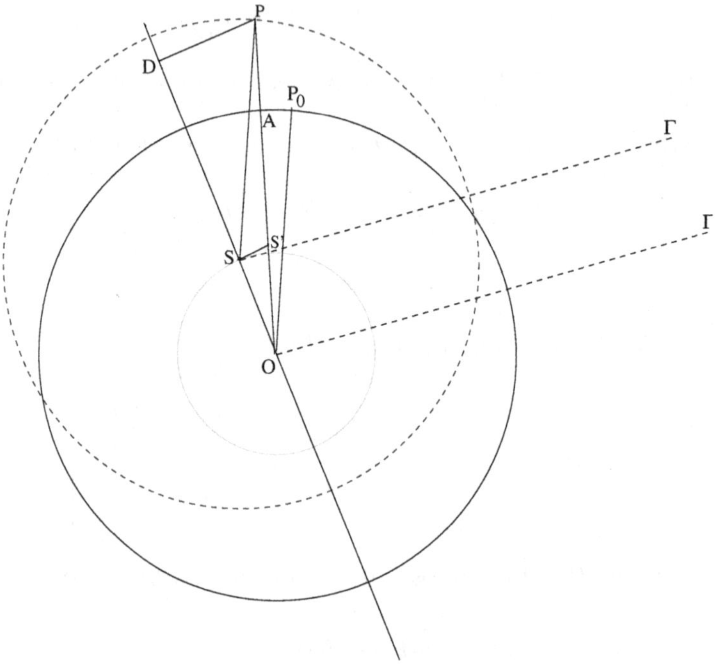

Fig. 7.5 Alternate method for obtaining the *śīghraphala*.

In Figure 7.5 the line SS' perpendicular to OP is the Rsine corresponding to the angle $S\hat{P}O$ as $SP = R$. Since the angles $S\hat{P}O$ and $A\hat{O}P_0$ are equal, the Rsine corresponding to the arc $\overparen{AP_0}$ is equal to SS'. That is, $R\sin(\Delta\theta_s) = SS'$. Now

$$PD = SP\sin(D\hat{S}P) = R\sin\theta_{sk}.$$

The triangles $SS'O$ and PDO are similar. Hence,

$$\begin{aligned} R\sin(\Delta\theta_s) = SS' &= PD.\frac{OS}{OP} \\ &= \frac{R\sin\theta_{sk}\times r_s}{K_s}. \end{aligned} \tag{7.48}$$

Therefore,

$$\Delta\theta_s = R\sin^{-1}\left(\frac{R\sin\theta_{sk} \times r_s}{K_s}\right). \tag{7.49}$$

This is the form in which the *śīghra* correction is given in most of the texts.

७.१० शीघ्रकर्णानयने प्रकारान्तरम्

7.10 An alternate method to obtain the *śīghrakarṇa*

अन्त्यं फलं कोटिगुणे द्विनिघ्ने कृत्वामुनैवान्त्यफलेन हत्वा।
संस्कृत्य तद् व्यासदलस्य वर्गे मूलीकृतो वा भवतीह कर्णः ॥ १४ ॥

antyaṃ phalaṃ koṭiguṇe dvinighne kṛtvāmunaivāntyaphalena hatvā |
saṃskṛtya tad vyāsadalasya varge mūlīkṛto vā bhavatīha karṇaḥ || 14 ||

Or else, having applied[7] the *antyaphala* to twice the Rcosine [of the *śīghrakendra*], and further having multiplied the result thus obtained by the *antyaphala* itself, and then having applied that to the square of the radius and taking the square root [of the resulting quantity], the *karṇa* is obtained.

Here, the author gives a different expression for *karṇa*. To derive this expression, consider the triangle ODP in Figure 7.5, which corresponds *makarādi-śīghrakendra*. $D\hat{S}P$ is the *śīghrakendra*, and $SP = R$, $SD = R\cos\theta_{sk}$ and $PD = R\sin\theta_{sk}$. Also, $OS = r_s$. Hence, the *karṇa* K_s is given by

$$\begin{aligned} K_s &= \sqrt{OD^2 + PD^2} \\ &= \sqrt{(OS + SD)^2 + PD^2} \\ &= \sqrt{(r_s + R\cos\theta_{sk})^2 + R^2\sin^2\theta_{sk}} \\ &= \sqrt{R^2 + r_s(r_s + 2|R\cos\theta_{sk}|)}. \end{aligned} \tag{7.50}$$

In case the *śīghrakendra* is *karkyādi*, we have from Figures 7.3 and 7.4,

$$\begin{aligned} OD' &= \pm(OS - SD') \\ &= \pm(r_s - |R\cos\theta_{sk}|), \end{aligned}$$

Hence,

[7] The term *kṛtvā* simply refers to application which can be positive or negative. Here, the *antyaphala* is to be subtracted from or added to depending on whether the *kendra* is *karkyādi* or *makarādi*.

$$\begin{aligned} K_s &= \sqrt{(OD')^2 + (P'D')^2} \\ &= \sqrt{(r_s - |R\cos\theta_{sk}|)^2 + R^2 \sin\theta_{sk}} \\ &= \sqrt{R^2 + r_s(r_s - 2|R\cos\theta_{sk}|)}, \end{aligned} \tag{7.51}$$

which is the expression for K_s as described in the verse.

७.११ शीघ्रकेन्द्रज्यातः शीघ्रवृत्तपरिधिगणना

7.11 Obtaining the circumference of the *śīghra* epicycle from the Rsine of the *śīghra*-anomaly

मृगकर्कटकादिदोःफलात् तद्भुजचापोनयुतात् भुजागुणो यः।
अमुना विभजेद् भुजाफलज्यां अजनिघ्नीं फलमत्र शीघ्रवृत्तम्॥ १५ ॥

mṛgakarkaṭakādidoḥphalāt
tadbhujacāponayutāt bhujāguṇo yaḥ |
amunā vibhajed bhujāphalajyāṃ
ajanighnīṃ phalamatra śīghravṛttam || 15 ||

The [arc] of the *mṛgakarkaṭādi-śīghraphala* has to be subtracted from or added to the *śīghrakendra* when the *kendra* is *makarādi* or *karkyādi* respectively, and Rsine of this is to be found. The product of 80 (*aja*) and the Rsine of the *bhujāphala* divided by the result [obtained above] would be the *śīghra-vṛtta* [*paridhi*].

The expression for the *śīghra-vṛttaparidhi* (C_s) given in the above verse may be written as

$$C_s = \frac{80 \times \textit{bhujāphalajyā}}{R\sin(\textit{śīghrakendra} \mp \text{arc of } \textit{doḥphala})}. \tag{7.52}$$

Here, in the denominator, '$-$' is to be used if *kendra* is *makarādi* and '$+$' if *karkyādi*.

The rationale behind this expression can be understood from Figure 7.6. Considering the triangle SOS',

$$SS' = OS.\sin(S\hat{O}P).$$

If we consider the triangle SPS',

$$SS' = SP.\sin(S\hat{P}S') = SP.\sin(S\hat{P}O).$$

Equating the two expressions for SS', we have

$$\textit{antyaphala}, \;\; OS = \frac{R\sin(S\hat{P}O) \times SP}{R\sin(S\hat{O}P)}. \tag{7.53}$$

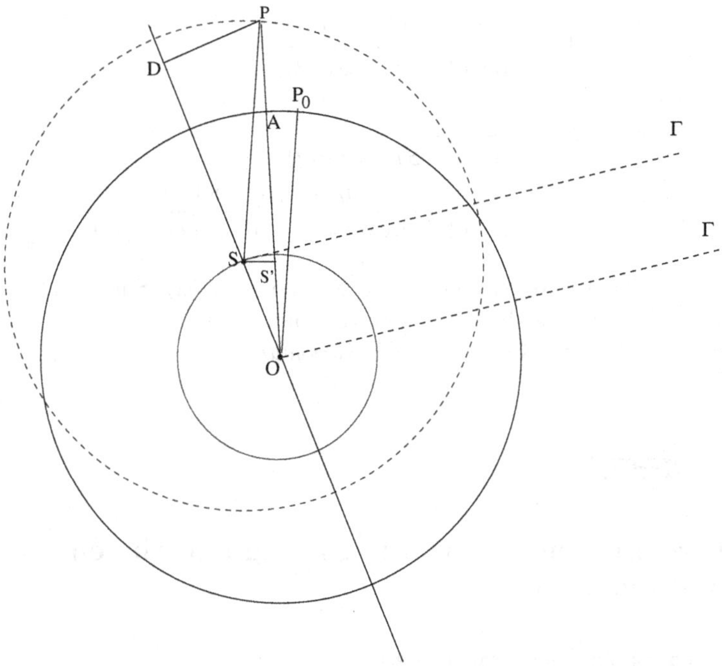

Fig. 7.6 Obtaining the circumference of the *śīghra* epicycle.

In the above expression $SP = R$, and $S\hat{P}O = A\hat{O}P_0$, the angle corresponding to the arc $\overset{\frown}{AP_0}$ which is nothing but the arc of the correction term (*śīghraphala* or *doḥphala*) used in *śīghrasaṃskāra*. That is,

$$R\sin(S\hat{P}O) = R\sin(AP_0) = \textit{bhujāphala}$$

$$\text{and} \qquad R\sin(S\hat{O}P) = R\sin(D\hat{S}P - S\hat{P}O).^{8}$$

Substituting the above relations in (7.53), we have

[8] In the triangle OSP, $S\hat{O}P + O\hat{S}P + S\hat{P}O = 180$ and $O\hat{S}P + D\hat{S}P$ is also 180. Hence, we have

$$R\sin(S\hat{O}P) = R\sin(D\hat{S}P - S\hat{P}O).$$

$$OS = \frac{R\sin(\overset{\frown}{AP_0}) \times R}{R\sin(D\hat{S}P - S\hat{P}O)},$$

$$\text{or} \qquad C_s = \frac{R\sin(\overset{\frown}{AP_0}) \times 80}{R\sin(D\hat{S}P - S\hat{P}O)}$$

$$= \frac{bhuj\bar{a}phalajy\bar{a} \times 80}{R\sin(\acute{s}\bar{\imath}ghrakendra - \text{arc of } do\d{h}phala)}, \tag{7.54}$$

as $\frac{C_s}{80} = \frac{OS}{R}$. The above expression is the same as (7.52). Similarly, it can be shown that there should be a '+' sign in front of the arc of *doḥphala* in the denominator, when the *śīghrakendra* is *karkyādi.*

७.१२ पदाद्यन्तपरिधी

7.12 Circumference at the beginning and the end of the odd quadrant

अन्त्यभुजाफलनीतं वृत्तं स्यादन्त्यवृत्तमोजपदे ।
एकभदोःफलनीतं द्विगुणितमन्त्योनितं भवेदाद्यम् ॥ १६ ॥

antyabhujāphalanītaṃ vṛttaṃ syādantyavṛttamojapade |
ekabhadoḥphalanītaṃ dviguṇitamantyonitaṃ bhavedādyam || 16 ||

The dimension of the epicycle (*vṛtta*) obtained by the *bhujāphala* at the end of the odd quadrant would be the *śīghravṛtta* at the end [of the odd quadrant or the beginning of the even quadrant]. Subtracting [this *vṛtta* at] the end [of the odd quadrant], from the product of 2 and the *vṛtta* obtained by the Rsine (*doḥphala*) at the end of first sign (*ekabha*) [from the beginning of odd quadrant], would result in the [*vṛttaparidhi* at the] beginning [of the odd quadrant].

If C_{se} and C_{so1} are the *śīghravṛtta-paridhis* at the end of the odd quadrant (beginning of the even quadrant) and at the end of the first *rāśi* from the beginning of the odd quadrant, then the *paridhi* at the beginning is given by

$$C_{so} = 2 \times C_{so1} - C_{se}. \tag{7.55}$$

We can understand the rationale behind the above expression as follows. To be specific, we consider the case $C_{se} > C_{so}$. At the end of the first *rāśi*, $\theta_{sk} = 30°$, and at the end of the first quadrant, $\theta_{sk} = 90°$. Then from (7.5), we have

$$C_{so1} = C_{so} + \frac{\Delta_s}{2},$$
$$C_{se} = C_{so} + \Delta_s, \tag{7.56}$$

where $\Delta_s = C_{se} - C_{so}$. Then it follows that

$$C_{so} = 2C_{so1} - C_{se}.$$

७.१३ रवेः व्यस्तकर्णानयनम्

7.13 Obtaining the *vyastakarṇa* of the Sun

राश्यन्तभानुस्फुटतो मृदूच्चं विशोध्य दोःकोटिगुणौ गृहीत्वा ।
त्रिसङ्गुणौ तावथ नन्दभक्तौ क्रमेण दोःकोटिफले भवेताम् ॥ १७ ॥

कोटीफलं कर्कमृगादिजातं त्रिमौर्विकायां स्वमृणं च कृत्वा ।
तद्वर्गतो दोःफलवर्गयुक्तात् मूलं विपर्यासकृतोऽत्र कर्णः ॥ १८ ॥

rāśyantabhānusphuṭato mṛdūccaṃ viśodhya doḥkoṭiguṇau gṛhītvā |
trisaṅguṇau tāvatha nandabhaktau krameṇa doḥkoṭiphale bhavetām || 17 ||

koṭīphalaṃ karkamṛgādijātaṃ trimaurvikāyāṃ svamṛṇaṃ ca kṛtvā |
tadvargato doḥphalavargayuktāt mūlaṃ viparyāsakṛto'tra karṇaḥ || 18 ||

Having subtracted the [longitude of the] *mandocca* from the true longitude of the Sun at the end of the *rāśi*, and having obtained the Rsine and Rcosine of that [result], and multiplying it by 3 and dividing by 80, the *doḥphala* and the *koṭiphala* are obtained successively.

The *koṭiphala* has to be added to or subtracted from the radius when [the *kendra* is] *karkyādi* or *makarādi* respectively. The square root of the sum of the squares of the result thus obtained and of the *doḥphala* would be the *viparītakarṇa* here.

The term *vyastakarṇa* or *viparītakarṇa* literally means 'inverse hypotenuse', and is nothing but the radius of the *kakṣyāvṛtta* when the measure of *mandakarṇa* is taken to be equal to the *trijyā*, R.[9]

The term *mṛdūcca* appearing in the first line of the verse is a synonym for *mandocca*. The *sphuṭa* that is referred to here should be understood as *manda-sphuṭa* or the *manda*-corrected longitude. Here onwards, we use the symbol θ for the *sphuṭa* or the true longitude of the Sun, instead of θ_{ms}, and r_m for the mean radius of the *manda* epicycle for convenience. If θ_m represents the longitude of the *mandocca* (of the Sun), then the *sphuṭa-doḥphala* and *sphuṭa-koṭiphala* are given by

$$\begin{aligned} doḥphala &= \frac{r_m}{R} R\sin(\theta - \theta_m), \\ koṭiphala &= \frac{r_m}{R} R\cos(\theta - \theta_m). \end{aligned} \tag{7.57}$$

[9] For a general discussion of the *viparītakarṇa* introduced by Mādhava in order to compute the itrated-*manda*-hypotenuse (*asakṛt-manda-karṇa*), see {TS 2011}, pp. 492–497.

The rationale behind the formula for *viparītakarṇa* which is mentioned in these verses, can be understood with the help of Figure 7.7a and 7.7b. In these figures P_0 and P represent the mean and the true planet respectively. N is the foot of the perpendicular drawn from the mean planet P_0 to the line joining the centre of the circle and the true planet. Let the radius of the *karṇavṛtta* OP be set equal to the *trijyā* R. Then the radius of the *uccanīcavṛtta* P_0P is r_m, as it is in the measure of the *karṇavṛtta*. In this measure, the radius of the *kakṣyāvṛtta* $OP_0 = R_v$ which is one of the sides of the triangle OP_0P, where other sides are $OP = R$ and $PP_0 = r_m$. In the triangle NPP_0, NP_0 is equal to the *doḥphala* and NP is equal to *koṭiphala* since the angle $P_0\hat{P}O = P\hat{O}U = \theta - \theta_m$. That is,

$$NP_0 = r_m \sin(\theta - \theta_m),$$
$$\text{and} \qquad NP = r_m \cos(\theta - \theta_m). \tag{7.58}$$

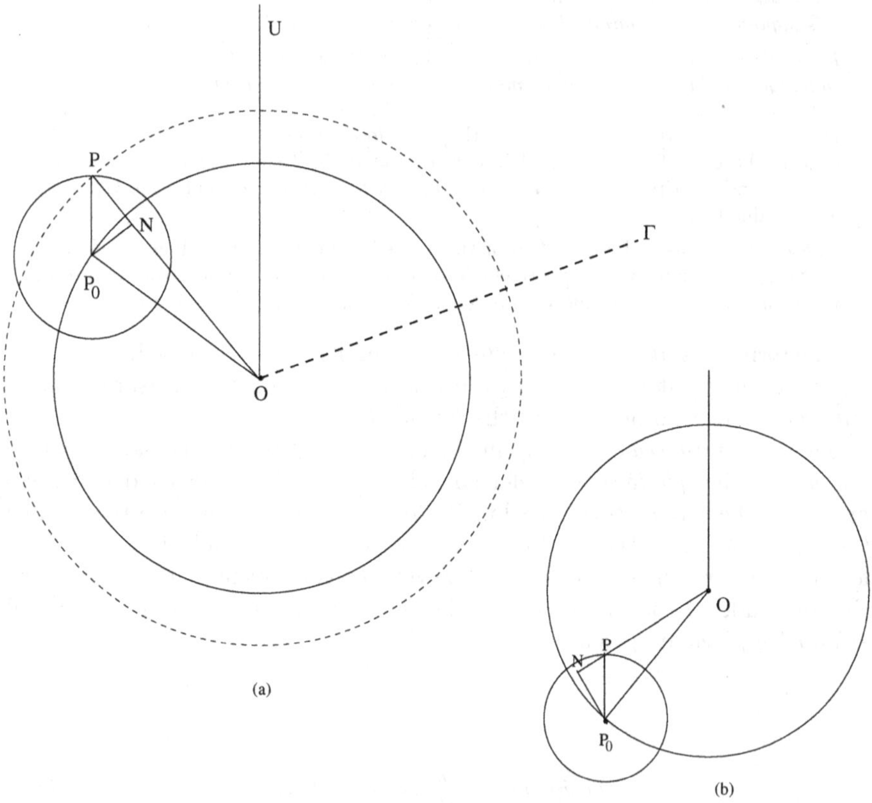

Fig. 7.7 Determination of the *viparītakarṇa* when the *kendra* is in (a) the first quadrant and (b) the third quadrant.

The *viparītakarṇa* (R_v) which is the hypotenuse (OP_0) of the triangle ONP_0 is given by

$$\begin{aligned} R_v &= \sqrt{ON^2 + NP_0^2} \\ &= \sqrt{(OP - NP)^2 + NP_0^2} \\ &= \sqrt{\left(R - \frac{r_m}{R} \times |R\cos(\theta - \theta_m)|\right)^2 + \left(\frac{r_m}{R} \times R\sin(\theta - \theta_m)\right)^2}. \end{aligned} \tag{7.59}$$

The above expression for the *vyastakarṇa* is applicable when the *kendra* is *makarādi* (in the first or fourth quadrant). If the *kendra* is *karkyādi* (in the second or third quadrant) as shown in Figure 7.7b, then

$$\begin{aligned} R_v &= \sqrt{ON^2 + NP_0^2} \\ &= \sqrt{(OP^2 + NP^2) + NP_0^2} \\ &= \sqrt{\left(R + \frac{r_m}{R} \times R\,|\cos(\theta - \theta_m)|\right)^2 + \left(\frac{r_m}{R} \times R\sin(\theta - \theta_m)\right)^2}. \end{aligned} \tag{7.60}$$

Equations (7.59) and (7.60) are the general expressions for the *viparītakarṇa*. The verse actually talks of the *viparītakarṇa* of the Sun at the end of various *rāśis*. For the Sun, $\frac{r_m}{R} = \frac{3}{80}$. The *viparītakarṇa* of the Sun at the end of various *rāśis* will be used in the next verse to calculate the mean longitudes of the Sun at the *saṅkramaṇas*.

७.१४ व्यस्तकर्णतः सङ्क्रमणकालिकमध्यार्कानयनम्

7.14 Finding the mean longitudes of the Sun at the zodiacal transits from the *vyastakarṇa*

त्रिज्याहताद् दोःफलतोऽमुनाप्तं चापीकृतं मेषतुलादितस्तत् ।
राश्यन्त्यभानौ स्वमृणं च कुर्यात् तदा भवेत् सङ्क्रमणार्कमध्यम् ॥ १९ ॥

trijyāhatād doḥphalato'munāptaṃ
cāpīkṛtaṃ meṣatulāditastat |
rāśyantyabhānau svamṛṇaṃ ca kuryāt
tadā bhavet saṅkramaṇārkamadhyam || 19 ||

The arc of [the quantity obtained] by multiplying the *doḥphala* by radius and dividing by this [*vyastakarṇa*], has to be added to or subtracted from the true longitude of the Sun when [the *kendra* is] *meṣādi* or *tulādi* respectively. The result would be the mean longitude of the Sun at the transit.

Having given the expression for the *viparītakarṇa* R_v in terms of the true anomaly $(\theta - \theta_m)$ in the previous verse, the text presents an application of it for obtaining the mean longitudes of the Sun from its true value at the transits (*saṅkramaṇas*).

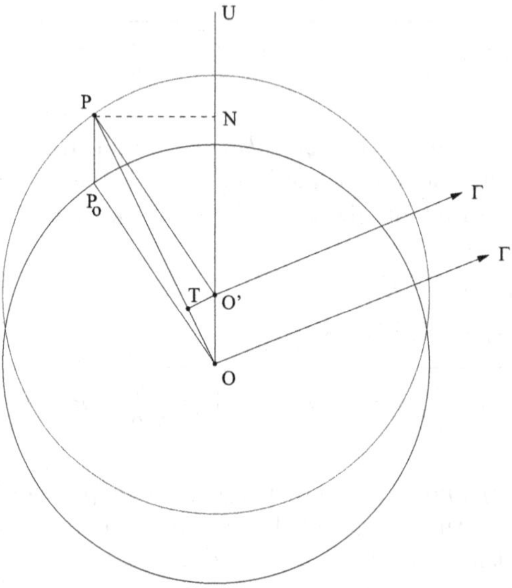

Fig. 7.8 Obtaining the *madhyama* (the mean position) from the *sphuṭa* (the true position).

We explain this with the help of Figure 7.8. Here O is the observer and P_0 is the mean planet (mean Sun). The point P represents the true Sun. The distance $P_0P = OO'$ represents the actual radius of the variable epicyle that may be denoted as r. The angle $P_0\hat{O}P = O\hat{P}O' = (\theta_0 - \theta)$. Considering the triangle $OO'P$, we draw a perpendicular from O' that intersects OP at T. Now, in the triangle $O'PT$,

$$\begin{aligned} O'T &= O'P \sin(O'\hat{P}T) \\ &= O'P \sin(P\hat{O}P_0) \\ &= R \sin(\theta_0 - \theta). \end{aligned} \tag{7.61}$$

Also

$$O'T = r \sin(\theta - \theta_m), \tag{7.62}$$

as $O'\hat{O}T = \theta - \theta_m$. Equating the two, we have

$$R \sin(\theta_0 - \theta) = r \sin(\theta - \theta_m). \tag{7.63}$$

Here, the term $R\sin(\theta_0 - \theta)$ can be positive or negative depending upon the quadrant in which $\theta - \theta_m$ is situated, so the *makarādi* and *karkyādi* cases are both considered here, and not discussed separately.

As noted in Appendix B, the *manda*-epicycle r, and the *mandakarṇa* K, vary in such a way that they always satisfy the equation $\frac{r}{K} = \frac{r_m}{R}$. Also, in the measure of the *karṇavṛtta*, the radius of the *pratimaṇḍala* is R_v, so that $\frac{K}{R} = \frac{R}{R_v}$.

Hence,

$$r = r_m.\frac{K}{R} = r_m.\frac{R}{R_v}. \tag{7.64}$$

Thus (7.63) reduces to

$$R\sin(\theta_0 - \theta) = r_m \sin(\theta - \theta_m)\frac{R}{R_v}. \tag{7.65}$$

$$\begin{aligned} \text{Or,} \qquad \theta_0 - \theta &= R\sin^{-1}\left[r_m \sin(\theta - \theta_m)\frac{R}{R_v}\right] \\ &= R\sin^{-1}\left[\frac{3}{80} \times R\sin(\theta - \theta_m)\frac{R}{R_v}\right], \end{aligned} \tag{7.66}$$

where we have used the tabulated value of the *manda*-epicycle of the Sun given by $\frac{r_m}{R} = \frac{3}{80}$.

In (7.66), since θ is known, the mean planet θ_0 can be obtained by adding the above difference $\theta_0 - \theta$ to it. The difference is positive when the *sphuṭakendra* $\theta - \theta_m$ is within the six signs beginning with *Meṣa*, viz., $0° < \theta - \theta_m \leq 180°$, and negative when the *kendra* is within the six signs beginning with *Tulā*, viz., $180° < \theta - \theta_m \leq 360°$.

For the purpose of illustration we shall work out an example. For this we assume that the longitude of the *mandocca* $\theta_m = 78°$ for the Sun.

Example: *Makara-saṅkramaṇa* (Transition to Capricorn, $\theta = 270°$)

$$\begin{aligned} R_v &= \sqrt{\left(R - \frac{3}{80}R\cos(270-78)\right)^2 + \left(\frac{3}{80}R\sin(270-78)\right)^2} \\ &= 3563.95', \\ \text{and} \quad \theta_0 - \theta &= \sin^{-1}\left[\frac{3}{80}\sin(270-78)\frac{R}{R_v}\right] \\ &= -0.431°. \end{aligned}$$

Therefore, $\theta_0 = 270° - 0.431° = 269.569° = 8^r 29° 34'$.

At the transit (*saṅkramaṇa*) into different *rāśis* the true longitudes of the Sun would be multiples of 30. That is,

$$\theta_i = 30 \times i,$$

where $i = 0, 1, \ldots, 11$ for *Meṣa*, *Vṛṣabha*, ..., and *Mīna* respectively. Substituting these 12 values for true longitudes (θ) in (7.66), we can obtain the mean longitudes of the Sun at each transit.

The mean longitudes at the transits known as '*saṅkramaṇārkamadhya*' are given as *vākyas* in the Commentary II of the *Karaṇapaddhati*. These are listed in Appendix Table D.3, and compared with the values computed as above.

Another important application of *vyastakarṇa* lies in finding the value of *mandakarṇa* which is described in the following verse.

७.१५ व्यस्तकर्णतः रविचन्द्रयोर्मन्दकर्णगणना

7.15 Finding the *mandakarṇas* of the Sun and the Moon from *vyastakarṇa*

सूर्येन्द्वोरेवमेवेष्टस्फुटात् तन्मध्यमानयेत् ।
तत्रोक्तव्यस्तकर्णात्तस्त्रिज्यावर्गो मृदुश्रुतिः ॥ २० ॥

sūryendvorevameveṣṭasphuṭāt tanmadhyamānayet |
tatroktavyastakarṇāttastrijyāvargo mṛduśrutiḥ || 20 ||

In the same manner, the mean longitudes of the Sun and the Moon have to be obtained from their true longitudes. Dividing the square of the radius by the corresponding *vyastakarṇa* gives *mandakarṇa*.

The expression for the *mandakarṇa*, or *manda* hypotenuse represented by K in the previous section, as described in the verse is given by

$$K = \frac{R^2}{R_v}.$$

This follows from the definition of the *vyastakarṇa* (R_v) itself. The radius of the *kakṣyāvṛtta* or the deferent is R_v, when the *karṇa* is taken to be R. Hence, when the radius of the deferent is the *trijyā* or R, the *karṇa* K can be found from the rule of proportions. That is,

$$\frac{K}{R} = \frac{R}{R_v} \qquad \text{or} \qquad K = \frac{R^2}{R_v}.$$

७.१६ व्यस्तकर्णतः रविचन्द्रयोः स्फुटयोजनकर्णज्ञानम्

7.16 Finding the true physical distances of the Sun and the Moon from *vyastakarṇa*

मृदुश्रुतिहतात् कक्ष्यावृत्ताच्चक्रकलाहृतम् ।
स्फुटयोजनकर्णाख्यं कक्ष्याव्यासदलं रवेः ॥ २१ ॥

mṛduśrutihatāt kakṣyāvṛttāccakrakalāhṛtam |
sphuṭayojanakarṇākhyaṃ kakṣyāvyāsadalaṃ raveḥ || 21 ||

The [instantaneous] radius of the orbit (*kakṣyāvyāsārdha*) which is [also] called the true hypotenuse in *yojanas* (*sphuṭayojanakarṇa*) can be obtained by multiplying *kakṣyā* by the *mandakarṇa* and dividing by 21600 (*cakrakalā*).

When the average distance of the Sun from the centre of the *bhagola* (essentially the centre of earth) is taken to be 21600 minutes, the distance of the Sun from this centre at any point of its orbit is given by K, in minutes. Hence, when the average distance of the Sun in *yojanas* is given by the *kakṣyāvyāsārdha* O_{sm} (see Section 1.20), the actual physical distance of the Sun from the earth known as the *sphuṭayojanakarṇa* is given by

$$O_{st} = \frac{K \times O_{sm}}{21600}.$$

७.१७ माससङ्क्रान्तिनक्षत्रवाक्यानयनम्

7.17 Obtaining the *māsavākyas, saṅkrāntivākyas* and *nakṣatravākyas*

Let d_i denote the time period that has elapsed from the beginning of the year (when the true longitude of the Sun is zero) to the end of the particular solar month (corresponding to the Sun transiting the ith *rāśi*). Obviously, d_i need not be an integer. A *māsavākya* is the integer closest to d_i. The fractional part, in terms of *nāḍikās* can be found from the *saṅkrāntivākyas*, which give the remainders when d_i are divided by 7. Similarly, *nakṣatra-saṅkrāntivākyas* give the time when the Sun crosses a *nakṣatra* division of the zodiac.

भागीकृतात् तदनु सङ्क्रमणार्कमध्यात्
अब्दान्तदोःफलयुताद्धरणीदिनघ्नात् ।
सौरैर्दिनैरपहृतं खलु मासवाक्यं
सङ्क्रान्तिवाक्यमिह तत्सुहृतावशिष्टम् ॥ २२ ॥

नक्षत्रान्तस्फुटोत्पन्नमध्यार्कादेवमेव च ।
नयेन्नक्षत्रसङ्क्रान्तिवाक्यं कविषुपूर्वकम् ॥ २३ ॥

bhāgīkṛtāt tadanu saṅkramaṇārkamadhyāt
abdāntadoḥphalayutāddharaṇīdinaghnāt |
saurairdinairapahṛtaṃ khalu māsavākyaṃ
saṅkrāntivākyamiha tatsuhṛtāvaśiṣṭam || 22 ||

nakṣatrāntasphuṭotpannamadhyārkādevameva ca |
nayennakṣatrasaṅkrāntivākyaṃ kaviṣupūrvakam || 23 ||

Having obtained the mean longitude of the Sun in degrees at [the time of] transit (*saṅkrānti*) and adding the *doḥphala* (difference between the mean and the true Sun) at the end of the year (*abdānta*) to it, multiply the result by the number of civil days (*bhūdina*) and divide by the number of solar days [in a *mahāyuga*]. [The result obtained gives] the *māsavākya*. The remainders obtained by dividing those (*māsavākyas*) by 7 (*su*) give the *saṅkrāntivākyas*.

In a similar manner, the *nakṣtravākyas* that commence with *kaviṣu* can be obtained by finding the mean longitudes of the Sun from its true longitudes at the end of the *nakṣatras*.

७.१७.१ मासवाक्यानि

7.17.1 The māsavākyas

Verse 22 gives the procedure for obtaining the *māsavākyas* and *saṅkrāntivākyas*. The true longitudes of the Sun at the end of each month are 30°, 60°..., 360°. At the end of the 12th month, which is the same as the beginning of the first month in the next year, the true longitude of the Sun is 360°. The mean longitude corresponding to the true longitude of 360° is found to be $357.883° = -2.117° \approx -2°7' = 11^r27°53'$. The difference between the true and the mean longitudes at the end of the year is termed the '*abdāntadoḥphala*' whose value is $2°7'$.

The *madhyamabhoga* (difference in the mean longitudes) reckoned from the *meṣa-saṅkramaṇa* to *iṣṭasaṅkramaṇa* (desired zodiacal transition) is the difference in the mean longitude at the desired zodiacal transit and the transit at *meṣādi* of the true Sun. It is found by adding $2°7'$ to the mean longitude at each transit. For example, the true longitude of the Sun at the *Makarasaṅkramaṇa* is 270°. The mean longitude corresponding to this is $269°34'$. Adding $2°7'$ to it, we obtain $271°41'$ as the *madhyamabhoga* from the *meṣasaṅkrama* to the *makarasaṅkrama*.

A mean solar day is the time interval corresponding to an increase of 1° in the mean longitude. This is slightly longer than a civil day, and is given in terms of the latter by $\frac{D_c}{D_s}$, where D_c and D_s represent the numbers of civil days and solar days in a *mahāyuga*. Note that the values given in the *Karaṇapaddhati* for D_c and D_s are 1577917500 and $360 \times 4320000 = 1555200000$ respectively. Let θ_{i0} represent the *madhyamabhoga* for the transit to the $(i+1)^{th}$

rāśi. Then

$$d_i = \theta_{i0} \times \frac{D_c}{D_s}.$$

For *makarasaṅkramaṇa*, $\theta_{90} = 271°41'$ and therefore

$$d_9 = \frac{271°41' \times 1577917500}{1555200000} = 275\ d\ 39\ n. \quad (7.67)$$

The *māsavākya* is the integer closest to d_i. Hence, 276 is the *māsavākya* at the *makarasaṅkrama*. The *māsavākyas* corresponding to all the transits as given in the Commentary I[10] are listed in Table 7.3 along with the computed d_i's.

Name of the *rāśi* transited (*saṅkramaṇa*)	*māsavākya* textual value of d_i in *kaṭapayādi*	in numerals	computed value of d_i (in days & *nāḍikās*)
Vṛṣabha	*kulīna*	31	30 *d* 56 *n*
Mithuna	*rūkṣajña*	62	62 *d* 20 *n*
Karkaṭaka	*vidhāna*	94	93 *d* 56 *n*
Siṃha	*mātrayā*	125	125 *d* 24 *n*
Kanyā	*kṣaṇasya*	156	156 *d* 26 *n*
Tulā	*siṃhasya*	187	186 *d* 54 *n*
Vṛścika	*suputra*	217	216 *d* 48 *n*
Dhanus	*catvarāṭ*	246	246 *d* 18 *n*
Makara	*tathādri*	276	275 *d* 39 *n*
Kumbha	*mīnāṅgi*	305	305 *d* 7 *n*
Mīna	*mṛgāṅgi*	335	334 *d* 55 *n*
Meṣa	*mātulaḥ*	365	365 *d* 15 *n*

Table 7.3 The *māsavākyās* given in the Commentary I and the computed values of d_i.

By finding the difference between the successive *māsavākyas*, the number of civil days corresponding to each month can be calculated.

७.१७.२ सङ्क्रान्तिवाक्यानि

7.17.2 The saṅkrāntivākyas

The instant at which the transit of true Sun from one *rāśi* to another occurs can be determined from the *saṅkrānti-vākyas*. By dividing d_i by 7, the

[10] {KP 1956}, p. 225.

remainders obtained are the *saṅkrānti-vākyas*. For instance, in the previous example

$$\frac{d_9}{7} = \frac{275\ d\ 39\ n}{7} = 39 + \frac{2\ d\ 39\ n}{7}.$$

The remainder is 2 *d* 39 *n*. Here the obtained day of the week corresponds to number 2 and the *nāḍikā* is 39. The *vākya* for this is *nṛvarāṭ*, which represents the day as 2 and *nāḍikā* as 40.

The *saṅkrāntivākyas* which are given in the Commentary II[11] for different transits are listed in Table 7.4, along with the computed values.

Name of the *rāśi*	*saṅkrāntivākya*				
	in *kaṭapayādi*	in numerals		computed value	
		day	*nāḍikā*	day	*nāḍikā*
Vṛṣabha	*timire*	2	56	2	55.5
Mithuna	*niratam*	6	20	6	19.5
Karkaṭaka	*camare*	2	56	2	56.0
Siṃha	*marutaḥ*	6	25	6	24.1
Kanyā	*surarāṭ*	2	27	2	26.1
Tulā	*ghṛṇibhaḥ*	4	54	4	53.5
Vṛścika	*javato*	6	48	6	47.7
Dhanus	*dhaṭakaḥ*	1	19	1	18.2
Makara	*nṛvarāṭ*	2	40	2	39.3
Kumbha	*sanibhaḥ*	4	7	4	6.7
Mīna	*maṇimān*	5	55	5	55.2
Meṣa	*cayakā*	1	16	1	15.5

Table 7.4 The *saṅkrāntivākyas* in the Commentary II and the computed values.

It is clear that the value of d_i corresponding to a *saṅkramaṇa* is obtained by adding a suitable multiple of 7, to the *saṅkrāntivākya*. For example, we have to add 91 to the day component of the *saṅkrāntivākya* for *karkaṭaka* (2+91) to obtain d_3 whose value is 93 days 56 *nāḍikās*.

७.१७.३ नक्षत्रवाक्यानि

7.17.3 The nakṣatravākyas

We know that the ecliptic (*rāśicakra*, 360°) is divided into 27 equal parts called *nakṣatras*, each part corresponding to 13°20′. The basis of this division is the Moon's sidereal period which is close to 27 days. The term *nakṣatra* also refers

[11] {KP 1956}, p. 226.

to the time spent by the Moon in any of these divisions. In the same vein, the time durations spent by the Sun to traverse through these divisions are called *mahānakṣatras*. The true longitudes of the Sun at the end of the 27 *nakṣatras* are 13°20′, 26°40′, 40°, 53°20′, ..., 360°. Converting these longitudes to the corresponding mean ones and adding 2°7′ to them, we obtain the increase in the mean longitude of the Sun at the end of each *nakṣatra* starting from *Aśvinī*. The number of civil days corresponding to these can be calculated by multiplying them by the civil days in a *mahāyuga* and dividing by the solar days in a *mahāyuga*. These values are presented in Table 7.5.

Name of the *nakṣatra*	No. of civil days elapsed before the *nakṣatra-saṅkramaṇa*
Bharaṇī	13.674
Kṛttikā	27.461
Rohiṇī	41.349
Mṛgaśirā	55.318
Ārdrā	69.343
Punarvasu	83.395
Puṣya	97.442
Āśleṣā	111.454
Maghā	125.401
Pūrvaphālgunī	139.260
Uttarāphālgunī	153.015
Hasta	166.654
Citrā	180.175
Svāti	193.581
Viśākhā	206.881
Anurādhā	220.090
Jyeṣṭhā	233.224
Mūla	246.304
Pūrvāṣāḍhā	259.352
Uttarāṣāḍhā	272.393
Śravaṇa	285.449
Dhaniṣṭhā	298.543
Śatabhiṣaj	311.697
Pūrvabhādrapadā	324.931
Uttarabhādrapadā	338.262
Revatī	351.702
Aśvinī	365.258

Table 7.5 Number of civil days elapsed at each *nakṣatra-saṅkramṇa*.

The instant at which the *nakṣatra-saṅkramaṇa* occurs can be obtained from the *nakṣatra-saṅkrāntivākyas*. When we divide the civil days at each transit by 7, the remainders obtained are the *nakṣatra-saṅkrāntivākyas*, similar to the *rāśi-saṅkrāntivākyas* discussed earlier. The *nakṣatra-saṅkrāntivākyas* as given in both the Commentaries[12] are tabulated along with the computed values in Table 7.6.

Nakṣatra transit (*saṅkramaṇa*)	*nakṣatra-saṅkrāntivākya*				
	in *kaṭapayādi*	in numerals		computed *saṅkrāntivākya*	
		day	*nāḍikā*	day	*nāḍikā*
Bharaṇī	*kaviṣu*	6	41	6	40.4
Kṛttikā	*hāriṣu*	6	28	6	27.7
Rohiṇī	*dīyata*	6	18	6	20.9
Mṛgaśirā	*dhīyate*	6	19	6	19.1
Ārdrā	*kariṣu*	6	21	6	20.6
Punarvasu	*māriṣu*	6	25	6	23.7
Puṣya	*sāriṣu*	6	27	6	26.5
Āśleṣā	*dūrataḥ*	6	28	6	27.2
Maghā	*smarati*	6	25	6	24.0
Pūrvaphālgunī	*duṣyati*	6	18	6	15.6
Uttaraphālgunī	*yoniṣu*	6	1	6	0.9
Hasta	*parvaṇā*	5	41	5	39.2
Citrā	*trikaśa*	5	12	5	10.5
Svāti	*tāṇḍava*	4	36	4	34.9
Viśākhā	*bhomṛga*	3	54	3	52.9
Anurādhā	*dhenugaḥ*	3	9	3	5.4
Jyeṣṭhā	*supura*	2	17	2	13.4
Mūla	*hāṭaka*	1	18	1	18.2
Pūrvāṣāḍhā	*nīrana*	0	20	0	21.1
Uttarāṣāḍhā	*bhāratā*	6	24	6	23.6
Śravaṇa	*caraṇa*	5	26	5	26.9
Dhaniṣṭhā	*gālava*	4	33	4	32.6
Śatabhiṣaj	*viśvagu*	3	44	3	41.8
Pūrvabhādrapadā	*carmarāṭ*	2	56	2	55.9
Uttarabhādrapadā	*cikura*	2	16	2	15.7
Revatī	*rāvaya*	1	42	1	42.1
Aśvinī	*markaṭa*	1	15	1	15.5

Table 7.6 The *nakṣatra-saṅkrāntivākyas* given in both the Commentaries and the computed values.

[12] {KP 1956}, p. 228.

७.१८ इष्टकालिकस्फुटार्कानयनार्थं योग्यादिवाक्यानि

7.18 The *yogyādivākyas*: True longitude of the Sun at any instant

मासादितोऽष्टाष्टदिनोत्थसूर्यस्फुटान्तरांशाष्टदिनान्तराणि ।
योग्यादिवाक्यानि धनर्णतैषां दिनाल्पताधिक्यवशादिनाप्तौ ॥ २४ ॥

māsādito'ṣṭāṣṭadinotthasūryasphuṭāntarāṃśāṣṭadināntarāṇi |
yogyādivākyāni dhanarṇataiṣāṃ dinālpatādhikyavaśādināptau || 24 ||

[First] the difference in the true longitudes of the Sun in degrees etc. at intervals of eight days from the beginning of the month [are found]. The difference between [these values] and eight degrees are [given by] the *yogyādivākyas*. These are [applied] positively or negatively, depending upon whether 8 degrees is lesser or greater [than the difference in longitudes at the 8 days interval], to obtain the [true] Sun [at any given instant].

The definition of *yogyādivākyas* and the method of applying them to obtain the true longitude of the Sun at intervals of 8 days in a solar month, are given in this verse.

Unlike the *rāśi* and *nakṣatra-saṅkramaṇavākyas* discussed earlier, whose nomenclature was based upon a certain time interval or phenomenon, the name *yogyādi-vākyas* stems from the fact that here we have a set of 48 *vākyas* beginning with the phrase *yogya*. These *vākyas* enable us to find the longitude of the Sun at any given instant. There are 4 *vākyas* corresponding to each solar month. Each month is divided into four parts with a maximum of 8 days per part. Now, the *sphuṭabhoga* of each part is the difference between the true longitudes of the Sun at the beginning and at the end of that part. The difference in minutes between the *sphuṭabhoga* of each part and 8° are the *yogyādivākyas*. If the longitudinal difference is greater (lesser) than 8°, then it will be notified as positive (negative).

७.१८.१ योग्यादिवाक्यानयनम्

7.18.1 Obtaining the yogyādivākyas

The *yogyādivākyas* as given in the edited version of the Commentary I[13] are listed in Table 7.7. Apart from the *vākyas* (here in the form of one word, which form part of meaningful sentences), the signs ('+' or '−') are also indicated in the commentary. Except in the case of *Tulā*, all the 4 *vākyas* corresponding to a particular *rāśi* have the same sign and this is indicated as such in the table. For *Tulā*, the sign for the first *vākya* is − and the signs for the other three are

[13] {KP 1956}, p. 229.

all +, as indicated in the table. The rationale behind these *yogyādivākyas* is best explained by taking up a couple of concrete examples.

Consider the solar month of *Makara*. The true longitude of the Sun at the beginning of the month is $\theta = 270°$. The corresponding mean longitude θ_0 can be determined using the method explained earlier and we find $\theta_0 = 269°34'9''$. Using the fact that the rate of motion of the mean longitude of the Sun is $59.136'$ per day, the mean longitude is $\theta_0 = 277°27'14''$ after 8 days in the month of *Makara*. The *mandaphala* $(\theta - \theta_0)$ corresponding to this value of θ_0 is found to be $42'56''$. Adding this to θ_0, we find the true longitude after 8 days to be $277°27'14'' + 42'56'' = 278°10'10''$. Hence the increase in the true longitude after the first 8 days of the month is $8°10'10''$. As the longitudinal increase is greater than $8°$, the difference is positive and is given by $+(8°10'10'' - 8°) = +10'10''$, compared with the value of $+11'$ as given by the *vākya* '*pūjya*' in the commentary.

After 16 days in the month of *Makara*, the mean longitude $\theta_0 = 269°34'9'' + 59'8'' \times 16 = 285°20'17''$. The true longitude corresponding to this is found to be $\theta = 286°19'29''$. Hence the difference between the true longitudes at the beginning and at the end of the second part is $286°19'29'' - 278°10'10'' = 8°9'19''$. Here again as the longitudinal increase is greater than $8°$, the difference is positive and is given by $+(8°9'19'' - 8°) = +9'19''$, which compares well with the value of $+9'$ as implied by the *vākya* '*dhenuḥ*' in the commentary.

७.१८.२ योग्यादिवाक्यात् स्फुटसूर्यानयनम्

7.18.2 Finding the true longitude of the Sun from the yogyādivākyas

One can obtain the true longitude of the Sun on any day using the *yogyādivākyas*, and linear interpolation. For example, suppose we would like to find the true longitude of the Sun after the lapse of 18 days in the *Makara* month. This comes in the third part (*khaṇḍa*). Therefore the approximate value of the true longitude of the Sun after 18 days elapsed would be

$$\theta' = 270° + 18° = 288°.$$

A correction which can be called *yogyādisaṃskāra* $\varDelta\theta'$ has to be applied to θ' in order to obtain the true longitude θ.

Now, the correction for 8 days of the third *khaṇḍa* is given as $8'$ (*dine*). Hence the correction for 2 days is $\frac{8\times 2}{8}$ minutes. Adding this to the sum of the first two *vākyas* (*pūjya* and *dhenuḥ*),

$$\varDelta\theta' = 11 + 9 + \frac{8 \times 2}{8} = 22'.$$

Month		*yogyādivākyas* (in minutes)							
Meṣa	−	*yogyo*	11	*vaidyaḥ*	14	*tapaḥ*	16	*satyam*	17
			(11′12″)		(13′3″)		(15′42″)		(17′42″)
Vṛṣabha	−	*dhanyaḥ*	19	*putraḥ*	21	*kharo*	22	*varaḥ*	24
			(19′18″)		(20′54″)		(22′18″)		(23′18″)
Mithuna	−	*vīraḥ*	24	*śūraḥ*	25	*śaro*	25	*vajrī*	24
			(24′6″)		(24′30″)		(24′36″)		(24′24″)
Karkaṭaka	−	*bhadram*	24	*gotro*	23	*ruruḥ*	22	*karī*	21
			(23′54″)		(23′6″)		(21′54″)		(20′30″)
Siṃha	−	*dhanyaḥ*	19	*sevyo*	17	*mayā*	15	*loke*	13
			(18′54″)		(17′0″)		(14′54″)		(12′42″)
Kanyā	−	*kāyo*	11	*dīnaḥ*	8	*stanām*	6	*ganā*	3
			(10′36″)		(8′12″)		(5′48″)		(3′18″)
Tulā		*yājño*	− 1	*yajñām*	+ 1	*ganā*	+ 3	*śūnā*	+ 5
			(−1′30″)		(+0′48″)		(3′0″)		(4′54″)
Vṛścika	+	*steno*	6	*dīno*	8	*dhunī*	9	*naṭaḥ*	10
			(6′12″)		(7′42″)		(8′54″)		(9′54″)
Dhanus	+	*āpaḥ*	10	*pāpaḥ*	11	*payaḥ*	11	*pathyam*	11
			(10′18″)		(10′42″)		(10′48″)		(10′30″)
Makara	+	*pūjyā*	11	*dhenuḥ*	9	*dine*	8	*rthinaḥ*	7
			(10.2)		(9.4)		(8.2)		(6.8)
Kumbha	+	*tanuḥ*	6	*bhinnā*	4	*khanī*	2	*jñānī*	0
			(5′42″)		(3′54″)		(1′54″)		(−0′18″)
Mīna	−	*ratnaṃ*	2	*bhānuḥ*	4	*sunīḥ*	7	*nayaḥ*	10
			(2′0″)		(4′24″)		(6′48″)		(9′18″)

Table 7.7 The 48 *yogyādivākyas* mentioned in the Commentary I along with the computed values in parentheses.

These corrections are indicated as positive in the listing of the *vākyas* in the commentary. Hence applying this result positively to θ' the true longitude of the Sun at the end of the 18th day of the solar month *Makara* is found to be

$$\theta = 288^\circ + 22' = 288^\circ 22'.$$

७.१९ मन्दस्फुटानयनप्रकारः

7.19 Procedure for obtaining the *mandasphuṭa* of the planets

भौमादेः कृतमन्ददोःफलदलात्तत्केन्द्रतो दोःफलं
नीतं केवलमध्यमे धनमृणं मन्दस्फुटस्याप्तये ।
विद्भृग्वोर्निजमध्यमे मृदुफलं स्वोच्चोनमध्योद्भवं
नन्दघ्नं स्फुटशीघ्रवृत्तविहृतं कुर्यात् स मन्दस्फुटः ॥ २५ ॥

bhaumādeḥ kṛtamandadoḥphaladalāttatkendrato doḥphalaṃ
nītaṃ kevalamadhyame dhanamṛṇaṃ mandasphuṭasyāptaye |
vidbhṛgvornijamadhyame mṛduphalaṃ svocconamadhyodbhavaṃ
nandaghnaṃ sphuṭaśīghravṛttavihṛtaṃ kuryāt sa mandasphuṭaḥ || 25 ||

In the case of Mars, etc. (Mars, Jupiter and Saturn)[14] having first applied half of the *mandaphala*, from the resulting value the *kendra* is obtained and from that, the *mandaphala* is [once again] obtained. The *mandaphala* is to be applied positively or negatively [depending on whether the *kendra* is *tulādi* or *meṣādi* respectively] to the original mean planet (*kevalamadhyama*) to obtain the *manda-sphuṭa*. In the case of Mercury and Venus, the *mandaphala* (*mṛduphala*)—obtained by subtracting the apogee from the mean longitude—is multiplied by 80 (*nanda*) and divided by the *sphuṭaśīghravṛtta* (C_s). The *mandasphuṭa* is obtained by applying [the result thus obtained] to its own mean (*nijamadhyama*).

The *mandaphala* obtained earlier can be written as

$$\Delta\theta = \sin^{-1}\left(\frac{C_m}{80}\sin(\theta_0 - \theta_m)\right). \tag{7.68}$$

The procedure for obtaining the *manda-sphuṭas* of the interior and the exterior planets are separately prescribed in the above verse, and they will be explained in the next two subsections.

७.१९.१ कुजादीनां मन्दस्फुटीकरणम्

7.19.1 Manda-sphuṭa of exterior planets

The procedure for obtaining *manda-sphuṭa* of Mars, Jupiter and the Saturn involves the following two steps:

[14] Here, the word "*bhaumādeḥ*" (of the Mars etc.), is to be understood in a restricted sense to refer to only three planets, namely Mars, Jupiter and Saturn, since the word "*vidbhṛgvoḥ*" used in the third quarter of the verse refers to the other two planets, Mercury and Venus .

Step 1: Having obtained the *mandaphala* $\Delta\theta$, half of it has to be applied to the mean planet (*madhyamagraha*) θ_0. Now we get the half *manda*-corrected mean planet θ_1.

$$\theta_1 = \theta_0 - \frac{\Delta\theta}{2} \qquad (\text{for } 0 < (\theta_0 - \theta_m) < 180), \tag{7.69}$$

$$= \theta_0 + \frac{\Delta\theta}{2} \qquad (\text{for } 180 < (\theta_0 - \theta_m) < 360). \tag{7.70}$$

Step 2: Then once again we have to compute *mandaphala*. In doing so, we have to find the *kendra* from the half-*manda*-corrected mean planet (θ_1). That is, the value of the *mandaphala* calculated for the second time is given by

$$\Delta\theta_2 = \sin^{-1}\left(\frac{C_m}{80}\sin(\theta_1 - \theta_m)\right). \tag{7.71}$$

It is said that this value $\Delta\theta_2$ is to be applied to *kevalamadhyama*[15] θ_0. Thus the *manda-sphuṭagraha* θ_{ms} is given by

$$\theta_{ms} = \theta_0 - \Delta\theta_2 \qquad (\text{for } 0 < (\theta_1 - \theta_m) < 180), \tag{7.72}$$

$$= \theta_0 + \Delta\theta_2 \qquad (\text{for } 180 < (\theta_1 - \theta_m) < 360). \tag{7.73}$$

७.१९.२ बुधशुक्रयोः मन्दस्फुटीकरणम्

7.19.2 *Manda-sphuṭa* of interior planets

In the case of interior planets (Mercury and Venus) the *manda-sphuṭa* is obtained by a single step process.

Multiply the *mandaphala* obtained from their respective *kendras* by 80 divided by *sphuṭaśīghravṛttaparidhi* (C_s). Thus the *doḥphala* is computed as

$$\frac{80}{C_s} \times \frac{C_m}{80}\sin(\theta_0 - \theta_m) = \frac{C_m}{C_s}\sin(\theta_0 - \theta_m).$$

The arc of the *doḥphala* is the *mandaphala*

$$\Delta\theta_1 = \sin^{-1}\left(\frac{C_m}{C_s}\sin(\theta_0 - \theta_m)\right). \tag{7.74}$$

Applying this to the mean planet, we obtain the *manda-sphuṭa*

[15] Usually the *mandaphala* is applied to the mean from which the *kendra* is computed. Here, though the *kendra* is computed from θ_1, the author specifically tells that the second *mandaphala* to be applied only to the original mean planet *kevala-madhyama* (θ_0).

$$\theta_{ms} = \theta_0 \pm \Delta\theta_1. \tag{7.75}$$

It is important to note that according to the Commentary II, the phrase "*vidbhṛgvor-nijamadhyame*" actually implies that, in the case of the interior planets, the *manda* correction is to be applied to the corresponding *śīghrocca*.[16] This is indeed a major departure from the traditional planetary theory where the mean Sun was taken to be the mean planet for Mercury and Venus.

It was Nīlakaṇṭha Somayājī who modified the traditional planetary model by suggesting that what was traditionally known as the *śīghrocca* in the case of the interior planets, should actually be taken as the corresponding mean planet to which the *manda* correction is to be applied. By interpreting the phrase *nijamadhyama* as explained above, the Commentary II is suggesting that the *Karaṇapaddhati*, by prescribing that the *manda* correction for the mean planets should be applied to their *śīghroccas* and not to the mean Sun, is also following the revised planetary model proposed by Nīlakaṇṭha in his *Tantrasaṅgraha* (c. 1500).[17]

In this context, it should be noted that the Commentary I does not attach any such special meaning to the phrase *nijamadhyama* for the interior planets. Further, barring a few indications here and there, there is no explicit statement in *Karaṇapaddhati* that it is following the revised model proposed by Nīlakaṇṭha. Therefore, the question as to whether Putumana Somayājī was indeed aware of and followed the modified planetary model of Nīlakaṇṭha is still an open question which could perhaps be settled by a study of the other works of Putumana Somayājī most of which are still unpublished.

There is also another peculiarity in the prescription for the *mandaphala* for the interior planets as given in the above verse. It seems to be suggesting that the stated value of the *mandaparidhi* is in the measure of the *śīghra-paridhi*. In other words, $\frac{C_m}{C_s}$ is being used as the ratio of the *manda*-epicycle and the deferent associated with the *mandasaṃskāra*.

---

[16] Here, the Commentary II states ({KP 1956}, p. 232):

"*budhaśukranmāruḍe tanṭe tanṭe nijamadhyamaṃ veccu ... mandasphuṭaññaḷ varuṃ. iviḍe 'nijamadhyama' mennu paraññadu śīghroccamākunnadu.*"

In terms of the respective mean longitudes (referred by the name *nijamadhyama*) of the Mercury and Venus ... *manda-sphuṭas* are obtained. Here, what is stated as the 'respective mean planet' is indeed the [corresponding] *śīghrocca*."

[17] For details of the revised planetary model of Nīlakaṇṭha, see {TS 2011}, pp. 508-522.

७.२० मन्दशीघ्रस्फुटाभ्यां मन्दशीघ्रकर्णज्ञानम्

7.20 Obtaining the *manda* and *śīghrakarṇas* from the *manda-sphuṭa* and *śīghra-sphuṭa*

स्वोपान्त्यान्त्यस्फुटाभ्यां निजनिजचलतुङ्गोनिताभ्यां भुजाज्ये
नीत्वा त्रिज्याहताभ्यां विभजतु परया शीघ्रकर्णस्तदा स्यात् ।
मन्दोच्चेनोनिताभ्यां श्रवणमपि नयेन्मध्यमन्दस्फुटाभ्यां
मान्दा यत्र स्फुटाप्तिः पठितभुजफलैस्तत्र कर्णाप्तिरेवम् ॥ २६ ॥

svopāntyāntyasphuṭābhyāṃ nijanijacalatuṅgonitābhyāṃ bhujājye
nītvā trijyāhatābhyāṃ vibhajatu parayā śīghrakarṇastadā syāt |
mandoccenonitābhyāṃ śravaṇamapi nayenmadhyamandasphuṭābhyāṃ
māndā yatra sphuṭāptiḥ paṭhitabhujaphalaistatra karṇāptirevam || 26 ||

Having obtained the Rsine of the *upāntyasphuṭa* and *antyasphuṭa* from which respective *tuṅgas* are subtracted, multiply [the former] by *trijyā* and divide by the latter. The result obtained then would be the *śīghrakarṇa*. [Similarly], the [*manda*] *karṇa* can be obtained by subtracting apogee from the mean and the *manda-sphuṭa*. The *karṇas* are to be obtained like this when the *sphuṭas* are obtained from the tabulated correction values.

The formula mentioned in the verse for obtaining the *karṇa* is

$$karṇa = \frac{R \times R\sin(upāntyasphuṭakendra)}{R\sin(antyasphuṭakendra)}. \tag{7.76}$$

Here, the terms '*upāntyasphuṭa*' (penultimate to the corrected planet) and '*antyasphuṭa*' (the corrected planet) have to be assigned meanings depending upon the process under consideration. If it is *śīghra*-process, then the *antyasphuṭa* refers to *śīghra-sphuṭa*, the *upāntyasphuṭa* refers to the *manda-sphuṭa*, and the associated *karṇa* is the *śīghrakarṇa*. On the other hand if the process under consideration is the *manda*-process, then the term *antyasphuṭa* refers to the *mandasphuṭa*, *upāntyasphuṭa* refers to *madhyamagraha* (mean planet), and the associated *karṇa* is *mandakarṇa*. As regards the *kendra*, it is obvious that depending upon whether the process considered is *manda* or *śīghra*, the *mandocca* and *śīghrocca* have to be employed in finding the anomaly.

Therefore, the expression for *śīghrakarṇa* can be written from (7.76) as

$$\begin{aligned} śīghrakarṇa &= \frac{R \times R\sin(mandasphuṭakendra)}{R\sin(śīghrasphuṭakendra)}, \\ \text{or} \qquad K_s &= \frac{R \times |R\sin(\theta_s - \theta_{ms})|}{|R\sin(\theta_s - \theta)|}, \end{aligned} \tag{7.77}$$

where θ_s is the *śīghrocca*, θ_{ms} is the *manda-sphuṭa* and θ is the *śīghra-sphuṭa*. The *mandakarṇa* can be obtained from

$$\begin{aligned} mandakarṇa &= \frac{R \times R\sin(mandakendra)}{R\sin(manda\text{-}sphuṭakendra)}, \\ \text{or} \qquad K_m &= \frac{R \times |R\sin(\theta_0 - \theta_m)|}{|R\sin(\theta_{ms} - \theta_m)|}. \end{aligned} \tag{7.78}$$

The rationale for obtaining the above expression (7.78) can be understood from Figure 7.9*a* where P_0 is the mean planet and P is the *manda-sphuṭa*.

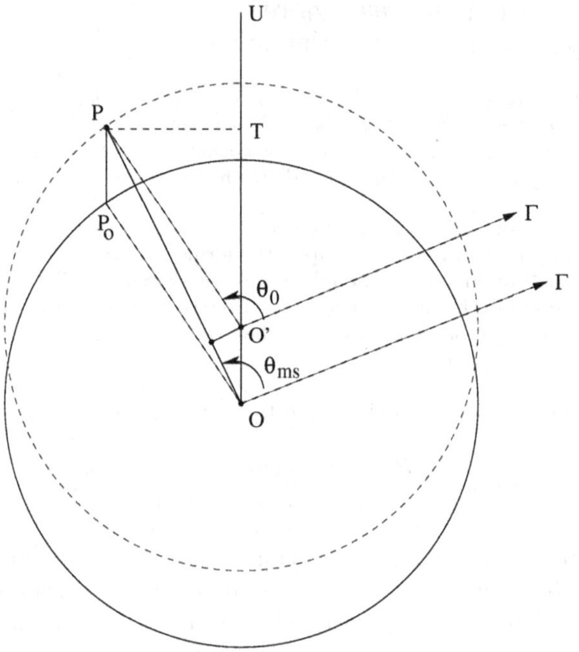

Fig. 7.9*a* Obtaining the *mandakarṇa*.

Draw PT perpendicular to OU. In triangle $PO'T$,

$$\begin{aligned} PT &= PO' \sin(T\hat{O}'P) \\ &= |R\sin(\theta_0 - \theta_m)|, \end{aligned} \tag{7.79}$$

as $\Gamma\hat{O}'P = \theta_0$, $\Gamma\hat{O}'U = \theta_m$ and $U\hat{O}'P = T\hat{O}'P = \theta_0 - \theta_m$. In triangle POT,

$$\begin{aligned} PT &= OP \sin(T\hat{O}P) \\ &= K_m |\sin(\theta_{ms} - \theta_m)|, \end{aligned} \tag{7.80}$$

as $P\hat{O}\Gamma = \theta_{ms}$, $\Gamma\hat{O}U = \theta_m$ and $U\hat{O}P = T\hat{O}P = \theta_{ms} - \theta_m$. Equating the two expressions (7.79) and (7.80), we find

$$K_m = \frac{R \times |R\sin(\theta_0 - \theta_m)|}{|R\sin(\theta_{ms} - \theta_m)|}, \tag{7.81}$$

which is the same as the expression given in (7.78).

Similarly, the expression for the *śīghrakarṇa*, K_s in (7.77) can be obtained by considering the Figure 7.9*b* where S is the *śīghrocca* and P is the *sphuṭa*.

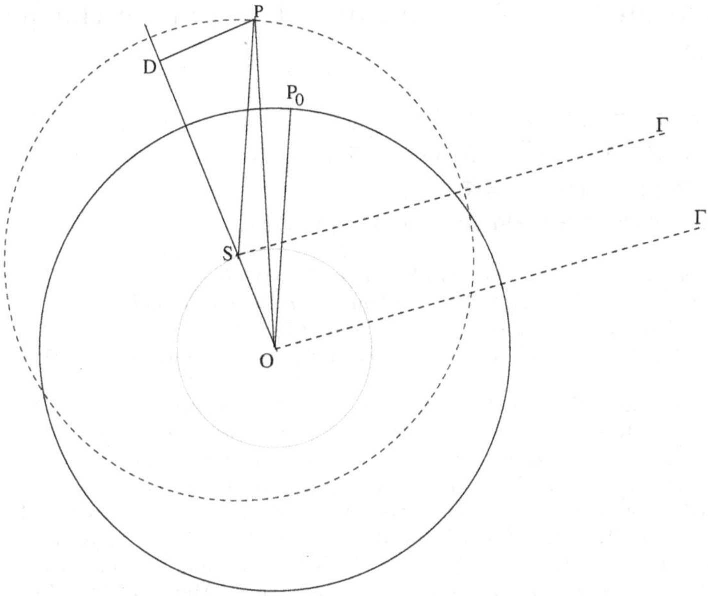

Fig. 7.9*b* Obtaining the *śīghrakarṇa*.

Draw PD perpendicular to OS extended. In triangle PSD,

$$\begin{aligned} PD &= SP\sin(P\hat{S}D) \\ &= R\sin(\Gamma\hat{S}D - \Gamma\hat{S}P) \\ &= R\sin(\Gamma\hat{O}S - \Gamma\hat{O}P_0) \\ &= R|\sin(\theta_s - \theta_{ms})|. \end{aligned} \tag{7.82}$$

In triangle POD,

$$\begin{aligned} PD &= OP\sin(P\hat{O}D) \\ &= K_s\sin(\Gamma\hat{O}S - \Gamma\hat{O}P) \\ &= K_s|\sin(\theta_s - \theta)|. \end{aligned} \tag{7.83}$$

Equating the two expressions (7.82) and (7.83), we find

$$K_s = \frac{R \times |R\sin(\theta_s - \theta_{ms})|}{|R\sin(\theta_s - \theta)|}. \tag{7.84}$$

७.२१ केन्द्रज्याभावे कर्णानयनम्

7.21 Obtaining the *karṇas* when the Rsine of the *kendra* is zero

दोर्ज्याभावे तु केन्द्रे सति मकरकुलीरादिके तत्र मान्दे
तद्वृत्तोनाढ्यनन्दैस्त्रिभगुणगुणितान्नन्दनान्मन्दकर्णः ।
शैघ्रे तद्वृत्तयुक्तोनितनदगुणिताद्विस्तरार्धान्नदाप्तो
भौमादेः शीघ्रकर्णः सततमपि विधोः शीघ्रकर्णोऽन्त्यकर्णः ॥ २७ ॥

dorjyābhāve tu kendre sati makarakulīrādike tatra mānde
tadvṛttonāḍhyanandaistribhaguṇaguṇitānnandanānmandakarṇaḥ |
śaighre tadvṛttayuktonitanadaguṇitādvistarārdhānnadāpto
bhaumādeḥ śīghrakarṇaḥ satatamapi vidhoḥ śīghrakarṇo'ntyakarṇaḥ || 27 ||

While obtaining the *mandakarṇa* if the Rsine [of the *kendra*] is zero, then the *ojavṛtta* has to be subtracted from or added to 80 (*nanda*) depending on whether [the *kendra*] is *makarādi* or *karkyādi* respectively. Dividing the product of the *trijyā* and 80 by this [result] the *mandakarṇa* is obtained. In the case of *śīghra*[*karṇa*] if the Rsine [of the *kendra*] is zero, then the *śīghrojaparidhi* has to be added to or subtracted from 80 (*nada*) [depending on whether the *kendra* is *makarādi* or *karkyādi* respectively]. Multiplying the result by *trijyā* and dividing by 80 (*nada*) would give the *śīghrakarṇa* in the case of the [planets] Mars etc. The *śīghrakarṇa* of the Moon is its *antyakarṇa* itself.

Recalling the expressions (B.4) and (B.5) for the *mandakarṇa* K, in appendix B, we have

$$K = \left[(R + r\cos(\theta_0 - \theta_m))^2 + (r\sin(\theta_0 - \theta_m))^2\right]^{\frac{1}{2}}, \tag{7.85}$$

which is valid for both *makarādi* and *karkyādi* cases. It is easily seen that, when the *kendra* $(\theta_0 - \theta_m) = 0$ or $180°$, then the expression for K reduces to

$$K = R + r \qquad (\text{for } (\theta_0 - \theta_m) = 0), \tag{7.86}$$

$$= R - r \qquad (\text{for } (\theta_0 - \theta_m) = 180°), \tag{7.87}$$

where r is the instantaneous radius of the epicycle. From the relation $\frac{r}{K} = \frac{r_m}{R} = \frac{C_m}{80}$, we have

$$r = \frac{K \times C_m}{80}. \tag{7.88}$$

Now, the *sphuṭamandaparidhi* C_m is simply *ojamandaparidhi* C_{mo} when $(\theta_0 - \theta_m) = 0$ or $180°$. Using the above relation for r in the expression for K, we have

$$K = R \pm \frac{K \times C_{mo}}{80},$$

Hence,
$$K = \frac{R \times 80}{80 \mp C_{mo}}$$
$$= \frac{trijy\bar{a} \times nanda}{nanda \mp ojaparidhi}, \qquad (7.89)$$

as stated in the text.

Similarly, using the expressions (B.13) and (B.14) for the *śīghrakarṇa* K_s in appendix B, we have,

$$K_s = \left[\left(R \pm \frac{C_s}{80}|R\cos\theta_{sk}|\right)^2 + \left(\frac{C_s}{80}R\sin\theta_{sk}\right)^2\right]^{\frac{1}{2}}, \qquad (7.90)$$

where we have used to relation, $\frac{r_s}{R} = \frac{C_s}{80}$. Here, we use '+' in the first term in RHS if *kendra* is *makarādi* and '−' if it is *karkyādi*. For $\theta_{sk} = 0$ and $\theta_{sk} = 180°$, $C_s = C_{so}$ and K_s reduces to

$$K_s = R \times \left(1 \pm \frac{C_{so}}{80}\right)$$
$$= R \times \left(\frac{80 \pm C_{so}}{80}\right). \qquad (7.91)$$

७.२२ भूमध्यखेटविवरम्

7.22 Earth–planet distance

कर्णद्वयस्य वधतस्त्रिगुणेन लब्धं
केन्द्रग्रहान्तरमिनस्य तु तत् स्वकर्णः ।
भूमध्यखेटविवरं च तदेव विद्यात्
प्रायेण शीतमहसः स्फुटमेव तत् स्यात् ॥ २८ ॥

karṇadvayasya vadhatastriguṇena labdhaṃ
kendragrahāntaraminasya tu tat svakarṇaḥ |
bhūmadhyakheṭavivaraṃ ca tadeva vidyāt
prāyeṇa śītamahasaḥ sphuṭameva tat syāt || 28 ||

The distance of the planet from the centre (*kendragrahāntara*) is given by the product of the two *karṇas* (*manda-karṇa* and *śīghrakarṇa*) with each other and divided by the radius (*triguṇa* or *trijyā*). For the Sun its *mandakarṇa* itself is the *kendragrahāntara*. That (*kendragrahāntara*) itself may generally be taken as the distance between the center of the earth and the planet (*bhūmadhyakheṭavivara*).

In the case of the Moon it is indeed the accurate distance from the centre of the earth.

In traditional Indian planetary theory, at least from the time of Āryabhaṭa, the mean planetary distances were obtained based on the hypothesis that all the planets move with the same linear velocity, that is, they all cover the same physical distance in a given period of time.

The term *kendragrahāntara* mentioned in the verse is the distance of the planets from the center of the celestial sphere (earth), and is given by

$$\begin{aligned} kendragrah\bar{a}ntara &= \frac{mandakar\d{n}a \times \acute{s}\bar{\imath}ghrakar\d{n}a}{trijy\bar{a}} \\ &= \frac{K_m \times K_s}{R}. \end{aligned} \tag{7.92}$$

This is the relation stated in *Āryabhaṭīya*. Nīlakaṇṭha in his *Āryabhaṭīyabhāṣya* explains that, since usually the *śīghrakarṇa* is evaluated with respect to a concentric of the standard radius (*trijyā*), the above prescription of *Āryabhaṭīya* implies that the earth-planet distance is actually given by the *śīghrakarṇa* which is evaluated with respect to a concentric circle whose radius is given by the *mandakarṇa*.[18]

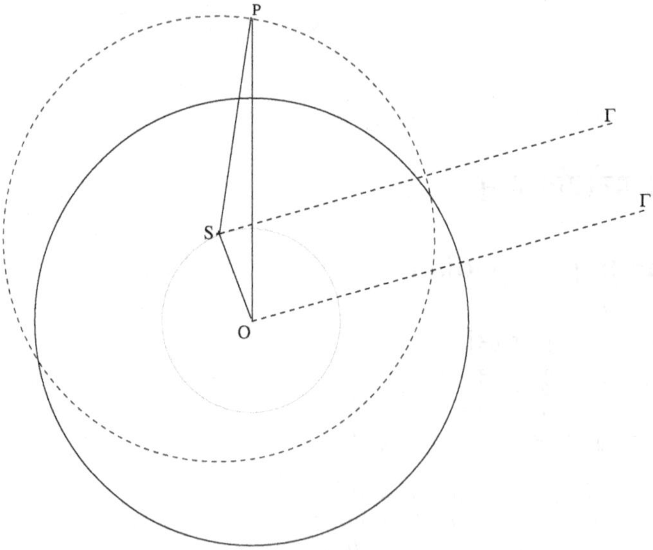

Fig. 7.10 Obtaining the distance of the planets from the centre of the celestial sphere (earth).

[18] {ABB 1931}, pp. 53-4.

In order to understand the conditions under which the *Āryabhaṭīya* relation holds, let us consider the case of exterior planets which is depicted in the Figure 7.10. The *kendragrahāntara* is the *śīghrakarṇa* OP itself if SP is taken as *trijyā*, R, that is P is taken to be *madhyamagraha* with respect to S. If we take P to be the *manda-sphuṭa* with respect to S and take SP, as the *mandakarṇa*, K, we would obtain

$$kendragrah\bar{a}ntara = \frac{K_m \times K_s}{R}, \tag{7.93}$$

only if OS the radius of the *śīghra* epicycle which is scaled by the factor $\frac{K_m}{R}$. There is no mention of such a scaling of the *śīghra* epicycle in *Karaṇapaddhati.* To that extent the prescription (7.92) for the earth–planet distance is somewhat *ad hoc.*

७.२३ ग्रहयोजनकर्णाः

7.23 *Yojanakarṇas* (physical distance in *yojanas*) of the planets

भूग्रहान्तरहता निजकक्ष्या शीघ्रवृत्तहतनिस्सरभक्तः ।
ज्ञाच्छयोर्भवति योजनकर्णो ज्ञानतत्परहृतश्च परेषाम् ॥ २९ ॥

bhūgrahāntarahatā nijakakṣyā śīghravṛttahatanissarabhaktaḥ |
jñācchayorbhavati yojanakarṇo jñānatatparahṛtaśca pareṣām || 29 ||

Multiply the earth-planet distance (*bhūgrahāntara*) by the orbit of the planet (*kakṣyā*). When we divide this by the product of the circumference of the *śīghra*-epicycle (*śīghravṛtta*) and 270 (*nissara*), we obtain the distance in *yojanas* (*yojanakarṇas*) of Mercury and Venus. Dividing [the product of (*bhūgrahāntara*) and *kakṣyā*] by 21600 (*jñānatatpara*) would result in the *yojanakarṇas* of other planets (the exterior planets)].

The *yojanakarṇas* of the interior planets are stated to be

$$\begin{aligned} yojanakar\d{n}a &= \frac{bh\bar{u}grah\bar{a}ntara \times kak\d{s}y\bar{a}paridhi}{nissara \times \acute{s}\bar{\imath}ghravr\d{t}ta} \\ &= \frac{bh\bar{u}grah\bar{a}ntara \times 2\pi \times O_p}{270 \times C_s}, \end{aligned} \tag{7.94}$$

where O_p is the radius of the orbit in *yojanas*. The rationale behind the expression in the denominator can be understood as follows. If r_s is the radius of the *śīghravṛtta* in minutes, $\frac{C_s}{80} = \frac{r_s}{R} = \frac{2\pi r_s}{21600}$, then $C_s \times 270 = 2\pi r_s$ (in minutes). Thus,

$$yojanakarṇa = \frac{bhūgrahāntara \times O_p}{r_s}. \quad (7.95)$$

This is essentially the same as the expression for the *yojanakarṇa* for Mercury and Venus in the *Tantrasaṅgraha* (verse VIII.37). However, in *Tantrasaṅgraha* what appears in the denominator is the planet's own orbital radius around the mean Sun and that is equal to the radius of the *śīghravṛtta*.[19]

Karaṇapaddhati does not present any geometrical model of planetary motion. There is no mention of the geometrical picture of planetary motion proposed by Nīlakaṇṭha that the interior planets are going around the mean Sun along their *śīghravṛttas*. Hence the above prescription (7.95) for the *yojanakarṇas* also seems to have been made in an adhoc manner.

For the other planets (*pareṣāṃ*), that is Mars, Jupiter and Saturn, the distance in *yojanas* is stated to be

$$\begin{aligned} yojanakarṇa &= \frac{bhūgrahāntara \times kakṣyāparidhi}{jñānatatpara} \\ &= \frac{bhūgrahāntara \times O_p}{R}. \end{aligned} \quad (7.96)$$

This can be understood by a simple rule of three. The mean *kakṣyā* in *yojana* O_p is the distance of the planet corresponding to the radius of the concentric given by R minutes. If the actual *karṇa* is given by *bhūgrahāntara* minutes, what is the distance of the planet in *yojanas*? The answer is as given in equation (7.96).

७.२४ मौढ्योपक्रमावसानकर्णानयनम्

7.24 Obtaining the hypotenuse at the heliacal rising and setting

भौमात् सेव्यगयापयोधनमयात् मौढ्योदिता ह्यंशकाः
तद्दोःकोटिगुणौ स्वशीघ्रपरिधिक्षुण्णौ नदाप्तौ फले ।
त्रिज्यादोःफलवर्गभेदजपदं कोटीफलेनान्वितं
कर्णः स्यादुदयास्तकालसविधे मन्दामरेड्यासृजाम् ॥ ३० ॥

bhaumāt sevyagayāpayodhanamayāt mauḍhyoditā hyaṃśakāḥ
taddoḥkoṭiguṇau svaśīghraparidhikṣuṇṇau nadāptau phale |
trijyādoḥphalavargabhedajapadaṃ koṭīphalenānvitaṃ
karṇaḥ syādudayāstakālasavidhe mandāmareḍyāsṛjām || 30 ||

The longitudinal difference of the planets in degrees for helical rising (*mauḍhyoditāṃśas*) [of the planets] starting from Mars etc. are 17 (*sevya*), 13 (*gayā*), 11

[19] {TS 2011}, pp. 434-437.

(*payaḥ*), 09 (*dhanaḥ*) and 15 (*mayā*). Multiply the Rsines and Rcosines of these [*aṃśas*] by the respective *śīghra-paridhis* and divide by 80 (*nada*). The square roots obtained by subtracting the square of the results obtained with Rsines from the square of the *trijyā* have to be added to the results obtained with Rcosines. [The results obtained] would be the hypotenuses (*karṇas*) near to the *udayāstamanakāla* of the Saturn (*manda*), Jupiter (*amaredya*) and Mars (*asṛk*).

The heliacal rising angles in degrees (*mauḍhyāmśas*) with Rsine and Rcosine values for five planets given in the verse are listed in Table 7.8. The value of the *mauḍhyāṃśas* stated here are the same as those in *Āryabhaṭīya* and *Tantrasaṅgraha*. The expression for the *karṇa* K_{md} at the *mauḍhyodaya* and

Planet	*mauḍhyāṃśa* (α) (in degrees)	$R\sin\alpha$ (in minutes)	$R\cos\alpha$ (in minutes)
Mars	17	1005.10	3287.53
Mercury	13	773.32	3349.64
Jupiter	11	655.95	3374.59
Venus	9	537.78	3395.42
Saturn	15	889.75	3320.61

Table 7.8 The *mauḍhyāṃśas* of the planets and their Rsines and Rcosines.

the *astamana* of the exterior planets is stated to be

$$K_{md} = \sqrt{R^2 - \left(\frac{C_s}{80} \times R\sin\alpha\right)^2} + \frac{C_s}{80} \times R\cos\alpha, \tag{7.97}$$

where α is the *mauḍhyāṃśa* and C_s the *śīghra-paridhi*.

This can be understood with the help of Figure 7.11. Here S' is the *śīghrocca* which is the centre of the *pratimaṇḍala* of radius R. The planet P is situated on this *pratimaṇḍala* such that $P\hat{S'}S$ is the *śīghrakendra*. $S\hat{O}P = \alpha$ is the angular separation between the planet P and the Sun. OS' is the radius of the *śīghravṛtta* and is given by

$$OS' = \frac{C_s}{80} \times R.$$

Then,

$$S'D = OS' \sin(S\hat{O}P) = \frac{C_s}{80} \times R\sin\alpha,$$

$$OD = \frac{C_s}{80} \times R\cos\alpha,$$

and $$PD = \sqrt{S'P^2 - SD^2} = \sqrt{R^2 - \left(\frac{C_s}{80} \times R\sin\alpha\right)^2}. \tag{7.98}$$

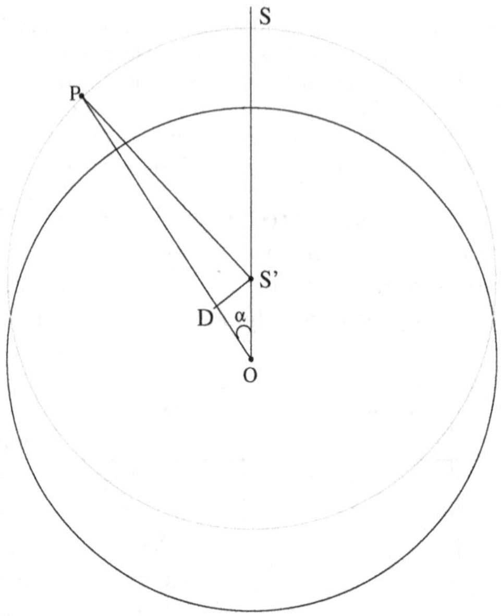

Fig. 7.11 Obtaining the *karṇa* of exterior planets at heliacal rising and setting.

Then OP, which is the *karṇa* K_{md} is given by

$$K_{md} = OP = PD + OD$$
$$= \sqrt{R^2 - \left(\frac{C_s}{80} \times R\sin\alpha\right)^2} + \frac{C_s}{80} \times R\cos\alpha. \qquad (7.99)$$

७.२४.१ बुधशुक्रयोः मौढ्योप्रक्रमावसानकर्णः

7.24.1 Hypotenuse at heliacal rising and setting of the interior planets

मौढ्योदितांशभुजकोटिगुणौ ज्ञभृग्वोः
नीत्वा भुजागुणहतात् चलवृत्तभेदात् ।
शीघ्रौजवृत्तहृतमन्त्यफलात् तदीयात्
संशोध्य शिष्टमिदमन्त्यफलं स्फुटं स्यात् ॥ ३१ ॥

तद्वर्गतो दोर्गुणवर्गहीनात् मूलं पुनः कोटिगुणे धनर्णम् ।
कुर्यात्तदा स्यान्मृगकर्कटाद्योः प्रायेण मौढ्याद्यवसानकर्णः॥ ३२ ॥

mauḍhyoditāṃśabhujakoṭiguṇau jñabhṛgvoḥ
nītvā bhujāguṇahatāt calavṛttabhedāt |

śīghraujavṛttahṛtamantyaphalāt tadīyāt
saṃśodhya śiṣṭamidamantyaphalaṃ sphuṭaṃ syāt || 31 ||
tadvargato dorguṇavargahīnāt mūlaṃ punaḥ koṭiguṇe dhanarṇam |
kuryāttadā syānmṛgakarkaṭādyoḥ prāyeṇa mauḍhyādyavasānakarṇaḥ || 32 ||

Having obtained the Rsine and Rcosine of the *mauḍhyāṃśas* of Mercury and Venus, multiply the Rsine by the difference between the odd and the even *śīghravṛttas* and divide by the *śīghravṛtta* at the beginning of odd quadrant (*śīghraujavṛtta*). The result when subtracted from the associated *antyaphala* would give the true *antyaphala*.

From the square of that (*sphuṭāntyaphala*), subtract the square of the Rsine [of the *mauḍhyāṃśa*] and square root of the result is added to or subtracted from the Rcosine in *mṛgādi* and *karkyādi* respectively. The result will be close to the *karṇa* at the beginning and ending of the *mauḍhya*.

The verse 31 expresses the *antyaphala* (radius of the epicycle) corresponding to the *mauḍhyāṃśa*, as the *antyaphala* at the beginning of the odd quadrant to which a correction term is added. The corrected *antyaphala* is further used to find the *karṇas* at the heliacal rising and setting of the interior planets. The correction term to be applied to the *antyaphala* in order to obtain the *sphuṭāntyaphala*, is stated to be

$$\Delta r_s = \frac{R \sin \alpha \times \Delta_s}{C_{so}}, \tag{7.100}$$

where Δ_s represents the difference between the odd and the even *śīghravṛttas*. Subtracting this from the *antyaphala*, we have

$$\textit{sphuṭāntyaphala} \quad r_s = r_{so} - \Delta r_s.$$

Now we present the rationale behind the above expression. In what follows, we ignore the correction due to eccentricity of the interior planet's orbit.

In Figure 7.12, O is the earth, P_0 is the Sun and P represents the interior planet. P_0P is parallel to OS_i, where OS_i is in the direction of the *śīghrocca*. Then $P_0\hat{O}S_i = \theta_{sk}$ is the *śīghrakendra*, and $P_0\hat{O}P$ is the angle between the true interior planet at P and the Sun at P_0, as viewed from the earth. $P_0\hat{O}P = \alpha$ at the beginning and end of the *mauḍhya*. We know that the *śīghra-paridhi* C_s varies with the *śīghrakendra* as

$$C_s = C_{so} - \Delta_s \frac{R \sin \theta_{sk}}{R}, \tag{7.101}$$

where Δ_s is the difference in the radius of the epicycle at the beginning and at the end of the odd quadrant. In the triangle P_0PO in Figure 7.12, $P_0\hat{O}P = \alpha$ and

$$P_0\hat{P}O = P\hat{O}S = P_0\hat{O}S_i - P_0\hat{O}P = \theta_{sk} - \alpha.$$

Now in the triangle P_0OD,

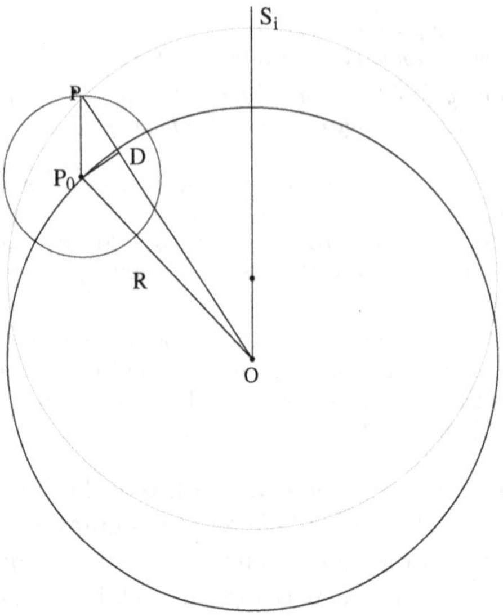

Fig. 7.12 Obtaining the *karṇa* at heliacal rising and setting of the interior planets.

$$P_0D = OP_0 \sin(P_0\hat{O}D) = R \sin\alpha.$$

Considering the triangle P_0DP,

$$P_0D = PP_0 \sin(P_0\hat{P}D) = r_s \sin(\theta_{sk} - \alpha),$$

as $PP_0 = r_s$ is the radius of the *śīghravṛtta*. Equating the two expressions for P_0D,

$$r_s \sin(\theta_{sk} - \alpha) = R \sin\alpha,$$

$$\text{or} \qquad \frac{\sin(\theta_{sk} - \alpha)}{R} = \frac{\sin\alpha}{r_s}.$$

Since α is small compared to the *śīghrakendra* θ_{sk}, and $r_s \approx r_{so}$, we have

$$\frac{\sin\theta_{sk}}{R} \approx \frac{\sin\alpha}{r_{so}}.$$

Hence

$$C_s \approx C_{so} - \Delta_s \frac{R\sin\alpha}{r_{so}}.$$

The above relation is equivalent to

$$r_s \approx r_{so} - \Delta_s \frac{R\sin\alpha}{C_{so}}, \tag{7.102}$$

since $\frac{r_s}{R} = \frac{C_s}{80}$.

According to verse 32, the *karṇas* at the rising and setting of the *mauḍhya* for *mṛgādi* and *karkyādi* are given by

$$K_{md} = (R\cos\alpha) \pm \sqrt{(r_s)^2 - (R\sin\alpha)^2}.$$

This is an exact result. The approximation is due to the fact that the only an approximate value of r_s was calculated in the previous verse. The expression for the *karṇa* can be understood as follows. In Figure 7.12,

$$OD = R\cos(P_0\hat{O}D) = R\cos\alpha,$$

$$\text{and} \qquad P_0D = R\sin\alpha.$$

Hence,

$$PD = \sqrt{P_0P^2 - P_0D^2} = \sqrt{r_s^2 - (R\sin\alpha)^2}.$$

The *karṇa* $OP = K_{md} = OD \pm DP$. Here, we should choose '+' for *makarādi* as in the figure, and '−' for *karkyādi*. Hence,

$$K_{md} = (R\cos\alpha) \pm \sqrt{(r_s)^2 - (R\sin\alpha)^2}. \qquad (7.103)$$

७.२५ मौढ्योपक्रमावसानयोः ग्रहविक्षेपः

7.25 Latitude of a planet at heliacal rising and setting

कर्णोऽयं क्षेपहारः[20] स्यान्मौढ्यारम्भावसानयोः ।
गुणो हि परमक्षेपो यद्वा तावपवर्तितौ ॥ ३३ ॥

आसुरन्नाळमारण्यमन्तमाश्रयमप्रियम् ।
चन्द्रादीनां क्रमादेताः परमक्षेपलिप्तिकाः ॥ ३४ ॥

karṇo'yaṃ kṣepahāraḥ syānmauḍhyārambhāvasānayoḥ |
guṇo hi paramakṣepo yadvā tāvapavartitau || 33 ||

āsurannāḷamāraṇyamantamāśrayamapriyam |
candrādīnāṃ kramādetāḥ paramakṣepaliptikāḥ || 34 ||

While obtaining the latitude (*kṣepa*) at the commencement and the end of *mauḍhya*, the divisor is this *mauḍhyārambhāvasāna-karṇa* (stated in the previous verse), and the multiplier is the maximum latitude (*paramakṣepa*). Or else, we can use these [*kṣepa* and *hāra*] after doing *apavartana*.

The maximum latitudes in minutes of the planets starting from the Moon etc., are successively *āsuram* (270), *nāḷam* (90), *āraṇyam* (120), *antam* (60), *āśrayam* (120) and *apriyam* (120).

[20] The term *kṣepahāraḥ* should be understood as "*kṣepasya* [*ānayane*] *yo hāraḥ saḥ*".

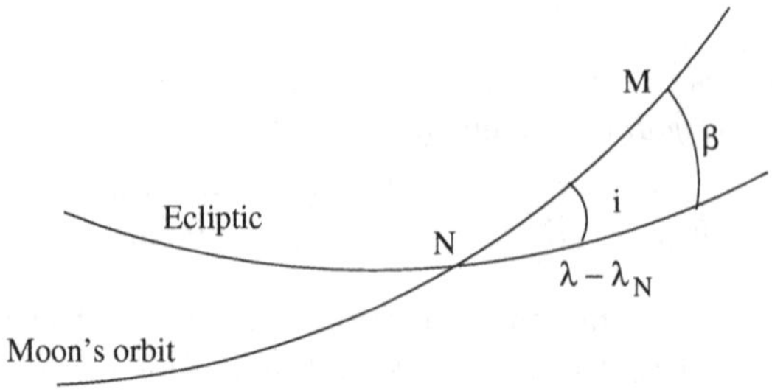

Fig. 7.13 Latitude of the Moon.

For the Moon, the latitude β (see Figure 7.13) for an arbitrary value of its longitude λ is given by $\beta \approx i\sin(\lambda - \lambda_N)$, where λ_N is the longitude of its node, and i is the inclination of its orbit with the ecliptic whose value is taken to be $270'$. For the planets, the latitude at the commencement and end of *mauḍhya* is stated to be

$$\textit{vikṣepa} = \frac{\textit{paramakṣepa}}{\textit{mauḍhyārambhāvasāna-karṇa}}.$$

Here, Putumana Somayājī has in view some approximate version of the expression for the latitudes of planets as in *Tantrasaṅgraha* (where also the *śīghrakarṇa* appears in the denominator). In Figures 7.14 and 7.15, P and

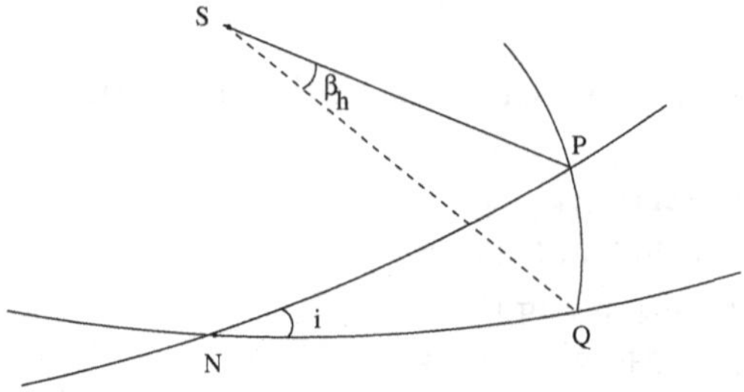

Fig. 7.14 Heliocentric latitude of a planet.

N refer to the true planet and the node, and S the mean Sun. The orbit of the planet is inclined at an angle i with respect to the ecliptic on which the mean Sun moves. Now consider Figure 7.14. Since the inclination of the orbit is small, the heliocentric latitude β_h may be written as

$$\beta_h \approx i\sin(\theta_{ms} - \theta_n), \tag{7.104}$$

where θ_{ms} and θ_n are the longitudes of the planet and its node respectively. The relation between the geocentric latitude, β_E, which is measured with respect to the Earth, and β_h is depicted in Figure 7.15. Here, the arc PQ may

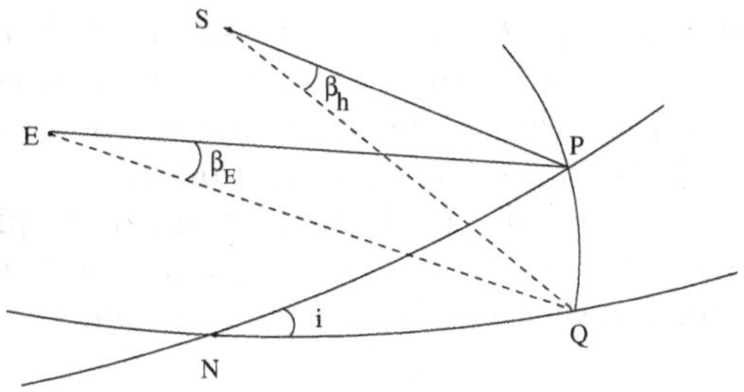

Fig. 7.15 Obtaining the geocentric latitude of a planet from its heliocentric latitude.

be expressed in two ways

$$PQ = \beta_E \times EP. \quad (7.105)$$

$$\text{and} \qquad PQ = \beta_h \times SP. \quad (7.106)$$

Hence,

$$\beta_E = \beta_h \frac{SP}{EP}, \quad (7.107)$$

$$\text{or} \qquad \beta_E = i \sin(\theta_{ms} - \theta_n) \, \frac{SP}{EP}. \quad (7.108)$$

For exterior planets, the mean Sun S is the *śīghrocca*, and P is the *manda-sphuṭa* which moves on the *pratimaṇḍala* centred at S, and $SP = R$ (the *trijyā*), and $EP =$ *śīghra-karṇa*. Then

$$\beta_E = \frac{iR \sin(\theta_{ms} - \theta_n)}{\textit{śīghra-karṇa}}. \quad (7.109)$$

For the interior planets, the mean Sun is the mean planet, and P is the true planet which moves on the *śīghravṛtta* or the *śīghra*-epicycle centred at S and is in the direction of the *śīghrocca*, with respect to S, and $SP = r_s$, the radius of the *śīghra* epicycle. Then

$$\beta_E = \frac{i \, r_s \sin(\theta_{ms} - \theta_n)}{\textit{śīghra-karṇa}} = \frac{i\left(\frac{r_s}{R}\right) \, R \sin(\theta_{ms} - \theta_n)}{\textit{śīghra-karṇa}}.$$

Here θ_{ms} is the longitude of the *śīghrocca* corrected by the *mandaphala*. Hence, for the interior planets, β_{max} should be identified with $i\left(\frac{rs}{R}\right)$. Thus for both interior and exterior planets, we can write the latitude as

$$\beta_E = \frac{\beta_{max} \times R \sin(\theta_{ms} - \theta_n)}{\textit{śīghra-karṇa}}.$$

This relation is stated in *Tantrasaṅgraha* (verse VII. 5).[21] However, it must be noted that in the above verse in *Karaṇapaddhati*, the maximum deflection β_{max} in the numerator and the *śīghrakarṇa* in the denominator are mentioned, whereas the factor $R\sin(\theta_{ms} - \theta_n)$ is not mentioned,

The above verse also gives the maximum latitudes of the planets in minutes using *kaṭapayādi* notation. These values are listed in Table 7.9. It may be noted that these values too are the same as in *Tantrasaṅgraha*.

Planet	Maximum latitude (β_{max} in minutes)	
	in *kaṭapayādi*	in numerals
Moon	*āsuram*	270
Mars	*nāḷam*	90
Mercury	*āraṇyam*	120
Jupiter	*antam*	60
Venus	*āśrayam*	120
Saturn	*apriyam*	120

Table 7.9 The maximum latitudes of planets.

७.२६ रविचन्द्रभुवां बिम्बव्यासयोजनानि

7.26 Diameter of the orbs of the Sun, Moon and the Earth in *yojanas*

अर्कोद्भवं रवेर्बिम्बव्यासः स्याद्योजनात्मकः ।
शश्यङ्गं शशिनस्तद्वद् भवेदात्मनयं भुवः ॥ ३५ ॥

arkodbhavaṃ raverbimbavyāsaḥ syādyojanātmakaḥ |
śaśyaṅgaṃ śaśinastadvad bhavedātmanayaṃ bhuvaḥ || 35 ||

The diameter of the Sun in *yojanas* is 4410 (*arkodbhavam*). Similarly that of the Moon and the Earth are 315 (*śaśyaṅgam*) and 1050 (*ātmanayam*) [respectively].

The diameters of the Sun, Moon and the Earth are specified to be 4410, 315, 1050 *yojanas* respectively. These values are the same as in *Tantrasaṅgraha*.

[21] {TS 2011}, pp. 391-394.

Chapter 8
छायाप्रकरणम्
Gnomonic shadow

८.१ मध्याह्नच्छायातः अक्षज्यालम्बज्यानयनम्

8.1 Obtaining the Rsine and Rcosine of the latitude from the midday shadow

विषुवद्दिनमध्यभाकृतिर्भवकाढ्या पदिता पलश्रुतिः ।
विषुवद्दिनमध्यभाहतात् त्रिगुणात् तद्विहृताक्षमौर्विका ॥ १ ॥
तत्त्रिज्यावर्गविश्लेषमूलं लम्बनमौर्विका ।
एके संस्कारमिच्छन्ति स्फुटत्वार्थं तयोर्मिथः ॥ २ ॥

viṣuvaddinamadhyabhākṛtirbhavakāḍhyā paditā palaśrutiḥ |
viṣuvaddinamadhyabhāhatāt triguṇāt tadvihṛtākṣamaurvikā || 1 ||
tattrijyāvargaviśleṣamūlaṃ lambanamaurvikā |
eke saṃskāramicchanti sphuṭatvārthaṃ tayormithaḥ || 2 ||

The hypotenuse of the gnomon (*palaśruti*) is obtained by taking the square root of the sum of 144 (*bhavaka*) and the square of the mid-day shadow [of the gnomon] on the equinoctial day (*viṣuvaddina*).[1] The mid-day shadow on the equinoctial day when multiplied by the radius (*triguṇa*) and divided by that [hypotenuse] would be the Rsine of the terrestrial latitude (*akṣamaurvikā*).

The square root of the square of that [*akṣajyā*] subtracted from the square of the radius is the Rcosine of the terrestrial latitude (*lambanamaurvikā* or *lambajyā*). Some [astronomers] prefer that their values be corrected in order to obtain accurate results.

Several quantities of physical interest can be found through observations using a *śaṅku* or a gnomon. The gnomon is placed on level ground, perpendicular to it, and shadow measurements are done with that. The set of verses given above present expressions for the Rsine and Rcosine of the latitude in

[1] The day on which the Sun passes through the equinox.

V. Pai et al., *Karaṇapaddhati of Putumana Somayājī*, Sources and Studies in the History of Mathematics and Physical Sciences, https://doi.org/10.1007/978-981-10-6814-0_8

terms of the equinoctial shadow. The height of the *śaṅku* (gnomon) is usually taken to be 12 units (*aṅgulas*) in the texts on Indian astronomy. The following are the expressions for the hypotenuse of the shadow cast by the *śaṅku*, the Rsine of the latitude (*akṣajyā*) and its Rcosine (*lambajyā*), as given in these verses:

$$\begin{aligned} karṇa &= \sqrt{chāyā^2 + 144}, \\ akṣajyā &= \frac{trijyā \times chāyā}{karṇa}, \\ lambajyā &= \sqrt{trijyā^2 - akṣajyā^2}. \end{aligned} \tag{8.1}$$

We now provide the rationale behind the above expressions using Figure 8.1. It can be easily seen from the figure that the triangle formed by the *śaṅku* (gnomon), the *chāyā* (shadow) and the *karṇa* (hypotenuse) is a right-angled triangle. OX represents the *śaṅku* (12 units), OY the *chāyā* and XY the *karṇa*. The expression for *karṇa* is straightforward. Considering the triangle OXY,

$$\begin{aligned} XY^2 &= OY^2 + OX^2, \\ \text{or} \qquad karṇa^2 &= chāyā^2 + śaṅku^2 \\ &= chāyā^2 + 144, \\ \text{or} \qquad karṇa &= \sqrt{chāyā^2 + 144}, \end{aligned}$$

as stated in the verse. On the equinoctial day, the Sun is almost on the equator throughout the day. Hence, the zenith distance of the Sun as it crosses the prime meridian (at noon) would be equal to the latitude (ϕ) of the place. That is, $O\hat{X}Y = \phi$. Hence,

$$\sin\phi = \frac{OY}{XY}, \qquad \cos\phi = \frac{OX}{XY}. \tag{8.2}$$

Now *akṣajyā* is $R\sin\phi$ and *lambajyā* or *lambaka* is $R\cos\phi$. Hence, multiplying the above equation by the radius (*trijyā*) we have

$$akṣajyā = \frac{trijyā \times chāyā}{karṇa}, \tag{8.3a}$$

$$\begin{aligned} \text{and} \qquad lambaka &= \frac{trijyā \times śaṅku}{karṇa}, \\ \text{or} \qquad lambaka &= \frac{trijyā \times \sqrt{karṇa^2 - chāyā^2}}{karṇa} \\ &= \sqrt{\frac{trijyā^2 \times (karṇa^2 - chāyā^2)}{karṇa^2}}. \end{aligned}$$

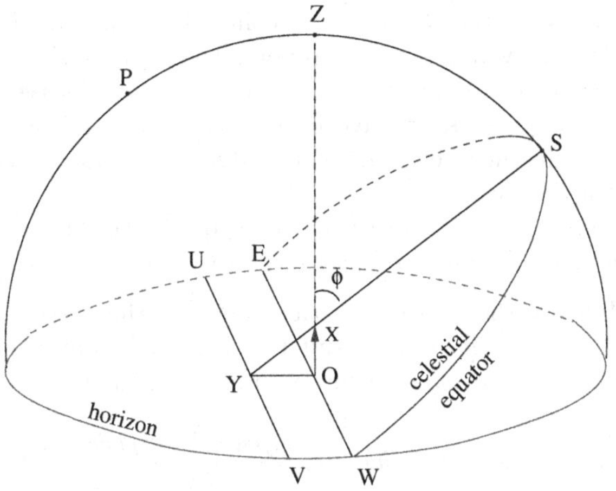

Fig. 8.1 Determination of the latitude from the equinoctial shadow of the *śaṅku*.

Using (8.3*a*) in the above, we have

$$lambaka = \sqrt{\left(trijy\bar{a}^2 - ak\d{s}ajy\bar{a}^2\right)}. \tag{8.3b}$$

This is what is mentioned in the verse.

८.२ लम्बाक्षज्यास्फुटीकरणम्

8.2 Corrections to the Rsine and Rcosine of the latitude

भूव्यासार्धसमाहताक्षगुणतस्त्रिज्याहृतं भास्वतो
व्यासार्धादपनीयशिष्टगुणितौ लम्बाक्षसंज्ञौ गुणौ ।
कक्ष्याव्यासदलेन चण्डमहसस्तात्कालिकेनाहरेत्
तत्राप्तौ क्रमशोऽक्षलम्बगुणयोः स्वर्णं तदाप्तौ स्फुटौ ॥ ३ ॥

bhūvyāsārdhasamāhatākṣaguṇatastrijyāhṛtaṃ bhāsvato
vyāsārdhādapanīyaśiṣṭaguṇitau lambākṣasaṃjñau guṇau |
kakṣyāvyāsadalena caṇḍamahasastātkālikenāharet
tatrāptau kramaśo'kṣalambaguṇayoḥ svarṇaṃ tadāptau sphuṭau || 3 ||

The radius of the earth is multiplied by the *akṣajyā* and divided by the *trijyā*, and the result is subtracted from the radius of the Sun. The remainder is multiplied by *lambajyā* and *akṣajyā* [separately and the results] are divided by the instantaneous (*tātkālika*) radius of the orbit (*kakṣyāvyāsadala*) of the Sun. The results obtained, when added to and subtracted from the *akṣajyā* and *lambajyā* respectively, would give the accurate [values of the same].

The expressions for the Rsine and Rcosine of the latitude in terms of the shadow stated in the verses 1 and 2 do not take the finite size of the Sun into account. Also in all the calculations, it is the shadow as observed at the centre of the earth which appears, whereas observations are made on the surface of the earth. The difference between them is due to 'parallax', which is due to the finite size of the earth.

The above verse gives the corrections to be applied to the observed '*akṣajyā*' and '*lambajyā*' due to these two factors via a two step process.

Step 1: Multiply the *akṣajyā* by the radius of the earth (*bhūvyāsārdha*) and divide by the *trijyā*. The result is subtracted from radius of the Sun (*arkavyāsārdha*). That is, we have to find the quantity,

$$\left(arkavy\bar{a}s\bar{a}rdha - \frac{ak\d{s}ajy\bar{a} \times bh\bar{u}vy\bar{a}s\bar{a}rdha}{trijy\bar{a}}\right).$$

Step 2: This quantity obtained in Step 1 is multiplied separately by *lambajyā* and *akṣajyā* and divided by the instantaneous *kakṣyāvyāsārdha* of the Sun. The results obtained are applied positively and negatively to the *akṣajyā* and *lambajyā* respectively. That is, the corrected values of the *akṣajyā* and *lambajyā* are

$$ak\d{s}ajy\bar{a} + \frac{lambajy\bar{a} \times \left(arkavy\bar{a}s\bar{a}rdha - \frac{ak\d{s}ajy\bar{a}\times bh\bar{u}vy\bar{a}s\bar{a}rdha}{trijy\bar{a}}\right)}{tatk\bar{a}la\text{-}kak\d{s}y\bar{a}vy\bar{a}s\bar{a}rdha},$$

and

$$lambajy\bar{a} - \frac{ak\d{s}ajy\bar{a} \times \left(arkavy\bar{a}s\bar{a}rdha - \frac{ak\d{s}ajy\bar{a}\times bh\bar{u}vy\bar{a}s\bar{a}rdha}{trijy\bar{a}}\right)}{tatk\bar{a}la\text{-}kak\d{s}y\bar{a}vy\bar{a}s\bar{a}rdha}$$

respectively.

These corrections have been discussed in the *chāyāprakaraṇa* chapter of *Tantrasaṅgraha*[2] also, but they are formulated differently here.

The correction which arises owing to the finite size of the Sun is illustrated in Fig. 8.2. Here OA is the *śaṅku* and PSQ represents the sectional view of the Sun, S being the centre. $r_s = PS$ is the *arkavyāsārdha* (radius of the Sun) and $O_s = SA$ is the *tatkāla-kakṣyāvyāsārdha* (distance of the Sun from the centre of the earth at that instant).

If the Sun was a point source of light, then the tip of the shadow of the *śaṅku* would fall at S' and $\phi' = O\hat{A}S'$ would be the latitude as measured by the observer. However, if P is the upper extremity of the Sun's disc, the tip of the shadow of the *śaṅku* would fall at P' and $\phi'' = O\hat{A}P'$ would be the actual value of the latitude observed. Now

$$\phi' = O\hat{A}S' = O\hat{A}P' + P'\hat{A}S' = \phi'' + \gamma,$$

[2] {TS 2011}, 131-214.

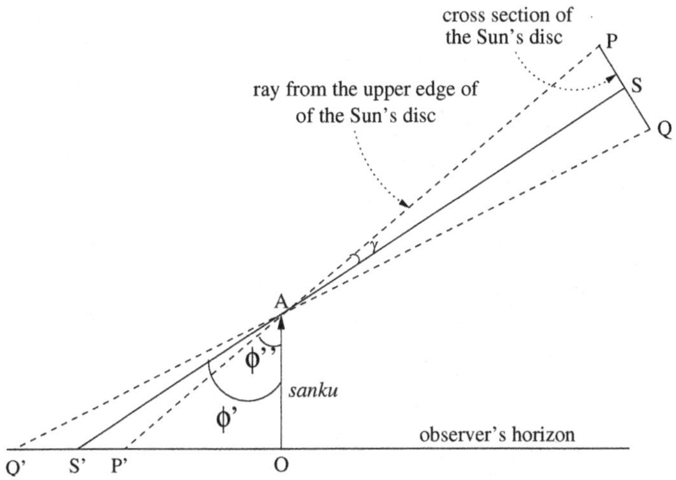

Fig. 8.2 Correction due to the finite size of the Sun.

where γ is the angular semidiameter of the Sun. Therefore,

$$\begin{aligned} R\sin\phi' &= R\sin(\phi''+\gamma) \\ &\approx R\sin\phi'' + \gamma.R\cos\phi'', \end{aligned} \tag{8.4}$$

since γ is very small, that is, $\gamma << \phi''$. Similarly,

$$R\cos\phi' = R\cos(\phi''+\gamma) \approx R\cos\phi'' - \gamma.R\sin\phi''$$

Also, since γ is small

$$\gamma \approx R\sin\gamma = \frac{PS}{SA} = \frac{r_s}{O_s} = \frac{arkavy\bar{a}s\bar{a}rdha}{tatk\bar{a}la\text{-}kak\d{s}y\bar{a}\text{-}vy\bar{a}s\bar{a}rdha}.$$

Therefore,

$$R\sin\phi' = R\sin\phi'' + \frac{r_s}{O_s}.R\cos\phi'', \tag{8.5}$$

and

$$R\cos\phi' = R\cos\phi'' - \frac{r_s}{O_s}.R\sin\phi''. \tag{8.6}$$

Since the observed value of the latitude ϕ'', the radius of the Sun r_s, and the distance of the Sun O_s are known, the first corrected value of the latitude ϕ' can be calculated.

Now we consider the correction due to the parallax as depicted in Figure 8.3. O represents the centre of the earth and A, the location of the observer. S is the Sun, $OS = O_s$ is the radius of the Sun's orbit at that instant and $OA = r_e$

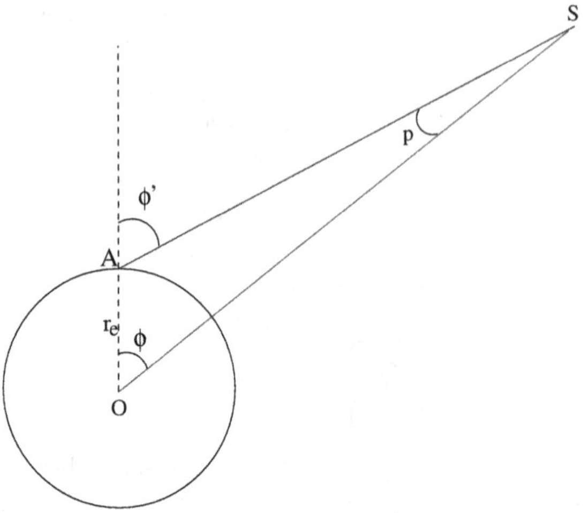

Fig. 8.3 Correction due to parallax.

is the *bhūvyāsārdha*, or the radius of the earth. ϕ' is the latitude including the correction due to the finite size of the Sun.

Considering the triangle OAS in Figure 8.3,

$$\frac{\sin p}{r_e} = \frac{\sin(180 - \phi')}{O_s},$$

or $$\sin p = \sin \phi' \frac{r_e}{O_s}. \tag{8.7}$$

Also, from the figure,

$$\phi = \phi' - p,$$

where $A\hat{S}O = p$ is the parallax. Therefore,

$$\begin{aligned} \sin \phi &= \sin(\phi' - p) \\ &= \sin \phi' \cos p - \cos \phi' \sin p. \end{aligned} \tag{8.8}$$

Using (8.7) in (8.8) and considering the fact that $p << \phi$ so that $\cos p \approx 1$, we have

$$R \sin \phi \approx R \sin \phi' - R \cos \phi' \frac{R \sin \phi'}{R} \frac{r_e}{O_s}.$$

Similarly

$$R \cos \phi \approx R \cos \phi' + R \sin \phi' \frac{R \sin \phi'}{R} \frac{r_e}{O_s}.$$

Using the expression for $R \sin \phi'$ and $R \cos \phi'$ given by (8.5) and (8.6) respectively, and neglecting higher order terms (of order $\frac{r_e r_s}{O_s^2}$), we have

$$R\sin\phi = R\sin\phi'' + R\cos\phi''\frac{\left(r_s - \dfrac{r_e.R\sin\phi''}{R}\right)}{O_s}, \quad (8.9)$$

and

$$R\cos\phi = R\cos\phi'' - R\sin\phi''\frac{\left(r_s - \dfrac{r_e.R\sin\phi''}{R}\right)}{O_s}. \quad (8.10)$$

These are the expressions stated in the above verse, as r_s is the *arkavyāsārdha*, r_e is the *bhūvyāsārdha*, O_s is the *tatkāla-kakṣyāvyāsārdha*, and $R\sin\phi''$ and $R\cos\phi''$ are the uncorrected *akṣajyā* and *lambajyā*.

८.३ इष्टकाले महाशङ्कुच्छायानयनम्

8.3 Obtaining *mahāśaṅku* and *chāyā* at any desired instant

अङ्गुलात्मिकयाभीष्टच्छायया तत्र भास्वतः ।
महाशङ्कुप्रभे कार्ये संस्कृते लम्बकाक्षवत् ॥ ४ ॥

aṅgulātmikayābhīṣṭacchāyayā tatra bhāsvataḥ |
mahāśaṅkuprabhe kārye saṃskṛte lambakākṣavat || 4 ||

Mahāśaṅku and *Mahācchāyā* at any instant can be obtained from the shadow of the Sun[3] in *aṅgulas* [at that instant]. [These] have to be corrected just like *lambaka* and *akṣa*.

We explain the concept of *mahāśaṅku* and *mahācchāyā* with the help of Figure 8.4. Here, S is the Sun and F is the foot of perpendicular drawn from the Sun to the horizon. The angle $O\hat{X}Y = F\hat{S}O = z$ is the zenith distance of the Sun, and $O\hat{Y}X = F\hat{O}S = 90 - z = a$ is the altitude of the Sun. The triangles OXY and FSO are similar. In the triangle OXY, OX represents the usual *śaṅku* of 12 units in height, and OY is its shadow, or the *chāyā*. $SF = R\cos z$ and $FO = R\sin z$ are referred to as the *mahāśaṅku* and the *mahācchāyā* respectively.

It is stated in the verse that the *mahāśaṅku* and *mahācchāyā* are to be corrected just like the *lambajyā* and *akṣajyā* in order to make them *sphuṭa*. Let $R\cos z$ and $R\sin z$ be the *sphuṭamahāśaṅku* and *sphuṭamahācchāyā* respectively, where z refers to the zenith distance at the centre of the earth. If z' is the observer's zenith distance,

---

[3] The third case *chāyayā* used here is '*hetau tṛtīyā*' and not '*karaṇe tṛtīyā*'. Hence it has been translated as, 'from the shadow'.

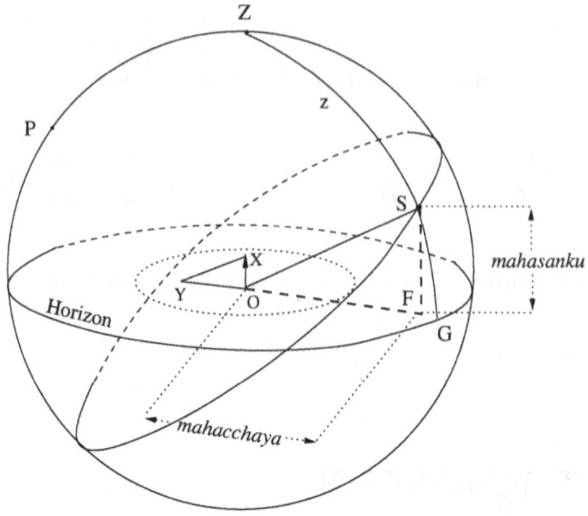

Fig. 8.4 The *mahāśaṅku* and the *mahācchāyā*.

$$R\cos z = R\cos z' - \frac{R\sin z' \times \left(arkavy\bar{a}s\bar{a}rdha - \frac{R\sin z' \times bh\bar{u}vy\bar{a}s\bar{a}rdha}{trijy\bar{a}}\right)}{tatk\bar{a}la\text{-}kak\d{s}y\bar{a}vy\bar{a}s\bar{a}rdha},$$

$$R\sin z = R\sin z' + \frac{R\cos z' \times \left(arkavy\bar{a}s\bar{a}rdha - \frac{R\sin z' \times bh\bar{u}vy\bar{a}s\bar{a}rdha}{trijy\bar{a}}\right)}{tatk\bar{a}la\text{-}kak\d{s}y\bar{a}vy\bar{a}s\bar{a}rdha}.$$

These relations can be derived in the same way as was done in the previous section for the case of the *akṣajyā* and *lambajyā*.

८.४ महाच्छायाभुजाकोट्यानयनम्

8.4 Obtaining the *bhujākoṭis* of *mahācchāyā*

छायाङ्गुलभुजाकोट्यौ हते स्फुटमहाभया ।
छायाङ्गुलहृते स्यातां महाभा बाहुकोटिके ॥ ५ ॥

याम्योत्तरा भुजा स्याच्छायायाः पूर्वपश्चिमकोटिः ।
सममण्डलगे भानौ नैव भुजा कोटिका न मध्याह्ने ॥ ६ ॥

chāyāṅgulabhujākoṭyau hate sphuṭamahābhayā |
chāyāṅgulahṛte syātāṃ mahābhā bāhukoṭike || 5 ||

yāmyottarā bhujā syācchāyāyāḥ pūrvapaścimakoṭiḥ |
samamaṇḍalage bhānau naiva bhujā koṭikā na madhyāhne || 6 ||

The *bhujā* and *koṭi* of *mahācchāyā* can be obtained by multiplying the *bhujā* and *koṭi* of *chāyāṅgula* by *sphuṭamahācchāyā* and dividing by the *chāyāṅgula*.

[The direction of] *bhujā* is along north-south line [whereas] that of *koṭi* is along east-west line. When the Sun is on the prime vertical, then there will not be *bhujā* [for *chāyā*], and there is no *koṭi* [for *chāyā* formed from] the noon [Sun].

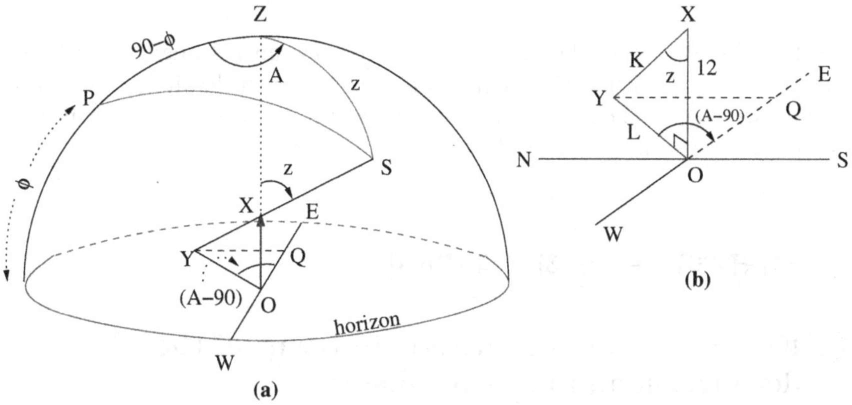

Fig. 8.5 Obtaining the *bhujā* and *koṭi* of *mahācchāyā*.

In Indian astronomical texts, the azimuthal angle (A') is measured with respect to the *samamaṇḍala* or the prime vertical (great circle passing through E, W and Z) $A' = A \sim 90°$, where A is the azimuth measured with respect to the meridian. In Figure 8.5, $OX = 12$ is the *śaṅku*. The Sun is at S, corresponding to a zenith distance z. A' is the azimuthal angle of the Sun, as measured from the prime vertical. OY is the *chāyā* (shadow), and $XY = K$ is the *karṇa*. We have $OX = K \cos z$ and $OY = K \sin z$, where z is the zenith distance. YQ is drawn perpendicular to the EW line from the tip of the shadow Y. Then

$$\begin{aligned} chāyābhujā &= YQ = OY \sin A' \\ &= K \sin z \sin A', \qquad (8.11) \\ \text{and} \quad chāyākoṭi &= OQ = OY \cos A' \\ &= K \sin z \cos A'. \qquad (8.12) \end{aligned}$$

Then

$$\begin{aligned} mahācchāyābhujā &= R \sin z \sin A' \\ &= R \sin z \frac{K \sin z \sin A'}{K \sin z} \\ &= mahācchāyā \times \frac{chāyābhujā}{chāyā}, \qquad (8.13) \end{aligned}$$

and

$$
\begin{aligned}
mah\bar{a}cch\bar{a}y\bar{a}ko\d{t}i &= R\sin z\cos A' \\
&= R\sin z\frac{K\sin z\cos A'}{K\sin z} \\
&= mah\bar{a}cch\bar{a}y\bar{a}\times\frac{ch\bar{a}y\bar{a}ko\d{t}i}{ch\bar{a}y\bar{a}}. \qquad (8.14)
\end{aligned}
$$

When the Sun is on the prime vertical, the azimuthal angle $A' = 0$, and hence the *chāyābhujā* and the *mahācchāyābhujā* also vanish. Similarly, at noon, the Sun will be on the meridian, and the zenith distance $z = 0$. Then, both the *chāyākoṭi* and *mahācchāyā-koṭi* are also zero.

८.५ क्रान्तिमध्याह्नभाभ्यां अक्षज्यानयनम्

8.5 Expression for the latitude in terms of the declination and mid-day shadow

त्रिज्याहतापक्रमतोऽक्षजीवा स्याच्छङ्कुभक्ता समवृत्तगेऽर्के ।
मध्याह्नभापक्रमचापयोर्वा भेदोऽक्षचापो विदिशोस्तु योगः ॥ ७ ॥

trijyāhatāpakramato'kṣajīvā syācchaṅkubhaktā samavṛttage'rke |
madhyāhnabhāpakramacāpayorvā bhedo'kṣacāpo vidiśostu yogaḥ || 7 ||

The Rsine of latitude (*akṣajīvā*) [of a place] is the product of Rsine of declination (*apakramajyā*) and the *trijyā* divided by *śaṅku* when the Sun is on the prime vertical (*samaśaṅku*). Or, the arc of the latitude is the difference between the arcs of mid-day shadow and the declination. The two arcs are to be added if they are in opposite directions.

In Figure 8.6, EZW is the *samamaṇḍala* or the prime vertical. Let z_0 be the zenith distance of the Sun S, when it is on the prime vertical. Then the *samaśaṅku* is $R\cos z_0$. If the declination of the Sun is δ, then the *apakramajyā* is $R\sin\delta$. Here it is stated that

$$R\sin\phi = \frac{apakramajy\bar{a}\times trijy\bar{a}}{sama\acute{s}a\dot{n}ku} = \frac{R\sin\delta\times R}{R\cos z_0}. \qquad (8.15)$$

In Figure 8.6, $SP = 90 - \delta$ and $ZP = 90 - \phi$. In the spherical triangle SZP, the spherical angle at Z is 90°. Applying the cosine formula to the side SP, we have

$$\cos(90-\delta) = \cos z_0 . \cos(90-\phi),$$

or
$$\sin\phi = \frac{\sin\delta}{\cos z_0}, \qquad (8.16)$$

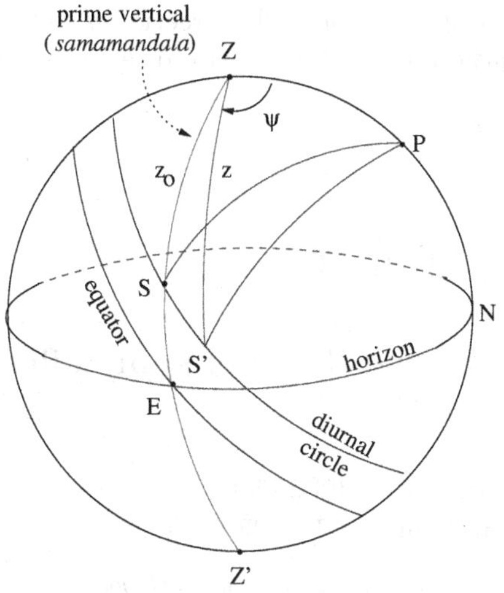

Fig. 8.6 Latitude in terms of the declination and zenith distance.

which is the same as (8.15).

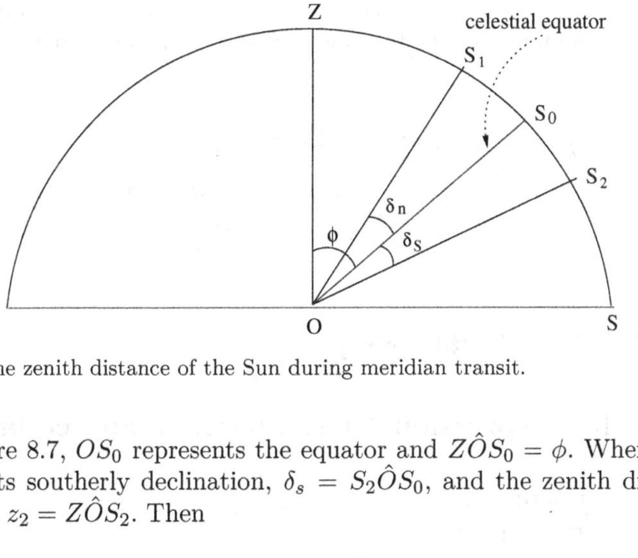

Fig. 8.7 The zenith distance of the Sun during meridian transit.

In Figure 8.7, OS_0 represents the equator and $Z\hat{O}S_0 = \phi$. When the Sun is at S_2, its southerly declination, $\delta_s = S_2\hat{O}S_0$, and the zenith distance at mid-day is $z_2 = Z\hat{O}S_2$. Then

$$\phi = z_2 - \delta_s.$$

Here, the zenith distance $ZS_2 = z_2$ and the declination $S_0S_2 = \delta_s$ are in the same direction, and the arc of the latitude is the difference between the arcs of the zenith distance and the declination.

When the Sun is at S_1, its northerly direction is $z_1 = Z\hat{O}S_1$. Here, z_1 and the northerly declination δ_n are in opposite directions. In this case,

$$\phi = z_1 + \delta_n.$$

८.६ अभीष्टचापयोगविवरयोः ज्या

8.6 Expression for the Rsine of sum/difference of two arcs

अन्योन्यकोटिहतयोरभिमतगुणयोः त्रिजीवया हृतयोः ।
योगवियोगौ स्यातामभिमतगुणचापयोगविवरगुणौ ॥ ८ ॥

anyonyakoṭihatayorabhimataguṇayoḥ trijīvayā hṛtayoḥ |
yogaviyogau syātāmabhimataguṇacāpayogavivaraguṇau || 8 ||

Multiply each *guṇa* (Rsine of an arc) by the other *koṭi* (Rcosine of the other arc) and divide them by the *trijyā*. Their sum or difference becomes the *guṇa* (Rsine) of the sum or difference of the arcs.

This verse essentially gives the rule for the Rsine of the sum or difference of two arcs (*yogavivaraguṇa*). That is, $\sin(A \pm B)$ formula. If α and β be the two arcs corresponding to the two angles θ and ϕ, then the rule given may be expressed as

$$\begin{aligned} jy\bar{a}\,(\alpha \pm \beta) &= \frac{jy\bar{a}\ \alpha\ ko\d{t}ijy\bar{a}\ \beta \pm ko\d{t}ijy\bar{a}\ \alpha\ jy\bar{a}\ \beta}{trijy\bar{a}}, \\ R\sin\,(\theta \pm \phi) &= \frac{R\sin\ \theta\ R\cos\ \phi \pm R\cos\ \theta\ R\sin\ \phi}{R}. \end{aligned} \tag{8.17}$$

८.७ लम्बाक्षज्यानयने प्रकारान्तरम्

8.7 Another expression for the latitude and co-latitude

भाकोटिका द्युगुणवर्गभिदा पदाढ्य-
च्छायाभुजापहृतशङ्कुपमैक्यवर्गः ।
हारे धनर्णमनयोर्महताल्पतोऽक्षः
त्रिज्या हताद् भवति लम्बगुणोऽस्य कोटिः ॥ ९ ॥

bhākoṭikā dyuguṇavargabhidā padāḍhya-
cchāyābhujāpahṛtaśaṅkvapamaikyavargaḥ |
hāre dhanarṇamanayormahatālpato'kṣaḥ
trijyā hatād bhavati lambaguṇo'sya koṭiḥ || 9 ||

> [Obtain] the square root of the difference between the squares of the *bhākoṭi* and day radius (*dyujyā*) from which the *chāyābhujā* is subtracted. By the resulting quantity divide the square of the sum of the gnomon (*śaṅku*) and the Rsine of declination (*apamajyā*), and the result obtained is added to and subtracted from [the initial result]. Among these, the one which is smaller in magnitude is multiplied by the radius (*trijyā*) and divided by the one which is larger in magnitude, in order to obtain the Rsine of latitude (*akṣajyā*). The associated Rcosine is the *lambajyā*.

To start with, we list a few technical terms employed here.

$$\begin{aligned}
\textit{śaṅku} &= R\cos z, \\
\textit{akṣajyā} &= R\sin\phi, \\
\textit{lambajyā} &= R\cos\phi, \\
\textit{dyuguṇa} = \textit{dyujyā} &= R\cos\delta, \\
\textit{bhākoṭikā} = \textit{chāyākoṭi} &= R\sin z\cos A', \\
\textit{bhādoḥchāyābhujā} &= R\sin z\sin A', \\
\textit{apamajyā} &= R\sin\delta.
\end{aligned}$$

Now, Let $x = \sqrt{R^2\cos^2\delta - R^2\sin^2 z\cos^2 A'} - R\sin z\sin A'$. The expression for the Rsine of the latitude (*akṣajyā*) stated in the verse is

$$\textit{akṣajyā} = R\sin\phi = R\times\left[\frac{\dfrac{(R\cos z + R\sin\delta)^2}{x} - x}{\dfrac{(R\cos z + R\sin\delta)^2}{x} + x}\right]. \tag{8.18}$$

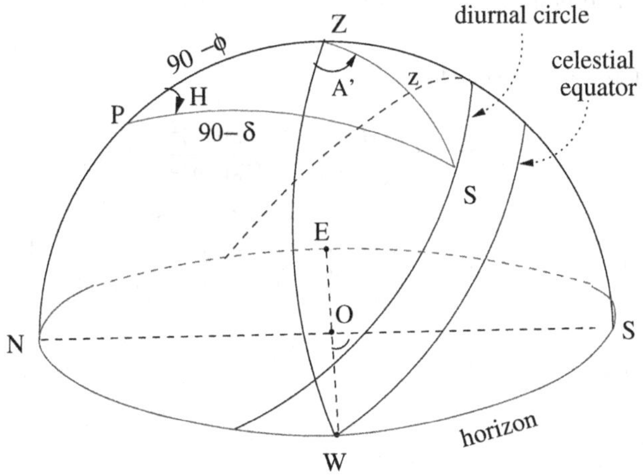

Fig. 8.8 Another method to obtain the latitude.

The rationale behind the above expression can be understood with the help of Figure 8.8. Here, in the spherical triangle $P\hat{Z}S$, $PZ = 90 - \phi$, $ZS = z$, $PS = 90 - \delta$, $P\hat{Z}S = 90 + A'$ and $Z\hat{P}S = H$, the hour angle. Applying the sine formula,

$$\frac{\sin(90 + A')}{\sin(90 - \delta)} = \frac{\sin H}{\sin z}.$$

Hence $\sin z \cos A' = \cos\delta \sin H$. From this it follows that

$$\begin{aligned} x &= \sqrt{R^2\cos^2\delta - R^2\sin^2 z\cos^2 A'} - R\sin z\sin A' \\ &= R(\cos\delta\cos H - \sin z\sin A'). \end{aligned} \tag{8.19}$$

Now applying the cosine formula to the sides $ZS = z$, and $PS = 90 - \delta$, we have

$$\begin{aligned} \cos z &= \sin\phi\sin\delta + \cos\phi\cos\delta\cos H, \\ \text{and} \qquad \sin\delta &= \sin\phi\cos z - \cos\phi\sin z\sin A'. \end{aligned} \tag{8.20}$$

Adding the two expressions, and rearranging the terms, we obtain

$$(\cos z + \sin\delta)(1 - \sin\phi) = \cos\phi(\cos\delta\cos H - \sin z\sin A').$$

Multiplying by R, and squaring, we get

$$\begin{aligned} (R\cos z + R\sin\delta)^2 &= \frac{\cos^2\phi}{(1-\sin\phi)^2}x^2 \\ &= \frac{1+\sin\phi}{1-\sin\phi}x^2. \end{aligned}$$

$$\text{Therefore,} \qquad \frac{1+\sin\phi}{1-\sin\phi} = \frac{(R\cos z + R\sin\delta)^2}{x^2}. \tag{8.21}$$

From this, it is easy to see that

$$\sin\phi = \frac{(R\cos z + R\sin\delta)^2 - x^2}{(R\cos z + R\sin\delta)^2 + x^2},$$

$$\text{or} \qquad R\sin\phi = R \times \left[\frac{\dfrac{(R\cos z + R\sin\delta)^2}{x} - x}{\dfrac{(R\cos z + R\sin\delta)^2}{x} + x}\right], \tag{8.22}$$

which is the same as (8.18).

८.८ स्वदेशहारकः

8.8 Obtaining the *svadeśahāraka*

व्यासार्धवर्गतो लब्धः स्फुटया लम्बजीवया ।
स्वदेशहारकः प्रोक्तो व्यस्तलम्बः स उच्यते ॥ १० ॥

vyāsārdhavargato labdhaḥ sphuṭayā lambajīvayā |
svadeśahārakaḥ prokto vyastalambaḥ sa ucyate || 10 ||

The result [obtained by] dividing the square of the *vyāsārdha* (*radius*) by the corrected *lambajyā* is known as *svadeśahāraka*. This is [also] referred to as the *vyastalamba.*

The *svadeśahāraka* defined in the above verse may be expressed as

$$svade\acute{s}ah\bar{a}raka = \frac{R^2}{R\cos\phi}. \tag{8.23}$$

Svadeśahāraka is used to calculate the time difference (*deśāntarakāla*) between two places on the same latitude circle. Now a distance d between two places on a latitudinal circle corresponding to a latitude ϕ, corresponds to a distance $\frac{d}{\cos\phi}$ on the equator.

The circumference of the earth is specified as 3300 *yojanas*. A distance of 3300 *yojanas* on the equator corresponds to a difference of one day or 60 *ghaṭikās*. Hence the time difference corresponding to a distance d on the latitudinal circle, or $\frac{d}{\cos\phi}$ on the equator would be

$$\begin{aligned} \delta t &= \frac{d}{3300\cos\phi} \times 60 \\ &= \frac{d}{3300} \times \frac{60}{R}\left(\frac{R^2}{R\cos\phi}\right) \quad gha\d{t}ik\bar{a}s. \end{aligned} \tag{8.24}$$

८.९ अर्कक्रान्तिज्याद्युज्ययोरानयनम्

8.9 Obtaining the *krāntijyā* of Sun and the *dyujyā*

गुणश्चतुर्विंशतिभागजातः परापमस्तेन हतेष्टजीवा।
त्रिज्या हृता क्रान्तिगुणोऽस्य कोटिः द्युमौर्विका स्याद् दिननायकस्य॥ ११ ॥

guṇaścaturviṃśatibhāgajātaḥ parāpamastena hateṣṭajīvā |
trijyā hṛtā krāntiguṇo'sya koṭiḥ dyumaurvikā syād dinanāyakasya || 11 ||

The Rsine of 24° is the Rsine of maximum declination (*paramāpamajyā*). [This] multiplied by the desired Rsine [of the longitude] and divided by the *trijyā* would be the *krāntijyā*. The associated Rcosine is the *dyujyā* of the Sun.

Let δ and λ be the declination and the longitude of the Sun respectively, then the relation stated in the text may be written as

$$R\sin\delta = \frac{R\sin(24°)R\sin\lambda}{R}, \tag{8.25}$$

where $R\sin(24°)$ is stated to be the Rsine of the maximum declination (*paramāpamajyā*) in the text.

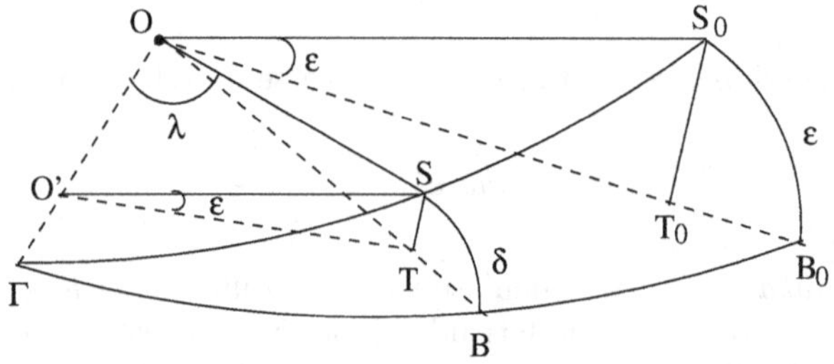

Fig. 8.9 Obtaining the *krāntijyā* of Sun in terms of the longitude.

The rationale for the above expression can be understood with the help of Figure 8.9. ΓBB_0 and ΓSS_0 are the quadrants of the equator and the ecliptic intersecting at the vernal equinox, Γ. Consider the situation when the Sun is at S on the ecliptic, when its longitude is $\lambda = \Gamma\hat{O}S$. Draw the arc $SB = \delta$ perpendicular to the equator, where δ is the declination. Draw ST perpendicular to OB, and $SO' = R\sin\lambda$ perpendicular to $O\Gamma$. Draw S_0T_0 perpendicular to OB_0. Now $S_0\hat{O}T_0 = S\hat{O'}T = \epsilon$, the obliquity of the ecliptic. Then, $S_0T_0 = R\sin\epsilon$ and $ST = R\sin\delta$.

Now the triangles $SO'T$ and S_0OT_0 are similar. Therefore,

$$\frac{ST}{S_0T_0} = \frac{SO'}{S_0O},$$

$$\text{or} \qquad \frac{R\sin\delta}{R\sin\epsilon} = \frac{R\sin\lambda}{R}. \tag{8.26}$$

Hence,

$$R\sin\delta = \frac{R\sin\epsilon . R\sin\lambda}{R}. \tag{8.27}$$

It is straightforward to see from the above relation that the declination has the maximum value when $\lambda = 90°$. That is, $\delta_{max} = \epsilon$. This is known as *paramāpama*. In all the Indian astronomical works the obliquity ϵ is taken to be 24°. Hence (8.27) reduces to

$$R \sin \delta = \frac{R \sin(24^\circ) R \sin \lambda}{R}.$$

The corresponding *koṭijyā*, $R \cos \delta$ is known as *dyujyā*. This is the radius of the diurnal circle, when the declination of the Sun is δ, and it figures in many relations related to diurnal problems.

८.१० प्राणकलान्तरगणनायां प्रकारवैविध्यम्

8.10 Different methods of obtaining *prāṇakalāntaras*

अन्त्यद्युजीवाहतबाहुजीवां इष्टद्युमौर्व्या विभजेदवाप्तम् ।
चापीकृतं बाहुगुणस्य चापाद् विशोधितं प्राणकलान्तरं स्यात् ॥ १२ ॥

कोटीगुणं व्यासदलेन संहत्येष्टद्युमौर्व्या विभजेदवाप्तम्।
चापीकृतात् कोटिगुणस्य चापे त्यक्तेऽथवा प्राणकलान्तरं स्यात् ॥ १३ ॥

दोः कोटिमौर्व्योर्वधतस्त्रिमौर्व्या लब्धं परापक्रमबाणनिघ्नम् ।
द्युज्याहृतं प्राणकलान्तरं तत् युग्मौजपादक्रमतो धनर्णम् ॥ १४ ॥

antyadyujīvāhatabāhujīvāṃ iṣṭadyumaurvyā vibhajedavāptam |
cāpīkṛtaṃ bāhuguṇasya cāpād viśodhitaṃ prāṇakalāntaraṃ syāt || 12 ||

koṭīguṇaṃ vyāsadalena saṃhatyeṣṭadyumaurvyā vibhajedavāptam |
cāpīkṛtāt koṭiguṇasya cāpe tyakte'thavā prāṇakalāntaraṃ syāt || 13 ||

doḥ koṭimaurvyorvadhatastrimaurvyā
labdhaṃ parāpakramabāṇanighnam |
dyujyāhṛtaṃ prāṇakalāntaraṃ tat
yugmaujapādakramato dhanarṇam || 14 ||

Multiply the last day-radius (*antyadyujyā*) by the Rsine of the longitude (*bāhujyā*) and divide by the desired day-radius (*dyujyā*). The arc of this, when subtracted from the longitude (the arc of the desired *bāhujyā*), would be the *prāṇakalāntara*.

Or else, the Rcosine of the longitude of the Sun is multiplied by the radius (*trijyā*) and divided by the desired day-radius (*dyujyā*). From the arc of [this], when the arc of the Rcosine of the longitude of the Sun is subtracted, [the result obtained] would be the *prāṇakalāntara*.

The product of the Rsine and Rcosine of the longitude is divided by the radius. The result is multiplied by the versine of the maximum declination (*paramāpakramabāṇa*) and divided by desired day-radius (*dyujyā*). [The associated arc] gives the *prāṇakalāntara*. This is positive for even quadrants and negative for odd quadrants.

We explain the rationale behind the three expressions for *prāṇakalāntara* given in the verses above, with the help of Figure 8.10. Here the point of intersection of the celestial equator and the ecliptic is the *sāyana meṣādi*. The Sun is situated on the ecliptic at S. The angle between the ecliptic and the celestial equator, $S\hat{\Gamma}D = \epsilon$. $\Gamma S = \lambda$ is the longitude of the Sun, measured along the ecliptic. The circle PSD is the meridian passing through S and is perpendicular to the celestial equator. Hence, the spherical angle $\Gamma\hat{D}S = 90°$ and, $SD = \delta$, is the declination of the Sun. Also, $\Gamma D = \alpha$, is the Right Ascension (R.A.) of the Sun, which is called the *natakāla*. The term *prāṇakalāntara* refers to the difference between the R. A. and the longitude of the Sun. That is,

$$prāṇakalāntara = \alpha - \lambda,$$

where α and λ are arcs expressed in minutes. This corresponds to part of the equation of time which is due to the obliquity of the ecliptic. It can be seen that the *prāṇakalāntara* is positive in the even quadrants and negative in the odd quadrants.

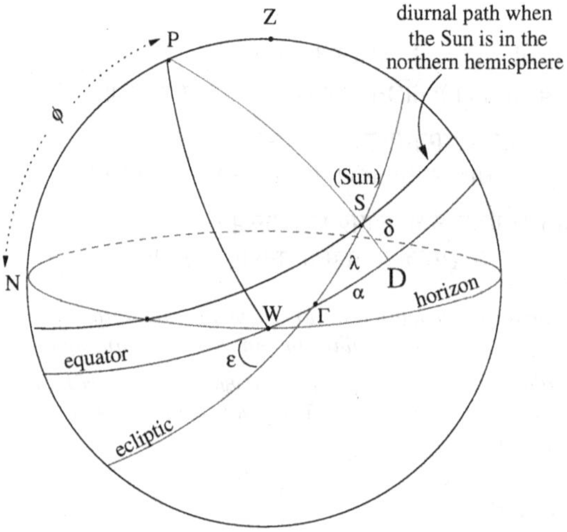

Fig. 8.10 Determination of *prāṇakalāntara*.

Verses 12, 13 and 14, essentially present the following three formulae to obtain the *prāṇakalāntara*:

$$\alpha \sim \lambda = \lambda - \sin^{-1}\left(\frac{\cos\epsilon \sin\lambda}{\cos\delta}\right),$$

$$\alpha \sim \lambda = \sin^{-1}(\cos\lambda) - \sin^{-1}\left(\frac{\cos\lambda}{\cos\delta}\right),$$

$$\alpha \sim \lambda = \sin^{-1}\left(\frac{\sin\lambda\cos\lambda(1-\cos\epsilon)}{\cos\delta}\right). \tag{8.28}$$

In what follows, we present a derivation of these relations. For this, consider the spherical triangle ΓPS. In this, $\Gamma\hat{P}S = \alpha$ and $PS = 90 - \delta$. Also, $P\hat{\Gamma}S = 90° - S\hat{\Gamma}D = 90 - \epsilon$. Applying the sine formula to this triangle,

$$\frac{\sin\alpha}{\sin\lambda} = \frac{\sin(90-\epsilon)}{\sin(90-\delta)}, \qquad \text{or} \qquad \sin\alpha = \frac{\cos\epsilon\sin\lambda}{\cos\delta}. \tag{8.29}$$

Hence

$$\alpha = \sin^{-1}\left(\frac{\cos\epsilon\sin\lambda}{\cos\delta}\right). \tag{8.30}$$

Thus, we obtain

$$\alpha - \lambda = \sin^{-1}\left(\frac{\cos\epsilon\sin\lambda}{\cos\delta}\right) - \lambda, \tag{8.31}$$

which is the first of the equations (8.28) for $\alpha \sim \lambda$.

Now, consider the spherical triangle ΓSD. Here, $\Gamma\hat{D}S = 90°$, and $SD = \delta$. Applying the cosine formula, we obtain

$$\begin{aligned} \cos\lambda &= \cos\alpha\cos\delta, \\ \text{or} \qquad \cos\alpha &= \frac{\cos\lambda}{\cos\delta}. \end{aligned} \tag{8.32}$$

Therefore,

$$\begin{aligned} \sin^{-1}\left(\frac{\cos\lambda}{\cos\delta}\right) &= \sin^{-1}(\cos\alpha) = 90° - \alpha \text{ or } 90° + \alpha, \\ \text{and} \qquad \sin^{-1}(\cos\lambda) &= 90° - \lambda \text{ or } 90° + \lambda. \end{aligned}$$

Hence

$$\begin{aligned} \alpha \sim \lambda &= 90° \pm \alpha - (90° \pm \lambda) \\ &= \sin^{-1}\left(\frac{\cos\lambda}{\cos\delta}\right) - \sin^{-1}(\cos\lambda), \end{aligned} \tag{8.33}$$

which is the second of the relations (8.28) for $\alpha \sim \lambda$.

Now consider the expression,

$$\begin{aligned} \frac{\sin\lambda\cos\lambda(1-\cos\epsilon)}{\cos\delta} &= \sin\lambda\frac{\cos\lambda}{\cos\delta} - \cos\lambda\frac{\sin\lambda\cos\epsilon}{\cos\delta} \\ &= \sin\lambda\cos\alpha - \cos\lambda\sin\alpha \\ &= \sin(\lambda-\alpha). \end{aligned} \tag{8.34}$$

Hence

$$\alpha \sim \lambda = \sin^{-1}\left[\frac{\sin\lambda\cos\lambda(1-\cos\epsilon)}{\cos\delta}\right], \tag{8.35}$$

which is the third of the relations (8.28) for $\alpha \sim \lambda$ given in the set of verses above.

८.११ चरज्यानयनम्

8.11 The expression for the ascensional difference or *carajyā*

पलज्यया हतापमात् स्वलम्बकेन भूगुणः ।
ततस्त्रिजीवयाहताद् द्युजीवया हृतं चरम् ॥ १५ ॥

palajyayā hatāpamāt svalambakena bhūguṇaḥ |
tatastrijīvayāhatād dyujīvayā hṛtaṃ caram || 15 ||

The Rsine of the latitude (*palajyā*), multiplied by Rsine of the declination (*apamajyā*) and divided by Rcosine of the latitude (*lambaka*) of the place, would be the *bhūguṇa* (generally referred to as earth-sine or *kṣitijyā*). That multiplied by the radius and divided by the day-radius (*dyujīvā*) would be the Rsine of ascessional difference (*carajyā*).

We first explain the concepts of the earthsine (*kṣitijyā*) and Rsine ascessional difference (*carajyā*), before presenting the derivation of the expressions for them given in the above verse.

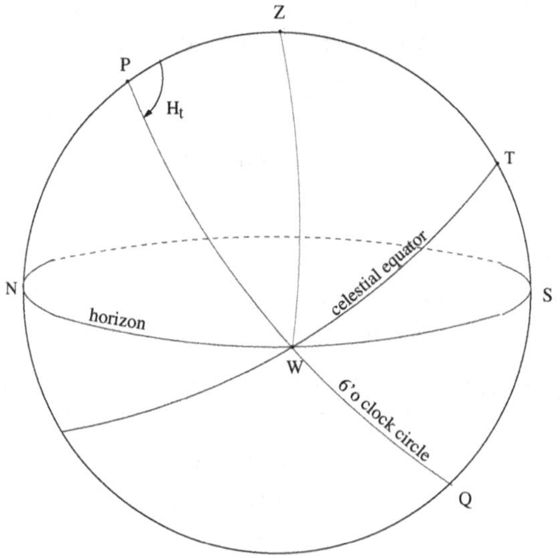

Fig. 8.11 Ascessional difference when the Sun is on the equator.

When the Sun is on the equator (declination, $\delta = 0$), the duration of the day (sunrise to sunset) is 12 hours irrespective of the latitude (ϕ) of the observer. Hence, the Sun takes six hours from meridian-transit to the setting on the horizon, and the hour angle at sunset would be 90°. When the declination of the Sun is northerly ($\delta > 0$), the Sun takes more than six hours from the meridian transit to the setting on the horizon, and correspondingly, the hour angle at sunset can be written as $90° + \Delta\alpha$. This ascessional difference $\Delta\alpha$ is called *cara* and $R\sin\Delta\alpha$ is called *carajyā*. Similarly the hour angle at sunset is less than 90° when the declination is southerly ($\delta < 0$) and can be expressed as $90° - \Delta\alpha$, as depicted in Figure 8.13.

In Figure 8.12, X represents the setting point of the Sun on the horizon, and $XV = \delta$ is Sun's declination. The great circle VWT is the celestial equator and the small circle XYT' which is parallel to it is the diurnal circle of the Sun on that day, whose radius is *dyujyā* or $R\cos\delta$. The great circle PYW is known as the *unmaṇḍala* (6 o' clock circle). $W\hat{P}Z = 90°$, and $H_t = X\hat{P}Z$ is the hour angle at sunset. $X\hat{P}Y = \Delta\alpha$ is the ascessional difference or *cara*.

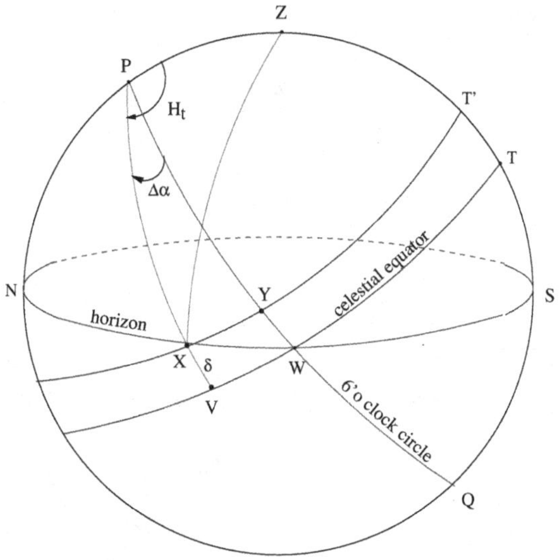

Fig. 8.12 Ascessional difference when the declination of the Sun is northerly.

Now in the spherical triangle XPZ, $XZ = 90°$, $PX = 90° - \delta$ and $PZ = 90° - \phi$. Applying the cosine formula, we have

$$\begin{aligned} \cos 90° &= \cos(90-\delta)\cos(90-\phi) + \sin(90-\delta)\sin(90-\phi)\cos H_t, \\ \text{or} \qquad \cos H_t &= -\tan\phi\tan\delta. \end{aligned}$$

As $H_t = (90 + \Delta\alpha)$, we obtain

$$\sin\Delta\alpha = \tan\phi\tan\delta. \tag{8.36}$$

Now the earthsine (*kṣitijyā*) is the sine of the ascessional difference on the diurnal circle and is given by

$$ksitijy\bar{a} = R\sin \Delta\alpha \cos\delta = \frac{R\sin\phi R\sin\delta}{R\cos\phi}$$
$$= \frac{palajy\bar{a} \times apamajy\bar{a}}{lambaka}, \qquad (8.37)$$

as stated in the verse. Hence, *carajyā* is given by

$$carajy\bar{a} = R\sin\Delta\alpha = R\tan\phi\tan\delta$$
$$= \frac{R\sin\phi R\sin\delta}{R\cos\phi} \times \frac{R}{R\cos\delta}$$
$$= ksitijy\bar{a} \times \frac{trijy\bar{a}}{dyujy\bar{a}}, \qquad (8.38)$$

as stated in the verse.

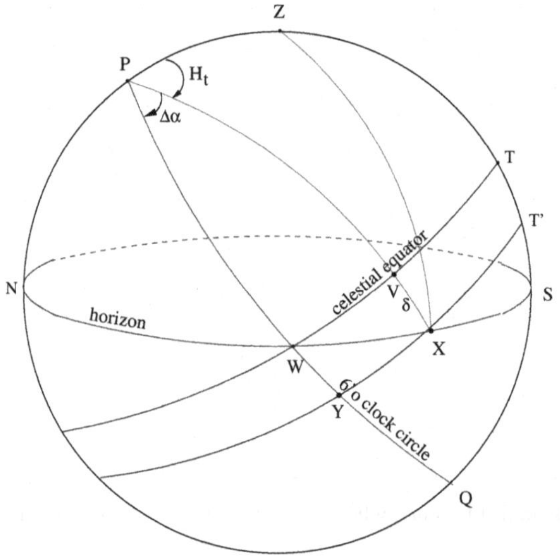

Fig. 8.13 Ascensional difference when the declination of the Sun is southerly.

८.१२ चरज्यानयने प्रकारान्तरम्

8.12 Alternate expressions for the *carajyā*

चरमद्युगुणाहतेष्टदोर्ज्या विहृतेष्टद्युगुणेन कालजीवा ।
चरमेण चरेण ताडिता सा त्रिभजीवापहृताथवा चरज्या ॥ १६ ॥

त्रिज्याक्षघातादवलम्बकाप्तेनाहत्य दोःक्रान्तिगुणं द्युमौर्व्या ।
हरेदवाप्तं चरमौर्विका स्यात् तदीयचापा हि चरासवः स्युः ॥ १७ ॥

त्रिज्याक्षघातलम्बांशेनाहतादिष्टदोर्गुणात् ।
तत्कोट्याप्तं चरज्या स्याद् ग्राह्या सा क्रान्तिचापतः ॥ १८ ॥

caramadyuguṇāhateṣṭadorjyā vihṛteṣṭadyuguṇena kālajīvā |
carameṇa careṇa tāḍitā sā tribhajīvāpahṛtāthavā carajyā || 16 ||

trijyākṣaghātādavalambakāptenāhatya doḥkrāntiguṇaṃ dyumaurvyā |
haredavāptaṃ caramaurvikā syāt tadīyacāpā hi carāsavaḥ syuḥ || 17 ||

trijyākṣaghātalambāṃśenāhatādiṣṭadorguṇāt |
tatkoṭyāptaṃ carajyā syād grāhyā sā krānticāpataḥ || 18 ||

The day-radius associated with the maximum declination (*caramadyujyā*), multiplied by the desired Rsine of longitude (*dorjyā*) and divided by the desired day-radius (*dyujyā*), would be the Rsine of Right Ascension (*kālajīvā*). That multiplied by the last *carajyā* and divided by the radius would be the *carajyā* alternatively.

The product of the radius and Rsine of the latitude (*akṣajyā*) divided by the Rcosine of the latitude (*lambaka*), when multiplied by Rsine of declination (*doḥkrāntiguṇa*) and divided by the Rcosine of declination (*dyumaurvī* or *dyujyā*), would give the *carajyā*. The arc of this would give the ascensional difference (*carāsavaḥ*).

The product of the radius and Rsine of latitude (*akṣa*) is divided by the Rcosine (*laṃbajyā*). The desired Rsine [of declination] multiplied by this and divided by the corresponding Rcosine would be the *carajyā*. This has to be obtained from the declination (*krānticāpa*).

The right ascension α is measured along the celestial equator, and is associated with 'time' or '*kāla*'. Hence $R\sin\alpha$ is called *kālajyā* or *kālajīvā*. In the first half of the verse 16 it is stated that

$$R\sin\alpha = \frac{R\cos\epsilon . R\sin\lambda}{R\cos\delta}. \tag{8.39}$$

This expression is the same as (8.29) which was derived earlier. In the latter half of the verse 16 it is stated that

$$carajy\bar{a} = k\bar{a}lajy\bar{a} \times \frac{caramacara}{trijy\bar{a}}. \tag{8.40}$$

We will now verify this relation. Recalling the expression (8.38) for *carajyā*, it may be noted that, once ϕ is fixed the variation is only due to δ, and it attains maximum (*carama*) value when $\delta = \epsilon$. Thus,

$$caramacara = \frac{R\sin\phi R\sin\epsilon}{R\cos\phi} \times \frac{R}{R\cos\epsilon}. \tag{8.41}$$

Hence,

$$
\begin{aligned}
k\bar{a}lajy\bar{a} \times \frac{caramacara}{trijy\bar{a}} &= \frac{R\cos\epsilon R\sin\lambda}{R\cos\delta}\frac{R\sin\phi R\sin\epsilon}{R\cos\phi}\frac{R}{R\cos\epsilon}\frac{1}{R} \\
&= \frac{R\sin\phi R\sin\delta}{R\cos\phi R\cos\delta}R \\
&= caraj y\bar{a}. \qquad (8.42)
\end{aligned}
$$

In verse 17 it is stated that

$$
caraj y\bar{a} = trijy\bar{a} \times \frac{ak\d{s}ajy\bar{a}}{lambaka} \times \frac{do\d{h}kr\bar{a}ntigu\d{n}a}{dyumaurv\bar{\imath}}. \qquad (8.43)
$$

Using the standard expressions for the various quantities in the above equation, we have

$$
\begin{aligned}
R\sin\varDelta\alpha &= R \times \frac{R\sin\phi}{R\cos\phi} \times \frac{R\sin\delta}{R\cos\delta} \\
&= R\tan\phi\tan\delta. \qquad (8.44)
\end{aligned}
$$

In the last quarter of verse 17, it is stated that the arc (*cāpa*) corresponding to the above *carajyā*, gives the ascensional difference (*carāsava*). Verse 18 presents another version of the same result given by (8.43). Towards the end of the verse it is mentioned that the *carajyā* is to be obtained from the declination. Perhaps this has been mentioned explicitly to indicate that, for a given observer, since the latitude is fixed, $\varDelta\alpha$ is essentially a function of the declination δ.

८.१३ चन्द्रविक्षेपानयनम्

8.13 Obtaining the declination of the Moon

अन्त्यद्युज्याहतादन्त्यक्षेपादन्त्यापमोद्धृते ।
अन्त्यक्षेपशराभ्यस्तां कोटिज्यां त्रिज्यया हृताम् ॥ १९ ॥

कर्किनक्रादितः स्वर्णं कुर्यादन्त्यफलाप्तये ।
तद्धते बाहुकोटिज्ये त्रिज्याप्ते बाहुकोटिजे ॥ २० ॥

व्यासार्धे कोटिजं स्वर्णं मृगकर्कादितः क्रमात् ।
तद्बाहुफलवर्गैक्यमूलं कर्णोऽत्र राहुजः ॥ २१ ॥

दोःफलं त्रिज्ययाभ्यस्तं राहुकर्णेन संहरेत् ।
लब्धचापं भवेदिन्दोः विक्षेपचलनाह्वयम् ॥ २२ ॥

परमापक्रमाभ्यस्तं राहुकर्णं त्रिजीवया ।
विभजेल्लब्धमिन्दोः स्यात् परमक्रान्तिमौर्विका ॥ २३ ॥

antyadyujyāhatādantyakṣepādantyāpamoddhṛte |
antyakṣepaśarābhyastāṃ koṭijyāṃ trijyayā hṛtām || 19 ||

karkinakrāditaḥ svarṇaṃ kuryādantyaphalāptaye |
taddhate bāhukoṭijye trijyāpte bāhukoṭije || 20 ||

vyāsārdhe koṭijaṃ svarṇaṃ mṛgakarkāditaḥ kramāt |
tadbāhuphalavargaikyamūlaṃ karṇo'tra rāhujaḥ || 21 ||

doḥphalaṃ trijyayābhyastaṃ rāhukarṇena saṃharet |
labdhacāpaṃ bhavedindoḥ vikṣepacalanāhvayam || 22 ||

paramāpakramābhyastaṃ rāhukarṇaṃ trijīvayā |
vibhajellabdhamindoḥ syāt paramakrāntimaurvikā || 23 ||

The Rsine of the maximum latitude of Moon (*antyavikṣepa*) is multiplied by the last day-radius (*antyadyujyā*) and divided by the Rsine of maximum declination (*antyāpama*). The result obtained by dividing the product of the Rversine of maximum latitude (*antyakṣepaśara*) and the *koṭijyā* [of the *Rāhu*] by the radius, has to be added to or subtracted from that, depending on whether the *rāhubhujā* is *karkyādi* or *makarādi* respectively, in order to obtain the *antyaphala*. That [*antyaphala*] multiplied by the *bhujājyā* and *koṭijyā* [of the *Rāhu*] separately and divided by the radius, would be the *bāhuphala* and *koṭiphala*[4] respectively. The *koṭiphala* is added to, or subtracted from the radius, depending upon whether [the *rāhubhujā* is] *mṛgādi* or *karkyādi* respectively. The square root of the sum of the squares of that and *bāhuphala* is *rāhukarṇa*. The arc of the result obtained by multiplying *bhujāphala* by the radius and dividing by the *rāhukarṇa*, is known as *vikṣepacalana* of the Moon. The product of *rāhukarṇa* and Rsine of maximum declination of the Sun (*paramāpakramajyā*), divided by the radius would be the Rsine of the maximum declination (*paramakrāntijyā*) of the Moon.

The above set of verses essentially presents the procedure for finding the declination (*krānti*) of the Moon. Today it is known that the inclination of Moon's orbit with the ecliptic varies over a period of time. However, the angle of inclination, i is taken to be a constant in Indian astronomy, and its value is taken as $270'$ or $4.5°$.

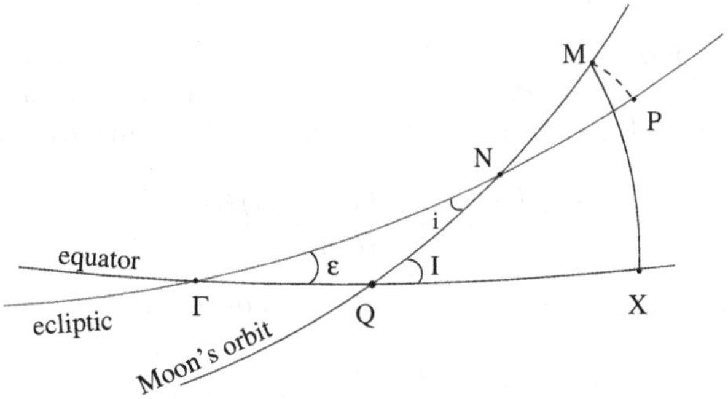

Fig. 8.14 Determination of the *vikṣepacalanajyā* and *paramakrāntijyā*.

[4] In the verse, the word "*bāhukoṭije*" should be understood as *bāhuja* and *koṭija*, or derived from *bāhu* and *koṭi*, namely, *bāhuphala* and *koṭiphala* respectively.

In Figure 8.14, $\varGamma$ is the vernal equinox, or *sāyana-meṣādi*. The Moon's orbit is inclined to the ecliptic and intersects it at N, at an angle i. N is the *Rāhu*, or the ascending node of the Moon. Consider the instant when the Moon is at M in its orbit. The arc MP is drawn perpendicular to the ecliptic from M. $M\hat{N}P = \beta$ is the latitude of the Moon and $\overset{\frown}{MP} = R\beta$. The *vikṣepa* is the perpendicular distance of the Moon from the plane of ecliptic, and is given by

$$\textit{vikṣepa} = R \sin\beta.$$

It can be seen that the maximum value of the *vikṣepa* or the *paramavikṣepa* is given by

$$\textit{paramavikṣepa} = R \sin i.$$

Now $\varGamma N = \lambda_N$ is the longitude of the node in minutes and $\varGamma P = \lambda_m$ is the longitude of the Moon (in minutes). MX is a part of the meridian circle passing through the Moon which intersects the equator at X. Then, $MX = \delta_m$ is the declination of the Moon in minutes. Then, it can be shown that

$$\sin\delta_m = \cos\epsilon\sin\beta + \sin\epsilon\cos\beta\sin\lambda_m. \tag{8.45}$$

In the *Karaṇapaddhati*, as in the *Tantrasaṅgraha* of Nīlakaṇṭha, an alternate method to find the declination is given. For this, an expression for the Rsine of maximum declination of the Moon, is given first. This maximum declination of the Moon, for a given position of the node, is indeed the instantaneous inclination I of the Moon's orbit with the equator.

First, a quantity called *antyaphala* (x) is defined through the relation

$$\begin{aligned} x &= \frac{\textit{paramavikṣepa} \times \textit{antyadyujyā}}{\textit{antyāpakramajyā}} \pm \frac{\textit{vikṣepaśara} \times \textit{rāhukoṭijyā}}{\textit{trijyā}} \\ &= \frac{R\sin i \times R\cos\epsilon}{R\sin\epsilon} \pm \frac{R(1-\cos i)\,|R\cos\lambda_N|}{R}, \end{aligned} \tag{8.46}$$

where the '+' sign is to be taken when the longitude of the node (*rāhubhujā*) is *karkyādi* or $90° \leq \lambda_N \leq 270°$, and '$-$' when the *rāhubhujā* is *mṛgādi* or $270° \leq \lambda_N \leq 90°$. Both these cases are taken into account by the equation

$$x = \frac{R\sin i R\cos\epsilon - R(1-\cos i) R\cos\lambda_N \sin\epsilon}{R\sin\epsilon}. \tag{8.47}$$

The *bāhuphala* B_p and the *koṭiphala*, K_p are now defined to be

$$B_p = \frac{xR|\sin\lambda_N|}{R} = x|\sin\lambda_N|, \tag{8.48a}$$

$$\text{and} \qquad K_p = \frac{xR|\cos\lambda_N|}{R} = x|\cos\lambda_N|, \tag{8.48b}$$

respectively, and the *rāhukarṇa* is given by

$$K_R = \sqrt{(R \pm K_p)^2 + B_p^2} \tag{8.49}$$

$$= \sqrt{(R \pm x|\cos\lambda_N|)^2 + x^2 \sin^2\lambda_N}$$

$$= R\sqrt{(1 + \frac{x}{R}\cos\lambda_N)^2 + \frac{x^2}{R^2}\sin^2\lambda_N}, \tag{8.50}$$

as $\cos\lambda_N = \pm|\cos\lambda_N|$, depending upon whether λ_N is *mṛgādi* or *karkyādi* respectively.

Then it is stated that the *paramakrāntijyā* of the Moon, $R\sin I$ (where I is the instantaneous inclination of the Moon's orbit as shown in the Figure (8.14)) is

$$R\sin I = \frac{\textit{rāhukarṇa} \times \textit{paramāpakramajyā}}{\textit{trijyā}}$$

$$= \frac{K_R \times R\sin\epsilon}{R} = K_R \sin\epsilon. \tag{8.51}$$

Now from equations (8.47) and (8.50)

$$\frac{K_R^2}{R^2} = 1 + \frac{x^2}{R^2} + 2\frac{x}{R}\cos\lambda_N$$
$$= \frac{1}{\sin^2\epsilon}[\sin^2\epsilon + (\sin i\cos\epsilon + \cos i\cos\lambda_N\sin\epsilon - \cos\lambda_N\sin\epsilon)^2$$
$$+2(\sin i\cos\epsilon + \cos i\cos\lambda_N\sin\epsilon - \cos\lambda_N\sin\epsilon)\cos\lambda_N\sin\epsilon].$$

By simplifying this, we get

$$\frac{K_R^2}{R^2} = \frac{1}{\sin^2\epsilon}\left[(\sin\epsilon\cos i + \cos\epsilon\sin i\cos\lambda_N)^2 + \sin^2 i\sin^2\lambda_N\right].$$

Using the above in (8.51) we have,

$$R\sin I = \sqrt{(R\sin\epsilon\cos i + R\cos\epsilon\sin i\cos\lambda_N)^2 + (R\sin i\sin\lambda_N)^2}. \tag{8.52}$$

Equation (8.52) is the same as the expression for the *paramakrāntijyā* given in *Tantrasaṅgraha*.[5] In *Yuktibhāṣā*, this expression for the maximum declination is derived.[6] We now provide a derivation based on modern spherical trigonometry. In Figure 8.14, let $NQ = y$. Then, ϵ, $\lambda_N = \Gamma N$, $i = \Gamma\hat{N}Q$, and $y = NQ$ are four adjacent parts in the spherical triangle ΓNQ. Using the "four-parts" formula, we find

$$\cos\lambda_N\cos i = \sin\lambda_N\cot y - \sin i\cot\epsilon. \tag{8.53}$$

[5] {TS 2011}, p. 363.

[6] {GYB 2008}, pp. 815-817.

Thus,

$$\cot y = \frac{\sin\epsilon\cos\lambda_N\cos i + \sin i\cos\epsilon}{\sin\epsilon\sin\lambda_N}. \tag{8.54}$$

After some straightforward manipulations, we get

$$\frac{\sin\epsilon}{\sin y} = \frac{\sqrt{(\sin\epsilon\cos i + \cos\epsilon\sin i\cos\lambda_N)^2 + (\sin i\sin\lambda_N)^2}}{\sin\lambda_N}. \tag{8.55}$$

Now applying the sine formula in the spherical triangle ΓNQ, we have

$$\frac{\sin\epsilon}{\sin y} = \frac{\sin I}{\sin\lambda_N}. \tag{8.56}$$

Using (8.55) in (8.56) we get,

$$\sin I = \sqrt{(\sin\epsilon\cos i + \cos\epsilon\sin i\cos\lambda_N)^2 + (\sin i\sin\lambda_N)^2},$$

which is the same as (8.52).

We now take up the expression for $R\sin\delta_m$, where δ_m is the declination of the Moon. From the spherical triangle MQX in Figure 8.14, the declination MX is given by

$$R\sin\delta_m = \frac{R\sin MQ.R\sin I}{R}. \tag{8.57}$$

Now

$$\begin{aligned} MQ &= MN + NQ \\ &= MN + \Gamma N + NQ - \Gamma N \\ &\approx NP + \Gamma N + NQ - \Gamma N, \end{aligned} \tag{8.58}$$

where we have assumed that $MN \approx NP$ as the inclination i is small. Now, $NP + \Gamma N = \lambda_m$, is the longitude of the Moon. Hence,

$$MQ \approx \lambda_m - (\Gamma N - NQ). \tag{8.59}$$

Here, $\Gamma N - NQ$ is called the '*vikṣepacalana*' and shall be denoted by A.[7] Then

$$MQ \approx \lambda_m - A.$$

Now the expression for the Moon's declination given in *Tantrasaṅgraha* is[8]

[7] In *Tantrasaṅgraha*, the *vikṣepacalana*, $A = \Gamma N - NQ$ is approximated by ΓQ. This is reasonable as i is small ({TS 2011}, pp. 368-369).

[8] Though *Karaṇapaddhati* does not give this formula for the declination of Moon, it gives all the necessary inputs for calculating it, such as the *vikṣepacalana* A and the maximum declination I.

$$R\sin\delta_m = \frac{R\sin MQ.R\sin I}{R} \approx \frac{R\sin(\lambda_m - A)R\sin I}{R}. \tag{8.60}$$

In the above verses of *Karaṇapaddhati*, the expression for the *vikṣepacalana* is given by the relation

$$\begin{aligned} R\sin A &= \frac{\textit{bāhuphala} \times \textit{trijyā}}{\textit{rāhukarṇa}} \\ &= \frac{x|\sin\lambda_N| \times R}{K_R} \\ &= \frac{x\sin\lambda_N}{R\sin I}.R\sin\epsilon. \end{aligned} \tag{8.61}$$

In arriving at the above equation, we have used (8.48b) and (8.51). Substituting for x from (8.47) in (8.61), the RHS reduces to

$$\frac{(R\sin iR\cos\epsilon - R(1-\cos i)R\cos\lambda_N\sin\epsilon)\sin\lambda_N}{R\sin I}. \tag{8.62}$$

We will now show that the above expression is the same as $R\sin(\Gamma N - NQ) = R\sin A$. Now $NQ = y$, and we had already noted that

$$\sin y = \frac{\sin\epsilon\sin\lambda_N}{\sin I}. \tag{8.63}$$

Also, from the "four-parts formula" involving ϵ, $\lambda_N = PN$, $i = P\hat{N}Q$ and $y = NQ$, we have

$$\begin{aligned} &\cos\lambda_N\cos i = \sin\lambda_N\cot y - \sin i\cot\epsilon, \\ \text{or} \qquad &\sin\lambda_N\frac{\cos y}{\sin y} = \sin i\frac{\cos\epsilon}{\sin\epsilon} + \cos\lambda_N\cos i. \end{aligned}$$

Multiplying this equation with the equation (8.63) for $\sin y$, we find

$$\cos y = \frac{\sin\epsilon}{\sin I}\left(\sin i\frac{\cos\epsilon}{\sin\epsilon} + \cos\lambda_N\cos i\right). \tag{8.64}$$

Hence,

$$\begin{aligned} R\sin A &= \sin(\lambda_N - NQ) \\ &= \sin\lambda_N\cos y - \cos\lambda_N\sin y \\ &= \frac{\sin\epsilon}{\sin I}\left[\sin\lambda_N\left(\sin i\frac{\cos\epsilon}{\sin\epsilon} + \cos\lambda_N\cos i\right) - \cos\lambda_N\sin\lambda_N\right] \\ &= \frac{[\sin i\cos\epsilon - (1-\cos i)\cos\lambda_N\sin\epsilon]\sin\lambda_N}{\sin I}. \end{aligned} \tag{8.65}$$

Using the above equations, Moon's latitude can be calculated from the relation

$$R\sin\delta_m = \frac{R\sin(\lambda_m - A) R\sin I}{R}. \tag{8.66}$$

This is an exact formula which does not involve any approximation (except for the relation $MN \approx NP$), unlike the expression in *Tantrasaṅgraha* where $A = \Gamma N - NQ$ is approximated by ΓQ.

८.१४ मान्यादि-इनादिज्यानयनम्

8.14 Obtaining the *mānyādijyās* and the *inādijyās*

त्रिज्यावर्गेणाहतादक्षकर्णात् द्युज्याभक्तास्त्रिज्यकाभक्तहीनाः ।
मान्यादिज्याः संभृताक्षेत्रदेशे देवाप्तास्ता हारजीवा इनाद्याः ॥ २४ ॥

trijyāvargeṇāhatādakṣakarṇāt dyujyābhaktāstrijyakābhaktahīnāḥ |
mānyādijyāḥ saṃbhṛtākṣetradeśe devāptāstā hārajīvā ināḍyāḥ || 24 ||

The hypotenuse of the equinoctial shadow (*akṣakarṇa*) multiplied by the square of the radius [is to be kept at two places]. [One is] divided by the day-radius (*dyujyā*) and [the other] by the radius. Subtracting the latter result from the former would give the *mānyādijyās*. At a place where the Rsine of the latitude (*akṣajyā*) is 647 (*saṃbhṛtaṃ*) [minutes], [these] divided by 48 would give the *hārajīvas* given by the *vākyas ina* etc.

For the standard *śaṅku* of 12 *aṅgulas*, the hypotenuse of the equinoctial shadow is given by

$$akṣakarṇa = \frac{12}{lambajyā} = \frac{12}{\cos\phi}. \tag{8.67}$$

The *mānyādijyās* and the *inādijyās* given in the above verse may be expressed as

$$\begin{aligned} mānyādijyās &= akṣakarṇa \times \left(\frac{R^2}{R\cos\delta} - R\right) \\ &= \frac{12}{\cos\phi} \times \left(\frac{R^2}{R\cos\delta} - R\right), \end{aligned} \tag{8.68}$$

and

$$\begin{aligned} inādijyās &= \frac{mānyādijyās}{48} \\ &= \frac{1}{4\times\cos\phi} \times \left(\frac{R^2}{R\cos\delta} - R\right). \end{aligned} \tag{8.69}$$

The above quantities are of relevance in the computation of solar eclipses. The verse also specifies that these quantities are to be calculated at a place where the Rsine of the latitude (*akṣajyā*) is given by 647′. This corresponds to a latitude of 10°50′50″.[9]

The *inādijyās*, given in the edition of *Karaṇapaddhati* along with Malayalam Commentaries[10] for the above value (647′) of the *akṣajyā*, are listed in Table 8.1. These *vākyas* can be put together in the verse form as given below.

इनः पुनर्गानमनाः सनी पटुर्भिया धिया ।
गात्रहरो बली हली विभुर्धवशर्मनटः ॥
शतं धृतं लसत् तथा धीस्थ-पदम् प्रभो गदः ।

inaḥ punargānamanāḥ sanī paṭurbhiyā dhiyā |
gātraharo balī halī vibhurdhavaśarmanaṭaḥ ||
śatam dhṛtaṃ lasat tathā dhīstha-padam prabho gadaḥ |

In Table 8.1, we also give the computed values, for the same latitude, of *inādijyās* by taking the declination values in (8.69) that correspond to longitudes which are multiples of 225′.

८.१५ लम्बनहारकः तस्योपयोगश्च

8.15 *Lambanahāraka* and its application

धूळीरागो भुवो वृत्तं तेन चक्रांशका हतात् ।
दिनयोजनभोगात् स्यात् लब्धो लम्बनहारकः ॥ २५ ॥
त्रिज्यातो लम्बहाराप्तं नाडिकाद्यन्तलम्बनम् ।
तदेवगतिभागघ्नं शोध्यं छायाविधौ नरात् ॥ २६ ॥

dhūḷīrāgo bhuvo vṛttaṃ tena cakrāṃśakā hatāt |
dinayojanabhogāt syāt labdho lambanahārakaḥ || 25 ||
trijyāto lambahārāptaṃ nāḍikādyantalambanam |
tadevagatibhāgaghnaṃ śodhyaṃ chāyāvidhau narāt || 26 ||

The circumference of the earth is 3299 (*dhūḷīrāga*) [*yojanās*]. The product of 360 (*cakrāṃśaka*) and the [common] daily motion of the planets in *yojanās* (*dinayojanabhoga*) is divided by this, and what is obtained is the divisor for parallax (*lambanahāraka*).

By dividing the *trijyā* by [this] *lambanahāraka*, the maximum parallax (*antyalambaka*) in *nāḍikā* etc., is obtained. The same (*antyalambaka*) multiplied by the daily motion of the Moon in degrees, has to be subtracted from the gnomon (*śaṅku*) in the computation of the shadow (*chāyā*).

---

[9] According to P. K. Koru, this could be a place close to Shoranur (latitude 10°46′12″) ({KP 1953}, p. 293). According to S. K. Nayar, the place could be Alattur ({KP 1956}, p. 272).

[10] {KP 1956}, p. 272.

i	longitude $\lambda = i \times 225'$	declination δ (in min.)	mnemonic in the commentary	tabulated value	comp. value *akṣakarṇa* $\times \frac{\left(\frac{R^2}{R\cos\delta} - R\right)}{48}$
1	225	91.46	*inaḥ*	0	0.3098
2	450	182.60	*punar*	1	1.2359
3	675	273.07	*gāna*	3	2.7682
4	900	362.57	*manāḥ*	5	4.8898
5	1125	450.75	*sanī*	7	7.5767
6	1350	537.27	*paṭur*	11	10.7977
7	1575	621.82	*bhiyā*	14	14.5139
8	1800	704.04	*dhiyā*	19	18.6787
9	2025	783.60	*gātra*	23	23.2372
10	2250	860.15	*haro*	28	28.1268
11	2475	933.36	*balī*	33	33.2762
12	2700	1002.88	*halī*	38	38.6068
13	2925	1068.38	*vibhur*	44	44.0323
14	3150	1129.53	*dhava*	49	49.4602
15	3375	1186.00	*śarma*	55	54.7931
16	3600	1237.48	*nataḥ*	60	59.9301
17	3825	1283.68	*śatam*	65	64.7694
18	4050	1324.34	*dhṛtam*	69	69.2105
19	4275	1359.19	*lasat*	73	73.1575
20	4500	1388.02	*tathā*	76	76.5220
21	4725	1410.65	*dhīstha*	79	79.2262
22	4950	1426.92	*padam*	81	81.2061
23	5175	1436.72	*prabho*	82	82.4134
24	5400	1440.0	*gadaḥ*	83	82.8196

Table 8.1 The *inādijyās* given in the commentary and the computed values.

These verses tell us how to obtain the parallax, which is used to find the corrections to the gnomon (*śaṅku*), the shadow (*chāyā*) as well as in the computation of eclipses. Let r_e be the radius of the earth and g the *dinayojanagati* (which is the same for all the planets) whose value is 7906 *yojanas*/day (see (1.93)). The circumference of the earth $C_e = 2\pi r_e$ is given as 3299 *yojanas*. The expression for *lambanahāraka* given in the verse is

$$l_h = \frac{360 \times g}{C_e} = \frac{360 \times 7906}{3299}. \tag{8.70}$$

The *antyalambana* in *nāḍikās*, l is stated to be

$$l = \frac{R}{l_h} = \frac{R \times C_e}{360 \times g} = \frac{3438 \times 3299}{360 \times 7906} = 3.99 \textit{ nāḍikās}. \tag{8.71}$$

Now, the latter half of verse 26 states that this value multiplied by the motion of the Moon is to be subtracted from the gnomon in the shadow measurement process. This is to take into account the effect of parallax. Thus the expression for the parallax in minutes is given to be

$$\begin{aligned} P \text{ (minutes)} &= l \times \textit{gati-bhāga} \text{ of the Moon} \\ &= l \times \frac{790.6}{60}. \end{aligned} \tag{8.72}$$

The word *gati-bhāga* used in the verse, as well as in the above equation, refers to the daily motion (*gati*) expressed in degrees (*bhāgas*).[11] Using (8.71) in the above equation we have,

$$\begin{aligned} P \text{ (minutes)} &= \frac{R \times C_e}{360 \times g} \times \frac{790.6}{60} \\ &= \frac{r_e}{10}, \end{aligned} \tag{8.73}$$

as $R \times C_e = R \times 2\pi r_e = 360 \times 60 \times r_e$ and $g = 7906$.

We now show that this is the value of the horizontal parallax or the maximum value of the parallax of the Moon, whose mean distance from the earth (in *yojanas*) is given by $O_m = 10R$. In Figure 8.15, the zenith distances of the Moon at the location of the observer A, and at center of the earth, O are z' and z respectively. The angle $A\hat{M}O = z' - z = p$, subtended by the radius $OA = r_e$ at M is termed parallax.

In the triangle AMO,

$$\begin{aligned} \frac{\sin p}{r_e} &= \frac{\sin(180 - z')}{O_m}, \\ \text{or} \qquad \sin p &= \frac{r_e}{O_m} \sin z'. \end{aligned} \tag{8.74}$$

When the Moon is on the horizon $z' = 90°$, the corresponding parallax, P is known as the horizontal parallax. Clearly,

$$\sin P = \frac{r_e}{O_m}. \tag{8.75}$$

Using (8.75) in (8.74), we get

[11] The compound *gati-bhāga* may be derived as: गतिः भागेषु निर्दिष्टा = गतिभागाः। *Gatiḥ bhāgeṣu nirdiṣṭā = gatibhāgāḥ* /

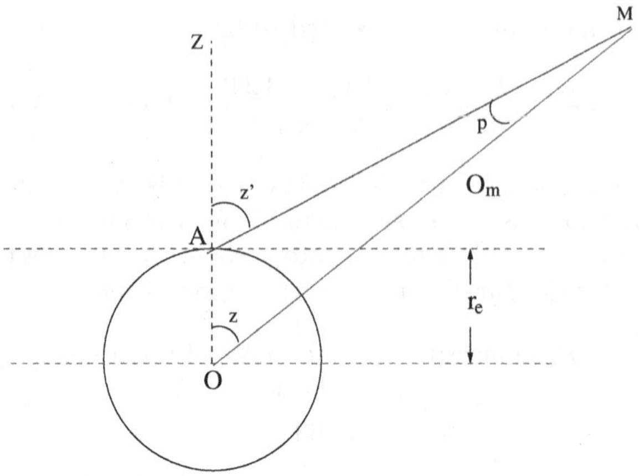

Fig. 8.15 Horizontal parallax.

$$\sin p = \sin P \sin z'. \tag{8.76}$$

Since $\frac{r_e}{O_m} \approx \frac{1}{60}$, both P and p are small angles, we can use the approximations, $\sin p \approx p$ and $\sin P \approx P = \frac{r_e}{O_m}$. Therefore,

$$p \approx P \sin z' = \frac{r_e}{O_m} \sin z'. \tag{8.77}$$

Here, p and P are in radians. The *trijyā*, R, is the number of minutes in a radian, and the distance of Moon $O_m = 10R$. Hence,

$$P \ \text{(minutes)} = \frac{r_e}{O_m} \times R = \frac{r_e}{10}, \tag{8.78}$$

as implied in the verses (see (8.73)).

Let us now consider the corrections to the *chāyā* and the *śaṅku* due to parallax. The actual zenith distance $z = z' - p$. Hence the corrected *śaṅku* (that is, the *śaṅku* at the centre of the earth) is given by

$$\begin{aligned} R\cos z &= R\cos(z' - p) \\ &\approx R\cos z' + R\sin z'.p \\ &= R\cos z' + R\sin z'.\frac{p(\text{in min.})}{R} \\ &= R\cos z' + R\sin z'.\frac{P(\text{in min.})}{R}.\frac{R\sin z'}{R}. \end{aligned} \tag{8.79}$$

Similarly, the corrected *chāyā* is given by

$$\begin{aligned} R\sin z &= R\sin(z' - p) \\ &\approx R\sin z' - R\cos z'.\frac{p(\text{in min.})}{R} \\ &= R\sin z' - R\cos z'.\frac{P(\text{in min.})}{R}.\frac{R\sin z'}{R}. \end{aligned} \tag{8.80}$$

८.१६ लम्बनज्यानयनम्

8.16 Obtaining the *lambanajyās*

सूर्येन्द्वोर्भुक्तिलिप्ताविवरविरहिता राशिचक्रस्य लिप्ता
भूव्यासार्धेन निघ्ना नृपहतहिमरुग्भुक्तिलिप्ता विभक्ताः ।
त्रिज्यान्त्यक्रान्तिकोट्योर्युतिदलगुणितास्त्रिज्ययाहृत्य लब्धं
लम्बघ्नं त्रिज्ययाप्तं चरमफलमिदं लम्बनोत्थं वदन्ति ॥ २७ ॥

त्यजेत् त्रिज्यावर्गाच्चरमफलवर्गेण सहितात्
द्विनिघ्नीं कोटिज्यां चरमफलसंवर्धिततनुम् ।
ततो मूलं कर्णस्त्रिभगुणहतं बाहुजगुणं
हरेत् कर्णेनाप्ता ग्रहणपठिता लम्बनगुणाः ॥ २८ ॥

sūryendvorbhuktiliptāvivaravirahitā rāśicakrasya liptā
bhūvyāsārdhena nighnā nṛpahatahimarugbhuktiliptā vibhaktāḥ |
trijyāntyakrāntikoṭyoryutidalaguṇitāstrijyayāhṛtya labdhaṃ
lambaghnaṃ trijyayāptaṃ caramaphalamidaṃ lambanotthaṃ vadanti || 27 ||

tyajet trijyāvargāccaramaphalavargeṇa sahitāt
dvinighnīṃ koṭijyāṃ caramaphalasaṃvardhitatanum |
tato mūlaṃ karṇastribhaguṇahataṃ bāhujaguṇaṃ
haret karṇenāptā grahaṇapaṭhitā lambanaguṇāḥ || 28 ||

[The result obtained by] subtracting the difference between the daily motions (*gatis*) of the Sun and the Moon in minutes from the 21600 (*rāśicakralipta*) has to be multiplied by the radius of the earth (*bhūvyāsārdha*) and divided by the product of ten and the rate of motion of the Moon expressed in minutes (*himarugbhuktilipta*). [The result] is then multiplied by half of the sum of the radius and the Rcosine of maximum declination (*antyakrānti-koṭi*) and divided by the radius. This result when multiplied by Rcosine of latitude (*lamba*) and divided by the radius is called as *lambanottha-caramaphala*.

Subtract the product of this *caramaphala* and Rcosine (*koṭijyā*) multiplied by two from the sum of the squares of the *caramaphala* and the radius. Square root [of this result] is the *karṇa*. The Rsine (*bāhuguṇa*) [of the zenith distance] multiplied by the radius (*tribha*) and divided by this *karṇa* is the *lambanaguṇa* tabulated in [the computation of] eclipse.

The effect of the parallax is to increase the zenith distance z, of a celestial object. That is, z as seen by an observer on the surface of the earth, would be greater than the zenith distance at the centre of the earth. So, effectively, if z be the zenith distance of a celestial object, it will be enhanced by an amount,

$p = P \sin z$, along the vertical passing through it, due to the parallax. The component of the parallax along the ecliptic, which represents the effective change in the longitude due to the parallax, is called '*lambana*'.

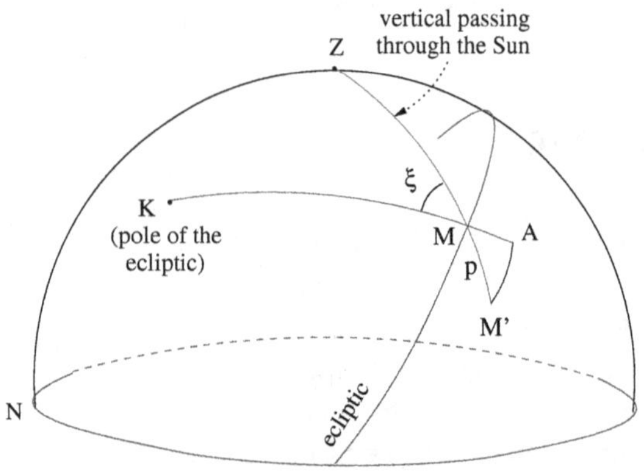

Fig. 8.16 Obtaining the parallax of a celestial body.

In Figure 8.16, M and M' represent the actual and the apparent positions of the Moon. That is, ZM and ZM' would be the zenith distances measured by the observers at the centre of the earth and on the surface respectively. Now, $MM' = p$ is the parallax, which is along the vertical passing through M. The ecliptic and the secondary to the ecliptic KMA passing through M are shown in the figure. Let ξ be the angle between the vertical through M and the secondary to the ecliptic. Then the *lambana*, $\Delta\lambda$, is given by

$$\Delta\lambda = M'A = MM' \sin\xi = p \sin\xi. \quad (8.81)$$

Hence, one needs to find the angle ξ to determine the *lambana*. The procedure for this, given in *Tantrasaṅgraha*, is fairly lengthy and involved. However, in *Karaṇapaddhati*, an approximate method to find the *lambana* is given, based on simpler considerations, and a different approach.

To start with, instead of considering the *lambana* as such, the text considers a related quantity called *lambanottha-caramaphala*. While, the *lambana* is the displacement along the ecliptic due to parallax, the *lambanotthaphala* is the westward displacement along the diurnal circle of the Moon in the time interval corresponding to the '*lambana*'. *Lambanottha-caramaphala* is the maximum value of this when $z = 90°$. The z-dependence comes in through '*lambanajyā*', which is considered in verse 28. In verse 27 the expression for *lambanottha-caramaphala* (l_c) is given as follows:

$$\begin{aligned} l_c &= \frac{[\textit{rāśicakraliptā} - (\textit{indugati} - \textit{sūryagati})(\textit{in liptās})]}{\textit{nṛpa} \times \textit{himarugbhukti}} \\ &\quad \times \textit{bhūvyāsārdha} \times \frac{1}{2}\frac{[\textit{trijyā} + \textit{antyakrāntikoṭi}]}{\textit{trijyā}} \times \frac{\textit{lamba}}{\textit{trijyā}} \\ &= \frac{21600 - (g_m - g_s)}{10 \times g_m} \times (r_e) \times \frac{\frac{1}{2}[R + R\cos\epsilon]}{R} \times \frac{R\cos\phi}{R}, \end{aligned} \quad (8.82)$$

where g_m and g_s are the daily rates of motion of the Moon and the Sun in minutes respectively, and ϕ is the latitude. The rationale behind this can be understood with the help of Figure 8.17.

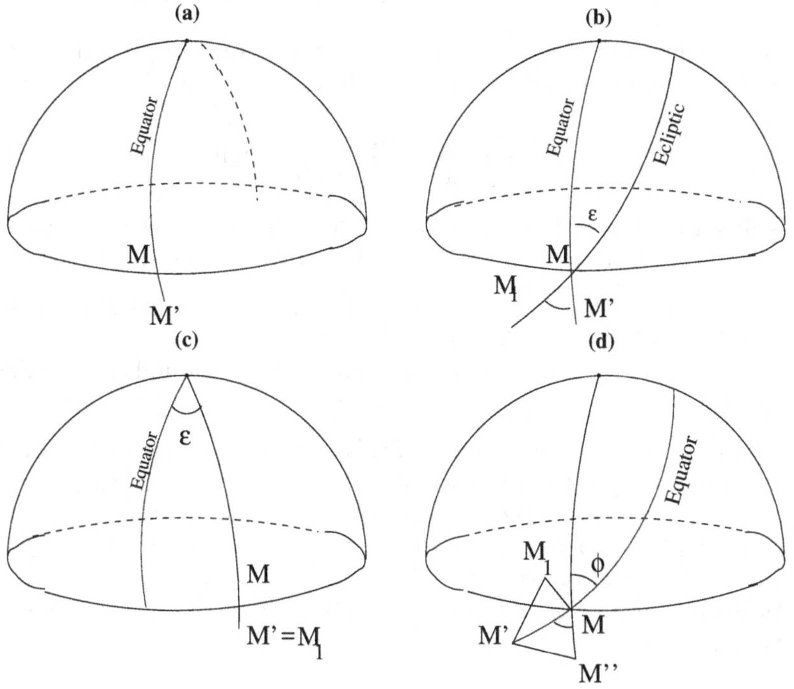

Fig. 8.17 Obtaining the *lambanottha-caramaphala.*

Consider an equatorial observer ($\phi = 0$). Then, the celestial equator is a vertical circle. In fact, it coincides with the prime vertical. For the moment, if we ignore the inclination of the Moon's orbit and also assume that it is close to the vernal or autumnal equinox, then the diurnal motion of the Moon would be more or less on the equator, which is the prime vertical. Then the horizontal parallax, MM' would be along the equator as shown in Figure 8.17 (a). Here,

$$MM' \text{ (minutes of arc)} = \frac{r_e}{10}, \quad (8.83)$$

as explained earlier. This is the change in the longitude $\Delta\lambda$ due to parallax. The time interval in days corresponding to this change in longitude is

$$\frac{r_e}{10 \times g_m}, \tag{8.84}$$

where g_m is the daily motion of the Moon. Now the east-west motion of the stars per civil day is $(21600 + g_s)$, where g_s is the daily motion of the Sun in minutes. The Moon moves eastwards with respect to the stars at the rate of g_m per day (g_m in minutes). Hence, the net westward motion of the Moon per day is

$$21600 + g_s - g_m = 21600 - (g_m - g_s) \text{ minutes/civil day.}$$

In the time interval given by (8.84), the net westward displacement of the Moon along the diurnal circle due to parallax, denoted by l, is

$$l = \frac{21600 - (g_m - g_s)}{g_m} \times \frac{r_e}{10} \text{ minutes.} \tag{8.85}$$

Assuming that the Moon moves along the ecliptic (taking $i = 0$) which is inclined to the equator at an angle ϵ, we have two extreme situations:

1. the ecliptic and the equator intersect at M, the east point of the horizon as indicated in Figure 8.17 (b), and,
2. the ecliptic and equator intersect at zenith and nadir as indicated in Figure 8.17 (c).

In the former case, the displacement MM' due to the parallax and the ecliptic make an angle ϵ, and the *lambana*, along the ecliptic will be, $MM_1 = \cos\epsilon \times MM'$. In the latter case, the displacement MM' is along the ecliptic itself, since the ecliptic happens to be a vertical circle, and hence *lambana* $= MM'$.

In a general situation, the multiplication factor is between $\cos\epsilon$ and 1. In his Malayalam exposition of the *Karaṇapaddhati*, P. K. Koru[12] suggests that, the factor

$$\frac{1}{2}(1 + \cos\epsilon) = \frac{R + R\cos\epsilon}{2R} \tag{8.86}$$

represents the average situation.

Now consider the situation for a place with latitude ϕ, as depicted in Figure 8.17 (d). In this case, the deflection due to parallax MM'' along the vertical has a component along the equator given by $M_1M' = MM''\cos\phi$, as the equator is inclined at an angle ϕ with the prime vertical. Now, we have to multiply this by $\frac{1}{2}(1+\cos\epsilon)$ as earlier, to take into account the obliquity of the ecliptic. Thus, the net displacement along the diurnal circle due to parallax

[12] {KP 1953}, p. 297.

corresponding to a zenith distance of 90°, or the *lambanottha-caramaphala*, is given by,

$$\begin{aligned} l_c &= \frac{r_e}{10} \times \frac{(1+\cos\epsilon)}{2}\cos\phi \qquad \text{(in min. of arc)} \\ &= \frac{21600-(g_m-g_s)}{g_m} \times \frac{r_e}{10} \times \frac{(R+R\cos\epsilon)}{2R}\frac{R\cos\phi}{R}, \end{aligned} \tag{8.87}$$

which is the same as (8.82) given in the text. The expression (8.87) will be in minutes westwards, along the diurnal circle.

Now, consider the *lambana* for an arbitrary zenith distance z. This is depicted in Figure 8.18. In this case,

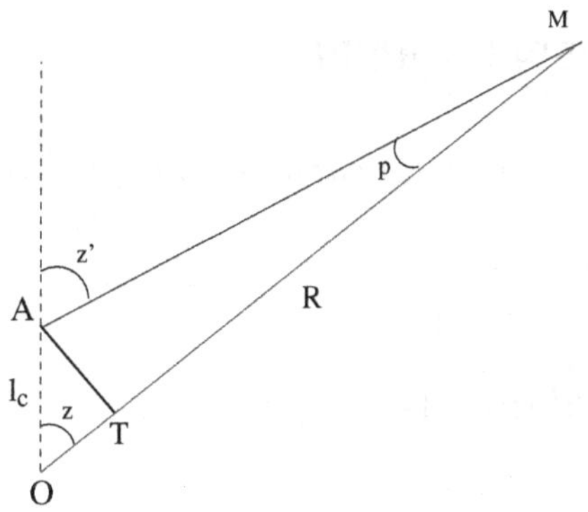

Fig. 8.18 Obtaining the *lambanajyā*.

$$\begin{aligned} \sin p \approx p \text{ (radians)} &= \frac{AT}{AM} \\ &= \frac{l_c \sin z}{\sqrt{AT^2+TM^2}} \\ &= \frac{l_c \sin z}{\sqrt{l_c^2\sin^2 z+(R-l_c\cos z)^2}} \\ &= l_c\frac{\sin z}{\sqrt{R^2+l_c^2-2l_cR\cos z}}. \end{aligned} \tag{8.88}$$

Hence

$$\begin{aligned} p\text{ (minutes)} &= p \text{ (radians)}.R \\ &= \frac{l_c}{R}\frac{R\sin z.R}{\sqrt{R^2+l_c^2-2l_cR\cos z}}. \end{aligned} \tag{8.89}$$

Now, the *karṇa* is given by

$$K = \sqrt{R^2 + l_c^2 - 2l_c R \cos z}. \tag{8.90}$$

Hence, the *lambanajyā* as defined in verse 28, is

$$lambanajy\bar{a} = \frac{R.R \sin z}{K}. \tag{8.91}$$

In verses 27 and 28, a method for obtaining *lambanajyās* at any place (i.e., for any value of ϕ) is described. Now, the author proceeds to explain the procedure for obtaining the 24 tabular values of *lambanajyās* at a particular place. These values are encoded in the 24 *vākyas*, *prītāṅganā* etc.

८.१७ प्रीताङ्गनादि-लम्बनज्यानयनम्

8.17 Obtaining the *prītāṅganādi-lambanajyās*

कोटीगुणाभिनिहतस्तनचित्रहीनं
स्त्रीकेळिमाल्यमलयं पदितं च कर्णः ।
त्रिज्याहतात् भुजगुणादमुना हृता वा
प्रीताङ्गनाद्युदितलम्बनमौर्विकाः स्युः ॥ २९ ॥

koṭīguṇābhinihatastanacitrahīnaṃ
strīkeḷimālyamalayaṃ paditaṃ ca karṇaḥ |
trijyāhatāt bhujaguṇādamunā hṛtā vā
prītāṅganādyuditalambanamaurvikāḥ syuḥ || 29 ||

The product of 2606 (*stanacitra*) and the Rcosine [of the zenith distance or the *mahāśaṅku*] is subtracted from 13515912 (*strīkeḷimālyamalayaṃ*). The square root [of the result obtained] is the *karṇa*. The *bhujāguṇa* (Rsine of the zenith distance or the *mahācchāya*) multiplied by the radius and divided by the *karṇa* would be the *lambanajyās* stated [by the *vākyas*] beginning with *prītāṅganā* etc.

In the above verse, the expression for the *lambanajyā* is stated to be

$$lambanajy\bar{a} = \frac{bhuj\bar{a}gu\d{n}a \times trijy\bar{a}}{kar\d{n}a}, \tag{8.92}$$

where the *karṇa* is given by

$$kar\d{n}a = \sqrt{13515912 - (2606 \times R \cos z)}. \tag{8.93}$$

Thus the complete expression for *lambanajyā* is

$$lambanajy\bar{a} = \frac{R \sin z \times R}{\sqrt{13515912 - (2606 \times R \cos z)}}. \tag{8.94}$$

The above equation was used to compute the 24 tabular *lambanajyās*, and the values obtained are tabulated for $z = 0$ to $90°$, at an interval of $225' = 3°45'$ along with the *vākyas* in Table 8.2.

z in minutes	computed value *lambanajyā*	vākya in the commentary[13]	tabulated value
225	361.2480	*prītāṅgana*	362
450	716.4680	*dhiyāsana*	719
675	1060.0057	*gītijñoyam*	1063
900	1386.8988	*naḷālaye*	1390
1125	1693.1000	*sudhātāpam*	1697
1350	1975.5867	*hasaddhānyam*	1978
1575	2232.3620	*bhṛguśreṣṭho*	2234
1800	2462.3700	*matirvarā*	2465
2025	2665.3570	*dantaaturo*	2668
2250	2841.7063	*vibhurjāro*	2844
2475	2992.2737	*vidhurdhīraḥ*	2994
2700	3118.2378	*parākulaḥ*	3121
2925	3220.9741	*parorogi*	3221
3150	3301.9565	*pranīlāṅgo*	3302
3375	3362.6835	*bhartālolaḥ*	3364
3600	3404.6261	*munirbhṛguḥ*	3405
3825	3429.1948	*dhīrobhṛguḥ*	3429
4050	3437.7181	*sulābhogaḥ*	3437
4275	3431.4332	*rudrobhargaḥ*	3422
4500	3411.4824	*priyobhṛguḥ*	3412
4725	3378.9150	*sosaulambaḥ*	3377
4950	3334.6914	*bhṛgorlīlā*	3334
5175	3279.6898	*dhīsthorāgi*	3279
5400	3214.7129	*mayūragaḥ*	3215

Table 8.2 The *Prītāṅganādi-lambanajyās* given in the commentary.

The rationale for the numbers appearing in the above equation can be understood by comparing (8.93) with (8.90). It is clear that the *lambanottha-caramaphala*, $l_c = 1303$. Hence,

$$R^2 + l_c^2 = \left(\frac{21600}{2\pi}\right)^2 + 1303^2 \approx 13515911.86 \approx 13515912,$$

[13] {KP 1956}, p. 277.

which is the number stated in the verse.

Now, we recall the expression for l_c given by the equation (8.82) from the previous section. Here, we know that $g_m = 790.6'$, $g_s = 59.1'$, $r_e = \frac{3299}{2\pi} =$ 525 *yojanas* and $\epsilon = 24°$. Hence,

$$\frac{l_c}{\cos\phi} = \frac{21600 - (790.6 - 59.1)}{7906} \times \frac{525(1+\cos(24°))}{2} = 1325.87,$$

$$\text{or} \qquad \phi = \cos^{-1}\left(\frac{l_c}{1325.87}\right) = 639.44'. \tag{8.95}$$

Therefore, the *lambanajyās* given in the Table 8.2 appear to be the ones for a place whose latitude is $\phi = 639.44'$.[14] It is seen that the tabulated values are close to the computed values, except when $z = 4275'$, where the discrepancy is nearly $9.4'$.

८.१८ योगीरक्तादि-लम्बनज्यानयनम्

8.18 Obtaining the *yogīraktādi-lambanajyās*

कोटीज्याघ्नेनाघनागेन हीनात् प्रज्ञामोहप्रायवाक्यात् पदं यत् ।
तेनावाप्ताद् दोर्गुणात् कातरघ्नात् योगीरक्तेत्यादिका लम्बनज्याः ॥ ३० ॥

koṭījyāghnenāghanāgena hīnāt
prajñāmohaprāyavākyāt padaṃ yat |
tenāvāptād dorguṇāt kātaraghnāt
yogīraktetyādikā lambanajyāḥ || 30 ||

The product of 3040 (*aghanāgaṃ*) and the Rcosine (of the zenith distance), is subtracted from 14128502 (*prajñāmoha-prāyavākyam*). 261 (*kātara*) multiplied by the *dorguṇa* (Rsine of the zenith distance or the *mahāśaṅku*) and divided by the the squareroot of the quantity obtained earlier would be the *lambanajyās* [given by the *vākyas*] *yogīrakta* etc.

In the above verse, the expression for a new *lambanajyā* is given to be

$$lambanajy\bar{a} = \frac{R\sin z \times 261}{kar\d{n}a}, \tag{8.96}$$

where the *karṇa* is

$$kar\d{n}a = \sqrt{14128502 - (3040 \times R\cos z)}. \tag{8.97}$$

[14] This is close to the latitude value 10°50′50″ which was referred to in verse 24 above.

Comparing the expression for *karṇa* given in the above equation with (8.90), it follows that the value of l_c works out to be

$$l_c = \frac{3040}{2} = 1520, \tag{8.98}$$

Using this value of l_c in (8.95) we have,

$$\frac{l_c}{1325.87} = \cos\phi > 1. \tag{8.99}$$

That is, the value of $l_c = 1520$ is not possible for any physical value of the latitude ϕ. Notwithstanding this, we compute the *lambanajyās* which are now defined through the relation

$$lambanajy\bar{a} = \frac{R\sin z \times 261}{\sqrt{l_c^2 + R^2 - 2l_c R\cos z}}. \tag{8.100}$$

It may also be mentioned here that the significance of the number 261 is also not clear. However, we note that

$$\begin{aligned} \frac{R \times 60}{261} = \frac{3438 \times 60}{261} &= 790.345 \\ &\approx 790.6 \simeq g_m, \\ \text{or} \qquad 261 &\approx \frac{R \times 60}{g_m}, \end{aligned} \tag{8.101}$$

where g_m is the rate of motion of the Moon per day in minutes.

We have computed the values of *lambanajyās* for $z = i.225'$ (i = 1, 2...24) and have listed them with the values given by the mnemonics *yogīrakta* in both the Commentaries,[15] in Table 8.3. We note that for the smaller values of the zenith distance, the errors are large. We can also find the value of l_c which fits the computed value with the tabulated value. That is, find l_c such that

$$lambanajy\bar{a}\ \ \text{(tabulated)} = \frac{R\sin z \times 261}{\sqrt{l_c^2 + R^2 - 2l_c R\cos z}}, \tag{8.102}$$

for different values of z. We find that $l_c = 1549, 1573$ and 1545 for $z = 225', 450'$ and $675'$ respectively. In fact if the *lambanajyā* is taken to be 61 for $z = 450'$, $l_c = 1542$. The computed values of the *lambanajyās* for $l_c = 1545$ are also displayed in Table 8.3. We see that the values represented by the *vākyas* are very close to these computed values, especially for smaller values of z.

[15] {KP 1956}, p. 279.

z in minute	computed value of *lambanajyā*	*vākya* in the commentary	tabulated value	computed value of *lambanajyā* for $l_c = 1545$
225	27.4296	*yogī*	31	30.7804
450	54.4015	*rakto*	62	60.8568
675	80.4864	*nidhanam*	90	89.5892
900	105.3074	*supaṭam*	117	116.4527
1125	128.5573	*kavikṛt*	141	141.0654
1350	150.0066	*vṛṣasya*	164	163.1938
1575	169.5036	*lajjaḍhyaḥ*	183	182.7381
1800	186.9681	*anakhaḥ*	200	199.7073
2025	202.3809	*svapure*	214	214.1894
2250	215.7712	*tarurāṭ*	226	226.3292
2475	227.2038	*tagaro*	236	236.2800
2700	236.7682	*vivare*	244	244.2367
2925	244.5690	*niśendra*	250	250.3735
3150	250.7180	*śoṇendraḥ*	255	254.8613
3375	255.3290	*himaruṅ*	258	257.8588
3600	258.5138	*nitarām*	260	259.5101
3825	260.3793	*kataro*	261	259.9447
4050	261.0264	*kataro*	261	259.2779
4275	260.5492	*nitarām*	260	257.6118
4500	259.0344	*himarug*	258	255.0366
4725	256.5615	*śoṇendra*	255	251.6322
4950	253.2036	*niśendra*	250	247.4691
5175	249.0273	*vivara*	244	242.6101
5400	244.0936	*tuṅgendraḥ*	236	237.1108

Table 8.3 *The Yogīraktādilambanajyās.*

८.१९ ग्रहबिम्बयोजनानां कलासु परिवर्तनम्

8.19 Obtaining the dimension of the disc of the planets in minutes

बिम्बादीनां योजनानि हतानि त्रिभजीवया ।
स्फुटयोजनकर्णेन भक्तान्येषां कलाः स्मृताः ॥ ३१ ॥

bimbādīnāṃ yojanāni hatāni tribhajīvayā |
sphuṭayojanakarṇena bhaktānyeṣāṃ kalāḥ smṛtāḥ || 31 ||

The diameters of the planets in *yojanas* multiplied by the radius and divided by the true distance in *yojanas* (*sphuṭayojanakarṇa*) are said to be the diameters [of the discs] of the planets in minutes.

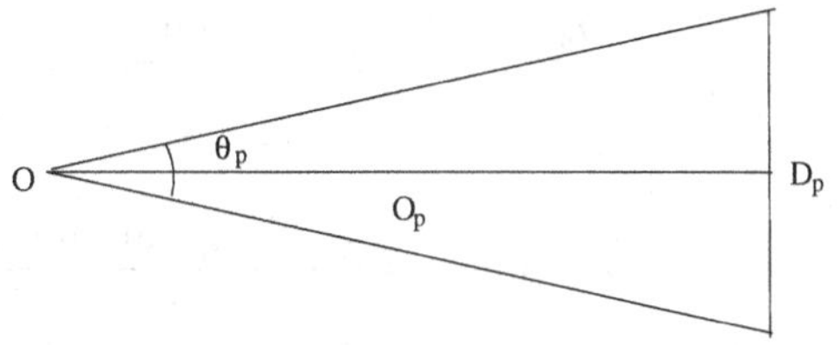

Fig. 8.19 Dimension of the disc of the planets.

Let D_p be the *bimbavyāsayojana* or the diameter of a planet in *yojanas*, and O_p, the true distance between the centres of the earth and the planet (*sphuṭayojanakarṇa*). Then, the angular diameter of the planet may be obtained from the relation

$$\tan\frac{\theta_p}{2} \approx \frac{\theta_p}{2} = \frac{D_p}{2O_p},$$

since θ_p is small. In the above relation both D_p and O_p are in *yojanas*, and θ_p is in radians. As the angular diameter in minutes D_p (min.) $= \theta_p R$, we are led to

$$D_p(\text{min.}) = \frac{D_p}{O_p} \times R, \tag{8.103}$$

which is the relation stated in the verse.

८.२० रवीन्दुबिम्बकलानयने प्रकारान्तरम्

8.20 Obtaining the dimension of the discs of the Sun and the Moon

अथवा स्फुटगतिलिप्ता बिम्बव्यासस्य योजनैर्गुणिताः ।
दिनयोजनगतिविहृतास्तस्य च लिप्ता भवन्ति रविशशिनोः ॥ ३२ ॥

athavā sphuṭagatiliptā bimbavyāsasya yojanairguṇitāḥ |
dinayojanagativihṛtāstasya ca liptā bhavanti raviśaśinoḥ || 32 ||

Or, the diameter of the discs of the Sun and the Moon in minutes can be obtained by multiplying their *sphuṭagatis* in minutes by the diameter of the orbs (*bimbavyāsa*) in *yojanas* and dividing by the daily motion (*dinayojanagatis*) in *yojanas*.

If g is the *dinayojanagati* or the mean rate of motion of a planet in *yojanas* per day, and g_{sp} (in min.) the true rate of motion (*sphuṭagati*) in minutes, it is clear that

$$g_{sp}(\text{in min.}) = \frac{g}{O_p} \times R.$$

Comparing this with equation (8.103),

$$\frac{R}{O_p} = \frac{D_p\ (\text{min.})}{D_p} = \frac{g_{sp}\ (\text{in min.})}{g}, \tag{8.104}$$

$$\text{or} \qquad D_p\ (\text{min.}) = \frac{D_p \times g_{sp}\ (\text{in min.})}{g} \tag{8.105}$$

$$= \frac{D_p \times sphu\d{t}agati}{dinayojanagati}, \tag{8.106}$$

which is what given in the verse. This applies to all the planets including the Sun and the Moon. It may be recalled here that the *dinayojanagati* is taken to be the same for all the planets.

८.२१ चन्द्रकक्ष्यास्थ-भूच्छायाव्यासानयनम्

8.21 Obtaining the diameter of earth's shadow on the Moon's orbit

भूव्यासयोजनहताः शशिभुक्तिलिप्ता
भूम्यर्कविस्तृतिभिदा हतभानुगत्या ।
हीनाः पुनर्दिवसयोजनभुक्तिभक्ता
लिप्ता भवन्ति तमसः शशिमार्गगस्य ॥ ३३ ॥

bhūvyāsayojanahatāḥ śaśibhuktiliptā
bhūmyarkavistṛtibhidā hatabhānugatyā |
hīnāḥ punardivasayojanabhuktibhaktā
liptā bhavanti tamasaḥ śaśimārgagasya || 33 ||

The product of the rate of motion of the Sun (*bhānugati*) and the difference between diameters (*vistṛti*) of the earth and the Sun, is subtracted from the product of the earth's diameter (*bhūvyāsa*) in *yojanas* and the rate of motion of the Moon (*śaśibhukti*) in minutes. The remainder divided by the daily rate of motion in *yojanas* (*divasayojanabhukti*), would be the diameter in minutes of the shadow (*tamas*) that moves on the Moon's orbit.

Figure 8.20 depicts a lunar eclipse. Here D_s and D_e are the diameters of the Sun and the earth in *yojanas*. O_s and O_m are the distances of the Sun and the Moon from the earth in *yojanas*. Let D_t be the diameter of the earth's shadow in the plane of the Moon's path in *yojanas* (or minutes). From the geometry of the figure, it is clear that

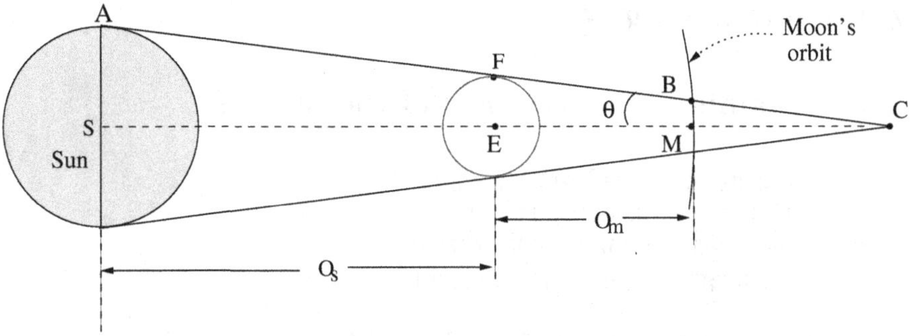

Fig. 8.20 Determination of the angular diameter of the Earth's shadow.

$$\frac{2SA - 2EF}{O_s} = \frac{2EF - 2MB}{O_m},$$

or $$\frac{D_s - D_e}{O_s} = \frac{D_e - D_t}{O_m}.$$

$$\therefore \quad D_t\ (yojanas) = D_e - \frac{O_m}{O_s}(D_s - D_e)$$

$$= \frac{O_s D_e - O_m(D_s - D_e)}{O_s}. \tag{8.107}$$

Let g_m and g_s be the rates of motion of the Moon and the Sun in minutes. Since the linear velocities of all the planets are the same, we have

$$O_s\ g_s = O_m g_m = g,$$

or $$\frac{O_s}{O_m} = \frac{g_m}{g_s}. \tag{8.108}$$

Using (8.108) in (8.107) we get

$$D_t\ (yojanas) = \frac{g_m D_e - g_s(D_s - D_e)}{g_m}. \tag{8.109}$$

Hence,

$$D_t\ (\text{minutes}) = \frac{D_t\ (yojanas) \times R}{O_m}$$

$$= \frac{g_m D_e - g_s(D_s - D_e)}{g\ (yojanas)}, \tag{8.110}$$

as stated in the verse.

८.२२ ग्रहबिम्बलिप्तानयनम्

8.22 Obtaining the *bimbaliptās* of the planets

हार्यं स्यान्मन्दशीघ्रोदितफलविवरे कर्किनक्रादियाते
स्वर्णं स्वर्णैक्यभेदान्निजवधशिखरांशाढ्यहीने शरीरे ।
तस्मात् पञ्चघ्नमौढ्योदितलवविहृता बिम्बलिप्ताः कुजादेः
स्थानोद्यन्नाथशीतक्षतपुलिनहृताः ताश्च कैश्चित् प्रदिष्टाः ॥ ३४ ॥

hāryaṃ syānmandaśīghroditaphalavivare karkinakrādiyāte
svarṇaṃ svarṇaikyabhedānnijavadhaśikharāṃśāḍhyahīne śarīre |
tasmāt pañcaghnamauḍhyoditalavavihṛtā bimbaliptāḥ kujādeḥ
sthānodyannāthaśītakṣatapulinahṛtāḥ tāśca kaiścit pradiṣṭāḥ || 34 ||

The product of differences of successive *mandaphalas* and *śīghraphalas* is divided by 225 (*śikhara*). [The result obtained is to be] added to [225 (*śikhara*)] itself, if both the differences are positive or negative and it is to be subtracted from [*śikhara*] if one [of them] is positive while the other is negative. The differences of successive *mandaphalas* and *śīghraphalas* are to be added [to the result obtained earlier] if they are *karkyādi* and are to be subtracted from [the result] if they are *makarādi*. That becomes the dividend (*hārya*). This *hārya* when divided by the product of 5 and the *mauḍhyāṃśas* of the planets, gives the *bimbaliptās* of the planets Mars etc. [There,] according to some, the *hārya* has to be divided by the *hārakas* 107 (*sthānodyat*), 70 (*nātha*), 65 (*śīta*), 60 (*kṣata*) and 91 (*puḷina*) in order to obtain the *bimbaliptās*.

The term *manda-śīghrodita-phalavivare* refers to the difference of successive *mandaphalas* and difference of successive *śīghraphalas*. Let Δ_{mi} and Δ_{si} denote these differences respectively. In the verse, the term *nijavadhaśikharāṃśa* refers to the product of them divided by 225 (*śikharāṃśa*), that is

$$\frac{\Delta_{mi} \times \Delta_{si}}{225}. \tag{8.111}$$

This quantity is added to or subtracted from 225 (*śarīra*). Then we obtain the result (denoted by x) as

$$x = 225 \pm \frac{\Delta_{mi} \times \Delta_{si}}{225}. \tag{8.112}$$

Here, the sign '+' is taken when both (the Δ_{mi} and Δ_{si}) are positive (*svarṇaikya*) and '$-$' is taken when one is positive and the other is negative (*svarṇabheda*). Both the Δ_{mi} and Δ_{si} are to be added or subtracted from the above result (x) obtained depending upon whether they are *karkyādi* or *makarādi*. Hence, the term *hārya* mentioned in the above verse can be written as

$$
\begin{aligned}
h\bar{a}rya &= x + \Delta_{mi} + \Delta_{si} && \text{(if both } \Delta_{mi} \text{ and } \Delta_{si} \text{ are } \textit{karkyādi}\text{)},\\
h\bar{a}rya &= x + \Delta_{mi} - \Delta_{si} && \text{(for } \Delta_{mi} \text{ is } \textit{karkyādi} \text{ and } \Delta_{si}\ \textit{makarādi}\text{)},\\
h\bar{a}rya &= x - \Delta_{mi} + \Delta_{si} && \text{(for } \Delta_{mi} \text{ is } \textit{makarādi} \text{ and } \Delta_{si}\ \textit{karkyādi}\text{)},\\
h\bar{a}rya &= x - \Delta_{mi} - \Delta_{si} && \text{(if both } \Delta_{mi} \text{ and } \Delta_{si} \text{ are } \textit{makarādi}\text{)}.
\end{aligned}
$$

This *hārya* when divided by the product of 5 and *mauḍhyāṃśas* of the planets would give the *bimbaliptās* of the respective planets. That is,

$$
\textit{bimbaliptās} = \frac{\textit{hārya}}{5 \times \textit{mauḍhyāṃśas}}.
$$

Chapter 9
मध्याह्नकाललग्नम्
Ascendent at the meridian transit

९.१ काललग्ननिर्वचनं तस्माद्राश्युदयास्वानयनञ्च

9.1 Defining *kāla-lagna* and obtaining the rising times of *rāśis* therefrom

कृतायनेष्टराश्यन्ते चरप्राणकलान्तरे ।
कुर्यात् तदिष्टराश्यन्तकाललग्नमुदाहृतम् ॥ १ ॥
इष्टतत्पूर्वराश्यन्तकाललग्नान्तरांशकाः ।
दशाहता भवन्तीष्टराशिमानविनाडिकाः ॥ २ ॥

kṛtāyaneṣṭarāśyante caraprāṇakalāntare |
kuryāt tadiṣṭarāśyantakālalagnamudāhṛtam || 1 ||
iṣṭatatpūrvarāśyantakālalagnāntarāṃśakāḥ |
daśāhatā bhavantīṣṭarāśimānavināḍikāḥ || 2 ||

The ascensional difference (*cara*) and the *prāṇakalāntara* are to be applied to the [longitude of the] end of the desired zodiacal sign (*rāśyanta*) which is corrected for the movement of equinox (*ayanacalana*). The result thus obtained is stated to be the *kālalagna* corresponding to the end of the desired zodiacal sign (*rāśyanta*).

The difference in degrees, between the *kālalagnas* of the desired *rāśyanta* and the one preceding it when multiplied by 10, would be the rising time of the sign (*rāśimāna*) in *vināḍikās*.

In the set of verses given above the author introduces the notion of *kālalagna* and gives the procedure for finding the rising time of *rāśis* using *kālalagna*. The *kālalagna* of a point on the ecliptic is the time interval between the rise of the vernal equinox, Γ, and the rise of the given point on the ecliptic. We explain the concepts with the help of Figure 9.1.

Here Γ is the vernal equinox, where the celestial equator and the ecliptic intersect, which is rising at the east point E on the horizon. R is a point on the ecliptic with tropical (*sāyana*) longitude λ, right ascension α, and declination δ. A section of the diurnal circle in which R moves is also depicted in the

V. Pai et al., *Karaṇapaddhati of Putumana Somayājī*, Sources and Studies in the History of Mathematics and Physical Sciences, https://doi.org/10.1007/978-981-10-6814-0_9

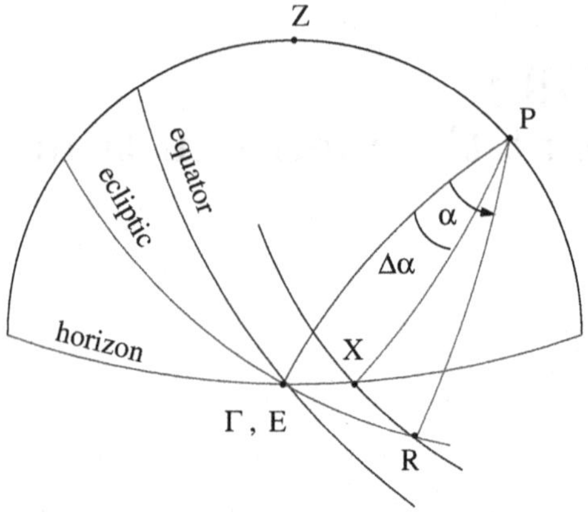

Fig. 9.1 Determination of the *kālalagna*.

figure. X represents the point on the diurnal circle at which the object with longitude λ rises. Consider the angle

$$R\hat{P}X = \alpha - \Delta\alpha. \tag{9.1}$$

Here $\Delta\alpha = E\hat{P}X$ denotes the *cara* which is given by

$$\sin \Delta\alpha = \tan\phi \tan\delta. \tag{9.2}$$

The expression (9.1) denotes the time between the rise of Γ and an object with tropical longitude λ, in sidereal units. This is essentially the expression for *kālalagna* which may also be written as

$$kālalagna = \lambda + (\alpha - \lambda - \Delta\alpha). \tag{9.3}$$

Here $\alpha - \lambda$ is the *prāṇakalāntara* and λ is the *sāyana* longitude[1] which is referred to as *kṛtāyana* in the verse. Hence,

$$kālalagna = \lambda + (prāṇakalāntara - cara),$$

which is the expression given in verse 1. What is of particular interest is the *iṣṭarāśyantakālalagna*, which is the *kālalagna* corresponding to the end point of the desired *rāśi*, given by $\lambda = i \times 30$, where $i = 1, 2, \ldots, 12$.

The term *rāśimāna* employed in verse 2 actually refers to "the time taken by the *rāśi*" to come up above the horizon. If K_i denotes the *kālalagna* corresponding to the ith *rāśi*, given by

[1] The term *sāyana* longitude refers to the sum of the *nirayaṇa* longitude (*sphuṭa*) and the *ayanāṃśa* (amount of precession of equinoxes.)

$$K_i = \lambda_i - [(\alpha_i - \lambda_i) - \Delta\alpha_i] \, ,$$

then $K_i - K_{i-1}$ is the *kālalagnāntara* (difference in *kālalagnas*) referred to in the first half of the verse 2. The beginning point of a *rāśi* is of course the end-point of the previous *rāśi*. Now the duration of a day is 60 *naḍikās* or 3600 *vināḍikās* corresponding to 360° in angular units. Hence 1° corresponds to 10 *vināḍikās* and the *rāśimāna* is given by

$$\text{Rising time of } i^{th} \ \textit{rāśi} = (K_i - K_{i-1}) \times 10 \ \ (\text{in } \textit{vināḍikās}).$$

Note: In *Tantrasaṅgraha* also, the *kālalagna* of a point on the ecliptic is the time interval between the rise of the vernal equinox Γ and the rise of the given point on the ecliptic. However, apart from this, the concepts considered in *Tantrasaṅgraha* and *Karaṇapaddhati* are somewhat different. In *Tantrasaṅgraha*, *madhyakāla* is the R.A. of a point on the equator which is situated on the meridian, and *madhyalagna* is the longitude of the meridian ecliptic point and an algorithm is presented for finding the *madhyalagna* from *madhyakāla*.[2]

In *Karaṇapaddhati*, *madhyakāla* and *madhyalagna* are not mentioned. Instead, as may be seen from Section 9.4, we have the concept of *madhyāhnakālalagna* which is the time interval between the rise of Γ and the instant when a star with a non zero latitude is on the meridian. Algorithms for finding the *madhyāhnakāla* given here have no equivalents in *Tantrasaṅgraha*. These algorithms involve very careful analysis of the properties of spherical triangles.

९.२ अश्विन्यादियोगताराणां स्फुटांशक्षेपौ

9.2 The longitudes and latitudes of the “junction stars” commencing with *aśvinī*

तटज्ञो गुणेन प्रसन्नो नृनम्यो वरेण्यो निविष्टो वदान्यो नृपेन्द्रः ।
हरिद्रा समुद्रः प्रहारो दिनाम्बुः स्तवाङ्गो निसर्गो विळङ्गो विरूढः ॥ ३ ॥

विभावस्तृणौघो गजाभो धनेशो बलेशो हताशः कळात्मा विपक्षः ।
क्षमाक्षो निळान्तो निरासः स्युरेते क्रमादर्धिता दस्रपूर्वाः स्फुटांशाः ॥ ४ ॥

नम्यः प्रियो मुनिर्मानी नित्यं पूज्यस्तनुर्ननु ।
स्थाने ज्ञानी प्रियो लोके सूनुः प्राज्ञः स्थले पुनः ॥ ५ ॥

लग्ने भानुर्जनस्थाने सूनुर्नागस्तिलं ननु ।
वक्रस्तत्र ननु क्षेपभागा दस्रादितः क्रमात् ॥ ६ ॥

दस्रान्तकार्यमभगादिति वह्निवायुवस्वेकपाद् द्विहरिशूर्पगताश्च सौम्याः ।
याम्याः परे वरुणनैर्ऋतशूर्पभानां क्षेपास्तु दिव्यनगनागकलान्वितास्ते ॥ ७ ॥

---

[2] {TS 2011}, pp. 248-254.

taṭajño guṇena prasanno nṛnamyo
vareṇyo niviṣṭo vadānyo nṛpendraḥ |
haridrā samudraḥ prahāro dināmbuḥ
stavāṅgo nisargo viḷaṅgo virūḍhaḥ || 3 ||

vibhāvastṛṇaugho gajābho dhaneśo
baleśo dṛtāśaḥ kaḷātmā vipakṣaḥ |
kṣamākṣo niḷānto nirāsaḥ syurete
kramādardhitā dasrapūrvāḥ sphuṭāṃśāḥ || 4 ||

namyaḥ priyo munirmānī nityaṃ pūjyastanurnanu |
sthāne jñānī priyo loke sūnuḥ prājñaḥ sthale punaḥ || 5 ||

lagne bhānurjanasthāne sūnurnāgastilaṃ nanu |
vakrastatra nanu kṣepabhāgā dasrāditaḥ kramāt || 6 ||

dasrāntakaryamabhagāditi vahnivāyu-
vasvekapād dviharisūrpagatāśca saumyāḥ |
yāmyāḥ pare varuṇanairṛtaśūrpabhānāṃ
kṣepāstu divyanaganāgakalānvitāste || 7 ||

The numbers 16, 53, 72, 100, 124, 140, 184, 210, 228, 257, 282, 308, 346, 370, 394, 424, 444, 456, 483, 509, 533, 569, 591, 614, 656, 690 and 720 when halved (*ardhita*) represent the values of the longitudes [of the stars], beginning with *aśvinī*, in degrees.

The numbers 10, 12, 5, 5, 10, 11, 6, 0, 7, 0, 12, 13, 7, 2, 37, 1, 3, 4, 8, 7, 7, 30, 36, 0, 24, 26 and 00 represent the values of the latitudes (*kṣepa*) [of the stars] in degrees beginning with *aśvinī*.

[The stars] *dasra* (*aśvinī*), *antaka* (*bharaṇī*), *aryama* (*pūrvāphalgunī*), *bhaga* (*uttarāphalgunī*), *aditi* (*punarvasu*), *vahni* (*kṛttikā*), *vāyu* (*svātī*), *vasu* (*dhaniṣṭhā*), *ekapādvī* (*pūrvabhādrapadā* and *uttarabhādrapadā*), *hari* (*śravaṇa*) and *śūrpa* (*viśākhā*) have northerly latitude (*Saumya*). The others have southerly latitude (*Yāmya*). To the aforementioned latitudes of *varuṇa* (*śatabhiṣaj*), *nairṛta* (*mūla*) and *śūrpa* (*viśākhā*), 18, 30 and 30 minutes respectively have to be added.

The verses 3 and 4, give the values of twice the longitudes of the junction stars in the 27 *nakṣatras* in degrees in *kaṭapayādi* notation. These values are specified commencing with the star *aśvinī* referred to as *dasra*, and are listed in Table 9.1.

Verses 5–6 present the magnitude of deflection (*vikṣepa*) of these stars from the ecliptic in degrees in *kaṭapayādi* notation. These values are tabulated in Table 9.2. Verse 7 gives the direction of this *vikṣepas*. It is stated that the *vikṣepas* of *aśvinī*, *bharaṇī*, *pūrvaphalgunī*, *uttaraphalgunī*, *punarvasu*, *kṛttikā*, *svātī*, *dhaniṣṭhā*, *pūrvabhādrapadā*, *uttarabhādrapadā*, *śravaṇa* and *viśākhā* are towards the north (*saumya*) of the ecliptic. These are denoted by introducing '+' sign in the table. The *vikṣepas* of the rest of the junction stars having southerly declination (*yāmya*) are denoted by '−' sign in the table. It is further stated that 18′ (*divya*), 30′ (*naga*) and 30′ (*nāga*) are to be added to the *vikṣepas* of *śatabhiṣaj*, *mūla* and *viśākhā* respectively. These corrrected values are indicated in the parentheses in Table 9.2.

Name of the *Nakṣatra*	Twice the longitude of stars in degrees	
	in *kaṭapayādi*	in numerals
Aśvinī	*taṭajñaḥ*	16
Bharaṇī	*guṇena*	53
Kṛttikā	*prasannaḥ*	72
Rohiṇī	*nṛnamyḥ*	100
Mṛgaśirā	*vareṇyaḥ*	124
Ārdrā	*niviṣṭaḥ*	140
Punarvasu	*vadānyaḥ*	184
Puṣya	*nṛpendraḥ*	210
Āśleṣā	*haridrā*	228
Maghā	*samudraḥ*	257
Pūrvaphalgunī	*prahāraḥ*	282
Uttaraphalgunī	*dināṃbuḥ*	308
Hasta	*tavāṅgaḥ*	346
Citrā	*nisarga*	370
Svāti	*viḷaṅgaḥ*	394
Viśākhā	*virūḍhaḥ*	424
Anurādhā	*vibhāvaḥ*	444
Jyeṣṭhā	*tṛṇaughaḥ*	456
Mūla	*gajābhaḥ*	483
Pūrvāṣāḍhā	*dhaneśaḥ*	509
Uttarāṣāḍhā	*baleśaḥ*	533
Śravaṇa	*dhṛtāśaḥ*	569
Dhaniṣṭhā	*kaḷātmā*	591
Śatabhiṣaj	*vipakṣaḥ*	614
Pūrvabhādrapadā	*kṣamākṣaḥ*	656
Uttarabhādrapadā	*niḷāntaḥ*	690
Revatī	*nirāsaḥ*	720

Table 9.1 Twice the longitudes of the "junction stars" in the 27 *nakṣatras*.

९.३ भानां स्फुटविक्षेपानयनम्

9.3 Obtaining the true declinations of the stars

नक्षत्राणां स्फुटाः कार्याः सकलाः संस्कृतायनाः ।
तेषां क्रान्तिगुणाः स्पष्टाः स्वस्वविक्षेपसंस्कृताः ॥ ८ ॥

Name of the *Nakṣatra*	The latitudes of stars in degrees	
	in *kaṭapayādi*	in numerals
Aśvinī	*namyaḥ*	+10
Bharaṇī	*priyaḥ*	+12
Kṛttikā	*muniḥ*	+05
Rohiṇī	*mānī*	−05
Mṛgaśirā	*nityam*	−10
Ārdrā	*pūjyaḥ*	−11
Punarvasu	*tanuḥ*	+06
Puṣya	*nanu*	00
Āśleṣā	*sthānī*	−07
Maghā	*jñānī*	00
Pūrvaphalgunī	*priyaḥ*	+12
Uttaraphalgunī	*loke*	+13
Hasta	*sūnuḥ*	−07
Citrā	*prājñaḥ*	−02
Svāti	*sthale*	+37
Viśākhā	*punaḥ*	+01(1.5)*
Anurādhā	*lagne*	−03
Jyeṣṭhā	*bhānuḥ*	−04
Mūla	*janaḥ*	08(8.5)*
Pūrvāṣāḍhā	*sthāne*	−07
Uttarāṣāḍhā	*sūnuḥ*	−07
Śravaṇa	*nāgaḥ*	+30
Dhaniṣṭhā	*tilam*	+36
Śatabhiṣaj	*nanu*	00(0.3)*
Pūrvabhādrapadā	*vakraḥ*	+24
Uttarabhādrapadā	*tatra*	+26
Revatī	*nanu*	00

Table 9.2 The latitudes of the "junction stars" in the 27 *nakṣatras*.

परमापक्रमकोट्या विक्षेपज्यां निहत्य तत्कोट्या ।
इष्टक्रान्तिं चोभे त्रिज्याप्ते योगविरहयोग्ये स्तः ॥ ९ ॥
सदिशोः संयुतिरनयोर्वियुतिर्विदिशोरपक्रमः स्पष्टः ।
स्पष्टापक्रमकोटिर्द्युज्या[3] विक्षेपमण्डले वसताम् ॥ १० ॥

[3] For getting the intended meaning (evident from the context), as well as to see that the sentence is gramatically correct we need to introduce a *'visarga'* after the word *'koṭi'*.

nakṣatrāṇāṃ sphuṭāḥ kāryāḥ sakalāḥ saṃskṛtāyanāḥ |
teṣāṃ krāntiguṇāḥ spaṣṭāḥ svasvavikṣepasaṃskṛtāḥ || 8 ||

paramāpakramakoṭyā vikṣepajyāṃ nihatya tatkoṭyā |
iṣṭakrāntiṃ cobhe trijyāpte yogavirahayogye staḥ || 9 ||

sadiśoḥ saṃyutiranayorvidiśorapakramaḥ spaṣṭaḥ |
spaṣṭāpakramakoṭidyujyā vikṣepamaṇḍale vasatām || 10 ||

The true longitudes (*sphuṭas*) of the *nakṣatras* have to be corrected by the amount of precession (*ayanacalana*). Their true declinations can be obtained by applying the corrections based on their respective latitudes (*vikṣepas*).

Take the product of the Rcosine of the maximum declination (*paramāpakramakoṭi*) and Rsine of the latitude (*vikṣepajyā*), and similarly obtain the product of Rcosine of the latitude (*vikṣepakoṭijyā*) and a desired Rsine of the declination (*iṣṭakrānti*); divide [both the products] by *trijyā*. [Now, the results obtained are] ready for additive and subtractive operations.

If the latitude (*vikṣepa*) and declination (*krānti*) are along the same direction, then take the sum [of the results obtained earlier] and if they are in opposite directions, then take their difference and that would be the true declination (*spaṣṭāpakrama*). The Rcosine of the true declination would be the radius of the diurnal circle (*dyujyā*) for objects that lie on a latitudinal circle (*vikṣepamaṇḍala*) [off the ecliptic].

It is the tropical or the *sāyana* longitude which figures in most of the diurnal problems. This is equal to the sum of the *nirayana* longitude and the amount of precession or the *ayanacalana*. The declination of a celestial object which lies on the plane of the ecliptic in terms of the tropical longitude λ is given by

$$\sin\delta = \sin\epsilon \sin\lambda,$$

where ϵ is the obliquity of the ecliptic. But when the celestial object is off the ecliptic, this relation is no longer valid, as we have to take the latitude β or *vikṣepa* of the object into account. Verse 8 simply states that the latitude of the stars have to be considered for obtaining the accurate value of declination without giving details of how it has to be applied. The next couple of verses give the procedure for obtaining the declination of a celestial object with latitude.

Let δ be the declination of a celestial object whose tropical longitude is λ and latitude is β. Let δ' be the declination of an imaginary object situated on the ecliptic ($\beta = 0$) with the same longitude λ. The *iṣṭakrāntijyā* of the celestial object is the declination of this imaginary object and is given by the formula

$$R\sin\delta' = R\sin\epsilon\sin\lambda. \tag{9.4}$$

Then it is stated that the Rsine of true declination (*spaṣṭāpakrama*) is the sum (or difference) of the two terms given below.

The reading in the existing editions is without *visarga*. That is स्पष्टापक्रमकोटिद्युज्या। (*Spaṣṭāpakramakoṭidyujyā*).

$$spaṣṭāpakramajyā = \frac{paramāpakramakoṭi \times vikṣepajyā}{trijyā} \sim \frac{vikṣepakoṭijyā \times iṣṭakrāntijyā}{trijyā}. \tag{9.5}$$

Considering only the '+' sign for the time being, the above relation may be expressed in the form

$$R \sin\delta = \frac{R\cos\epsilon R \sin|\beta|}{R} + \frac{R\cos\beta R\sin\delta'}{R}. \tag{9.6}$$

Using (9.4) in the above equation and simplifying we have

$$\sin\delta = \cos\epsilon \sin|\beta| + \cos\beta \sin\epsilon \sin\lambda. \tag{9.7}$$

In Figure 9.2(a) the star is to the north of both the ecliptic and the equator. Hence the declination is given by the sum of the two terms in the RHS of (9.5). However when the star lies to the south of the ecliptic and to the north of the equator as indicated in Figure 9.2(b), then the difference of the two terms in RHS of (9.5) is to be considered for obtaining the declination of the star. In other words, the true declination is given by

$$\sin\delta = \cos\epsilon \ \sin|\beta| \ \sim \ \cos\beta \ \sin\epsilon \ \sin\lambda. \tag{9.8}$$

The rationale behind (9.7) and (9.8) can be understood with the help of Figure 9.2.

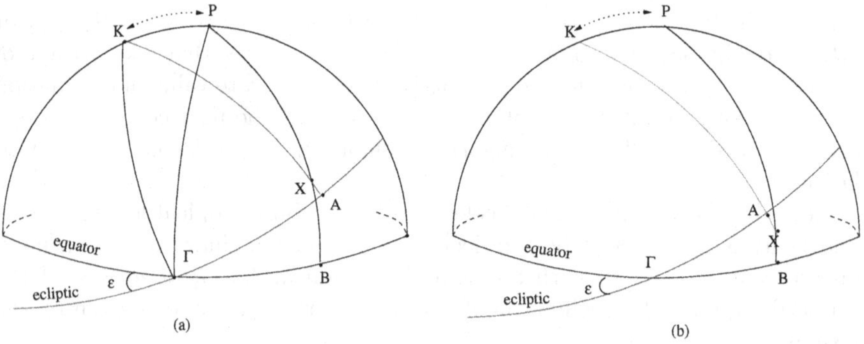

Fig. 9.2 Finding the declination of a celestial object with a non-zero latitude.

Let X be the celestial object whose longitude and latitude are given by $\Gamma A = \lambda$ and $XA = |\beta|$ respectively in Figure 9.2(a). Here, P and K are the poles of the equator and ecliptic respectively. Consider the spherical triangle PKX. The sides of the triangle are given by $PX = 90 - \delta$, $KX = 90 - |\beta|$ and $KP = \epsilon$. The spherical angle $P\hat{K}X = 90 - \lambda$. Now, applying the cosine formula, we have

$$\cos(90-\delta) = \cos\epsilon\cos(90-|\beta|) + \sin\epsilon\sin(90-|\beta|)\cos(90-\lambda),$$

or

$$\sin\delta = \cos\epsilon\sin|\beta| + \cos\beta\sin\epsilon\sin\lambda, \tag{9.9}$$

which is the same as (9.2) given in the text. In a similar manner by applying the cosine formula to the triangle PKX in Figure 9.2(b), we get

$$\sin\delta = \cos\beta\ \sin\epsilon\ \sin\lambda\ - \cos\epsilon\ \sin|\beta|, \tag{9.10}$$

which is the same as (9.8).

Finally it is said that for objects lying on the *vikṣepamaṇḍala*, which is a small circle parallel to the ecliptic with radius $R\cos\beta$, the Rcosine of the true declination thus determined would be the *dyujyā*. That is for determining the radius of the diurnal circle, the value of declination (δ) obtained from (9.9) is to be used and not the one given by (9.4).

९.४ मध्याह्नकाललग्नानयनम्

9.4 Obtaining the *madhyāhnakālalagna*

स्फुटक्रान्तिगुणो भूयः केवलक्षेपसंस्कृतः ।
परक्रान्तिशराभ्यस्तः परक्रान्तिहृतो गुणः ॥ ११ ॥

गुणहतात् त्रिभयुक्स्फुटदोर्गुणात्[4] दिनगुणाप्तधनुस्त्रिभयुक्स्फुटे ।
ऋणधनं समभिन्नदिशो भवेत् दिवसमध्यगकालविलग्नकम् ॥ १२ ॥

sphuṭakrāntiguṇo bhūyaḥ kevalakṣepasaṃskṛtaḥ |
parakrāntiśarābhyastaḥ parakrāntihṛto guṇaḥ || 11 ||

guṇahatāt tribhayuksphuṭadorguṇāt
dinaguṇāptadhanustribhayuksphuṭe |
ṛṇadhanaṃ samabhinnadiśo bhavet
divasamadhyagakālavilagnakam || 12 ||

The Rsine of the true declination (*sphuṭakrānti*), again corrected by the latitude, multiplied by the Rversine of maximum declination (*para-krānti*) and divided by the Rsine of maximum declination would be the multiplier (*guṇa*).

[This] multiplier has to be multiplied by the Rsine of the sum of the true longitude (*sphuṭa*) and 3 signs, and divided by the day-radius (*dina-guṇa*). The arc [of the result obtained] is added to or subtracted from the sum of the true longitude and 3 signs, depending on whether the directions of the multiplier and the declination are the same or different respectively.

---

[4] In the earlier editions of the text, the reading found is "त्रिभयुक् स्फुटदोर्गुणात्" (*"Tribhayuk sphuṭadorguṇāt"*). The word-splitting here is unintended and also misleading. This would mean $(90^\circ + R\sin\lambda)$. What is intended is $R\sin(90+\lambda)$. Hence we have presented the phrase in its compound form.

The *madhyāhnakālalagna* corresponding to any celestial object is the *kālalagna* when the object is on the prime meridian. It is the time interval between the rise of the vernal equinox Γ, and the instant when the celestial object is on the meridian. It can be seen that the *madhyāhnakālalagna* is equal to $90+\alpha$, where α is the *natakāla* or the right ascention (R.A). In Figure 9.3, X denotes

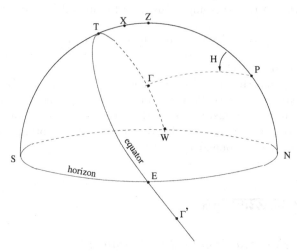

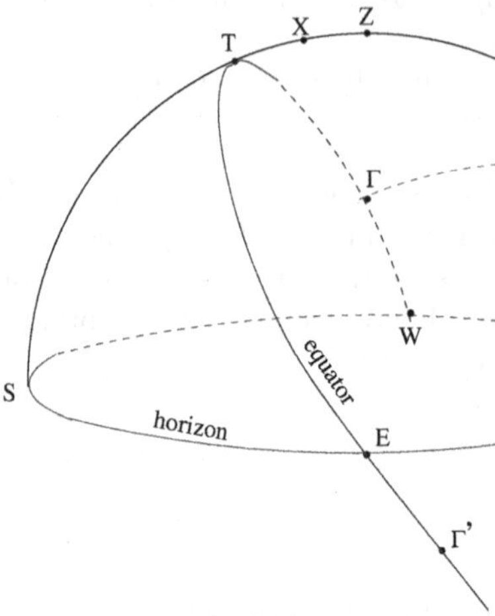

Fig. 9.3 Determination of the *madhyāhnakālalagna*.

the celestial object on the prime meridian, and Γ the vernal equinox. The R.A. of the object is $\Gamma T = \alpha$. Clearly, the time (in angular measure) after the rise of Γ at the eastern point is arc $ET + T\Gamma = 90 + \alpha$. This is the *madhyāhnakālalagna* for the object X, as it is on the prime meridian.

Verses 11 and 12 prescribe the following procedure for the determination of the *madhyāhnakālalagna*. The text first introduces a quantity called *guṇa* (the multiplier) which is given by the relation:[5]

$$\begin{aligned} guṇa &= \frac{(sphuṭakrānti \pm kevalavikṣepa) \times parakrāntiśara}{parakrānti} \\ &= \frac{(|R\sin\delta + R\sin\beta|) \times R(1-\cos\epsilon)}{R\sin\epsilon}. \end{aligned} \quad (9.11)$$

With this *guṇa* we are asked to find the *phala* given by

[5] Here onwards, we include the sign in β, so that it is negative when the declination is south.

$$\begin{aligned}
phala &= \frac{guṇa \times tribhayuk\text{-}sphuṭaguṇa}{dinaguṇa} \\
&= \frac{guṇa \times R|\sin(90+\lambda)|}{R\cos\delta} \\
&= \left| \frac{R(\sin\delta + \sin\beta) \times (1-\cos\epsilon)}{\sin\epsilon} \times \frac{\cos\lambda}{\cos\delta} \right|.
\end{aligned} \tag{9.12}$$

Then,

$$madhyāhnakālalagna = (\lambda + 90^\circ) \mp R\sin^{-1}(phala). \tag{9.13}$$

The latter half of the verse 12 talks about $\mp$ (*ṛṇadhanaṃ*). When $(\sin\delta + \sin\beta)$ and $\cos\lambda$ have the same sign or have the same directions, then their product is positive and we should subtract the angle corresponding to the *phala* from $90 + \lambda$. If they have the opposite sign, then we should add the angle corresponding to the *phala* to $90 + \lambda$.

The rationale behind the relation (9.13) can be understood as follows. It was already shown (9.9) that

$$\sin\delta = \cos\epsilon\sin\beta + \cos\beta\sin\epsilon\sin\lambda. \tag{9.14}$$

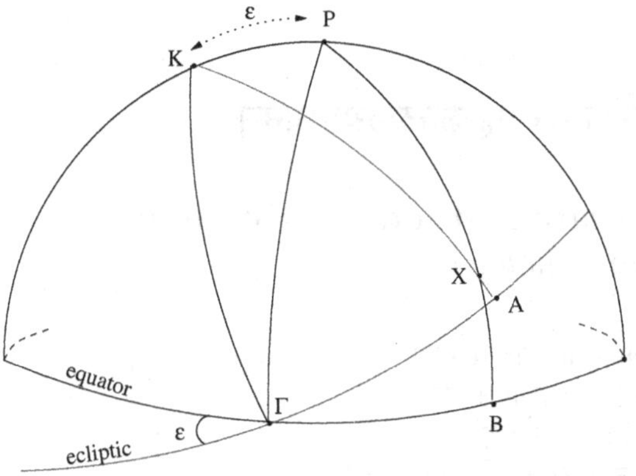

Fig. 9.4 The relation between declination, true longitude, latitude and R.A.

It is to be noted in the Figure 9.4 that, in the triangle KPX, the spherical angle $K\hat{P}X = 90 + \alpha$. Applying the cosine formula to the side $KX = 90 - \beta$, we have

$$\begin{aligned}
\cos(90-\beta) &= \cos\epsilon\cos(90-\delta) + \sin\epsilon\sin(90-\delta)\cos(90+\alpha), \\
\text{or} \qquad \sin\beta &= \cos\epsilon\sin\delta - \sin\epsilon\cos\delta\sin\alpha.
\end{aligned} \tag{9.15}$$

Adding (9.14) and (9.15), we have

$$\sin\delta + \sin\beta = \cos\epsilon(\sin\delta + \sin\beta) - \sin\epsilon\cos\delta\sin\alpha + \cos\beta\sin\epsilon\sin\lambda,$$

or
$$\frac{(\sin\delta + \sin\beta)(1 - \cos\epsilon)}{\sin\epsilon} = -\cos\delta\sin\alpha + \cos\beta\sin\lambda. \tag{9.16}$$

Multiplying this by $\frac{\cos\lambda}{\cos\delta}$ and using $\frac{\cos\beta\cos\lambda}{\cos\delta} = \cos\alpha$ (which follows from the application of the sine formula to the spherical triangle KPX), we have

$$\frac{(\sin\delta + \sin\beta)(1 - \cos\epsilon)}{\sin\epsilon} \times \frac{\cos\lambda}{\cos\delta} = -\cos\lambda\sin\alpha + \cos\alpha\sin\lambda = \sin(\lambda - \alpha). \tag{9.17}$$

The LHS of the above equation is nothing but the *phala* given in (9.12). Using (9.17) in (9.12) we have,

$$\lambda + 90 - \sin^{-1}\left[\frac{(\sin\delta + \sin\beta) \times (1 - \cos\epsilon)}{\sin\epsilon}.\frac{\cos\lambda}{\cos\delta}\right] = \lambda + 90 - (\lambda - \alpha) = 90 + \alpha, \tag{9.18}$$

which is the desired result.

९.५ प्रकारान्तरेण मध्याह्नकाललग्नानयनम्

9.5 An alternate method for obtaining the *madhyāhnakālalagna*

यद्वा स्फुटे निजकलासुभिदां[6] च कृत्वा
तस्मिन् पुनस्त्रिभयुते स्वभुजापमज्याम् ।
क्षेपाहतामपमकोटिहृतां पुरोवत्
कुर्यात् स्वमध्यदिनकालविलग्नसिद्ध्यै ॥ १३ ॥

yadvā sphuṭe nijakalāsubhidāṃ ca kṛtvā
tasmin punastribhayute ṣvabhujāpamajyām |
kṣepāhatāmapamakoṭihṛtāṃ purovat
kuryāt svamadhyadinakālavilagnasiddhyai || 13 ||

[6] The reading in the earlier editions of the text is: निजकलासु भिदां। (*Nijakalāsu bhidāṃ*). (This is not acceptable as the word is a compound word and is a synonym of प्राणकलान्तर (*Prāṇakalāntara*), which is also a compound word denoting the difference between the longitude and the *R.A.* ($\lambda \sim \alpha$).

Alternatively, having applied the difference between the true longitude and the right ascension (*nijakalāsubhidā*) to the true longitude (*sphuṭa*), adding 3 signs to it, and finding the Rsine of the declination corresponding to the resulting longitude (*svabhujāpamajyā*), multiply [the result obtained] by the Rsine of the latitude (*vikṣepajyā*). Dividing this by the Rcosine of the declination (*apakramakoṭi*), [computation] has to be done, as was done before in order to obtain *madhyāhnakālalagna*.

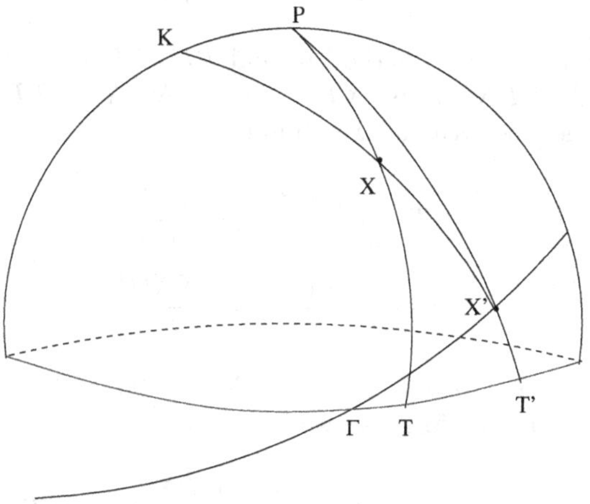

Fig. 9.5 Obtaining the *madhyāhnakālalagna*.

Let λ be the true longitude of the celestial object. It is said that this has to be first corrected by the *prāṇakalāntara*. It should be noted that the *prāṇakalāntara* ($\Delta\alpha$) for a celestial object having a non-zero latitude is

$$\Delta\alpha = \lambda \sim \alpha',$$

where λ is the longitude and α' is the R.A. of an object on the ecliptic whose longitude is λ. Let $\lambda' = \lambda - \Delta\alpha$ be the *sāyana* longitude of the celestial object corrected by its *prāṇakalāntara*, and $R\sin\delta''$ be the Rsine of declination associated with a point on the ecliptic, corresponding to a longitude $\lambda' + 90$. Then the *madhyāhnakālalagna* is stated to be

$$\alpha + 90 = \lambda' + 90 - R\sin^{-1}\left[\frac{R\sin\delta'' \times R\sin\beta}{R\cos\delta}\right]. \tag{9.19}$$

The rationale for the above expression is as follows. In Figure 9.5, X is the position of a star with a latitude β, and longitude λ. X' is the point on the ecliptic with the same longitude. From the figure $\lambda = \Gamma X'$, and $\alpha' = \Gamma T'$ is the R.A., corresponding to the point X' on the ecliptic, and, $\Delta\alpha = \lambda - \alpha' = \lambda - \Gamma T'$, is the *prāṇakalāntara*. Hence,

$$\begin{aligned}\alpha = \Gamma T &= \Gamma T' - TT' \\ &= \lambda - (\lambda - \Gamma T') - TT' \\ &= \lambda - \Delta\alpha - TT' \\ &= \lambda' - TT'. \end{aligned} \tag{9.20}$$

Now the *madhyāhnakālalagna* can be written as

$$\alpha + 90 = \lambda' + 90 - TT'. \tag{9.21}$$

Verse 13 essentially gives a method to find TT'. TT' can be found from the spherical triangle XPX'. Here, $XP = 90 - \delta$, $X\hat{P}X' = TT'$ and $XX' = \beta$. Now applying the sine formula, we obtain

$$\begin{aligned}\sin TT' &= \frac{\sin\beta \times \sin X\hat{X}'P}{\sin(90-\delta)} \\ &= \frac{\sin\beta \times \sin X\hat{X}'P}{\cos\delta}. \end{aligned} \tag{9.22}$$

In the spherical triangle KPX', $KP = \epsilon$, $KX' = 90$, and $K\hat{P}X' = 90 + \Gamma T'$.
Now applying the sine formula, we have

$$\begin{aligned}\sin X\hat{X}'P &= \frac{\sin K\hat{P}X}{\sin(KX')} \times \sin KP \\ &= \frac{\sin(90 + \Gamma T')}{\sin(90)} \times \sin\epsilon, \end{aligned}$$

$$\text{or} \qquad \sin X\hat{X}'P = \cos\Gamma T' \sin\epsilon. \tag{9.23}$$

The quantity $\Gamma T'$ is the RA of the object at X' and is equal to the *sāyana* longitude corrected by *prāṇakalāntara*. That is, $\Gamma T' = \lambda' = \lambda - \Delta\alpha$. Hence, the Rsine of declination (*apakramajyā*) corresponding to $\lambda' + 90$ is $\sin\delta'' = \cos\Gamma T' \sin\epsilon$, which is numerically equal to $\sin X\hat{X}'P = \cos\Gamma\hat{X}'T'$. Therefore, (9.22) becomes

$$\begin{aligned}\sin TT' &= \frac{\sin\beta \times \cos\Gamma T' \sin\epsilon}{\cos\delta} \\ &= \frac{\sin\beta \times \sin\delta''}{\cos\delta}. \end{aligned} \tag{9.24}$$

Substituting (9.24) in (9.21), we obtain the *madhyāhnakālalagna*

$$\alpha + 90 = \lambda' + 90 - R\sin^{-1}\left[\frac{R\sin\delta'' \times R\sin\beta}{R\cos\delta}\right],$$

which is the same as (9.19).

९.६ मध्याह्नकाललग्नानयने अन्यतरः प्रकारः

9.6 Yet another method for obtaining the *madhyāhnakālalagna*

यद्वा क्रान्तितदीयकोट्यसुकलाभेदांस्त्रिभोनस्फुटात्
आनीयासुकलान्तरं त्रिभयुते कुर्यात् स्फुटे तत्र तु ।
तत्क्रान्त्योर्वधतोऽर्धविस्तृतिहताद् द्युज्यावधाप्ताद् धनुः
क्रान्त्योर्भिन्नसमाशयोर्धनमृणं मध्याह्नकालाप्तये ॥ १४ ॥

yadvā krāntitadīyakoṭyasukalābhedāṃstribhonasphuṭāt
ānīyāsukalāntaraṃ tribhayute kuryāt sphuṭe tatra tu |
tatkrāntyorvadhato'rdhavistṛtihatād dyujyāvadhāptād dhanuḥ
krāntyorbhinnasamāśayordhanamṛṇaṃ madhyāhnakālāptaye || 14 ||

Or, for the true longitude diminished by three signs (*tribhonasphuṭa*) obtain the Rsine of declination, the corresponding Rcosine, and the *prāṇakalāntara*. This *prāṇakalāntara* has to be applied to the sum of the true longitude (*sphuṭa*) and three *rāśis*. The product of these declinations [of the *sphuṭa* and the one diminished by three *rāśis*] has to be multiplied by the radius (*ardhavistṛti*) and divided by the product of the [corresponding] day-radii (*dyujyās*). The arc [of the result obtained] is to be applied positively or negatively, depending upon whether the direction of the declinations are opposite or the same, in order to obtain the *madhyāhnakālalagna*.

Let δ_1 and δ_2 be the declinations of two celestial objects whose longitudes are λ and $\lambda - 90$ respectively. Then, the *madhyāhnakālalagna* given in the verse is

$$\lambda + 90 - \Delta\alpha_{\lambda-90} - \sin^{-1}\left[\frac{\sin\delta_1 \sin\delta_2}{\cos\delta_1 \cos\delta_2}\right], \tag{9.25}$$

where $\Delta\alpha_{\lambda-90}$ is the *prāṇakalāntara* at the longitude $\lambda - 90$.

We give the rationale behind the above expression with the help of Figure 9.6. Here X and X' are the objects whose longitudes are λ and $\lambda - 90$ respectively. If α and $\Delta\alpha_\lambda = \lambda - \alpha$ are the R.A. and the *prāṇakalāntara* of the star when it is at X (whose longitude is λ), then the *madhyāhnakālalagna* $(\alpha + 90)$ can be written as

$$\alpha + 90 = \lambda + 90 - \Delta\alpha_\lambda. \tag{9.26}$$

Now, PX' is the great circle arc which intersects the equator at R. Therefore, the *prāṇakalāntara* when the star is at X' is

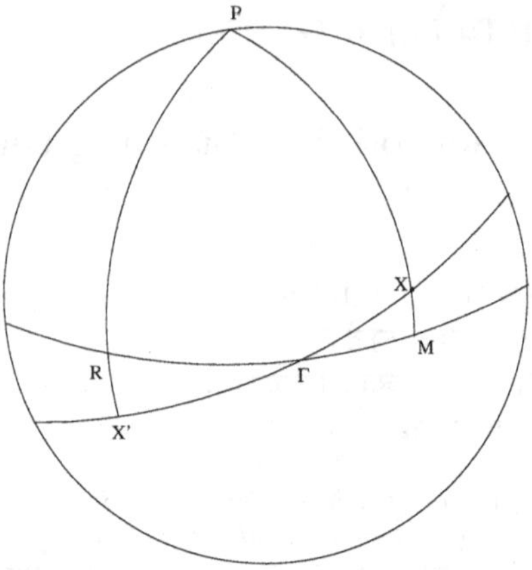

Fig. 9.6 Obtaining the *madhyāhnakālalagna* from the declinations of two celestial objects whose longitudinal difference is 90°.

$$\begin{aligned} \Delta\alpha_{\lambda-90} &= (\lambda - 90) - (\alpha - RM) \\ &= \lambda - 90 - \alpha + RM \\ &= (\lambda - \alpha) - (90 - RM) \\ &= \Delta\alpha_{\lambda} - \sin^{-1}(\sin(90 - RM)) \\ &= \Delta\alpha_{\lambda} - \sin^{-1}(\cos(RM)), \\ \text{or} \qquad \Delta\alpha_{\lambda} &= \Delta\alpha_{\lambda-90} + \sin^{-1}(\cos(RM)). \end{aligned} \tag{9.27}$$

Substituting (9.27) in (9.26), we have

$$\alpha + 90 = \lambda + 90 - \Delta\alpha_{\lambda-90} - \sin^{-1}(\cos(RM)). \tag{9.28}$$

The term $\sin^{-1}(\cos(RM))$ in (9.28) can be found from the spherical triangle PXX'. Here $PX = 90 - \delta_1$, $PX' = 90 + \delta_2$, and $X\hat{P}X' = RM$. Here δ_1 and δ_2 are the magnitudes of the declinations. Now applying the cosine formula to the side $XX'(= 90°)$, we get

$$\begin{aligned} \cos(90) &= \cos(90 - \delta_1)\cos(90 + \delta_2) \\ &\quad + \sin(90 - \delta_1)\sin(90 + \delta_2)\cos(RM), \\ \text{or} \qquad 0 &= -\sin\delta_1 \sin\delta_2 + \cos\delta_1 \cos\delta_2 \cos(RM). \\ \text{Hence,} \qquad \cos(RM) &= \frac{\sin\delta_1 \sin\delta_2}{\cos\delta_1 \cos\delta_2}. \end{aligned} \tag{9.29}$$

Substituting (9.29) in (9.28), we get

$$\alpha + 90 = \lambda + 90 - \Delta\alpha_{\lambda-90} - \sin^{-1}\left[\frac{\sin\delta_1 \sin\delta_2}{\cos\delta_1 \cos\delta_2}\right],$$

which is the expression (9.25) given in the text.

The significance of this result is not clear and it corresponds to a rather round about procedure. After all, $\alpha + 90 = \lambda + 90 - \Delta\alpha_{\lambda}$. Hence, the need of expressing $\alpha + 90$ in terms of $\Delta\alpha_{\lambda-90}$ and then applying a correction to it is not clear.

९.७ नतकाल-मध्याह्नकाललग्नानयनम्

9.7 Obtaining the *natakāla* (RA) and the *madhyāhnakālalagna*

विक्षेपकोटीपरमद्युजीवाघातात् परक्रान्तिहृतो हरः स्यात् ।
क्षेपाहते स्वस्फुटकोटिदोर्ज्ये हाराहृते कोटिभुजाफले स्तः ॥ १५ ॥

भुजाफलं व्यासदले धनर्णं विक्षेपदोर्ज्याहरिदैक्यभेदात् ।
तद्वर्गकोटीफलवर्गयोगात् मूलं भवेदायनसंज्ञकर्णः ॥ १६ ॥

कोटीफलं व्यासदलेन हत्वा कर्णाहृतं दृक्फलमायनं स्यात् ।
स्वर्णं स्फुटे तन्मृगकर्कटाद्योः क्षेपे तु मेषादिगतेऽन्यथा स्यात् ॥ १७ ॥

तस्मिन् पुनः प्राणकलान्तरं च कुर्यात् तदा स्यान्नतकाल एषः ।
नताख्यकालस्त्रिभसंयुतोऽयं माध्याह्निकं कालविलग्नकं स्यात् ॥ १८ ॥

vikṣepakoṭīparamadyujīvāghātāt parakrānti hṛto haraḥ syāt |
kṣepāhate svasphuṭakoṭidorjye hārāhṛte koṭibhujāphale staḥ || 15 ||

bhujāphalaṃ vyāsadale dhanarṇaṃ vikṣepadorjyāharidaikyabhedāt |
tadvargakoṭīphalavargayogānmūlaṃ bhavedāyanasaṃjñakarṇaḥ || 16 ||

koṭīphalaṃ vyāsadalena hatvā karṇāhṛtaṃ dṛkphalamāyanaṃ syāt |
svarṇaṃ sphuṭe tanmṛgakarkaṭādyoḥ kṣepe tu meṣādigate'nyathā syāt || 17 ||

tasmin punaḥ prāṇakalāntaraṃ ca kuryāt tadā syānnatakāla eṣaḥ |
natākhyakālastribhasaṃyuto'yaṃ mādhyāhnikaṃ kālavilagnakaṃ syāt || 18 ||

The product of the Rcosine of the (celestial) latitude (*vikṣepakoṭijyā*) and the Rcosine of maximum declination (*paramadyujyā*), divided by the Rsine of maximum declination (*paramakrāntijyā*), is the divisor. When the Rcosine and Rsine of the true longitude (*sphuṭa*) are multiplied by the Rsine of the latitude (*vikṣepajyā*) and divided by the divisor, the *koṭiphala* and the *bhujāphala* respectively are obtained.

Depending on whether the directions of the Rsine of latitude (*vikṣepa*) and the Rsine of longitude (*dorjyā*) are the same or different, the *bhujāphala* has to be added to or subtracted from the radius (*vyāsārdha*), respectively. Square root of the sum of the squares of this and the *koṭiphala* is known as the *āyanakarṇa*.

[The arc of] this *koṭiphala* multiplied by the radius and divided by the *karṇa* is the *āyanadṛkphala*. This is added to or subtracted from the true longitude for *karkyādi* or *makarādi* if the *kṣepa* is *meṣādi*, and vice versa if it is *tulādi*.

Then let the *prāṇakalāntara* of this [corrected true longitude] be applied to that. This gives the Right Ascension (*natakāla*). When three *rāśis* are added to the *natakāla* that will be the *madhyāhnakālalagna*.

The method to obtain the *madhyāhnakālalagna* as given in verses 15–18 may be outlined as follows. Let $|\beta|$ and ϵ be the latitude of the star and obliquity of the ecliptic respectively. Then the *hāra* H, defined in the verse is

$$H = \frac{R\cos\beta \; R\cos\epsilon}{R\sin\epsilon}.$$

The *koṭiphala* K_p, and the *bhujāphala* B_p, are defined as

$$K_p = \frac{|R\cos\lambda \; R\sin\beta|}{H},$$
$$B_p = \frac{|R\sin\lambda \; R\sin\beta|}{H}. \qquad (9.30)$$

The *āyanakarṇa*, denoted by K, and the *āyanadṛkphala*, A_p are defined by the relations

$$K = \sqrt{K_p^2 + (R \pm B_p)^2},$$
$$A_p = R\sin^{-1}\left(\frac{K_p \times R}{K}\right). \qquad (9.31)$$

Now the *madhyāhnakālalagna* can be found by using

$$\alpha + 90 = \lambda \pm A_p - \Delta\alpha + 90, \qquad (9.32)$$

where $\Delta\alpha$ is the *prāṇakalāntara* corresponding to $\lambda \pm A_p$. We explain the rationale for this relation in what follows.

In Figure 9.7, λ $(= \Gamma X')$ and β $(= XX')$ are the longitude and the latitude of the star X. It will be seen that $X'Y$ is the *āyanadṛkphala*, where Y is the point of intersection between the ecliptic and the meridian passing through X. Now, considering the spherical triangle YXX', and applying the sine formula, we find

$$\sin X'Y = \frac{\sin\beta}{\sin X\hat{Y}X'} \times \sin Y\hat{X}X'. \qquad (9.33)$$

In the spherical triangle KXY,

$$\frac{\sin(K\hat{X}Y)}{\sin(KY)} = \frac{\sin(180 - Y\hat{X}X')}{\sin 90} = \frac{\sin(X\hat{Y}K)}{\sin(KX)},$$
$$\text{or} \qquad \sin(Y\hat{X}X') = \frac{\sin(X\hat{Y}K)}{\cos\beta}. \qquad (9.34)$$

Applying the sine formula to the spherical triangle KYP,

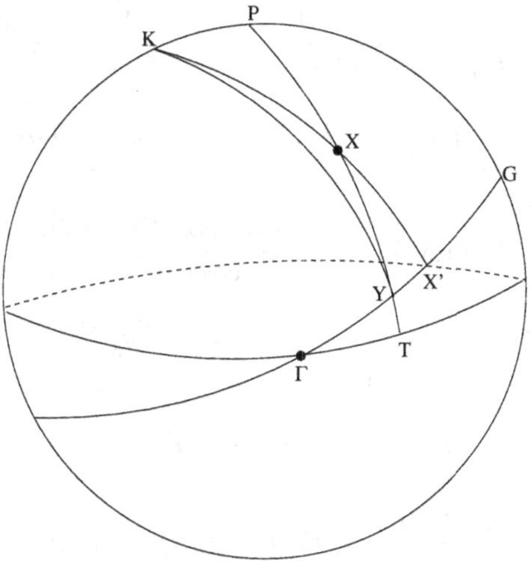

Fig. 9.7 Obtaining the *natakāla* and the *madhyāhnakālalagna*.

$$\frac{\sin(P\hat{Y}K)}{\sin(KP)} = \frac{\sin(X\hat{Y}K)}{\sin\epsilon} = \frac{\sin(K\hat{P}Y)}{\sin(KY)} = \frac{\sin(90+\Gamma\hat{P}T)}{\sin 90},$$

$$\text{or} \qquad \sin(X\hat{Y}K) = \sin\epsilon\cos(\Gamma\hat{P}T).$$

Using this in (9.34), we have

$$\sin(Y\hat{X}X') = \frac{\cos(\Gamma\hat{P}T)\sin\epsilon}{\cos\beta}. \tag{9.35}$$

Now, applying the sine formula to the spherical triangle ΓPY,

$$\frac{\sin(P\hat{Y}\Gamma)}{\sin(P\Gamma)} = \frac{\sin(P\hat{\Gamma}Y)}{\sin(PY)},$$

$$\text{or} \qquad \frac{\sin(180-X\hat{Y}X')}{\sin 90} = \frac{\sin(90-\epsilon)}{\sin(PY)},$$

$$\text{so that,} \qquad \sin X\hat{Y}X' = \frac{\cos\epsilon}{\sin(PY)}. \tag{9.36}$$

Using (9.35) and (9.36) in (9.33), we obtain

$$\sin X'Y = \frac{\sin\beta\cos(\Gamma\hat{P}T)\sin(PY)\sin\epsilon}{\cos\beta\cos\epsilon}. \tag{9.37}$$

In the spherical triangle PYG, $Y\hat{P}G = 90° - \Gamma\hat{P}T, YG = 90° - \Gamma Y$, and $P\hat{G}Y = 90°$. Applying the sine formula,

$$\frac{\sin(Y\hat{P}G)}{\sin(YG)} = \frac{\cos(\Gamma\hat{P}T)}{\cos(\Gamma Y)} = \frac{\sin(P\hat{G}Y)}{\sin(PY)} = \frac{\sin 90}{\sin(PY)}.$$

Thus, $\cos(\Gamma\hat{P}T)\ \sin(PY) = \cos(\Gamma Y)$. Using this in (9.37), we have

$$R\sin X'Y = \frac{R\sin\beta R\cos(\Gamma Y)R\sin\epsilon}{R\cos\beta R\cos\epsilon}. \tag{9.38}$$

In (9.38), $\Gamma Y = \Gamma X' - X'Y = \lambda - X'Y$, and $\frac{R\cos\beta R\cos\epsilon}{R\sin\epsilon}$ is referred to as the *hāra* (H) in the verse. Therefore,

$$\begin{aligned} R\sin X'Y &= \frac{R\sin\beta . R\cos(\lambda - X'Y)}{H} \\ &= \frac{R\sin\beta}{H} \times R\left[\cos\lambda\cos X'Y + \sin\lambda\sin X'Y\right]. \end{aligned}$$

Thus,

$$\sin X'Y\left[R - \frac{R\sin\beta R\sin\lambda}{H}\right] = \cos X'Y\left[\frac{R\sin\beta R\cos\lambda}{H}\right]. \tag{9.39}$$

Recalling the definition of *bhujāphala* and *koṭiphala* (9.30) given in the verse, and also squaring the equation (9.39) and adding $\sin^2 X'Y \times K_p^2$ on both sides, we get

$$\begin{aligned} \sin^2 X'Y\left[(R - B_p)^2 + K_p^2\right] &= K_p^2, \\ \text{or} \qquad \sin X'Y\sqrt{\left[(R - B_p)^2 + K_p^2\right]} &= K_p, \\ \text{or} \qquad \sin X'Y = \frac{K_p \times R}{K} &= \sin A_p, \end{aligned} \tag{9.40}$$

where $K = \sqrt{\left[(R - B_p)^2 + K_p^2\right]}$ is referred to as *āyanakarṇa* in the verse, and A_p is the *āyanadṛkphala*.[7]

Hence, $X'Y = A_p$. When this is applied to the *sphuṭa* or the longitude, $\lambda = \Gamma X'$, we have

$$\lambda' = \Gamma Y = \Gamma X' - X'Y = \lambda - A_p.$$

Now, when $\Delta\alpha$, or the *prāṇakalāntara* corresponding to λ', is applied to this, we obtain ΓT, the R.A. corresponding to Y, which is the *natakāla* or the R.A. corresponding to the object X also, as the secondary to the equator, PT passes through both X and Y. Hence, the *natakāla* is

[7] In Figure 9.7, $\sin\beta\sin\lambda$ is positive and $B_p = \frac{R\sin\lambda\sin\beta}{H}$. When $\sin\beta\sin\lambda$ is negative $B_p = \frac{-R\sin\lambda\sin\beta}{H}$ and $K = \sqrt{(R + B_p)^2 + K_p^2}$.

$$\alpha = \lambda' - \Delta\alpha = \lambda - A_p - \Delta\alpha,$$

and the *madhyāhnakālalagna* is

$$\alpha + 90 \approx \lambda - A_p - \Delta\alpha + 90, \tag{9.41}$$

which is the relation (9.32) mentioned in the verse.

९.८ प्रकृतराशेः व्यतीतकालानयनम्

9.8 The time elapsed in the current *rāśi*

> नक्षत्रमध्याह्नजकाललग्नात् स्वासन्नराश्यन्तजकाललग्नम् ।
> त्यक्त्वावशेषस्य लवाः षडाप्ताः तद्राशियाता घटिका भवन्ति ॥ १९ ॥
>
> *nakṣatramadhyāhnajakālalagnāt svāsannarāśyantajakālalagnam |*
> *tyaktvāvaśeṣasya lavāḥ ṣaḍāptāḥ tadrāśiyātā ghaṭikā bhavanti || 19 ||*
>
> Having subtracted the *kālalagna* of the nearest end of a *rāśi* (*svāsanna-rāśyanta*) from the *madhyāhnakālalagna* of the desired star, divide the remainder obtained, in degrees (*lavas*), by six. [The result] would be the *ghaṭikās* that have elapsed in the [next] *rāśi*.

The *madhyāhnakālalagna* corresponding to a star with R.A. α is $90 + \alpha$. This means that the time interval between the rise of Γ and the instant when the star is on the meridian is $90+\alpha$. Let the *kālalagna* of the *rāśyanta* nearest to the star be $\alpha' - \Delta\alpha'$. This means that the nearest *rāśyanta* rises $\alpha' - \Delta\alpha'$ time units after the rise of Γ. Hence, at the instant when the star is on the meridian, a time interval $90 + \alpha - (\alpha' - \Delta\alpha')$ in degrees has elapsed after the rise of the *rāśyanta* or, in other words, in the current *rāśi*. Now,

$$60 \text{ } ghaṭikās = 360 \text{ degrees} = 360 \text{ } lavas.$$

Therefore, 1 *ghaṭikā* = 6 *lavas*. Hence, the time elapsed in the current *rāśi* is

$$\frac{90 + \alpha - (\alpha' - \Delta\alpha')}{6} \quad ghaṭikās,$$

as stated in the verse.

Chapter 10
नतकालाद्यानयनम्
Obtaining the Right Ascension, etc.

१०.१ यन्त्रवेधस्य प्रामुख्यम्

10.1 Importance of observations with instruments

भानां स्फुटाश्च विक्षेपाः बहुधोक्ता बुधैस्ततः ।
स्वछायाद्यैः परीक्ष्यैते निर्णेया यन्त्रसाधितैः ॥ १ ॥

bhānāṃ sphuṭāśca vikṣepāḥ bahudhoktā budhaistataḥ |
svachāyādyaiḥ parīkṣyaite nirṇeyā yantrasādhitaiḥ || 1 ||

The longitudes (*sphuṭas*) and latitudes (*vikṣepas*) of the planets have been instructed in various ways by different scholars. Therefore, they have to be established after examination of their shadows etc. (*chāyādi*), as observed by the instruments.

Here the author emphasizes the importance of actual observations of the celestial objects through the measurement of their shadows etc. in determining their longitudes and latitudes. In the case of the Sun, the measurements associated with the shadow at noon and other times pose no difficulty, in principle at least, as these are done during the day. Observations pertaining to the Moon's shadow are also possible during the night. But what about the planets and stars?

It is indeed possible to measure the zenith distance of the planets and stars at the meridian transit and other times during the night using sights and quadrants, and the longitude and latitude can be determined from these observations. But the text or the commentaries do not give any further details concerning measurements done with instruments.

V. Pai et al., *Karaṇapaddhati of Putumana Somayājī*, Sources and Studies in the History of Mathematics and Physical Sciences, https://doi.org/10.1007/978-981-10-6814-0_10

१०.२ नतकालस्य वायुकालस्य वानयनम्

10.2 Obtaining the *natakāla* or the *vāyukāla* (RA)

सौम्ये शङ्क्वक्षघाते स्वमृणमिह दिशा भाभुजां लम्बकघ्नीं
कृत्वास्मात् त्रिज्ययाप्तं विदुरपमगुणस्तस्य कोटिर्द्युजीवा ।
त्रिज्याभाकोटिघाताद् दिनगुणविहृतं चापितं काललग्ने
स्वर्णं पूर्वापराह्णे त्रिभवनरहितं तद्भवेत् वायुकालः ॥ २ ॥

saumye śaṅkvakṣaghāte svamṛṇamiha diśā bhābhujāṃ lambakaghnīṃ
kṛtvāsmāt trijyayāptaṃ vidurapamaguṇastasya koṭirdyujīvā |
trijyābhākoṭighātād dinaguṇavihṛtaṃ cāpitaṃ kālalagne
svarṇaṃ pūrvāparāhṇe tribhavanarahitaṃ tadbhavet vāyukālaḥ || 2 ||

When [the declination is] north, the product of Rcosine of altitude (*śaṅku*) and Rsine of latitude (*akṣajyā*), and the product of *bhābhujā* or *chāyā-bhujā* and Rcosine of latitude (*lambaka*), have to be added to or subtracted from each other depending upon their directions. [The result obtained] divided by the radius is known to be the Rsine of declination (*apamajyā*). The Rcosine of that [declination] is day-radius (*dyujyā*). The product of *bhākoṭi* and the radius is divided by *dyujyā* and the arc of that is added to or subtracted from *kālalagna* before noon or after noon respectively. The right ascension (*natakāla* or *vāyukāla*) is obtained by subtracting 3 *rāśis* from the result.

This verse, in *sragdharā* metre, essentially presents an expression for the right ascension α (*yāyukāla*) of a celestial object in terms of its azimuth A, and the zenith distance z.

In Figure 10.1, X is the position of a celestial body whose zenith distance is z. Then, $XF = R\cos z$, is the *śaṅku* and, $OF = R\sin z$, is the *mahācchāyā* or just *chāyā*. The *chāyābhujā* is the projection of the *chāyā*, $R\sin z$, along the north-south line (or perpendicular to the east-west line) and is given by $RF = R\sin z \sin A'$, where A' is the angle between the prime vertical and the vertical passing through the celestial body.

Then it is stated that

$$apamajy\bar{a} = \frac{\acute{s}a\dot{n}ku \times ak\d{s}ajy\bar{a} \pm ch\bar{a}y\bar{a}bhuj\bar{a} \times lambaka}{trijy\bar{a}},$$

$$\text{or} \qquad R\sin\delta = \frac{R\cos z R\sin\phi \pm R\sin z \sin A' R\cos\phi}{R}. \tag{10.1}$$

This is obtained by applying the cosine formula to the side PX in the spherical triangle PZX, which gives

$$\sin\delta = \cos z \sin\phi - \sin z \cos\phi \sin A'. \tag{10.2}$$

Note that in this figure $P\hat{Z}X = 90 + A'$, so that $\cos(90 + A') = -\sin A'$. However, when X lies to the north of the prime vertical, $P\hat{Z}X = 90 - A'$ (not shown in the figure). Then, $\cos(90 - A') = \sin A'$, and hence

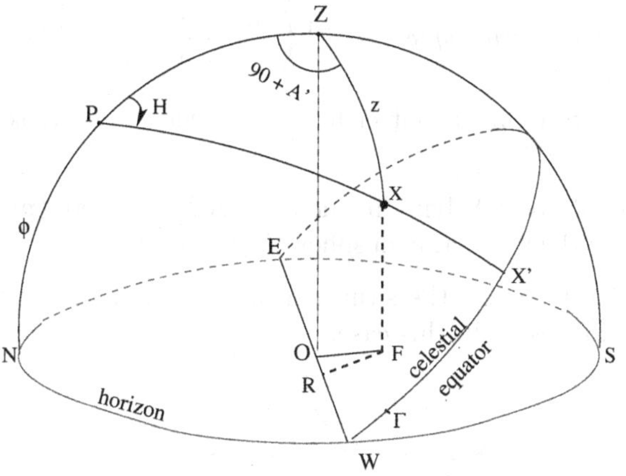

Fig. 10.1 Obtaining the *natakāla* (R.A.) from the *śaṅku* and *chāyā*.

$$\sin\delta = \cos z \sin\phi + \sin z \cos\phi \sin A'. \qquad (10.3)$$

The two possible cases, given by (10.2) and (10.3), have been stated in the verse. Having given the expression for $R\sin\delta$ (*apamajyā*) it is said that *dyujyā* ($R\cos\delta$) may be obtained from that.

Now applying the sine formula to the triangle $P\hat{Z}X$ (where H is the hour angle),

$$\frac{\sin H}{\sin z} = \frac{\sin(90 + A')}{\sin(90 - \delta)} = \frac{\cos A'}{\cos\delta},$$

$$\text{or} \qquad H = \sin^{-1}\left(\frac{\sin z \cos A'}{\cos\delta}\right)$$

$$= R\sin^{-1}\left(\frac{R\sin z \cos A'.R}{R\cos\delta}\right)$$

$$= R\sin^{-1}\left(\frac{bh\bar{a}ko\d{t}i \times trijy\bar{a}}{dyujy\bar{a}}\right), \qquad (10.4)$$

as *bhākoṭi* or *chāyākoṭi* is projection of the *chāyā* along the east-west line and is given by $R\sin z \cos A'$.

Now in Figure 10.1, the time interval after the rise of Γ, referred to as the *kālalagna*, is given by

$$k\bar{a}lalagna = \Gamma E = 90 + H + \alpha,$$

$$\text{or} \qquad \alpha = k\bar{a}lalagna - H - 90. \qquad (10.5)$$

where α is the *natakāla* or *vāyukāla* which is the R.A. of X. Hence, using (10.4) in (10.5) we have

$$\alpha = k\bar{a}lalagna - \sin^{-1}\left(\frac{\sin z \cos A'}{\cos\delta}\right) - 90°. \tag{10.6}$$

This is what is stated in the latter half of the verse. The verse also considers two cases,

(i) the celestial body X lying in the eastern hemisphere (*pūrvāhṇa*) and
(ii) X lying in the western hemisphere (*aparāhṇa*).

While Figure 10.1 depicts the situation in the afternoon, Figure 10.2 corresponds to the forenoon. In this case,

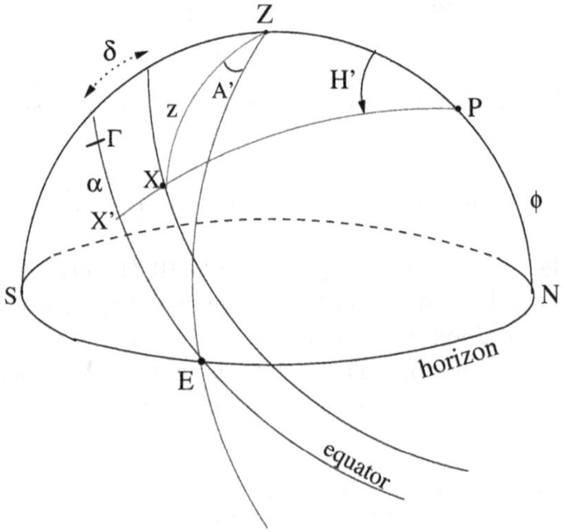

Fig. 10.2 Obtaining the *vāyukāla* (R.A.) from the *śaṅku* and *chāyā*.

$$\begin{aligned} H' &= 90 - X'E \\ &= 90 - (\Gamma E - \Gamma X') \\ &= 90 - k\bar{a}lalagna + \alpha. \end{aligned} \tag{10.7}$$

Therefore,

$$\alpha = k\bar{a}lalagna + \sin^{-1}\left(\frac{\sin z \cos A'}{\cos\delta}\right) - 90, \tag{10.8}$$

as stated before. Here H' is measured eastwards whereas H in Figure 10.1 is measured westwards. *Kālalagna* is always measured westwards, whereas α is measured eastwards in both Figure 10.1 and Figure 10.2.

१०.३ वायुकालानयने प्रकारान्तरम्

10.3 An alternate method for obtaining the *vāyukāla*

यद्वा याम्योत्तरापक्रमपलगुणयोर्घाततस्त्रिज्ययाप्तं
शङ्कौ संस्कृत्य तस्मात् त्रिगुणकृतिहताल्लम्बकक्रान्तिकोट्योः ।
घाताप्तं काललग्ने क्षिपतु कृतधनुस्त्यज्यतां प्राक् कपाले
कालो माध्याह्निकोऽयं भवति पुनरसौ वायुकालस्त्रिभोनः ॥ ३ ॥

yadvā yāmyottarāpakramapalaguṇayorghātatastrijyayāptaṃ
śaṅkau saṃskṛtya tasmāt triguṇakṛtihatāllambakakrāntikoṭyoḥ |
ghātāptaṃ kālalagne kṣipatu kṛtadhanustyajyatāṃ prāk kapāle
kālo mādhyāhniko'yaṃ bhavati punarasau vāyukālastribhonaḥ || 3 ||

Alternatively, the Rsine of the declination (*apakramajyā*) along the south or north is multiplied by Rsine of the latitude (*akṣajyā*) and divided by the radius. Having applied this to the Rcosine of zenith distance (*śaṅku*), multiply the result by the square of the radius and divide by the product of the Rcosine of the latitude (*lambaka*) and the Rcosine of declination (*krāntikoṭi*). The result is added to the *kālalagna* [when the Sun is in the western hemisphere] and is subracted from it when [the Sun] is in the eastern hemisphere (*prākkapāla*). And again three *rāśis* are added to or subtracted respectively, in order to obtain the *mādhyāhnika-kālalagna*. Subtracting 3 *rāśis* from this would result in the *vāyukāla*.

Consider the situation in the afternoon, as shown in the Figure 10.1. Applying the cosine formula to the side $ZX = z$ in the spherical triangle PZX,

$$\cos z = \sin\phi \sin\delta + \cos\phi\cos\delta\cos H,$$

$$\text{or} \qquad R\sin(90-H) = R\cos H = R\frac{\cos z - \sin\phi\sin\delta}{\cos\phi\cos\delta}$$

$$= R^2.\frac{R\cos z - \dfrac{R\sin\delta.R\sin\phi}{R}}{R\cos\phi.R\cos\delta}$$

$$= \left[(trijy\bar{a})^2 \times \frac{\acute{s}a\dot{n}ku - \dfrac{apakramajy\bar{a}\times ak\d{s}ajy\bar{a}}{trijy\bar{a}}}{lambaka \times kr\bar{a}ntiko\d{t}i}\right],$$

as indicated in the verse. Therefore,

$$90 - H \text{ (degrees)} = \sin^{-1}\left[\frac{\cos z - \sin\phi\sin\delta}{\cos\phi\cos\delta}\right] \text{ (degrees)}.$$

It is stated in the verse that this has to be applied to *kālalagna*, either positively or negatively depending on whether X lies in the eastern on western hemisphere. It may be recalled from (10.5) that when X is in the western hemisphere, *kālalagna* $= 90 + H + \alpha$. Hence,

$$
\begin{aligned}
\textit{mādhyāhnika-kālalagna} &= 90 + \alpha \\
&= \textit{kālalagna} + (90 - H) - 90, \\
\text{and,} \qquad \textit{vāyukāla} = \alpha &= \textit{mādhyāhnika-kālalagna} - 90, \qquad (10.9)
\end{aligned}
$$

as stated in the verse.

Consider the situation in the forenoon, when X is in the eastern hemisphere as depicted in the Figure 10.2. In this case,

$$
90 - H' \text{ (degrees)} = \sin^{-1}\left[\frac{\cos z - \sin\phi \sin\delta}{\cos\phi\cos\delta}\right] \text{ (degrees)},
$$

and, *kālalagna* $= E\Gamma = EX' + X'\Gamma = 90 - H' + \alpha$, as we saw earlier. Hence *kālalagna* $-(90 - H') = \alpha$, and we have to subtract the arc from the *kālalagna*. Again,

$$
\begin{aligned}
\textit{madhyāhna-kālalagna} &= 90 + \alpha \\
&= \textit{kālalagna} - (90 - H') + 90, \\
\text{and,} \qquad \textit{vāyukāla} = \alpha &= \textit{madhyāhna-kālalagna} - 90, \qquad (10.10)
\end{aligned}
$$

as stated in the verse.

१०.४ नतकालनिर्वचनम्

10.4 Definition of the *natakāla*

घटिकामण्डले यत्र स्पृष्टं तन्नतमण्डलम् ।
तत्प्रदेशो वायुकालो नतकालः स चोच्यते ॥ ४ ॥

सर्वेषामपि मध्याह्ने महाच्छायैव दोःप्रभा ।
काललग्नं त्रिराश्यूनं नतकालस्तदा भवेत् ॥ ५ ॥

ghaṭikāmaṇḍale yatra spṛṣṭaṃ tannatamaṇḍalam |
tatpradeśo vāyukālo natakālaḥ sa cocyate || 4 ||

sarveṣāmapi madhyāhne mahāchāyaiva doḥprabhā |
kālalagnaṃ trirāśyūnaṃ natakālastadā bhavet || 5 ||

Vāyukāla corresponds to the place on the equator where the secondary to the equator (*tannatamaṇḍala*) [passing through the celestial object] touches it. And this is [also] referred to as the *natakāla*.

In the noon *mahācchāyā* itself is the *chāyābhujā* for all [planets]. Three signs subtracted from the *kālalagna* at that time would be the *natakāla*.

In Figure 10.1 or 10.2, the secondary to the equator (*ghaṭikānatavṛtta*) passing through the object X intersects the equator at X'. This point corresponds to the *natakāla* or *vāyukāla*. In fact *natakāla*, as used in the earlier verses corresponds to the Right Ascension, $\Gamma X' = \alpha$.

As may be seen from Figure 10.1, at noon, $A' = 90$, as the Sun (X) will be on the prime meridian. In this case, *chāyābhujā* will be

$$R \sin z \sin A' = R \sin z = mah\bar{a}cch\bar{a}y\bar{a}.$$

This is true of any celestial object X on the meridian. Hence the use of the word *'sarveṣām'* (for all). Also, $H = H' = 0$ at noon. Then $k\bar{a}lalagna = 90 + \alpha$ and hence at noon, the $nat\bar{a}k\bar{a}la = \alpha = k\bar{a}lalagna - 90$, as stated in Verse 5.

१०.५ वायुकालाद्विक्षेपानयनम्

10.5 Obtaining the latitude from the *vāyukāla*

नतकालभुजाक्रान्तिं[1] व्यस्तदिक्कं प्रकल्प्यताम् ।
स्फुटापक्रमकोटिघ्नीं स्फुटक्रान्तिगुणे पुनः ॥ ६ ॥

परमक्रान्तिकोटिघ्ने कृत्वा तस्मात् त्रिजीवया ।
लब्धो भवति विक्षेपः तत्कोटिं च समानयेत् ॥ ७ ॥

natakālabhujākrāntiṃ vyastadikkaṃ prakalpyatām |
sphuṭāpakramakoṭighnīṃ sphuṭakrāntiguṇe punaḥ || 6 ||

paramakrāntikoṭighne kṛtvā tasmāt trijīvayā |
labdho bhavati vikṣepaḥ tatkoṭiṃ ca samānayet || 7 ||

Obtain the Rsine of declination (*krānti*) associated with the *natakāla*, which is taken to be in the opposite direction and multiply it by the Rcosine of the true declination. [The result] has to be applied to the product of Rcosine of maximum declination (*paramakrānti*) and the Rsine of the true declination (*sphuṭakrānti*). This divided by the radius is the Rsine of the latitude (*vikṣepa*). The corresponding Rcosine may [also] be obtained.

This verse presents the expression for the latitude of a celestial object in terms of its *R.A.* α, and declination δ. The formula given in the verse may be expressed as

$$R \sin \beta = \frac{R \sin \delta \times R \cos \epsilon - R \sin \alpha \sin \epsilon \times R \cos \delta}{R}. \tag{10.11}$$

Here the Rsine declination (*krāntijyā*) of the R.A. (*natakāla*) α is $R \sin \alpha \sin \epsilon$ (as the declination of a point on the ecliptic with longitude λ is $R \sin \lambda \sin \epsilon$) and this is what is referred to as *'natakālabhujākrānti'* in the verse. Also the

[1] The term नतकालभुजाक्रान्ति (*natakālabhujākrānti*) should be understood as नतकालसम्बन्धिनी या क्रान्तिः, तस्याः भुजा (*natakālasambandhinī yā krāntiḥ, tasyāḥ bhujā*). Mathematically it represents the quantity $R \sin \alpha \sin \epsilon$.

term *vikṣepa* used in last quarter of the verses refers to the Rsine of the latitude ($R \sin \beta$) and not just the arc β. Also in (10.11) the choice of the negative sign for the second term in RHS is as per the prescription in the verse that the *krāntijyā* of the *natakāla* is taken in the opposite direction.

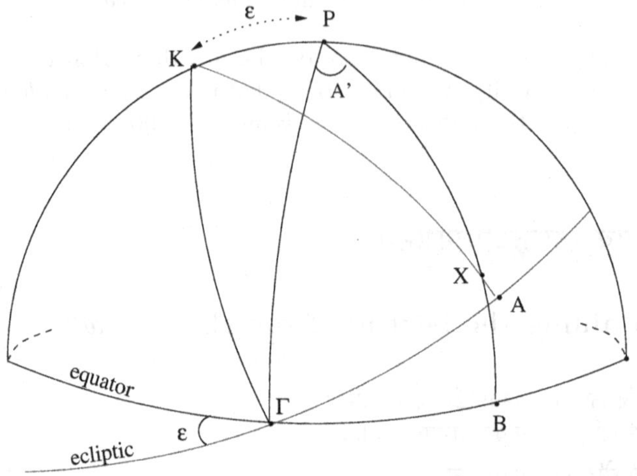

Fig. 10.3 Obtaining the Rsine of the latitude from the R.A.

By considering the spherical triangle KPX as shown in Figure 10.3, and applying the cosine formula for the side $KX = 90 - \beta$, we get,

$$\cos(90 - \beta) = \cos\epsilon \cos(90 - \delta) + \sin\epsilon \sin(90 - \delta)\cos(90 + A'),$$

$$\text{or} \qquad \sin\beta = \cos\epsilon \sin\delta - \sin\alpha \sin\epsilon \cos\delta, \qquad (10.12)$$

which is the same as (10.11) given in the verse.

१०.६ वायुकालात् स्फुटानयनम्

10.6 Obtaining the longitude from the *vāyukāla*

भूयः क्षेपगुणं स्फुटामपगुणे कृत्वामुना ताडिता
दोर्ज्यासत्रिभवायुकालजनिता विक्षेपकोट्या हृता ।
अन्त्यक्रान्तिशराहता परमया क्रान्त्या हृता चापिता
स्वर्णं तुल्यभिदाशघातवशतः स्याद्वायुकालस्फुटः ॥ ८ ॥

bhūyaḥ kṣepaguṇaṃ sphuṭāmapaguṇe kṛtvāmunā tāḍitā
dorjyāsatribhavāyukālajanitā vikṣepakoṭyā hṛtā |

antyakrāntiśarāhatā paramayā krāntyā hṛtā cāpitā
svarṇaṃ tulyabhidāśaghātavaśataḥ syādvāyukālasphuṭaḥ || 8 ||

Apply the Rsine of the latitude (*vikṣepa*) to the Rsine of the true declination (*sphuṭāpakrama*), and that is to be multiplied by the Rsine of the sum of three *rāśis* and the R.A. (*natakāla*), and divided by the Rcosine of the latitude (*vikṣepa*). [This, again] has to be multiplied by the versine of maximum declination (*paramakrānti*) and divided by Rsine of maximum declination (*antyakrānti*). The arc of the [result obtained] has to be added to or subtracted from the R.A. (*natakāla*) depending on the similarity or otherwise of [the directions of declination and latitude], since there is a product. This would give the true longitude (*sphuṭa*) from the R.A. (*vāyukāla*).

In the previous verse an expression for the latitude β of a celestial object was presented as a function of α and δ. This verse gives an expression for the longitude λ (*sphuṭa*) as a function of β, α and δ. This is actually done by means of a correction term to be applied to the R.A. or *natakāla*. The correction term given in the verse may be expressed as:

$$R\sin^{-1}\left[\frac{R\sin(90+\alpha)}{R\cos\beta}\times(R\sin\delta+R\sin\beta)\frac{R(1-\cos\epsilon)}{R\sin\epsilon}\right].$$

This term has to be applied to the *natakāla* positively or negatively depending on whether the directions of *vikṣepa* and *krāntijyā* are opposite or the same respectively. Therefore, the true longitude is given by

$$\lambda=\alpha\pm\sin^{-1}\left[\frac{\cos\alpha}{\cos\beta}\times(\sin\delta+\sin\beta)\times\frac{(1-\cos\epsilon)}{\sin\epsilon}\right]. \tag{10.13}$$

The rationale for the above expression is as follows. Rewriting equation (9.18) we have,

$$\lambda=\alpha+\sin^{-1}\left[\frac{(\sin\delta+\sin\beta)\times(1-\cos\epsilon)}{\sin\epsilon}.\frac{\cos\lambda}{\cos\delta}\right]. \tag{10.14}$$

Now consider the spherical triangle PKX in Figure 10.3. Here $P\hat{K}X=90-\lambda$, $PX=90-\delta$, $K\hat{P}X=90+\alpha$, and $KX=90-\beta$. Applying the sine formula we have,

$$\frac{\cos\lambda}{\cos\delta}=\frac{\cos\alpha}{\cos\beta}. \tag{10.15}$$

Using (10.15) in (10.14), we get

$$\lambda=\alpha+\sin^{-1}\left[\frac{\cos\alpha}{\cos\beta}\times(\sin\delta+\sin\beta)\times\frac{(1-\cos\epsilon)}{\sin\epsilon}\right], \tag{10.16}$$

which is the result (10.13) given in the verse. Here the quantity in the bracket is positive or negative depending upon the sign of the product $\cos\alpha(\sin\delta+\sin\beta)$.

१०.७ मध्याह्नकाललग्नात् स्फुटानयने प्रकारान्तरम्

10.7 An alternate method for obtaining the longitude from the *madhyāhnakālalagna*

यद्वा स्वमध्याह्नगकाललग्ने कृतासुलिप्ताविवरे स्वदोर्ज्याम् ।
क्षेपान्तिमक्रान्तिवधेन हत्वा तत्कोटिघातेन विभज्य लब्धम् ॥ ९ ॥
चापीकृतं च स्वमृणं प्रकुर्यात् विक्षेपदोर्ज्याहरिदैक्यभेदात् ।
त्रिभोनितेऽस्मिन् पुनरायनांशं व्यस्तं च कुर्यात् स निजस्फुटः स्यात् ॥ १० ॥

yadvā svamadhyāhnagakālalagne
kṛtāsuliptāvivare svadorjyām |
kṣepāntimakrāntivadhena hatvā
tatkoṭighātena vibhajya labdham || 9 ||

cāpīkṛtaṃ ca svamṛṇaṃ prakuryāt
vikṣepadorjyāharidaikyabhedāt |
tribhonite'smin punarāyanāṃśaṃ
vyastaṃ ca kuryāt sa nijasphuṭaḥ syāt || 10 ||

Or, having obtained the Rsine of *prāṇakalāntara* from the *sva-madhyāhna-kālalagna* and applying the result to itself, [the Rsine of the result obtained] has to be multiplied by the product of the Rsine of latitude (*kṣepajyā*) and Rsine of maximum declination (*antyakrānti*) and divided by the product of their Rcosines. The arc of [the result obtained] is added to or subtracted from the corrected (*saṃskṛta*)-*madhyāhnakālalagna* depending on whether the directions of Rsine of latitude (*vikṣepa*) and Rsine of longitude (*dorjyā*) are the same or different respectively. Then, by subtracting 3 signs and the amount of precession (*ayanāṃśa*), the true longitude (*sphuṭa*) is obtained.

The two verses above essentially give yet another method to obtain the true longitude (*sphuṭa*) from the *madhyāhnakālalagna*. Let λ' be the *madhyāhnakālalagna* corrected by the *prāṇakalāntara*, $\Delta\alpha$. That is,

$$\lambda' = madhy\bar{a}hnak\bar{a}lalagna + \Delta\alpha.$$

Then the expression for the true longitude λ given by the verse is

$$\lambda = \lambda' \pm R\sin^{-1}\left[\frac{R\sin\beta\ R\sin\epsilon\ R\sin\lambda'}{R\cos\beta R\cos\epsilon}\right] - 90. \qquad (10.17)$$

The *ayanāṃśa* has to be subtracted from λ to obtain the *nirayana* longitude.

In Figure 10.4, $\Gamma X' = \lambda$ is the true longitude, and $\Gamma T = \alpha$, is the right ascension. And $\Gamma X' - \Gamma T' = \lambda - \alpha' = \Delta\alpha$ is the *prāṇakalāntara*, where α' is the R.A. of X' which has the same longitude as X and lies on the ecliptic. Now the *madhyāhnakālalagna* $(\alpha + 90)$ corrected by *prāṇakalāntara* is given by

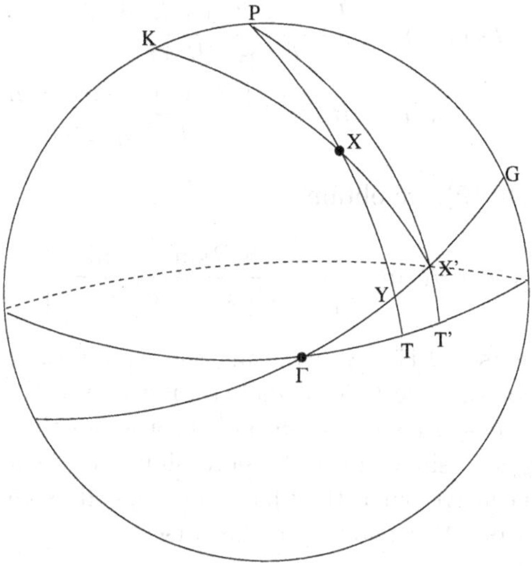

Fig. 10.4 Obtaining the true longitude from the *madhyāhnakālalagna*.

$$\begin{aligned} \lambda' &= \mathit{\Gamma}T + 90 + \mathit{\Gamma}X' - \mathit{\Gamma}T' && (10.18)\\ &= (\mathit{\Gamma}T - \mathit{\Gamma}T') + 90 + \mathit{\Gamma}X' \\ &= -TT' + 90 + \lambda, \\ \text{or} \qquad \lambda &= \lambda' + TT' - 90 \\ &\approx \lambda' + X'Y - 90, && (10.19) \end{aligned}$$

as $TT' \approx X'Y$. Now recalling the equation (9.38), we have

$$R\sin X'Y = \frac{R\sin\beta\, R\sin\epsilon\, R\cos(\mathit{\Gamma}Y)}{R\cos\beta R\cos\epsilon}. \qquad (10.20)$$

Now, again from the figure, $\mathit{\Gamma}X' = \mathit{\Gamma}Y + X'Y$. Substituting this for $\mathit{\Gamma}X'$ in (10.18), we have

$$\begin{aligned} \lambda' &= (\mathit{\Gamma}T - \mathit{\Gamma}T') + 90 + \mathit{\Gamma}Y + X'Y \\ &= -TT' + X'Y + \mathit{\Gamma}Y + 90, \\ \text{or} \qquad \mathit{\Gamma}Y &= \lambda' + TT' - X'Y - 90 \\ &\approx \lambda' - 90. && (10.21) \end{aligned}$$

Substituting (10.21) in (10.20), we have

$$R\sin X'Y = \frac{R\sin\beta \; R\sin\epsilon \; R\sin\lambda'}{R\cos\beta \; R\cos\epsilon},$$

or
$$X'Y = R\sin^{-1}\left[\frac{R\sin\beta \; R\sin\epsilon \; R\sin\lambda'}{R\cos\beta \; R\cos\epsilon}\right]. \qquad (10.22)$$

Using (10.22) in (10.19), we obtain

$$\lambda = \lambda' + R\sin^{-1}\left[\frac{R\sin\beta \; R\sin\epsilon \; R\sin\lambda'}{R\cos\beta \; R\cos\epsilon}\right] - 90, \qquad (10.23)$$

which is the same as (10.17). Subtracting the *ayanāṃśa* from the above expression, we will obtain the true *nirayana* longitude of the celestial body as mentioned in the verses. Here, the second term is positive, when the *vikṣepa* ($R\sin\beta$) and *dorjyā* ($R\sin\lambda'$) have the same sign, and the arc has to be added to λ'; while it is negative when they have the opposite signs, and the arc has to be subtracted from λ', as stated in the verse.

१०.८ उपसंहारवचनम्

10.8 Concluding remarks

गणितमिदमशेषं युक्तियुक्तं पठन्तः
भुवि गणितजनानां अग्रगण्या भवेयुः ।
अपि च गतिविशेषात् कालरूपस्य विष्णोः
सुभृशमनुभवन्तो यान्ति तद्धाम शुद्धम् ॥ ११ ॥

इति शिवपुरनामग्रामजः कोऽपि यज्वा
किमपि करणपद्धत्याह्वयं तन्त्ररूपम् ।
व्यधितगणितमेतत् सम्यगालोक्य सन्तः
कथितमिह विदन्तः सन्तु सन्तोषवन्तः ॥ १२ ॥

gaṇitamidamaśeṣaṃ yuktiyuktaṃ paṭhantaḥ
bhuvi gaṇitajanānāṃ agragaṇyā bhaveyuḥ |
api ca gativiśeṣāt kālarūpasya viṣṇoḥ
subhṛśamanubhavanto yānti taddhāma śuddham || 11 ||

iti śivapuranāmagrāmajaḥ ko'pi yajvā
kimapi karaṇapaddhatyāhvayaṃ tantrarūpam |
vyadhitagaṇitametat samyagālokya santaḥ
kathitamiha vidantaḥ santu santoṣavantaḥ || 12 ||

Those who study and comprehend (*paṭhantaḥ*) all the mathematical principles supported by rationales (*yuktiyuktam*) enunciated here, would become the foremost leaders in the community of mathematicians in this world. Moreover, blessed with the compassion of Lord Viṣṇu, who also manifests in the form of time, they would attain His pristine abode.

> Thus someone, who has performed sacrifices (*yajvā*) hailing from the village by name *Śivapura*, has composed a work called *Karaṇapaddhati*, in the form of a *Tantra*. By assiduously going through this mathematical work and understanding whatever has been set out here, may the noble ones become happy.

It was conjectured by Whish that the phrase "*gaṇitametadsamyak*" may be encoding the *ahargaṇa* of the time of composition of this text (Whish 1834). This *ahargaṇa*, which works out to be 1765653, corresponds to the year 1733 CE.[2]

[2] On the date of *Karaṇapaddhati*, see the discussion in the Introduction.

Appendix A
Vallyupasaṃhāra and continued fractions

Ever since the work of Āryabhaṭa on the *kuṭṭaka* procedure for solving linear indeterminate equations, Indian astronomers and mathematicians have been using this method to solve a variety of problems. The method, also referred to as *kuṭṭākāra*, basically makes use of a technique called *vallyupasaṃhāra* which is analogous to the continued fraction expansion of a ratio of integers. The *vallī* introduced by Āryabhaṭa is nothing but the column composed of the quotients which arise in the mutual division of the integers. The *vallyupasaṃhāra* method of transforming the *vallī* is essentially the recursive process of calculating the successive convergents of the associated continued fraction.

In *Karaṇapaddhati* Putumana Somayājī displays a very sophisticated understanding of the mathematical properties of the continued fraction expansion of a ratio of two integers G, H.[1] Usually, G is the *guṇa* or *guṇakāra* and H is the *hāra* or *hāraka*, and their ratio $(\frac{G}{H})$ is the rate of motion of a particular planet or its apogee or node etc. Thus, G being the corrected revolution number and H the total number of civil days, they are indeed very huge numbers. Chapter 2 of *Karaṇapaddhati* essentially presents the method of approximating the ratio $\frac{H}{G}$ by the successive convergents of the associated continued fraction. *Karaṇapaddhati* also reveals a very sophisticated understanding of the properties of the convergents including a very interesting "remainder theorem", as we shall explain in this appendix.

A.1 Simple continued fraction and its convergents

We start with the ratio of two integers $\frac{G}{H}$, where G is the *guṇakāra* and H is the *hāraka*. Normally H is larger than G and it is useful to consider the ratio $\frac{H}{G}$. We now discuss the continued fraction expansion of $\frac{H}{G}$. Dividing H by G, we get

[1] For an introduction to continued fractions, see Khinchin 1964.

V. Pai et al., *Karaṇapaddhati of Putumana Somayājī*, Sources and Studies in the History of Mathematics and Physical Sciences, https://doi.org/10.1007/978-981-10-6814-0

$$\frac{H}{G} = q_1 + \frac{r_1}{G}, \tag{A.1}$$

where q_1 is the quotient and $r_1 < G$ is the remainder. Now dividing G by r_1, we get

$$\frac{G}{r_1} = q_2 + \frac{r_2}{r_1}. \tag{A.2}$$

Continuing in this manner, we obtain a series of quotients and remainders

$$\begin{aligned} \frac{r_1}{r_2} &= q_3 + \frac{r_3}{r_2}, \\ \vdots &= \vdots \\ \frac{r_{i-2}}{r_{i-1}} &= q_i + \frac{r_i}{r_{i-1}}. \end{aligned} \tag{A.3}$$

As these quotients (q_i) and remainders (r_i) are obtained by the mutual division of the numbers H, G, we can write

$$\frac{H}{G} = q_1 + \cfrac{1}{q_2 + \cfrac{1}{q_3 + \cfrac{1}{q_4 + \ldots + \cfrac{1}{q_i + \cfrac{r_i}{r_{i-1}}}}}}. \tag{A.4}$$

Since $\frac{H}{G}$ is a ratio of two integers, the process will terminate for some n, when $r_n = 0$. We thus have

$$\frac{H}{G} = q_1 + \cfrac{1}{q_2 + \cfrac{1}{q_3 + \cfrac{1}{q_4 + \ldots + \cfrac{1}{q_n}}}}. \tag{A.5}$$

This process of mutual division is also the well known process (so called Euclidean algorithm) for finding GCD of the numbers H, G which is in fact given by r_{n-1}.

The above equation gives the simple continued fraction expansion of $\frac{H}{G}$. If we truncate the above process at any intermediate stage $k < n$, then we get one of the so called convergents of the continued fraction given by

$$\frac{H_k}{G_k} = q_1 + \cfrac{1}{q_2 + \cfrac{1}{q_3 + \cfrac{1}{q_4 + \ldots + \cfrac{1}{q_k}}}}. \tag{A.6}$$

In particular, we have

$$\frac{H_1}{G_1} = \frac{q_1}{1} \qquad (H_1 = q_1,\ G_1 = 1), \tag{A.7}$$

$$\frac{H_2}{G_2} = q_1 + \frac{1}{q_2}$$

$$= \frac{q_1 q_2 + 1}{q_2} \qquad (H_2 = q_1 q_2 + 1, G_2 = q_2), \tag{A.8}$$

and so on.

A.2 Properties of the convergents

Consider the simple continued fraction expansion of $\frac{H}{G}$:

$$q_1 + \cfrac{1}{q_2 + \cfrac{1}{q_3 + \cfrac{1}{q_4 + \dots}}} \, . \tag{A.9}$$

The successive convergents are

$$\frac{H_1}{G_1} = \frac{q_1}{1}, \qquad \frac{H_2}{G_2} = \frac{q_2 q_1 + 1}{q_2}, \qquad \frac{H_3}{G_3} = \frac{q_3(q_2 q_1 + 1) + q_1}{q_3 q_2 + 1}, \dots \tag{A.10}$$

We shall now proceed to explain the properties of convergents. First we shall show that the following recursion relations are satisfied by H_k, G_k for $k > 2$:

$$H_k = q_k H_{k-1} + H_{k-2}, \tag{A.11}$$

$$G_k = q_k G_{k-1} + G_{k-2}. \tag{A.12}$$

The proof is by induction on k. Clearly from (A.10), we see that (A.11) and (A.12) hold when $k = 3$. Assuming that these equations hold for k, we shall show that they hold for $k + 1$. From (A.6), it is clear that $\frac{H_{k+1}}{G_{k+1}}$ is same as $\frac{H_k}{G_k}$ with q_k replaced by $q_k + \frac{1}{q_{k+1}}$. Therefore,

$$\begin{aligned}\frac{H_{k+1}}{G_{k+1}} &= \frac{(q_k + \frac{1}{q_{k+1}})H_{k-1} + H_{k-2}}{(q_k + \frac{1}{q_{k+1}})G_{k-1} + G_{k-2}} \\ &= \frac{(q_{k+1}q_k + 1)H_{k-1} + q_{k+1}H_{k-2}}{(q_{k+1}q_k + 1)G_{k-1} + q_{k+1}G_{k-2}} \\ &= \frac{q_{k+1}(q_k H_{k-1} + H_{k-2}) + H_{k-1}}{q_{k+1}(q_k G_{k-1} + G_{k-2}) + G_{k-1}} \\ &= \frac{q_{k+1}H_k + H_{k-1}}{q_{k+1}G_k + G_{k-1}}.\end{aligned}$$

Thus, we have shown that the recurrence relations (A.11) and (A.12) are valid for all $k > 2$.

We shall now show another important property of H_k's and G_k's, namely

$$H_k G_{k+1} - H_{k+1} G_k = (-1)^k. \tag{A.13}$$

From the recurrence relations (A.11) and (A.12), we see that

$$\begin{aligned}H_k G_{k+1} - H_{k+1} G_k &= H_k(q_{k+1}G_k + G_{k-1}) - (q_{k+1}H_k + H_{k-1})G_k \\ &= H_k G_{k-1} - H_{k-1} G_k \\ &= -(H_{k-1} G_k - H_k G_{k-1}) \\ &\;\;\vdots \\ &= (-1)^{k-1}(H_1 G_2 - H_2 G_1) \\ &= (-1)^k. \end{aligned} \tag{A.14}$$

From the above relation we can also derive yet another interesting property of the convergents, namely

$$\left| \frac{H_{k+1}}{G_{k+1}} - \frac{H_k}{G_k} \right| = \frac{1}{G_k G_{k+1}}. \tag{A.15}$$

We can easily see that

$$\begin{aligned}\frac{H_{k+1}}{G_{k+1}} - \frac{H_k}{G_k} &= \frac{q_{k+1}H_k + H_{k-1}}{q_{k+1}G_k + G_{k-1}} - \frac{H_k}{G_k} \\ &= \frac{(H_{k-1}G_k - H_k G_{k-1})}{G_{k+1}G_k} \\ &= \frac{(-1)^{k-1}}{G_{k+1}G_k}. \end{aligned} \tag{A.16}$$

A.3 Remainder theorem of *Karaṇapaddhati*

The *Karaṇapaddhati* states and makes extensive use of a "remainder theorem" which gives the difference between the number $\frac{H}{G}$ and its convergents $\frac{H_i}{G_i}$ in terms of the remainder r_i which is obtained in the mutual division of H, G.

Now from the previous discussions, we know that the ratio

$$\frac{H}{G} = q_1 + \frac{r_1}{G} = \frac{Gq_1 + r_1}{G}. \tag{A.17}$$

The difference between the actual ratio and its first approximation (A.7) can be written as

$$\begin{aligned} \frac{H}{G} - \frac{H_1}{G_1} &= \frac{Gq_1 + r_1}{G} - q_1 \\ &= \frac{r_1}{G}. \end{aligned} \tag{A.18}$$

Therefore,

$$GH_1 - HG_1 = Gq_1 - H = -r_1. \tag{A.19}$$

Now, from (A.4)

$$\begin{aligned} \frac{H}{G} &= q_1 + \cfrac{1}{q_2 + \cfrac{r_2}{r_1}} \\ &= q_1 + \frac{r_1}{q_2 r_1 + r_2} \\ &= \frac{(q_1 q_2 + 1) r_1 + q_1 r_2}{q_2 r_1 + r_2}. \end{aligned} \tag{A.20}$$

Using (A.8) in the above we have,

$$\frac{H}{G} = \frac{H_2 r_1 + q_1 r_2}{G_2 r_1 + r_2}. \tag{A.21}$$

Therefore,

$$\begin{aligned} G(H_2 r_1 + q_1 r_2) &= HG_2 r_1 + Hr_2, \\ \text{or} \quad (GH_2 - HG_2) r_1 &= -(Gq_1 - H) r_2 \\ &= r_1 r_2, \end{aligned} \tag{A.22}$$

where we have used (A.19). Hence,

$$GH_2 - HG_2 = r_2. \tag{A.23}$$

Now we present the general version of the above remainder relation which the *Karaṇapaddhati* states and makes extensive use of.

Theorem

If $\frac{H_i}{G_i}$ is the ith convergent of $\frac{H}{G}$ and r_i is the ith remainder in the mutual division of H and G, then

$$HG_i - GH_i = (-1)^{i-1} r_i. \tag{A.24}$$

Proof:

We have already seen that (A.24) is valid for $i = 1, 2$. Now, we shall assume that relation (A.24) is true for some i and then show that it is true for $i + 1$. To be specific let i be even so that

$$(HG_i - GH_i) = -r_i. \tag{A.25}$$

Using the recursion relations

$$\begin{aligned} G_{i+1} &= G_i q_{i+1} + G_{i-1}, \\ \text{and} \qquad H_{i+1} &= H_i q_{i+1} + H_{i-1}, \end{aligned}$$

we obtain

$$\begin{aligned} HG_{i+1} - GH_{i+1} &= (HG_i - GH_i) q_{i+1} + HG_{i-1} - GH_{i-1} \\ &= -r_i q_{i+1} + r_{i-1}. \end{aligned} \tag{A.26}$$

In the expression for the continued fraction of $\frac{H}{G}$ considered earlier, the remainder r_{i+1} is obtained by dividing r_{i-1} by r_i. The corresponding quotient is q_{i+1} (see section A.1). Hence, r_{i-1} can be written as

$$\begin{aligned} r_{i-1} &= r_i q_{i+1} + r_{i+1}, \\ \text{or} \qquad r_{i+1} &= -(r_i q_{i+1} - r_{i-1}). \end{aligned} \tag{A.27}$$

Thus from (A.26), we have

$$r_{i+1} = (HG_{i+1} - GH_{i+1}), \tag{A.28}$$

thereby proving the above theorem.

A.4 Some applications of the Remainder theorem

The above Remainder theorem is used in the computation of *dvitīyahāras* described in the second chapter of *Karaṇapaddhati*. The third chapter of *Karaṇapaddhati* also introduces what are called *kendraphalas* which are nothing but the remainders (r_{ij}) which arise when we mutually divide the *alpahāras* H_i and *alpaguṇakāras* G_i. Let r_{ij} be the jth remainder when we mutually divide H_i and G_i. This is called the jth *kendraphala* of the *hāraka* H_i. Applying the above remainder theorem (A.24) to $\frac{H_i}{G_i}$, we clearly obtain

$$r_{ij} = (-1)^{j+1}(H_i G_j - G_i H_j). \tag{A.29}$$

This is what is referred to as *hāraśeṣa* (when j is odd) and *guṇaśeṣa* (when j is even) in the Section 3.4.

Appendix B
Epicycle and eccentric models for *manda* and *śīghra* corrections

Chapter 7 of the text describes the procedures for finding the true geocentric longitudes of the planets beginning with the mean longitudes. Here, we explain the epicycle and eccentric models which form the basis for these computations.[1]

B.1 Equation of centre and the *manda-sphuṭa* of planets

As explained in Section 1.10 of Chapter 1, the mean longitude is calculated for the desired day by computing the number of mean civil days elapsed since the epoch (called the *ahargaṇa*) and multiplying it by the mean daily motion of the planet. Having obtained the mean longitude, a correction known as *manda-phala* is applied to it. In essence, this correction takes care of the eccentricity of the planetary orbit around the Sun. The equivalent of this correction is termed the 'equation of centre' in modern astronomy, and it is a consequence of the eccentricity of the orbit of the planet. The longitude of the planet obtained by applying the *mandaphala* is known as the *manda-sphuṭa-graha,* or simply the *manda-sphuṭa.*

The procedure for finding the *mandaphala* can be explained with the help of an epicycle or an eccentric circle model represented in Figure B.1. Here the mean planet P_0 moves at a uniform rate on the 'deferent' circle or the *kakṣyāvṛtta*, of radius R, whose circumference is usually taken to be 21600 minutes, so that $R = \frac{21600}{2\pi} \approx 3438'$. This circle is centered around O, which is *bhagolamadhya* (centre of the celestial sphere), and $O\Gamma$ is in the direction of *meṣādi*, or the first point of Aries. This is the reference direction for measuring the longitudes. The longitude of the mean planet P_0 is given by $\Gamma\hat{O}P_0 = \theta_0$. OU is in the direction of the '*mandocca*' or the apside. The '*manda-sphuṭa*' P

[1] For a more detailed overview of planetary models in Indian Astronomy, the reader is referred to {TS 2011}, Appendix F, pp. 487-535.

V. Pai et al., *Karaṇapaddhati of Putumana Somayājī*, Sources and Studies in the History of Mathematics and Physical Sciences, https://doi.org/10.1007/978-981-10-6814-0

is situated on an epicycle which is a small circle of radius r around P_0, such that P_0P is parallel to OU. The longitude of the *mandocca* is $\Gamma\hat{O}U = \theta_m$.

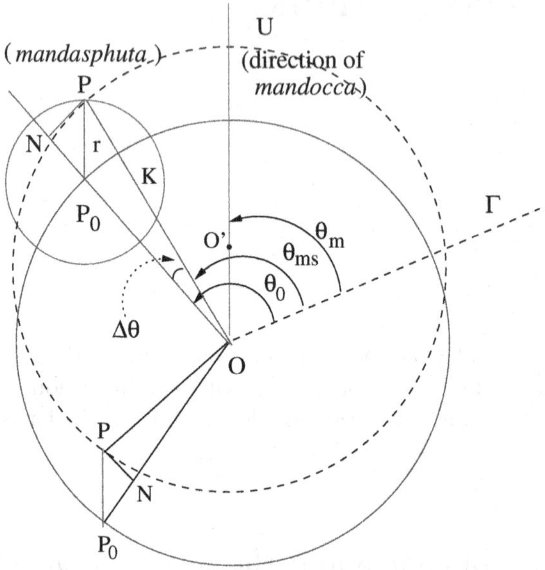

Fig. B.1 Obtaining the *manda-sphuṭa* in the epicycle and eccentric circle models.

We have another equivalent picture of this in the eccentric circle model. Here, O' is a point at a distance r from O, in the direction of OU. Then the *manda-sphuṭa* P moves uniformly around O' at the same rate as P_0 around O in a circle of radius R, the '*pratimaṇḍala*' (eccentric circle) or the *grahavṛtta*, which is represented by a dashed circle. The motion of P around O would not be uniform.

Now, draw a line PN perpendicular to OP_0 which is extended. The difference between the mean longitude and the *mandocca* is

$$U\hat{O}P_0 = P\hat{P_0}N = \theta_0 - \theta_m = \theta_{mk}, \tag{B.1}$$

which is known as the *mandakendra*. The *doḥphala* and the *koṭiphala* are defined as

$$\textit{doḥphala} = PN = |r\sin(\theta_{mk})|, \tag{B.2}$$

and

$$\textit{koṭiphala} = P_0N = |r\cos(\theta_{mk})|. \tag{B.3}$$

Now, the *mandakarṇa* K is the distance between the planet P and the center of the deferent circle O. When the *mandakendra* θ_{mk} is *makarādi*, that is, when $270° \leq \theta_{mk} \leq 360°$ and $0° \leq \theta_{mk} \leq 90°$, then the *mandakarṇa* is given by

$$\begin{aligned} K &= OP \\ &= \left[(ON)^2 + (PN)^2\right]^{\frac{1}{2}} \\ &= \left[(R + |r\cos(\theta_0 - \theta_m)|)^2 + |r\sin(\theta_0 - \theta_m)|^2\right]^{\frac{1}{2}}, \end{aligned} \tag{B.4}$$

as shown in the figure. Similarly, when the *mandakendra* θ_{mk} is *karkyādi*, that is, when $90° \leq \theta_{mk} \leq 270°$, then the *mandakarṇa* is given by

$$\begin{aligned} K &= OP \\ &= \left[(ON)^2 + (PN)^2\right]^{\frac{1}{2}} \\ &= \left[(R - |r\cos(\theta_0 - \theta_m)|)^2 + |r\sin(\theta_0 - \theta_m)|^2\right]^{\frac{1}{2}}, \end{aligned} \tag{B.5}$$

as shown in the figure.

In Figure B.1, the longitude of the planet, generally referred to as *manda-sphuṭa*, is given by $\Gamma\hat{O}P = \theta_{ms}$. Denoting the difference between the mean and true planets ($P\hat{O}P_0 = |\theta_0 - \theta_{ms}|$) by $\Delta\theta$, we have

$$PN = OP\sin(P\hat{O}P_0) = K\sin(\Delta\theta). \tag{B.6}$$

Considering the triangle PP_0N, PN is also given by

$$PN = PP_0\sin(P\hat{P_0}N) = |r\sin(\theta_{mk})|. \tag{B.7}$$

Equating the two expressions for PN,

$$\begin{aligned} K\sin(\theta_0 - \theta) &= |r\sin(\theta_{mk})|, \\ \text{or} \qquad \sin(\Delta\theta) &= \frac{|r\sin(\theta_{mk})|}{K}. \end{aligned} \tag{B.8}$$

In most of the Indian astronomical texts, the epicycle radius associated with the equation of centre, r, is stated to be proportional to the *mandakarṇa*, K, so that

$$\frac{r}{K} = \frac{r_m}{R},$$

where r_m is the specified value of the radius in the text.[2] Using this in (B.8), we have

$$\sin(\Delta\theta) = \frac{|r_m\sin(\theta_{mk})|}{R}. \tag{B.9}$$

It may be noted that (B.9) does not involve the *mandakarṇa* K.

Now the *manda-sphuṭa* of the planet θ_{ms} can be obtained from the mean planet θ_0 by applying $\Delta\theta$ to it. It is clear that $\theta_{ms} < \theta_0$, when the *man-*

[2] In fact, as discussed in Section 7.1, the specified value of epicycle radius r_m also depends upon the *mandakendra*, θ_{mk}.

dakendra is *meṣādi* ($0 \leq \theta_{mk} \leq 180°$), and $\theta_{ms} > \theta_0$, when it is *tulādi* ($180° \leq \theta_{mk} \leq 360°$). That is,

$$\theta_{ms} = \theta_0 \pm \Delta\theta,$$

where the value of $\Delta\theta$ is obtained using (B.9), and the '+' and '−' signs are applicable when the *mandakendra* is *tulādi* and *meṣādi* respectively.

B.2 *Śīghraphala* and the *śīghra-sphuṭa* or the true longitude of planets

While the *mandaphala* is the only correction that needs to be applied in the case of the Sun and the Moon for obtaining their true longitudes (*sphuṭa-grahas*), in the case of the other five planets, two corrections, namely the *manda-saṃskāra* and *śīghra-saṃskāra*, are to be applied in order to obtain their true longitudes. The application of the *śīghraphala* essentially converts the heliocentric longitude into the geocentric longitude. The true longitude of the planet obtained by applying the *śīghraphala* is known as the *śīghrasphuṭa-graha*, or simply *śīghra-sphuṭa*, or just *sphuṭa*.

For obtaining the *śīghraphala* also, epicycle/eccentric circle models are used, and the procedure is similar to that for the *mandaphala*. This is illustrated in Figure B.2. The *manda-sphuṭa* P_0 moves on the 'deferent' circle or the *kakṣyāvṛtta* of radius $R \approx 3438$ around the centre of the celestial sphere O. Its longitude is given by $\theta_{ms} = \Gamma\hat{O}P_0$. OS is in the direction of the *śīghrocca*, S, whose longitude is $\theta_s = \Gamma\hat{O}S$. We will discuss its physical significance later. Then, the *śīghra-sphuṭa* P is situated on a *śīghra*-epicycle of radius r_s around P_0 such that P_0P is parallel to OS.

Here also, we have an alternate picture of this in the eccentric circle model. Let $OS = r_s$, the radius of *śīghra*-epicycle. Draw a circle of radius R around S, which is the *pratimaṇḍala* (dashed circle). Then, the *śīghra-sphuṭa* P is the point on this circle, such that its longitude with respect to S is $\theta_{ms} = \Gamma\hat{S}P$.

Now, draw a line PN perpendicular to OP_0 which is extended. The difference between the *śīghrocca* and the *manda-sphuṭa* is

$$S\hat{O}P_0 = P\hat{P_0}N = \theta_s - \theta_{ms} = \theta_{sk}, \tag{B.10}$$

and is known as the '*śīghrakendra*'. Again, the *doḥphala* and the *koṭiphala* are defined as

$$doḥphala = PN = |r_s \sin(\theta_{sk})|, \tag{B.11}$$

and

$$koṭiphala = P_0N = |r_s \cos(\theta_{sk})|. \tag{B.12}$$

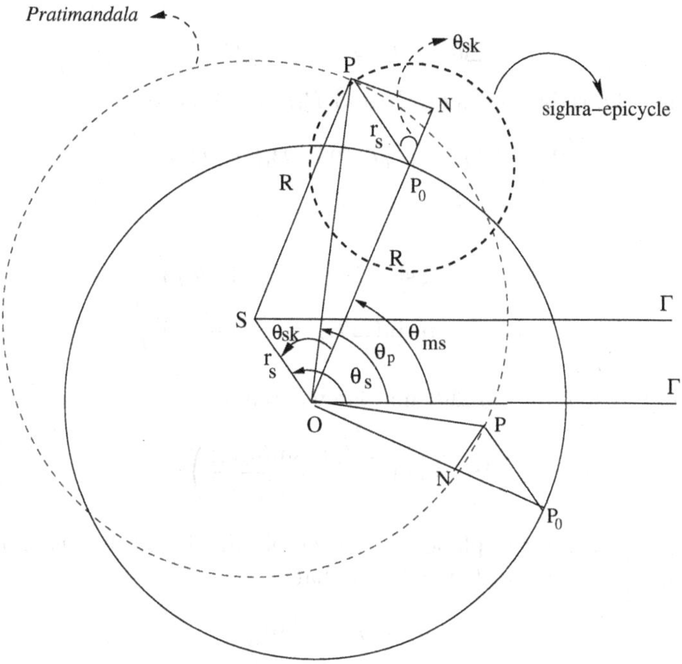

Fig. B.2 Obtaining the *śīghra-sphuṭa* in the epicycle and eccentric models.

Now, the *śīghrakarṇa* K_s is the distance between the planet P and the center of deferent circle, O. When the *śīghrakendra*, θ_{sk} is *makarādi*, that is, when $270° \leq \theta_{sk} \leq 360°$ and $0° \leq \theta_{sk} \leq 90°$, then the *śīghrakarṇa* is given by

$$\begin{aligned} K_s &= OP \\ &= \left[(ON)^2 + (PN)^2\right]^{\frac{1}{2}} \\ &= \left[(R + |r_s \cos(\theta_{sk})|)^2 + |r \sin(\theta_{sk})|^2\right]^{\frac{1}{2}}. \end{aligned} \tag{B.13}$$

Similarly, when the *śīghrakendra* θ_{sk} is *karkyādi*, that is, when $90° \leq \theta_{sk} \leq 270°$, then the *śīghrakarṇa* is given by

$$\begin{aligned} K_s &= OP \\ &= \left[(ON)^2 + (PN)^2\right]^{\frac{1}{2}} \\ &= \left[(R - |r_s \cos(\theta_{sk})|)^2 + |r \sin(\theta_{sk})|^2\right]^{\frac{1}{2}}. \end{aligned} \tag{B.14}$$

Now, the *śīghrasphuṭa* or the true longitude is the geocentric longitude of planet P with respect to O, and is given by $\theta_p = \Gamma\hat{O}P$. The *śīghraphala*, $\Delta\theta_s$ is the difference between the true longitude θ_p and the *manda-sphuṭa* θ_{ms}.

Hence, we have

$$\Delta\theta_s = \theta_p \sim \theta_{ms} = P_0\hat{O}P.$$

Considering the right angles OPN and P_0PN, we have

$$PN = OP\sin(P_0\hat{O}P) = P_0P\sin(P\hat{P}_0N), \tag{B.15}$$

or,

$$K_s\sin(\Delta\theta_s) = |r_s\sin(\theta_{sk})|, \tag{B.16}$$

$$\text{or,} \qquad R\sin(\Delta\theta_s) = |r_s\sin(\theta_{sk})|\frac{R}{K_s}. \tag{B.17}$$

Therefore, the *śīghraphala* $\Delta\theta_s$ may be written as

$$\Delta\theta_s = \sin^{-1}\left(\frac{|r_s\sin(\theta_{sk})|}{K_s}\right).$$

The *śīghra-sphuṭa* of the planet θ_p can be obtained from the *manda-sphuṭa*, θ_{ms}, by applying $\Delta\theta_s$ to it. It is clear that

$$\theta_p = \theta_{ms} \pm \Delta\theta_s,$$

where '+' and '−' signs are applicable, when the *śīghrakendra* is *meṣādi* and *tulādi* respectively.

Note that unlike in the case of *mandaphala*, the *karṇa*, K_s, occurs explicitly in the expression for the *śīghraphala*, $\Delta\theta_s$. For the exterior planets, Mars, Jupiter and Saturn, the *manda-sphuṭa* is the true heliocentric planet, and the *śīghrocca* is the mean Sun. For the interior planets, Mercury and Venus, the *karaṇapaddhati* seems to follow the traditional planetary model where the mean planet is taken to be the mean Sun to which the equation of centre is applied, and the *śīghrocca* is the mean heliocentric planet.[3]

[3] For the first time in the history of astronomy, a correct formulation of the equation of centre and the latitudinal motion of the interior planets was presented by Nīlakaṇṭha Somayājī based on a revised planetary theory outlined in his *Tantrasaṅgraha* (c. 1500) and other works (See for instance {TS 2011}, pp. 508-523). *Karaṇapaddhati* does not discuss this model explicitly. But, as noted in Section 7.19, the Commentary II suggests that the prescription in the verse 26 of Chapter VII that the *manda* correction for the interior planets should be applied to their "*nijamadhyama*" implies that the correction is actually to be applied to their *śīghroccas*. If this interpretation is adopted, then Putumana Somayājī is also following the modified planetary model of Nīlakaṇṭha according to which what was traditionally known as the *śīghroccas* of the interior planets are indeed the *madhyama-grahas* or the mean planets. However, barring a few such instances, there is no explicit statement in *Karaṇapaddhati* to the effect that the text is following the revised planetary model proposed by Nīlakaṇṭha.

Appendix C
Alpaguṇakāras and *alpahārakas* of the planets

In this appendix we present the tables of *alpaguṇakāras* and *alpahārakas*, which have been computed for various planets following the procedure discussed in Section 2.5.2.

alpaguṇakāras G_i	*alpahārakas* H_i
1	27
3	82
28	765
115	3142
143	3907
38868	1061939
116747	3189724
155615	4251663
1050437	28699702
2256489	61651067
10076393	275303970
12332882	336955037
34742157	949214044
81817196	2235383125

Table C.1 The *alpaguṇakāras* and *alpahārakas* of Moon.

alpaguṇakāras G_i	*alpahārakas* H_i
1	3232
1	3233
2	6465
3	9698
8	25861
27	87281
170	549547
197	636828
43904	141925363
131909	426412917
175813	568338280
483535	1563089477
1142883	3694517234
1626418	5257606711
10901391	35240157500

Table C.2 The *alpaguṇakāras* and *alpahārakas* of Moon's apogee.

V. Pai et al., *Karaṇapaddhati of Putumana Somayājī*, Sources and Studies in the History of Mathematics and Physical Sciences, https://doi.org/10.1007/978-981-10-6814-0

alpaguṇakāras G_i	*alpahārakas* H_i
1	686
1	687
85	58394
171	117475
256	175869
427	293344
1537	1055901
6575	4516948
14687	10089797
153445	105414918
475022	326334551
1103489	758084020
1578511	1084418571
8996044	6180176875

Table C.3 The *alpaguṇakāras* and *alpahārakas* of Mars.

alpaguṇakāras G_i	*alpahārakas* H_i
1	87
1	88
31	2727
63	5542
94	8269
251	22080
11640	1023949
23531	2069978
176357	15513795
199888	17583773
42152837	3708106125

Table C.4 The *alpaguṇakāras* and *alpahārakas* of Mercury.

alpaguṇakāras G_i	*alpahārakas* H_i
1	4332
1	4333
3	12998
4	17331
123	532928
127	550259
250	1083187
377	1633446
627	2716633
12917	55966106
91046	394479375

Table C.5 The *alpaguṇakāras* and *alpahārakas* of Jupiter.

alpaguṇakāras G_i	*alpahārakas* H_i
1	224
1	225
3	674
7	1573
10	2247
37	8314
47	10561
84	18875
131	29436
477	107183
608	136619
1693	380421
2301	517040
3994	897461
6295	1414501
29174	6555465
35469	7969966
738554	165954785
5205347	1169653461
11149248	2505261707
27503843	6180176875

Table C.6 The *alpaguṇakāras* and *alpahārakas* of Venus.

alpaguṇakāras G_i	*alpahārakas* H_i
1	10764
1	10765
5	53824
11	118413
16	172237
27	290650
97	1044187
221	2379024
2307	24834427
4835	52047878
7142	76882305
111965	1205282453
231072	2487447211
574109	6180176875

Table C.7 The *alpaguṇakāras* and *alpahārakas* of Saturn.

alpaguṇakāras G_i	*alpahārakas* H_i
1	6792
2	13585
3	20377
8	54339
59	400750
67	455089
126	855839
445	3022606
1016	6901051
2477	16824708
5970	40550467
14417	97925642
49221	334327393
309743	2103890000

Table C.8 The *alpaguṇakāras* and *alpahārakas* of Moon's node.

Appendix D
An introduction to the *Vākya* method of Indian astronomy

D.1 Introduction

The term *vākya* literally means a sentence consisting of one or more words. In the context of astronomy, it refers to a phrase or a string of letters in which numerical values associated with various astronomical parameters are encoded. The *vākyas* are composed using the *kaṭapayādi* system[1] of numeration. The strings used in composing the *vākyas* are chosen so that they not only represent numerical values, but are also in the form of beautiful meaningful phrases and sentences that convey worldly wisdom and moral values.

The *vākya* method of finding the true longitude of the Sun, Moon and the planets (*sphuṭagraha*) is a brilliantly designed simplified version of the methods outlined in the various *Siddhāntas*.[2] As per the *Siddhāntas*, we first find the mean longitudes of the planets and then apply a few *saṃskāras*[3] to get their true positions. On the other hand, the *vākya* method, by making use of a few series of *vākyas* presents a shortcut directly leading to the true longitudes of the planets at certain regular intervals,[4] starting from a certain instant in the past. We will discuss about this instant, which is also closely linked with other notions such as *khaṇḍa* and *dhruva*, during the course of our discussion. At this stage it would suffice to mention that this *vākya* method provides a simple elegant method for computing the true longitudes without

---

[1] For the *kaṭapayādi* system, see Section 1.2.

[2] This appendix presents an introductory overview of the *vākya* method of Indian astronomy. For further details see {CV 1948}, {KP 1956}, {VK 1962}, {SC 1973}, Hari 2001, 2003, Madhavan 2012, Pai 2011, 2013, Pai et al 2009, 2015, 2017, Sriram 2014, 2017, Sriram and Pai 2012.

[3] The *mandasaṃskāra* is to be applied in the case of the Sun and the Moon, whereas both the *mandasaṃskāra* and *śīghrasaṃskāra* are to be applied in the case of the other five planets.

[4] The interval is usually one day for the Moon, and in the case of planets it varies widely and depends on several factors which include their rates of motion with respect to their *mandocca* and *śīghrocca*.

V. Pai et al., *Karaṇapaddhati of Putumana Somayājī*, Sources and Studies in the History of Mathematics and Physical Sciences, https://doi.org/10.1007/978-981-10-6814-0

having to resort to the normal procedure of calculating a whole sequence of corrections involving sine functions etc., which would be quite tedious and time consuming. Therefore, the *vākya* method became very popular in south India and even today some *pañcāngas* are brought out using the *vākya* method in the southern states of India.

Ancient Indian astronomers were aware of various kinds of periodicities in the motions of celestial bodies. One such periodicity is the 248-day cycle during which the Moon's anomaly completes nearly 9 revolutions.[5] They used this cycle to find the true longitude of the Moon at the sunrise for each day of the cycle and expressed them as phrases or "*vākyas*", from early times. These are the 248 '*Vararuci-vākyas*' which are attributed to an astronomer Vararuci, who is also credited with the invention of the letter-numeral system of numeration, known as the *kaṭapayādi* system. Vararuci probably hailed from Kerala and is usually dated prior to the 4th century CE.

The canonical text of the *Parahita* system, *Grahacāranibandha* of Haridatta (7th Century), introduces *vākyas* for the *manda* and *śīghra* corrections which are referred to as the *manda-jyās* and *śīghra-jyās*. The fully developed *vākya* system is presented in the famous *karaṇa* text of the 13th century, *Vākyakaraṇa* which gives the method of directly computing the true longitudes of the Sun, the Moon and the planets using '*vākyas*'. Manuscripts of this work are available in various manuscript libraries of south India, especially Tamilnadu. Kuppanna Sastri and K. V. Sarma estimate that it was composed between 1282 and 1306 CE . The author of this work is not known, but probably hailed from the Tamil speaking region of south India. It has a commentary called *Laghuprakāśikā* by Sundararāja who hailed from *Kāñcī* near Chennai. The work is based on '*Mahābhaskarīya*' and '*Laghubhāskarīya*' of Bhāskara I belonging to the Āryabhaṭa School, and the *Parahita* system of Haridatta prevalent in Kerala.

Mādhava of *Saṅgamagrāma* (c. 1360-1420), the founder of the Kerala school of mathematics and astronomy, composed two works namely, '*Veṇvāroha*' and '*Sphuṭa-candrāpti*' which describe the *vākya* method for the Moon, and are appended by accurate *candra-vākyas* which give the true longitudes of the Moon correct to a second. Mādhava also gave an ingenious method by which the *vākyas* can be used to determine the true longitudes of the Moon at nine instants during the course of a day, thereby significantly reducing the error in calculating the true longitude at any instant using normal interpolation. Mādhava was also the originator of the *Agaṇita* system of computation of planetary positions. Mādhava's *Agaṇitagrahacāra* is yet to be edited, though manuscripts of the work are available.

Vākyakaraṇa and the above works of Mādhava only present the lists of *vākyas* and the computational procedures for obtaining the longitudes of the planets using these *vākyas*. It is indeed the *Karaṇapaddhati* of Putumana Somayāji which explains the basis of the entire *vākya* system. It outlines in

[5] This cycle had also been noticed by the Babylonians and Greeks (Jones 1983).

detail the procedure for composing a *karaṇa* text (*Karaṇapaddhati*) for any desired epoch, by presenting systematic procedures for the calculation of the *khaṇḍas*, *maṇḍalas* and the associated *dhruvas*, as well as the methods for arriving at an accurate set of *vākyas* for the Sun, Moon and the planets. In this process, *Karaṇapaddhati* also explains how to make use of the continued fraction techniques for approximating the rates of motion of the planets and their anomalies.

D.2 *Vākyas* related to the true motion of the Sun

D.2.1 *Vākyakaraṇa* method

The text *Vākyakaraṇa* presents a fairly accurate method for computing the true longitude of the Sun based on a set of *vākyas*.[6] This computation involves the knowledge of an accurate value of the mean motion of the Sun per day, which in turn depends on the knowledge of the duration of the sidereal year. The length of the sidereal year used in *Vākyakaraṇa* is $365 + 1/4 + 5/576$ days $= 365$ days (d) 15 *nāḍis* (n) and 31 $\frac{1}{4}$ *vināḍis* (v), where a *nāḍi* is one-sixtieth of a day, and a *vināḍi* is one sixtieth of a *nāḍi*. In what follows, we at times abbreviate the day, *nāḍi* and *vināḍi* by 'd', 'n' and 'v' respectively. Then, the mean motion of the Sun is $59'8'12''$ per day, which is slightly less than 1° per day.

Suppose n days have elapsed since the Sun was at the first point of Aries, which is termed the '*meṣa saṅkramaṇa*'.[7] Then, the true longitude of the Sun is n°, minus certain number of minutes, as deduced from the table of *vākyas*. The *vākyas* beginning with *bhūpajña* which stands for 14 in the *kaṭapayādi* system, are given for multiples of ten days upto 370 days, in the text. The text also specifies that the correction in minutes to be applied for an intermediate value of the time interval, is to be found by linear interpolation, which is a common technique employed in finding the longitude of any celestial body at an arbitrary instant.

In the following, we present a representative sample of the deductive minutes, at an interval of 50 days. We also compare the values, with those obtained by computing the true longitude of the Sun using a simple epicycle model discussed below.

In the epicycle model for the Sun, implicit in the text, the ratio of the radii of the epicycle to the deferent is $13.5/360$. Sun's apogee is fixed at $\theta_A = 78^\circ$. Then the true longitude of the Sun θ_t is obtained from its mean longitude θ_0, using the relation,

---

[6] {VK 1962}, pp. 10-18, 251-253.

[7] Here, n can include a fractional part also.

$$\theta_t = \theta_0 - \sin^{-1}\left[\frac{13.5}{360}\sin(\theta_0 - \theta_m)\right],$$

where $\theta_0 - \theta_m = \theta_0 - 78°$ is the mean anomaly. When the true longitude, $\theta_t = 0$, $\theta_0 = -2°6'58''$. Given this, the mean longitude after n days is given by

$$\theta_0 = -2°6'58'' + (59'8'') \times n, \tag{D.1}$$

from which the true Sun θ_t is computed. For instance, for $n = 100$ days, we find

$$\theta_0 = 96°26'42'',$$

$$\text{and} \qquad \theta_t = 100° - 254'5'' = 95°45'55''.$$

So the computed value of the deductive minutes is $254'5''$, compared with the '*vākya*' value $254'$.

No. of days	50	100	150	200	250	300	350
Vākya with corresponding number	*munīḍya* मुनीड्य 105′	*bhūmīndra* भूमीन्द्र 254′	*tathāmbu* तथाम्बु 376′	*tāpavān* तापवान् 416′	*vāsāṅga* वासाङ्ग 374′	*puṇyāṅga* पुण्याङ्ग 311′	*kunīla* कुनील 301′
As computed in the epicyle model	104′3″	254′5″	375′4″	414′30″	373′24″	311′0″	300′48″

Table D.1 Deductive minutes for finding the true Sun for some specified number of days.

The text also gives the *vākyas* for the time-intervals between the entry of the Sun into different zodiacal signs (*rāśi-saṅkramaṇa*) and the entry into Aries (*Meṣa-saṅkramaṇa*). These are presented in the format: day-*nāḍi*-*vināḍi* taking the entry into Aries as the reference time 0-0-0. Here the day refers only to the weekday and an appropriate multiple of 7 has to be added to it to obtain the instant of entry into the particular sign. These *vākyas* have been listed in Table D.2.

Illustrative example

From Table D.1, after 150 days, the true longitude of the Sun is $150° - 376'$. So, 150 days after the entry into the Aries, the Sun has to traverse $376'$ more to enter Virgo. The rate of motion of the Sun for the longitude around 150° can be estimated to be $58'18''$ per day from the table of deductive minutes (see Table D.1). Hence, it would take $\frac{376}{58'18''}$ days = 6 d 26 n 38 v to cover $376'$. Hence the *Kanyā-saṅkramaṇa* (entry into Virgo) is 156 *d* 26 *n* 38 *v* after the '*Meṣa-saṅkramaṇa*' (entry into Aries). This can be compared with the value

Name of the *rāśi*		Instant of entry of the Sun in *kaṭapayādi*		in numerals
वृषभ	*Vṛṣabha*	श्रीर्गुणमित्रा	*śrīrguṇamitrā*	2-55-32
मिथुन	*Mithuna*	भूर्विधिपक्षा	*bhūrvidhipakṣā*	6-19-44
कर्कटक	*Karkaṭaka*	स्त्री रतिशूरा	*strī ratiśūrā*	2-56-22
सिंह	*Siṃha*	भोगवरा ते	*bhogavarā te*	6-24-34
कन्या	*Kanyā*	भावचरोरिः	*bhāvacaroriḥ*	2-26-44
तुला	*Tulā*	तेन वशत्त्वं	*tena vaśattvaṃ*	4-54-6
वृश्चिक	*Vṛścika*	लोकजभीतिः	*lokajabhītiḥ*	6-48-13
धनुष्	*Dhanuṣ*	स्थूलहयोऽयम्	*sthūlahayo'yam*	1-18-37
मकर	*Makara*	अङ्गधिगारः	*aṅgadhigāraḥ*	2-39-30
कुम्भ	*Kumbha*	स्तम्भितनाभिः	*stambhitanābhiḥ*	4-06-46
मीन	*Mīna*	नित्यशशीशो	*nityaśaśīśo*	5-55-10
मेष	*Meṣa*	यागमयोऽयम्	*yāgamayo'yam*	1-15-31

Table D.2 *Vākyas* giving the instants of entry of the Sun into different zodiacal signs.

using the *vākya* method, 156 d 26 n 44 v, where we have added $22 \times 7 = 154$ days to the *vākya* value 2-26-44 in Table D.2.

D.2.2 *Vākyas* pertaining to the Sun according to *Karaṇapaddhati*

We now proceed to discuss the method for obtaining the *vākyas* pertaining to the Sun, as presented in the *Karaṇapaddhati*. An important ingredient in this method is the determination of the mean longitude corresponding to a specified true longitude of the Sun. We explain the methods for obtaining the '*māsavākyas*', '*saṅkrāntivākyas*' and '*nakṣatra-saṅkramaṇavākyas* (*nakṣatra*-transition sentences)', and the method for obtaining the true longitude at any instant, using the '*yogyādi-vākyas*'.

D.2.2.1 The mean longitude of the Sun at the *saṅkramaṇas*

The word *saṅkramaṇa* or *saṅkrānti* refers to 'cross over' or 'transit' of an object from one division to another. According to the solar calendrical system followed in many parts of India, a solar year is the time interval between successive transits of the Sun across the beginning point of the *Meṣarāśi* (First point of Aries). The solar year is divided into 12 solar months (*sauramāsas*). The durations of these months are equal to the time spent by the Sun in each of the twelve *rāśis* (zodiacal signs), namely *Meṣa* (Aries), *Vṛṣabha* (Taurus), *Mithuna* (Gemini), etc. In other words, it is the time interval between two successive *rāśi* transits (*rāśisaṅkramaṇa*), which occur when the Sun crosses the intersticial point between the two *rāśis*. For example, when the Sun is at the beginning of the *Siṃharāśi* (Leo), transiting from *Karkaṭaka* (Cancer) to *Siṃha* (Leo), it is *Siṃhasaṅkramaṇa*. Similarly, a *nakṣatra-saṅkramaṇa* (transition to the next *nakṣatra*) occurs when the Sun transits from one *nakṣatra* (27th part of the zodiac, with the names *Aśvinī*, *Bharaṇī*, etc.) to the other.

The calculations related to *saṅkramaṇas* (transitions) are based on the true longitudes of the Sun. For instance, a *rāśi-saṅkramaṇa* (zodiacal transit) occurs when the true longitude is an integral multiple of 30°. The true longitude of the Sun does not increase uniformly with time. However, the variation of the mean longitude is proportional to time. Conversely, the time-intervals are proportional to the difference in mean longitudes. The mean longitude of the Sun θ_0 is obtained from true longitude θ, using the relation (see equation (7.66) in Chapter 7)

$$\theta_0 - \theta = \sin^{-1}\left[\frac{3}{80}\sin(\theta - \theta_m)\frac{R}{R_v}\right],$$

where θ_m is the longitude of the Sun's apogee (taken to be 78° in the text), R is the *trijyā* (3438′), and R_v is the *viparyāsakarṇa* (inverse hypotenuse) given by (see equation (7.59))

$$R_v = \sqrt{\left(R - \frac{3}{80}R\cos(\theta - \theta_m)\right)^2 + \left(\frac{3}{80}R\sin(\theta - \theta_m)\right)^2}.$$

By definition, at the *saṅkramaṇa* (transit), the true longitudes of the Sun are multiples of 30. That is, $\theta_i = 30 \times i$, where $i = 0, 1, \ldots, 11$ for *Meṣa*, *Vṛṣabha*, ..., and *Mīna* respectively. We now illustrate the procedure for obtaining the mean longitude from the true longitude, for two transits namely *Mithuna-saṅkramaṇa* (transition to Gemini, $\theta = 60°$) and *Kanyā-saṅkramaṇa* (transition to Virgo, $\theta = 150°$).

Example 1: *Mithuna-saṅkramaṇa* (Transition to Gemini, $\theta = 60°$)

For this transit, the *viparītakarṇa* is given by

$$R_v = \sqrt{\left(R - \frac{3}{80}R\cos(60 - 78)\right)^2 + \left(\frac{3}{80}R\sin(60 - 78)\right)^2} = 3321.52',$$

and hence,

$$\theta_0 - 60 = \sin^{-1}\left[\frac{3}{80}\sin(360 - (78 - 60))\frac{R}{R_v}\right] = -0.687°.$$

Therefore, the mean longitude is

$$\begin{aligned}\theta_0 &= 60^\circ - 0.687^\circ \\ &= 59.313^\circ \\ &= 1^r 29^\circ 19'.\end{aligned}$$

Example 2: *Kanyā-saṅkramaṇa* **(Transition to Virgo, $\theta = 150^\circ$)**

For this transit, the *viparītakarṇa* is given by

$$R_v = \sqrt{\left(R - \frac{3}{80} R\cos(150-78)\right)^2 + \left(\frac{3}{80} R\sin(150-78)\right)^2} = 3398.14',$$

and hence,

$$\theta_0 - 150 = \sin^{-1}\left[\frac{3}{80}\sin(150-78)\frac{R}{R_v}\right] = 2.068^\circ.$$

Therefore, the mean longitude is

$$\begin{aligned}\theta_0 &= 150^\circ + 2.068^\circ \\ &= 152.068^\circ \\ &= 5^r 02^\circ 04'.\end{aligned}$$

The mean longitudes of the Sun at the transits known as '*saṅkramaṇārka-madhya*' are given as *vākyas* in the Commentary II of the *Karaṇapaddhati*.[8] These are listed in Table D.3, and compared with the values computed as above. Here, the *Vṛṣabha* (Taurus) appears first, as the transit into that *rāśi* corresponds to the end of the first solar month, and the *Meṣa* (Aries) appears last as the transit into that *rāśi* marks the end of the twelfth solar month, and also of the solar year itself. It may be noted that the computed and the *vākya* values differ only in three cases, namely the transit into *Karkaṭaka*, *Makara* and *Kumbha-rāśis*, and the difference is only $1'$.

D.3 Obtaining *māsavākyas*, *saṅkrāntivākyas* and *nakṣatravākyas*

Let d_i denote the time period that has elapsed since the beginning of a year to the end of a particular solar month (corresponding to the Sun transiting the ith *rāśi*). Obviously, d_i need not be an integer. A *māsavākya* is the integer

[8] {KP 1956}, p. 223.

Name of the *rāśi*	*Saṅkramaṇārka-madhya-vākya* (*r*)	(°)	(′)	in *kaṭapayādi*		comp. val. (*r*)	(°)	(′)
Vṛṣabha	0	28	22	श्रेष्ठं हि रत्नम्	*śreṣṭhaṃ hi ratnam*	0	28	22
Mithuna	1	29	19	धान्यधरोऽयम्	*dhānyadharo'yam*	1	29	19
Karkaṭaka	3	00	27	सुखी अनिलः	*sukhī anilaḥ*	3	00	28
Siṃha	4	01	29	धरण्यां नभः	*dharaṇyāṃ nabhaḥ*	4	01	29
Kanyā	5	02	04	वानरा अमी	*vānarā amī*	5	02	04
Tulā	6	02	05	मुनीन्द्रोऽनन्तः	*munīndro'nantaḥ*	6	02	05
Vṛścika	7	01	33	बलाढ्यो नाथः	*balāḍhyo nāthaḥ*	7	01	33
Dhanus	8	00	38	जले निनादः	*jale ninādaḥ*	8	00	38
Makara	8	29	35	शूलधरो हि	*śūladharo hi*	8	29	34
Kumbha	9	28	37	साम्बो हि प्रधानः	*sāmbo hi pradhānaḥ*	9	28	36
Mīna	10	27	59	धर्मसूक्ष्मं नित्यम्	*dharmasukhaṃ nityam*	10	27	59
Meṣa	11	27	53	लक्ष्मी सुरपूज्या	*lakṣmī surapūjyā*	11	27	53

Table D.3 The *vākyas* given in the Commentary II for the mean longitudes of the Sun at *saṅkramaṇas*, compared with the computed values.

closest to d_i. The fractional part, in terms of *nāḍikās* can be found from the *saṅkrāntivākyas*, which give the remainders when d_i are divided by 7.

D.3.1 The Māsavākyas

Verse 22 in chapter 7 of *Karaṇapaddhati* gives the procedure for obtaining the *māsavākyas* and *saṅkrāntivākyas*. The true longitudes of the Sun at the end of each month are 30°, 60°, ..., 360°. At the end of the 12th month, which is the same as the beginning of the first month in the next year, the true longitude of the Sun is 360°. The mean longitude corresponding to the true longitude of 360° is found from (7.66) to be $357.883° \;=\; -2.117° = \;-2°7' = 11^r27°53'$. The difference between the true and the mean longitudes at the end of the year is termed the '*abdāntadoḥphala*' (the difference between the true and mean longitudes at the year-end), whose value is $2°7'$.

The *madhyamabhoga* (difference in the mean longitudes) reckoned from the *meṣa-saṅkramaṇa* to *iṣṭasaṅkramaṇa* (desired zodiacal transition) is the difference in the mean longitudes at the desired zodiacal transit and the transit at *meṣādi* of the true Sun. It is found by adding $2°7'$ to the mean longitude at each transit. For example, the true longitude of the Sun at the *siṃhasaṅkra-*

maṇa is 120°. The mean longitude corresponding to this is 121°29′ (see Table D.3). Adding 2°7′ to it, we obtain 123°36′ as the *madhyamabhoga* from the *meṣasaṅkrama* to the *siṃhasaṅkrama*.

A mean solar day is the time interval corresponding to an increase of 1° in the mean longitude. This is slightly longer than a civil day, and is given by $\frac{D_c}{D_s}$, where D_c and D_s represent the numbers of civil days and solar days in a *mahāyuga*. Note that the values given in the *Karaṇapaddhati* for D_c and D_s are 1577917500 and 360 × 4320000 = 1555200000 respectively. Let θ_{i0} represent the *madhyamabhoga* for the transit to the $(i+1)^{th}$ rasi. Then

$$d_i = \theta_{i0} \times \frac{D_c}{D_s}.$$

For *siṃhasaṅkramaṇa*, $\theta_{40} = 123°36'$ and therefore

$$d_4 = \frac{123°36' \times 1577917500}{1555200000} = 125\,d\;24\,n. \tag{D.2}$$

The *māsavākya* is the integer closest to d_i. Hence, 125 is the *māsavākya* at the *siṃhasaṅkrama*. The *māsavākyas* corresponding to the transits as given in the Commentary I[9] as also the computed d_i's are listed in Table D.4.

Rāśi transited to (*saṅkramaṇa*)	*Māsavākya* (textual value) in *kaṭapayādi*		in numerals	computed value of d_i
Vṛṣabha	कुलीन	*kulīna*	31	30 d 56 n
Mithuna	रूक्षज्ञ	*rūkṣajña*	62	62 d 20 n
Karkaṭaka	विधान	*vidhāna*	94	93 d 56 n
Siṃha	मात्रया	*mātrayā*	125	125 d 24 n
Kanyā	क्षणस्य	*kṣaṇasya*	156	156 d 26 n
Tulā	सिंहस्य	*siṃhasya*	187	186 d 54 n
Vṛścika	सुपुत्र	*suputra*	217	216 d 48 n
Dhanus	चत्वराट्	*catvarāṭ*	246	246 d 18 n
Makara	तथाद्रि	*tathādri*	276	275 d 39 n
Kumbha	मीनाङ्गि	*mīnāṅgi*	305	305 d 07 n
Mīna	मृगाङ्गि	*mṛgāṅgi*	335	334 d 55 n
Meṣa	मातुलः	*mātulaḥ*	365	365 d 15 n

Table D.4 The *māsavākyās* given in the Commentary I compared with the computed values of d_i.

[9] {KP 1956}, p. 225.

By finding the difference between the successive *māsavākyas*, the number of civil days corresponding to each month can be calculated.

D.3.2 The Saṅkrāntivākyas

The instant at which the *rāśisaṅkramaṇas* occur can be determined from the *saṅkrānti-vākyas*. By dividing d_i by 7, the remainders obtained are the *saṅkrānti-vākyas*. For instance, in the previous example

$$\frac{125\,d\,24\,n}{7} = 17 + \frac{6\,d\,24\,n}{7}.$$

The remainder is $6\,d\;24\,n$. In this, the integral part represents the day and the fractional part multiplied by 60 would give the *nāḍikās*. Here the obtained day of the week corresponds to number 6 and the *nāḍikā* is 24. The *vākya* for this is *marutaḥ*, which represents the day as 6 and *nāḍikā* as 25.

The *saṅkrāntivākyas* which are given in the Commentary II[10] for different transits are listed in Table D.5, along with the computed values.

Name of the *rāśi*	*Saṅkrāntivākya* (textual value) in *kaṭapayādi*		in numerals		computed value day	*nāḍikā*
Vṛṣabha	तिमिरे	*timire*	2	56	2	55.5
Mithuna	निरतम्	*niratam*	6	20	6	19.5
Karkaṭaka	चमरे	*camare*	2	56	2	56.0
Siṃha	मरुतः	*marutaḥ*	6	25	6	24.1
Kanyā	सुरराट्	*surarāṭ*	2	27	2	26.1
Tulā	घृणिभः	*ghṛṇibhaḥ*	4	54	4	53.5
Vṛścika	जवतो	*javato*	6	48	6	47.7
Dhanus	धटकः	*dhaṭakaḥ*	1	19	1	18.2
Makara	नृवराट्	*nṛvarāṭ*	2	40	2	39.3
Kumbha	सनिभः	*sanibhaḥ*	4	07	4	06.7
Mīna	मणिमान्	*maṇimān*	5	55	5	55.2
Meṣa	चयका	*cayakā*	1	16	1	15.5

Table D.5 The *saṅkrāntivākyas* in the Commentary II and the computed values.

It is clear that the value of d_i corresponding to a *saṅkramaṇa* is obtained by adding a suitable multiple of 7 to the *saṅkrāntivākya*. For example, we have

[10] {KP 1956}, p. 226.

to add 91 to the day component of the *saṅkrāntivākya* for *karkaṭaka* (2+91) to obtain d_3 whose value is 93 days 56 *nāḍikā*.

D.3.3 The Nakṣatravākyas

We know that the ecliptic (*rāśicakra* of 360°) is divided into 27 equal parts called *nakṣatras*, each part corresponding to 13°20′. The basis of this division is the fact that Moon's sidereal period is around 27 days. The term *nakṣatra* also refers to the time spent by the Moon in any of these divisions. In the same vein, the time durations spent by the Sun to traverse through these divisions are called *mahānakṣatras*. The true longitudes of the Sun at the end of the 27 *nakṣatras* are 13°20′, 26°40′, 40°, 53°20′, ... 360°. Converting these longitudes to the corresponding mean ones and adding 2°7′ to them, we obtain the increase in the mean longitude of the Sun at the end of each *nakṣatra* starting from *Aśvinī*. The number of civil days corresponding to these can be calculated by multiplying them by the *bhūdinas* and dividing by the solar days in a *mahāyuga*. These values are presented in Table D.6.

The instant at which the *nakṣatra-saṅkramaṇa* occurs can be obtained from the *nakṣatra-saṅkrāntivākyas*. When we divide the civil days at each transit by 7, the remainders obtained are the *nakṣatra-saṅkrāntivākyas*, similar to the *rāśi-saṅkrāntivākyas* discussed earlier. The *nakṣatra-saṅkrāntivakyas* as given in both the Commentaries[11] are tabulated along with the computed values in Table D.7.

D.4 The *Yogyādivākyas*

Unlike the *vākyas* discussed earlier, wherein the nomenclature was based upon a certain time interval or phenomenon, here the name *yogyādivākyas* stems from the fact that the set of 48 *vākyas* begin with the word *yogya*. These *vākyas* enable us to find the longitude of the Sun at any given instant. There are 4 *vākyas* corresponding to each solar month. Each month is divided into four parts with a maximum of 8 days per part. Now, the *sphuṭabhoga* of each part is the difference between the true longitudes of the Sun at the beginning and at the end of that part. The difference in minutes between the *sphuṭabhoga* of each part and 8° are the *yogyādivākyas*. If the longitudinal difference is greater (lesser) than 8°, then it is to be taken as positive (negative).

The definition of *yogyādivākyas* and the method of applying them to obtain the true longitude of the Sun at an interval of 8 days in a solar month, are given in verse 24, chapter 7 of *Karaṇapaddhati*.

[11] {KP 1956}, p. 228.

Name of the *Nakṣatra*	No. of civil days elapsed before the *Nakṣatra-saṅkramaṇa*
Bharaṇī	13.674
Kṛttikā	27.461
Rohiṇī	41.349
Mṛgaśirā	55.318
Ārdrā	69.343
Punarvasu	83.395
Puṣya	97.442
Āśleṣā	111.454
Maghā	125.401
Pūrvaphālgunī	139.260
Uttaraphālgunī	153.015
Hasta	166.654
Citrā	180.175
Svātī	193.581
Viśākhā	206.881
Anurādhā	220.090
Jyeṣṭhā	233.224
Mūla	246.304
Pūrvāṣāḍhā	259.352
Uttarāṣāḍhā	272.393
Śravaṇa	285.449
Dhaniṣṭhā	298.543
Śatabhiṣaj	311.697
Pūrvabhādrapadā	324.931
Uttarabhādrapadā	338.262
Revatī	351.702
Aśvinī	365.258

Table D.6 Number of civil days elapsed at each *Nakṣatra-saṅkramaṇa*.

How to obtain the *yogyādivākyas*?

The *yogyādivākyas* as given in the edited version of the Commentary I[12] are listed in Table D.8. Apart from the *vākyas* (here in the form of one word, which form part of meaningful sentences), the signs ('+' or '−') are also indicated in the commentary. Except in the case of *Tulā*, all the 4 *vākyas* corresponding to

[12] {KP 1956}, p. 229.

Name of the *Nakṣatra*	*Nakṣatra-saṅkrāntivākya* in *kaṭapayādi*		in numerals		computed value day	*nāḍikā*
Bharaṇī	कविषु	*kaviṣu*	6	41	6	40.4
Kṛttikā	हारिषु	*hāriṣu*	6	28	6	27.7
Rohiṇī	दीयत	*dīyata*	6	18	6	20.9
Mṛgaśirā	धीयते	*dhīyate*	6	19	6	19.1
Ārdrā	करिषु	*kariṣu*	6	21	6	20.6
Punarvasu	मारिषु	*māriṣu*	6	25	6	23.7
Puṣya	सारिषु	*sāriṣu*	6	27	6	26.5
Āśleṣā	दूरतः	*dūrataḥ*	6	28	6	27.2
Maghā	स्मरति	*smarati*	6	25	6	24.0
Pūrvaphālgunī	दुष्यति	*duṣyati*	6	18	6	15.6
Uttaraphālgunī	योनिषु	*yoniṣu*	6	01	6	00.9
Hasta	पर्वणा	*parvaṇā*	5	41	5	39.2
Citrā	त्रिकश	*trikaśa*	5	12	5	10.5
Svātī	ताण्डव	*tāṇḍava*	4	36	4	34.9
Viśākhā	भोमृग	*bhomṛga*	3	54	3	52.9
Anurādhā	धेनुगः	*dhenugaḥ*	3	09	3	05.4
Jyeṣṭhā	सुपुर	*supura*	2	17	2	13.4
Mūla	हाटक	*hāṭaka*	1	18	1	18.2
Pūrvāṣāḍhā	नीरन	*nīrana*	0	20	0	21.1
Uttarāṣāḍhā	भारता	*bhāratā*	6	24	6	23.6
Śravaṇa	चरण	*caraṇa*	5	26	5	26.9
Dhaniṣṭhā	गालव	*gālava*	4	33	4	32.6
Śatabhiṣaj	विश्वगु	*viśvagu*	3	44	3	41.8
Pūrvabhādrapadā	चर्मराट्	*carmarāṭ*	2	56	2	55.9
Uttarabhādrapadā	चिकुर	*cikura*	2	16	2	15.7
Revatī	रावय	*rāvaya*	1	42	1	42.1
Aśvinī	मर्कट	*markaṭa*	1	15	1	15.5

Table D.7 The *nakṣatra-saṅkrāntivākyas* given in both the Commentaries and the computed values.

a partcular *raśi* have the same sign and this is indicated as such in the table. For *Tulā*, the sign for the first *vākya* is – and the signs for the other three are all +, as indicated in the table. The rationale behind these *yogyādivākyas* is best explained by taking up a couple of concrete examples.

Consider the solar month of *Milhuna*. Thc true longitnde of the Sun is $\theta = 60°$ at the beginning of the month. The corresponding mean longitude θ_0 can be determined using the method explained earlier and we find $\theta_0 = 59°18'42''$. Using the fact that the rate of motion of the mean longitude of the Sun is $59.136'$ per day, the mean longitude is $\theta_0 = 67°11'48''$ after 8 days in the month of *Mithuna*. The *mandaphala* $(\theta - \theta_0)$ corresponding to this value of θ_0 is found to be $24'6''$. Adding this to θ_0, we find the true longitude after 8 days to be $67°11'48'' + 24'6'' = 67°35'54''$. Hence the increase in the true longitude after the first 8 days of the month is $7°35'54''$. The longitudinal difference is less than $8°$, and is given by $-(8° - 7°35'54'') = -24'6''$, which compares well with with the value of $-24'$ as given by the *vākya* '*vīraḥ*' in the commentary.

After 16 days in the month of *mithuna*, the mean longitude $\theta_0 = 59°18'42'' + 59'8'' \times 16 = 75°4'48''$. The true longitude corresponding to this is found to be $\theta = 75°11'24''$. Hence the difference between the true longitudes at the beginning and at the end of the second part is $75°11.4 - 67°35'54'' = 7°35'30''$. Here again as the longitude difference is less than $8°$, and is given by $-(8° - 7°35'30'') = -24'30''$, compared with the value of $-25'$ as implied by the *vākya* '*śūraḥ*' in the commentary.

Solar month		*yogyādivākyas* (in minutes)							
Meṣa	–	*yogyo*	11 (11′12″)	*vaidyaḥ*	14 (13′3″)	*tapaḥ*	16 (15′42″)	*satyam*	17 (17′42″)
Vṛṣabha	–	*dhanyaḥ*	19 (19′18″)	*putraḥ*	21 (20′54″)	*kharo*	22 (22′18″)	*varaḥ*	24 (23′18″)
Mithuna	–	*vīraḥ*	24 (24′6″)	*śūraḥ*	25 (24′30″)	*śaro*	25 (24′36″)	*vajrī*	24 (24′24″)
Karkaṭaka	–	*bhadram*	24 (23′54″)	*gotro*	23 (23′6″)	*ruruḥ*	22 (21′54″)	*karī*	21 (20′30″)
Siṃha	–	*dhanyaḥ*	19 (18′54″)	*sevyo*	17 (17′0″)	*mayā*	15 (14′54″)	*loke*	13 (12′42″)
Kanyā	–	*kāyo*	11 (10′36″)	*dīnaḥ*	8 (8′12″)	*stanām*	6 (5′48″)	*ganā*	3 (3′18″)
Tulā		*yājño*	− 1 (−1′30″)	*yajñām*	+ 1 (+0′48″)	*ganā*	+ 3 (3′0″)	*śūnā*	+ 5 (4′54″)
Vṛścika	+	*steno*	6 (6′12″)	*dīno*	8 (7′42″)	*dhunī*	9 (8′54″)	*naṭaḥ*	10 (9′54″)
Dhanus	+	*āpaḥ*	10 (10′18″)	*pāpaḥ*	11 (10′42″)	*payaḥ*	11 (10′48″)	*pathyam*	11 (10′30″)
Makara	+	*pūjyā*	11 (10.2)	*dhenuḥ*	9 (9.4)	*dine*	8 (8.2)	*arthinaḥ*	7 (6.8)
Kumbha	+	*tanuḥ*	6 (5′42″)	*bhinnā*	4 (3′54″)	*khanī*	2 (1′54″)	*jñānī*	0 (−0′18″)
Mīna	–	*ratnaṃ*	2 (2′0″)	*bhānuḥ*	4 (4′24″)	*sunīḥ*	7 (6′48″)	*nayaḥ*	10 (9′18″)

Table D.8 The 48 *yogyādivākyas* mentioned in the Commentary I along with the computed values in parentheses.

D.4.1 Finding the true longitude of the Sun from the *yogyādivākyas*

One can obtain the true longitude of the Sun on any day using the *yogyādivākyas*, and linear interpolation. For example, suppose we would like to find the true longitude of the Sun after the lapse of 18 days in the *Vṛṣabha* month. This comes in the third part (*khaṇḍa*). Therefore the approximate value of the true longitude of the Sun after 18 days elapsed would be

$$\theta' = 30^\circ + 18^\circ = 48^\circ.$$

A correction called *yogyādisaṃskāra* $\Delta\theta'$ has to be applied to θ' in order to obtain the true longitude θ.

Now, the correction for 8 days of the third *khaṇḍa* is given as $22'$ (*khara*). Hence the correction for 2 days is $\frac{22\times2}{8}$ minutes. Adding this to the sum of the first two *vākyas* (*dhanya* and *putra*),

$$\Delta\theta' = 19 + 21 + \frac{22 \times 2}{8} = 45'30''.$$

These corrections are indicated as negative in the listing of the *vākyas* in the commentary. Hence applying this result negatively to θ' the true longitude of the Sun at the end of the 18th day of the solar month *Vṛṣabha* is found to be

$$\theta = 48^\circ - 45'30'' = 47^\circ 14'30''.$$

D.4.2 Some observations

It is clear from the examples given above, that this method can be used to determine the true longitude at any instant during the day using interpolation. In Table D.8, our computed values for the difference between 8° and the actual angular distance converted by the Sun in 8 days (i.e., the difference between the true longitudes computed after a separation of 8 days) are given in parentheses below the *vākya* values. It is clear from these figures that the *yogyādivākyas* are very accurate.

More importantly, what is noteworthy here is the phenomenal simplification that has been achieved in computing the true longitudes of the Sun at any moment using the *yogyādivākyas*. The *yogyādivākyas* are easily remembered through the following verses:

योग्यो वैद्यः तपः सत्यं धन्यः पुत्रः खरो वरः।
वीरः शूरः शरो वज्री भद्रं गोत्रो रुरुः करी॥
धन्यः सेव्यो मया लोके कायी दीनः स्तनाङ्गना।
याज्ञी यज्ञाङ्गना शूना स्तेनो दीनो धुनी नतः॥

आपः पापः पयः पथ्यं पूज्या धेनुर्दिनेऽर्थिनः।
तनुर्भिन्ना खनी ज्ञानी रत्नं भानुः सुनिर्नयः॥

yogyo vaidyaḥ tapaḥ satyaṃ dhanyaḥ putraḥ kharo varaḥ |
vīraḥ śūraḥ śaro vajrī bhadraṃ gotro ruruḥ karī ||
dhanyaḥ sevyo mayā loke kāyī dīnaḥ stanāṅganā |
yājñī yajñāṅganā śūnā steno dīno dhunī naṭaḥ ||
āpaḥ pāpaḥ payaḥ pathyaṃ pūjyā dhenurdine'rthinaḥ |
tanurbhinnā khanī jñānī ratnaṃ bhānuḥ sunirnayaḥ ||

The literal translation of the above verse is:

A qualified doctor; [Speaking] truth [by itself] is austerity; A blessed son; A donkey is better; A skilful warrior; Indra's arrow; This clan is safe; The antelope and elephant; In the world only the blessed are to be served by me; The one with a bulky body is pitiable; A lady with big breasts; The wife of the Yajamana and performer if the sacrifice is swollen; The thief is miserable; The river is a dancer; The water is the culprit; Milk is good; Cow is to be worshiped during the day by those desirous of becoming wealthy; The body has been split; The wise is like a mine; The Sun is a pearl; The one who is completely unscrupulous.

By simply memorizing the above verses, one can find out the longitude of Sun on any given day at any given instant with reasonable accuracy. In fact, for all practical purposes, the inaccuracies noted in Table 7.7 are negligible. This is a very small price paid for the enormous simplification and fun involved in computing the longitudes by simple arithmetical calculations.

D.5 *Vākya* method of finding the longitude of the Moon

All the texts dealing with the *vākya* method, be it the *Vākyakaraṇa* or *Sphuṭacandrāpti* or *Veṇvāroha*, first introduce what is known as the *khaṇḍa*, before proceeding to describe the procedure for the computation of planetary longitudes. Essentially, a *khaṇḍa* refers to *kalidinasaṅkhyā* (certain number of civil days elapsed from the beginning of *kaliyuga*), which is close to the *ahargaṇa* under consideration. The value of *khaṇḍa* will be varying from planet to planet. The choice of the *khaṇḍa* is dictated by certain conditions to be satisfied by the longitudes of the planets in relation with those of their *mandoccas* and *śīghroccas*. In the case of the Moon, the *khaṇḍa* is the number of days from the beginning of the *kali* to a day near to the chosen *ahargaṇa* such that the *mandakendra* or the mean anomaly is zero at sunrise on that day. It may be recalled that

$$\begin{aligned} \textit{mandakendra} &= \textit{madhyamagraha} - \textit{mandocca} \\ &= \text{mean longitude} - \text{longitude of apogee.} \end{aligned}$$

In Table D.9 we have listed the values of *khaṇḍas* for the Moon as given in different texts.

The *vākya* method for the Moon is based on the fact that the number of revolutions of the *mandakendra* (anomaly) or anomalistic revolutions is very close to an integer for certain specific number of days. For instance, the number of anomalistic revolutions in 55, 248, 3031, 12372 or 188611 days is very close to an integer (2, 9, 110, 449 and 6845 in the above cases). These are called the cycles of anomaly or anomalistic cycles. Larger the value of the anomalistic cycle, closer is the number of revolutions to an integral value.

Text	the *khaṇḍa* for Moon		
	in *kaṭapayādi*		in numerals
Vākyakaraṇa	वेदेद्धज्ञानतुष्टः	*vededdhajñānatuṣṭaḥ*	1600984
Sphuṭacandrāpti	दीननम्रानुशास्यः	*dīnanamrānuśāsyaḥ*	1502008
Veṇvāroha	दीननम्रानुशास्यः	*dīnanamrānuśāsyaḥ*	1502008

Table D.9 The *khaṇḍas* for Moon given in different texts.

Let A be the *ahargaṇa* corresponding to the day on which the mean longitude of the Moon is to be calculated. The algorithm for finding its true longitude involves the following steps.[13]

- Initially we have to find the difference between the *ahargaṇa* and the *khaṇḍa* which is generally referred to as *khaṇḍaśeṣa*. That is

$$\textit{khaṇḍaśeṣa} = A - K.$$

- This *khaṇḍaśeṣa* is to be divided by 12372. The quotient and the remainder thus obtained are to be stored separately.

$$\frac{(A-K)}{12372} = q_1 + \frac{r_1}{12372}. \tag{D.3}$$

- Now, we have to divide the remainder (r_1) by 3031.

$$\frac{r_1}{3031} = q_2 + \frac{r_2}{3031}. \tag{D.4}$$

  Having stored the quotient (q_2) and remainder (r_2) separately, the process has to be repeated with r_2 and the number 248. Thus we have

$$\frac{r_2}{248} = q_3 + \frac{r_3}{248}. \tag{D.5}$$

- Let D_0 be the true longitude of the Moon at the end of the *khaṇḍa* or *khaṇḍāntya* K. This is known as the *dhruva* at the *khaṇḍāntya* K. Let

[13] {VK 1962}, pp. 18-21, 253-254.

D_1, D_2 and D_3 denote the *dhruvas* at the end of 12372, 3031 and 248 days respectively, which represent the change in the true longitude of the Moon corresponding to these *hārakas*. Then the true longitude of Moon (θ_A) corresponding to the *ahargaṇa* (A) is given by

$$\theta_A = [D_0 + (D_1 \times q_1) + (D_2 \times q_2) + (D_3 \times q_3)] + V_{r_3}, \qquad \text{(D.6)}$$

where V_{r_3} represents the *vākya* corresponding to the number r_3 which is to be read off from the table of Moon's *vākyas*. The table contains 248 *vākyas* which are generally known as *gīrnaśreyādi-vākyas* composed by Vararuci (prior to 4th cent. CE). Mādhava has composed a more accurate set of 248 *cāndravākyas* commencing with "*śīlaṃ rājñaḥ śriye*". Both the Vararuci and Mādhava *vākyas* have been tabulated in Table E.1 of Appendix E. Their derivation is also discussed in the same Appendix.

It may be noted that the expression for the longitude θ_A given by (D.5), essentially consists of two parts, which may be called as the "*dhruva-part*" and the "*vākya-part*". The expression corresponding to the former is enclosed in the square bracket ("[ ]") in order to distinguish it from the latter. We note that it depends on D_i's ($i = 0, 1, 2, 3$) and q_i's ($i = 0, 1, 2, 3$). While q_i's are to be obtained from the equations (D.3) – (D.5), the values of D_i's are to be taken from the text.

The *dhruvas* D_0, D_1, D_2 and D_3 given in the text *Vākyakaraṇa* are listed in Table D.10.[14]

i	*Dhruvas* (D_i's) given in the text *Vākyakaraṇa*			
	in *kaṭapayādi*		in numerals	in degrees
0	सेनानारिनाथनम्	*senānārināthanam*	$07^r 02° 00' 7''$	212.00194
1	नित्यदेवसुखाधीनम्	*nityadevasukhādhīnam*	$09^r 27° 48' 10''$	297.80278
2	यज्ञयोगसुनायकम्	*yajñayogasunāyakam*	$11^r 07° 31' 01''$	337.51694
3	तनुर्विश्वसुखम्	*tanurviśvasukham*	$0^r 27° 44' 06''$	27.73500

Table D.10 The *dhruvas* as given in *Vākyakaraṇa*.

Now Moon's anomaly at the end of $K + 12372 \times q_1 + 3031 \times q_2 + 248 \times q_3$ days is taken to be zero in the first instance, as it is zero for the *khaṇḍa* K and the anomaly is supposed to complete an integral number of revolutions in 12372, 3031 and 248 days. Hence, the anomaly is solely determined by r_3.

However, the lunar anomaly does not complete an exactly integral number of revolutions in 248, 3031 or 12372 days. At the end of 248 days the anomaly

[14] {VK 1962}, pp. 18-19.

would have increased by 7′. Similarly, it would have decreased by 1.75′ at the end of 3031 days. The text *Vakyakaraṇa* prescribes a correction for taking into account this minute motion in anomaly in the following manner.

Find $32q_3 - 8q_2$ (where q_2 and q_3 are the quotients corresponding to the divisors 3031 and 248 respectively, in equations (D.3)–(D.5)). Find the true daily motion of Moon's longitude, say g_{mt} by subtracting the longitude corresponding to the *vākya* $(r_3 - 1)$ from that corresponding to r_3. Subtract the mean daily motion of the Moon, $g_{m0} = 13°11'$ from this. Then the correction to the true longitude obtained from the *vākya* in seconds is $(g_{mt} - g_{m0}) \times (32q_3 - 8q_2)$.

In *Vākyakaraṇa*, the rationale for the *dhruvas* D_0, D_1, D_2, D_3, or for the *vākyas*, or for the finer correction mentioned above, are not presented. The origin of these *dhruvas* can be understood as follows. The *dhruva* corresponding to 12372 days is mentioned as $9^r 27°48'10''$. Now there are 449 anomalistic cycles in 12372 days. As it is known that the sidereal period of the Moon is around 27.32 days, this will mean that the Moon's longitude increases by 452 revolutions and, $9^r 27°48'10''$ or 452.8272299383 revolutions. Hence the sidereal period of Moon employed in *Vākyakaraṇa* may be inferred to be $\frac{12372}{452.8272299383} = 27.3216785167$ days. Incidentally, this differs only in the 8th decimal place with the *Karaṇapaddhati*'s *śakābda* corrected revolution rate of 27.3216785014.

According to *Vākyakaraṇa*, the mean longitude of the Moon increases by $\frac{3031}{27.3216785167}$ revolutions in 3031 days. Subtracting the integral number of revolutions from this, the *dhruva* for 3031 days would be $11^r 7°31'1''$, which is precisely the value specified in the text. Similarly, the mean longitude increases by $\frac{248}{27.3216785167}$ revolution in 248 days. From this, we find the *dhruva* for 248 days to be $0^r 27°44'5''36'''$ compared to the value $0^r 27°44'6''$ specified in the text.

Now, Moon's dhruva at the *khaṇḍa* of 1600984 days is stated to be $7^r 2°0'7''$. This means that the Moon's mean longitude at the beginning of *kaliyuga* is

$$7^r 2°0'7'' - \left[\frac{1600984}{27.32167852}\right]_{frc} = 7^r 2°0'7'' - 6^r 25°36'2''$$
$$= 0^r 6°24'5''.$$

This *kalyādidhruva* does not figure in the calculation of Moon's true longitude. Still, it is interesting to compare the above *kalyādidhruva* with the *kalyādidhruva* in *Karaṇapaddhati* which is $0^r 6°23'37''$ (see Table 2.5). These two differ only by 28″ .

It appears that in *Vākyakaraṇa* the rate of motion of the anomaly is taken to be $\frac{449}{12372}$ revolutions/day. Let x be the longitude of the apogee at the beginning of the *kali*. As the Moon's mean longitude at that time was $0^r 6°24'5''$, the *kendra* will be $0^r 6°24'5'' - x$. The motion of the anomaly in 1600984 days is $1600984 \times \frac{449}{12372}$ revolutions. Hence the longitude of the anomaly at the *khaṇḍa* is given by

$$0^r 6°24'5'' - x + \left[1600984 \times \frac{449}{12372}\right]_{frc} \times 360,$$

where '$[\ ]_{frc}$' denotes the fractional part. At the *khaṇḍa* the anomaly is supposed to be zero. Hence we find

$$x = 6°24'5'' + 112°40'2'' = 3^r 29°4'7''.$$

This is the *kalyādidhruva* of Moon's apogee in *Vākyakaraṇa.* This can be compared with the *Karaṇapaddhati* value of $3^r 29°17'26''$. The difference is $13'19''$.

We now take up the finer correction to the 'true longitude' due to the minute motion in anomaly at the end of 248 and 3031 days. The anomaly at the end of 248 days is not zero but $\frac{449}{12372} \times 248$ revolutions if the rate of motion of the anomaly is taken to be $\frac{449}{12372}$ revolutions/day. This works out to be $6.98' \approx 7'$ after subtracting 9 integral number of revolutions. Similarly the anomaly at the end of 3031 days is

$$\begin{aligned} \frac{449}{12372} \times 3031 &= 110 - 0.0000808276 \text{ revolutions} \\ &= -1'44''45''' \approx -1.75', \end{aligned}$$

dropping the integral number of revolutions. Hence, the correction to the anomaly is $\delta = q_3 \times 7' - q_2 \times 1.75' = \frac{(q_3 \times 7' - q_2 \times 1.75')}{60}$ in degrees, when the quotients corresponding to the *hārakas* 3031 and 248 are q_2 and q_3.

Let X be the anomaly without the above correction. Hence the actual anomaly would be $X + \delta$. The equation of center would now be $-\sin^{-1}\left[\frac{7}{80}\sin(X+\delta)\right]$ instead of $-\sin^{-1}\left[\frac{7}{80}\sin(X)\right]$. Hence the correction to the equation of centre is given by

$$\begin{aligned} &-\sin^{-1}\left[\frac{7}{80}\sin(X+\delta)\right] + \sin^{-1}\left[\frac{7}{80}\sin(X)\right] \\ &\approx -\frac{d}{dX}\left[\sin^{-1}\left(\frac{7}{80}\sin X\right)\right] \times \delta. \end{aligned}$$

The true longitude is given by

$$\theta_t = \theta_m - \sin^{-1}\left(\frac{7}{80}\sin X\right).$$

Now in one day, X increases by $13°4'$. Therefore,

$$g_{mt} \approx g_{m0} - \frac{d}{dX}\left[\sin^{-1}\left(\frac{7}{80}\sin X\right)\right] \times 13°4',$$

or

$$-\frac{d}{dX}\left[\sin^{-1}\left(\frac{7}{80}\sin X\right)\right] \approx \frac{g_{mt}-g_{m0}}{13°4'},$$

where g_{mt} is the rate of motion of the true longitude of Moon/day and g_{m0} is rate of motion of the mean longitude of Moon/day, in degrees. Hence,

$$\begin{aligned}
-\frac{d}{dX}\left[\sin^{-1}\left(\frac{7}{80}\sin X\right)\right] \times \delta &\approx \frac{g_{mt}-g_{m0}}{13°4'} \times \delta \quad \text{(in degrees)} \\
&= (g_{mt}-g_{m0}) \times \frac{\delta \times 3600}{13.06} \quad \text{in seconds} \\
&= (g_{mt}-g_{m0}) \times \frac{(q_3 \times 7 - q_2 \times 1.75) \times 60}{13.06} \\
&= (g_{mt}-g_{m0}) \times (q_3 \times 32.159 - q_2 \times 8.03)
\end{aligned}$$

In the text, the correction to the true longitude is stated to be $(g_{mt}-g_{m0}) \times (32q_3 - 8q_2)$ seconds, where g_{mt} and g_{m0} are rates of motion of the true and mean longitudes of the Moon per day.

The longitude of the Moon thus obtained is to be further refined by taking into account the *saṃskāras* which are mentioned in *Vākyakaraṇa* such as the *bhujāntara*, *cara*, *deśāntara* and so on.

Illustrative example

We shall now illustrate the procedure outlined above by means of a numerical example. For this, we choose the *Kalyahargaṇa* $A = 1851974$ corresponding to *Nija-āṣāḍha-kṛṣṇa-navamī*, *Śaka* 1891 (August 6, 1969 CE).

$$\begin{aligned}
\textit{ahargaṇa} \quad A &= 1851974, \\
\textit{khaṇḍa} \quad K &= 1600984, \\
\textit{khaṇḍaśeṣa} \quad A-K &= 250990, \\
\text{First quotient of} \quad \frac{250990}{12372}, q_1 &= 20, \\
\text{First remainder of} \quad \frac{250990}{12372}, r_1 &= 3550, \\
\text{Second quotient of} \quad \frac{3550}{3031}, q_2 &= 1, \\
\text{Second remainder of} \quad \frac{3550}{3031}, r_2 &= 519
\end{aligned}$$

$$\text{Third quotient of} \quad \frac{519}{248}, q_3 = 2,$$

$$\text{Third remainder of} \quad \frac{519}{248}, r_3 = 23.$$

From the quotients obtained above and the values of the *dhruvas* given in the Table D.10, we get the net *dhruva* of the Moon to be

$$D_0 + (D_1 \times q_1) + (D_2 \times q_2) + (D_3 \times q_3) = 81.04448°.$$

Now, the *vākya* corresponding to the number 23 (r_3) is "*goraso nanu syāt*" (see Table E.1). The longitude encoded by this phrase is $10^r\ 07°\ 23' = 307.38333°$.

Adding 307.38333° to the above *dhruva* 81.04448°, we get the uncorrected true longitude of the Moon as

$$\theta_A = 28.4278 = 0^r 28° 25' 40''.$$

Now the 22nd *vākya* in the cycle is $9^r\ 24°\ 42'$. Subtracting this from the 23rd, namely $10^r\ 7°\ 23'$, we obtain the daily rate of motion of the true longitude, $g_{mt} = 12°41'$. As the daily motion of the mean Moon is $g_{m0} = 13°11'$, we get $g_{mt} - g_{m0} = -30' = -0.5°$. Hence the correction to the true longitude is

$$-0.5(32q_3 - 8q_2) = -.5(64 - 8)'' = -28''.$$

Hence the corrected true longitude is

$$0^r 28° 25' 40'' - 28'' = 0^r 28° 25' 12''.$$

D.6 *Vākya* method for finding the true longitudes of the planets

In the case of the Moon, only one correction, corresponding to the equation of center or *mandasaṃskāra*, is considered in *Vākyakaraṇa*. The *khaṇḍa* corresponds to a day which is close to the *ahargaṇa* when the anomaly or the *mandakendra* is zero. Then use is made of periods such as 248 days, 3031 days and 12372 days etc. in which the anomaly of *mandakendra* makes (very nearly) integral number of revolutions. Finally for the remainder of days, use is made of the tables of Moon's longitude in the form of *vākyas* in a cycle of 248 days. As the anomaly does not complete exactly integral number of revolutions in 248 or 3031 days, a correction term is also prescribed to take into account, the small positive or negative change in the anomaly after these cycles.

In the case of the five planets, Mars, Mercury, Jupiter, Venus and Saturn, two corrections (*saṃskāras*) have to be applied in order to obtain their true longitudes from the mean longitudes. One of them is the equation of centre or *mandasaṃskāra* involving the *mandocca*. The other is the *śīghrasaṃskāra* which converts the heliocentric longitude to the geocentric longitude. The *śīghrasaṃskāra* involves the *śīghrocca*, and the *śīghrakendra* or the *śīghra* anomaly. In the modified planetary model of Nīlakaṇṭha Somayājī as enunciated in his *Tantrasaṅgraha*, *Siddhāntadarpaṇa* etc., the mean Sun will be the *śīghrocca* for all the planets. However, *Vākyakaraṇa* and, to a large extent, *Karaṇapaddhati* are based on the traditional planetary model according to which the Sun is the *śīghrocca* for the exterior planets, Mars, Jupiter and Saturn, whereas the heliocentric mean planet itself is called the *śīghrocca* in the case of Mercury and Venus. Further, in the case of Mercury and Venus, the mean Sun is taken to be the mean planet.

Tabulation of *vākyas* for the planets would have been greatly simplified if there were small periods of time in which both the *mandakendra* and the *śīghrakendra* become zero. On the contrary it is only after a very large interval of time, of the order of millions of years, that both the *mandakendra* and *śīghrakendra* become zero at the same time. Such periods would not be convenient for the purpose of tabulations. Instead *Vākyakaraṇa* invokes the concept of *maṇḍalas* which is closely linked with the synodic period of the planet as we shall explain below.[15]

D.6.1 *Maṇḍalas, dhruvas* and *śodhyas*

The synodic period is the time interval between two successive conjunctions of the mean planet and the *śīghra*. Having known this, we find the integral multiples of the synodic period during which both the *śīghra* and the mean planet return to an angular position close to the *mandocca*, if they started near the *mandocca*. It is obvious that this would be possible only for certain specified time durations. These time units are the *maṇḍalas*. It is important to note that the *maṇḍalas* are not given by an integral number of civil days.

Let us consider an instant at which both the *mandakendra* and the *śīghrakendra* are zero. Physically this would correspond to a situation in which the planet and its *śīghrocca* are both along the direction of the *mandocca* (U) as shown in the Figure D.1.

In the figure, E represents the Earth and $E\Gamma$ the direction of *Meṣādi* from where the longitudes are measured. S is the *śīghrocca* (the mean Sun) which moves around E and P is an exterior planet which moves around S. $\Gamma\hat{S}U$ represents the longitude of *mandocca* of the planet which may be taken to be fixed

[15] For more details on the *Vākyakaraṇa* method of computing the true longitudes of planets, see {VK 1962}, pp. 30-62, 257-266.

planet	*śodhyas*			*dhruvas*	
	in *kaṭapayādi*	in numerals		in *kaṭapayādi*	in numerals
		days	*nāḍikās*		(in minutes)
Mars	*śīlasārahāramāṇikyo* शीलसारहारमाणिक्यो	1552827	35	*ratnavit* रत्नवित्	−402
Mercury	*rātrirnivāsaruddhāṃśako* रात्रिर्निवासरुद्धांशको	1592740	22	*rāgajñaḥ* रागज्ञः	−032
Jupiter	*satyaṃ śūravanasamayaiḥ* सत्यं शूरवनसमयैः	1570425	17	*pitaraḥ* पितरः	−261
Venus	*viśvaṃ sadgandhikāntimayaiḥ* विश्वं सद्गन्धिकान्तिमयैः	1561937	44	*satyajñāḥ* सत्यज्ञाः	+17
Saturn	*harirvāsavadhīdamāḍhyaḥ* हरिर्वासवधीदमाढ्यः	1589474	28	*tarugaḥ* तरुगः	−326

Table D.11 The *śodhyas* and corresponding *dhruvas* for different planets as given in *Vākyakaraṇa* ({VK 1962}, p. 30).

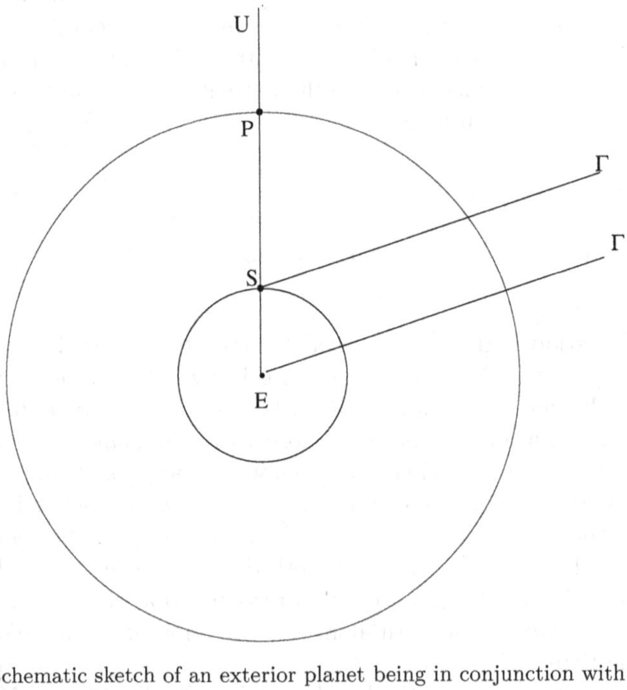

Fig. D.1 Schematic sketch of an exterior planet being in conjunction with its *śīghrocca* along the direction of its *mandocca*.

as the rate at which the *mandocca* moves is extremely small in comparison with that of the planets.

Both the planet (P) and the *śīghrocca* (S) keep moving continuously around S and E respectively. It may so happen that after a certain period of time when both of them complete certain integral number of revolutions once again they may fall in conjunction, along the direction of the *mandocca* of the planet. But for this to happen the time period would be extremely large. However, we can think of smaller periods in which, after a certain integral number of synodic revolutions, the planet could be fairly close to its *mandocca*. These periods are referred as *maṇḍalas*. Since we are considering a situation in which the planet, being in conjunction with *śīghrocca*, is only close to the direction of *mandocca*, there will be an angular separation between the two. In other words, though the *śīghrakendra* is zero, the *mandakendra* will not be zero. This difference between the longitude of the planet and its *mandocca* when the *śīghrakendra* is zero is what is referred to as *maṇḍaladhruva* or simply *dhruva*. Conventionally, *dhruva* would be taken to be positive if the longitude of the planet is greater than that of its *mandocca* and negative otherwise. It can be easily seen that if the *maṇḍala* is large, then the magnitude of *dhruva* will be small and vice versa.

In the context of the planets, we also have the notion of the *śodhya* which is roughly the equivalent of the *khaṇḍa* in the case of the Moon. This is the time instant which is close to the desired *ahargaṇa*, when the *śīghrakendra* is zero and the *mandakendra* has a small value. Again the *śodhya* does not correspond to an integral number of civil days. The *śodhyas* and the corresponding *dhruvas* for the planets, as given in the *Vākyakaraṇa*, are presented in Table D.11. The *maṇḍalas* and the corresponding *dhruvas* as given in *Vākyakaraṇa* are listed in Tables G.1 – G.5. In what follows, we shall briefly explain these tables in the case of the planet Mars with some specific examples.

Maṇḍalas and *dhruvas* of Mars[16]

The *maṇḍalas* and *dhruvas* of Mars as given in *Vākyakaraṇa* are compiled in Table D.12.

Based on the rough value of 780 days for the synodic period of Mars, it is clear that the last or fifth *maṇḍala* (M_5) of 11699 days 4 *nāḍikās* corresponds to 15 synodic periods of Mars (complete revolutions of *śīghrakendra*) and the fourth *maṇḍala* (M_4) of 17158 days 37 *nāḍikās* corresponds to 22 synodic revolutions. It is seen that the duration of the third *maṇḍala* (M_3) is exactly the sum of these two and corresponds to 37 revolutions.

$$M_3 = M_4 + M_5.$$

If the *dhruvas* corresponding to the *maṇḍala* M_i are denoted by D_i, we see that they also satisfy the relation

$$D_3 \approx D_4 + D_5.$$

[16] {VK 1962}, p. 259.

maṇḍalas (M_i)			*dhruvas* (D_i)	
in *kaṭapayādi*	in numerals		in *kaṭapayādi*	in numerals
	days	*nāḍikās*		(in minutes)
dhanadhīdānabhaṅgārtaiḥ धनधीदानभङ्गार्तैः	634089	09	*vanānām* वनानाम्	+004
putradhīhāmarālayaiḥ पुत्रधीहामरलयैः	132589	21	*satrajñāḥ* सत्रज्ञाः	+027
kavissammodahāraiḥ कविस्सम्मोदहारैः	28857	41	*gaṅgārcyā* गङ्गार्च्या	+133
saṅgajanmapathikaiḥ सङ्गजन्मपथिकैः	17158	37	*vaneśaḥ* वनेशः	−504
vanadhīdhṛtapuṇyaiḥ वनधीधृतपुण्यैः	11699	04	*jalārtā* जलार्ता	+638

Table D.12 The *maṇḍalas* and corresponding *dhruvas* for Mars as given in *Vākyakaraṇa* ({VK 1962}, p. 51).

The second *maṇḍala* (M_2) and the *dhruva* (D_2) correspond to 170 revolutions and satisfy the relations:

$$M_2 \approx 5M_3 - M_5,$$
$$D_2 \approx 5D_3 - D_5.$$

The first *maṇḍala* (M_1) and the *dhruva* D_1 correspond to 813 revolutions and satisfy the relations

$$M_1 \approx 5M_2 - M_3,$$
$$D_1 \approx 5D_2 - D_3.$$

The *dhruva* for the first *maṇḍala* is very small (4′). This means that if the longitude of the Sun and the Mars are both equal to the longitude of the *mandocca* of Mars at some instant, then after 813 synodic revolutions, the duration of which is 634089 days 9 *nāḍikās*, they will differ from the *mandocca* only by 4′.

From the first *maṇḍala*, the exact synodic period of Mars can be computed. Its value is 779.9374539 days. Using this and the sidereal period of the *śīghrocca*, (the Sun), which is the sidereal year assumed in *Vākyakaraṇa* equal to 365.258681 days, the sidereal period of Mars works out to be 686.9869988 days.

We can compute the increase in the longitude of the Sun or the Mars after each *maṇḍala* using the above inferred sidereal year and the sidereal period of Mars. The values obtained are listed in Table D.13.

maṇḍalas			*dhruvas*	
M_i	days	*nāḍikās*		
M_5	11699	−4	D_5'	638′
M_4	17158	−37	D_4'	−505.1′
M_3	28857	−41	D_3'	132.9′
M_2	132589	−21	D_2'	26.5′
M_1	634089	−9	D_1'	4.7′

Table D.13 Computed values of *maṇḍalas* and *dhruvas* of Mars.

Comparing these values with those given in Table D.12, it is noted that the computed values are very close to the *dhruvas* D_i listed in the *Vākyakaraṇa*.

Computation of the *śodhya* and planetary longitude

Suppose we want to find a time period that is close to, let us say 1560000 *kali* days, at which the *śīghrakendra* is zero. For this, if the longitude of Mars at epoch were to be zero, then we could have chosen this to be

$$2M_1 + 2M_2 + 2M_5 = 1556755.133333 \text{ days},$$

which amounts 1996 revolution of the *śīghrakendra*. However, the *śīghrakendra* is not zero at the *kali* beginning as the longitude of Mars at the epoch is actually $11^r 17^\circ 7.4' = 347^\circ 7.4'$. Taking this into consideration, we have to find the time at which the longitudes of both the Sun and the Mars are close to each other. If we take the time period as 1556727.236 days, the longitude of Sun would be $354.8134661^\circ = 354^\circ 48.8'$, and the longitude of Mars would be $354.8248167^\circ = 354^\circ 49.5'$ which are nearly equal.

Now at this time, both the Sun and Mars have a longitude of $\approx 354^\circ 49'$ which is far from the apogee of Mars which has the value 118°. Now in one synodic revolution, the longitude of Sun/Mars increases by 48.70858°. Hence in five synodic revolutions they increase by 243.54292°. Hence if we take the time period to be

$$\begin{aligned} 1556727.236 - 5 \times 779.9374539 &= 1556827.549 \\ &= 1556827 \text{ days } \; 33 \; nāḍikās, \end{aligned}$$

then the longitude of the Sun and Mars for this *ahargaṇa* would be $111^\circ 16'$. This is $6^\circ 44'$ less than the longitude of the *mandocca* of Mars which is 118°. In other words, for a value of *śodhya* equal to 1556827 days 33 *nāḍikās*, the *dhruva* is $-6^\circ 44' = -404'$. The value of *śodhya* and *dhruva* found in this manner are very close to the *Vākyakaraṇa* values.

For Mars, the *vākyas* are given for 15 synodic cycles or *parivṛtti* of nearly 780 days each. The period of $15 \times 780\,\text{days} = 11700\,\text{days}$, is very close to $M_5 = 11699d4n$. 38 *vākyas* are given for each synodic cycle of 780 days. These are not at equal intervals. For an arbitrary day, the longitude of Mars is to be found by interpolation. The *vākya* of the 0th day of a cycle is that of the last day of the previous cycle. The *vākya* of the 0th day of the first cycle is simply the same as the apogee of the planet, which is 118° in the case of Mars. The *vākyas* also incorporate a correction term due to the '*maṇḍaladhruvas*'.

Suppose we want to compute the true longitude of Mars for the day corresponding to the *ahargaṇa*, 18,44,004. Now,

$$\begin{aligned}18,44,004 &= 15,52,827d\ 35n + 2 \times 1,32,589d21n \\ &\quad + 1 \times 17,1587d\ 37n + 11 \times 780d + 259d\ 6n.\end{aligned}$$

This means that after the *śodhyadina* of $15,52,827d35n$, 2 *maṇḍalas* of $1,32,589d21n$, 1 *maṇḍala* of $17,158d37n$ and 11 cycles of 780 days in the *maṇḍala* of $11,699d4n$ are complete and $259d6n$ are over in the 12th cycle. In the table provided in the text, the *vākyas* for 250 days and 270 days are specified. The longitude corresponding to $259d6n$ is to be determined by interpolation from these two *vākyas*. There is a '*dhruva*', of

$$-402' + 2 \times 27' + 1 \times -504' = -852' = -14°12',$$

at the beginning of the '*maṇḍala*' of $11,699d4n$. This, as well as the correction for the longitude corresponding to 259 days and 6 *nāḍis* in the 12th cycle should be added to the latter, in order to obtain the true longitude of Mars, at the mean sunrise of the desired day (whose *ahargaṇa* is 18,44,404 days).

The true longitudes of the other planets are also to be found in this manner, using their '*śodhyadina*' '*maṇḍalas*', *dhruvas*, and the *vākyas* for the shortest *maṇḍala*.

Appendix E
The *candravākyas* of Vararuci and Mādhava

In Table E.1 below we present the 248 *candravākyas* of Vararuci and their refinement due to Mādhava. The *candravākyas* are nothing but the phrases encoding the true longitudes of the Moon for 248 successive days. The *gīranaḥ śreyādi vākyas* of Vararuci give the true longitude values correct to a minute. The *vākyas* of Mādhava, *śīlaṃ rājñaḥ śriye* etc., give the true longitudes correct to a second. We have taken the *Vararuci-vākyas* from the edition of *Vākyakaraṇa* due to K. V. Sarma and Kuppanna Sastri,[1] and we have taken the *Mādhava-vākyas* from the edition of *Veṇvāroha* due to K. V. Sarma.[2] As regards the latter, we have corrected the *vākyas* corresponding to the *vākya* numbers 25, 174, 181, 234 and 242 mostly following the variant readings given in the footnotes of the same edition. These variant readings have to be adopted because, it is the values encoded by them which actually coincide with the values of the true longitude of Moon computed accurately upto seconds. As we have explained below, these variant readings are also obtained following the traditional error correction procedure prescribed for correcting the *Mādhava-vākyas*.

E.1 Computing the *candravākyas* of Vararuci and Mādhava

The computation of the true longitude of the Moon depends on two parameters, namely the rates of motion of the mean Moon and the anomaly. We shall adopt the values for these as given in *Karaṇapaddhati*. In their prescrip-

[1] {VK 1962}, pp. 125-134. An earlier edition of the *Vararuci-vākyas* is due to Kunhan Raja {CV 1948}.

[2] {VR 1956} pp. 23-28. The same *vākyas* are also published in K. V. Sarma's edition of *Sphuṭacandrāpti* ({SC 1973}, pp. 46-59). There is a typographic error in the latter version, as regards *vākya* 98, which is noted in fn. 13.

V. Pai et al., *Karaṇapaddhati of Putumana Somayājī*, Sources and Studies in the History of Mathematics and Physical Sciences, https://doi.org/10.1007/978-981-10-6814-0

tions for the calculation of the true longitude of the Moon, the *Vākyakaraṇa* and the *Sphuṭacandrāpti* seem to use rates of motion of the Moon and the anomaly which are approximations to the *śakābda* corrected rates given in *Karaṇapaddhati*.[3]

To start with, let us compute the true longitudes of the Moon by using the approximate daily rate of motion of the anomaly, given by $\frac{9}{248}$ revolutions. Then, the true longitude of the Moon on the ith day of a 248-day anomalistic cycle can be expressed as

$$\frac{360^\circ \times i}{\frac{134122987500}{4909031760}} - \sin^{-1}\left[\frac{7}{80} \times \sin\left(\frac{9 \times i}{248} \times 360\right)\right]. \qquad \text{(E.1)}$$

In the above equation, the first term represents the mean longitude as the sidereal period of the Moon is $\frac{134122987500}{4909031760}$ days. The second term represents the equation of center, as the anomaly is $\frac{9i}{248} \times 360$ at the end of the ith day. Equation (E.1) gives the value of the true longitude of the Moon at the end of the first day to be $0^r 12^\circ 02' 34'' 50'''$ correct to the thirds. This is also the value encoded by the Vararuci *vākya*, *gīrnaḥ śreyaḥ* ($0^r 12^\circ 03'$) correct to a minute and also the value encoded by the Mādhava *vākya*, *śīlaṃ rājñaḥ śriye* ($0^r 12^\circ 02' 35''$). However, for several values of $i > 1$ the values resulting from the equation (E.1) are not exactly in agreement with the *candravākyas* of Mādhava. There is always a discrepancy of a few seconds. For instance, for the 52nd day ($i = 52$), the above expression (E.1) gives $10^r 28^\circ 26' 15'' 11'''$. This is in agreement with the 52nd Vararuci *vākya*, *tāḥ prajāḥ prājñāḥ syuḥ* which encodes the value $10^r 28^\circ 26'$. However, it differs by $6''$ from the value $10^r 28^\circ 26' 09''$ encoded in 52th *candravākya* of Mādhava, *dhanaṃ coro haren-nityam.*

The expression (E.1) for the true longitude of the Moon on i^{th} day is obviously not accurate as it uses the crude rate of $\frac{9}{248}$ for the daily motion of the anomaly. The correct expression using the exact rate $\frac{4574211340428709}{126040405436547500}$ for the motion of the anomaly, according to *Karaṇapaddhati* (Table 2.1), is given by

$$\frac{360^\circ \times i}{\frac{134122987500}{4909031760}} - \sin^{-1}\left[\frac{7}{80} \times \sin\left(\frac{4574211340428709 \times i}{126040405436547500} \times 360\right)\right]. \qquad \text{(E.2)}$$

Using the above expression (E.2), we have computed the values for the true longitude of the Moon for the 248 day cycle. The computed values accurate upto thirds are tabulated along with the values encoded in the *candravākyas* of Mādhava in Table E.2. For the sake of comparison, we have also computed the longitude values obtained by using the approximate value for the daily rate of motion of the anomaly given by $\frac{9}{248}$. These have also been included in Table E.2.

[3] See for instance, the discussion in {VK 1962}, pp. xi-xii.

From Table E.2, it may be seen that the computed values of Moon's true longitude using (E.2) with the exact rate of motion of the anomaly, coincide (except for a difference of $1''$ in a few places) with the values encoded in the *candravākyas* of Mādhava as tabulated in Table E.1, where, as noted earlier, we have carried out emendations, from the published version of the *vākyas* ({VR 1956} pp. 23-28), to the five *vākyas* bearing the numbers 25, 174, 181, 234 and 242. On the other hand, the longitudes calculated with the crude rate of motion $\frac{9}{248}$ differ from the accurate values (and also from the Mādhava *vākya* values)—the difference could be as much as $33''$ as seen in the case of 233rd *vākya*.

E.2 Error correction procedure for *candravākyas*

It is quite interesting to note that an error correcting mechanism (*vākyaśodhana*) has also been enunciated in the tradition to ensure that the 248 *candravākyas* do not get corrupted over a period of time. In the case of *candravākyas* of Vararuci, we have the following verses cited in the edition of *Vākyakaraṇa*, which present a method for checking the *candravākyas*.[4] However, the actual source of these verses is not known.

भवेत् सुखस्य राशीनां अर्धं वाक्यं तु मध्यमम् ।
आदिवाक्यमुपान्त्यं च भवतीति 'भवेत् सुखम्' ॥
यत्राप्यक्षरसन्देहः तत्र संस्थाप्य 'देवरम्' ।
त्यजेत् तद्गतवाक्यानि, शिष्टं शोध्यं 'भवेत् सुखात्' ॥

bhavet sukhasya rāśīnāṃ ardhaṃ vākyaṃ tu madhyamam |
ādivākyamupāntyaṃ ca bhavatīti '*bhavet sukham*' ||
yatrāpyakṣarasandehaḥ tatra saṃsthāpya '*devaram*' |
tyajet tadgatavākyāni, śiṣṭaṃ śodhyaṃ '*bhavet sukhāt*' ||

Six *rāśis* plus half '*bhavet sukham*' is the middle (124th) *vākya*. The first *vākya* plus the penultimate [*vākya*] is '*bhavet sukham*'. [Therefore] whenever there is a doubt regarding the letters of any *vākya*, deduct its serial number from 248 (*devara*) and take the *vākya* corresponding to the remainder; deduct this *vākya* from '*bhavet sukham*'. [The *vākya* in doubt should agree with this.]

Bhavet sukham is the last *vākya* of Vararuci which encodes the value $0^r 27° 44'$. The 124th *vākya* is *rāmā gīyate* which encodes the value $6^r 13° 52'$. As the above verses note,

$$\frac{1}{2}(360° + 0^r 27° 44') = 6^r 13° 52'.$$

More importantly, the above verses state that the following equation is satisfied by the 248 *candravākyas* given by Vararuci, which can be used to check

[4] {VK 1962}, p. 134.

any of the *vākyas* when in doubt.

$$V_i + V_{248-i} = V_{248} \qquad \text{(modulo } 360°\text{)}.$$

There is also a procedure for correcting the *Mādhava-vākyas* which is cited in the editions of *Veṇvāroha* and *Sphuṭacandrāpti*.[5] The procedure is described in the following verses, whose authorship, again, is unknown.

इष्टसंख्योनदेवेन्द्रवाक्यं तत्प्रतियोगिकम् ।
तस्याध-उर्ध्वविवराद् द्विगुणाद्यविवर्जितात् ॥
शिखराप्तकलाहीनं प्रतियोग्यन्ततस्त्यजेत् ।[6]
इष्टवाक्यं भवेच्छिष्टम् एवं स्याद् वाक्यशोधनम् ॥

iṣṭasaṃkhyonadevendravākyaṃ tatpratiyogikam |
tasyādha-ūrdhvavivarād dviguṇādyavivarjitāt ||
śikharāptakalāhīnaṃ pratiyogyantatastyajet |
iṣṭavākyaṃ bhavecchiṣṭam evaṃ syād vākyaśodhanam ||

The *vākya*, associated with the desired number subtracted from 248 (*devendra*), is the complementary [*vākya*] (*tatpratiyogika*). Find the difference of the *vākyas* above and below it, and subtract the difference from twice the first *vākya*. Divide the result by 225 (*śikhara*) and subtract the quotient, in minutes etc., from the complementary *vākya*. Subtract the end-result from the 248th *vākya*. The remainder will be the desired *vākya*. This is the procedure for the correction of *vākyas*.

We now explain the procedure outlined in the verses to verify the correctness of any of the *Mādhava-vākyas*.[7] Let V_i be the ith *vākya* which is to be checked. Then, the procedure is as follows:

- First, find out the complementary *vākya* V_j, where $j = 248 - i$.
- Then, find the *vākyas* above (V_{j-1}) and below (V_{j+1}) the complementary *vākya* (V_j) and find their difference, $\Delta V_j = V_{j+1} - V_{j-1}$.
- Subtract twice the first *vākya* (V_1) from ΔV_j, and divide the result by 225. That is, find
$$\delta_j = \frac{\Delta V_j - (2 \times V_1)}{225}.$$
- Subtract δ_j (in minutes, etc.) from the complementary *vākya*. This gives the transformed complementary *vākya*
$$V_j' = V_j - \delta_j.$$
- The desired *vākya*, if correct, should be equal to the result found by subtracting V_j' from the final *vākya* (V_{248}). That is, the desired *vākya* should satisfy

[5] {VR 1956}, pp. 27-28; {SC 1973} pp. 57-58.

[6] The *padaccheda* here is *pratiyogi+antataḥ+tyajet*. This is the version found in {VR 1956}, p. 27. The version found in {SC 1973}, p. 57 is erroneous.

[7] The rationale behind this ingenious *Vākyaśodhana* procedure is explained in Sriram (2017). See next section for details.

$$V_i = V_{248} - V'_j.$$

We shall now apply the above correction procedure for *vākya* numbers 25, 174, 181, 234, and 242, and show that this procedure actually leads to the variant readings of these *vākyas* that we have adopted based on our computation.

Correcting the 25th *vākya*

Since the desired *vākya* number $i = 25$, the number of the complementary *vākya* $j = 248 - 25 = 223$. The *vākyas* below (V_{j+1}) and above (V_{j-1}) the complementary *vākya* are:

$$V_{224} = 2^r07°51'36'' \quad \text{and} \quad V_{222} = 1^r13°23'59''.$$

Their difference is given by

$$\Delta V_{223} = 0^r24°27'37''.$$

By subtracting twice the first *vākya* ($2 \times V_1 = 0^r24°05'10''$), from the above, we get $22'27''$. Dividing this by 225, we obtain $\delta_{223} = 6''$. Subtracting this from the complementary *vākya* (V_{223}), we have

$$\begin{aligned} V'_{223} &= V_{223} - \delta_{223} \\ &= 1^r25°33'31'' - 6'' = 1^r25°33'25''. \end{aligned}$$

Now the desired *vākya* V_{25} is given by

$$\begin{aligned} V_{25} &= V_{248} - V'_{223} \\ &= 12^r27°43'29'' - 1^r25°33'25'' \\ &= 11^r02°10'04''. \end{aligned} \tag{E.3}$$

The *vākya* number 25 found in the editions of *Veṇvāroha* and *Sphuṭacandrāpti*[8] is *"dhananikaro niryayau"*. This when decoded corresponds to the value $11^r02°10'09''$ which varies from the value obtained above by $5''$. The correct reading of the 25th *vākya* should be *"ghananikaro niryayau"* which is in conformity with (E.3) and also the computed value given in Table E.2.

Correcting the 174th *vākya*

Since the desired *vākya* number $i = 174$, the number of the complementary *vākya* $j = 248 - 174 = 74$. The *vākyas* below (V_{j+1}) and above (V_{j-1}) the complementary *vākya* are:

$$V_{75} = 9^r03°10'03'' \quad \text{and} \quad V_{73} = 8^r05°55'11''.$$

---

[8] {VR 1956}, p. 23; {SC 1973}, p. 47.

Their difference is given by

$$\Delta V_{74} = 27°14'52''.$$

By subtracting twice the first *vākya* ($2 \times V_1 = 0^r 24°05'10''$), from the above and dividing the result obtained by 225, we obtain $\delta_{74} = 51''$. Subtracting this from the complementary *vākya* (V_{74}), we have

$$\begin{aligned} V'_{74} &= V_{74} - \delta_{74} \\ &= 8^r 19°39'48'' - 51'' = 8^r 19°38'57''. \end{aligned}$$

Now the desired *vākya* V_{174} is given by

$$\begin{aligned} V_{174} &= V_{248} - V'_{74} \\ &= 12^r 27°43'29'' - 8^r 19°38'57'' \\ &= 4^r 08°04'32''. \end{aligned} \tag{E.4}$$

The *vākya* number 174 found in the editions of *Veṇvāroha* and *Sphuṭacandrāpti* is "*bālye'vajño jano vai*". This when decoded corresponds to the value $4^r 08°04'13''$ which varies from the value obtained above by 19″. The *vākya* "*khāṇḍavaghno jano vai*" which corresponds to the value given by (E.4) is actually given as a variant reading in the editions of *Veṇvāroha* and *Sphuṭacandrāpti*,[9] and this should be taken as the correct one as it is also in conformity with the computed value given in Table E.2.

Correcting the 181th *vākya*

Since the desired *vākya* number $i = 181$, the number of the complementary *vākya* $j = 248 - 181 = 67$. The *vākyas* below (V_{j+1}) and above (V_{j-1}) the complementary *vākya* are:

$$V_{68} = 5^r 24°59'07'' \quad \text{and} \quad V_{66} = 4^r 26°34'16''.$$

Their difference is given by

$$\Delta V_{67} = 28°24'51''.$$

By subtracting twice the first *vākya* ($2 \times V_1 = 0^r 24°05'10''$), from the above and dividing the result obtained by 225, we obtain $\delta_{67} = 1'\ 10''$. Subtracting this from the complementary *vākya* (V_{67}), we have

$$\begin{aligned} V'_{67} &= V_{67} - \delta_{67} \\ &= 5^r 10°43'28'' - 1'10'' = 5^r 10°42'18''. \end{aligned}$$

[9] {VR 1956}, p. 25; {SC 1973}, p. 53.

Now the desired *vākya* V_{181} is given by

$$\begin{aligned} V_{181} &= V_{248} - V'_{67} \\ &= 12^r 27°43'29'' - 5^r 10°42'18'' \\ &= 7^r 17°01'11''. \end{aligned} \tag{E.5}$$

The *vākya* number 181 found in the editions of *Veṇvāroha* and *Sphuṭacandrāpti* is "*phalajñānecchā katham*". This when decoded corresponds to the value $7^r 17°00'32''$ which varies from the value obtained above by 39″. The *vākya* "*payasyanicchā katham*" which corresponds to the value given by (E.5) is actually given as a variant reading in the editions of *Veṇvāroha* and *Sphuṭacandrāpti*,[10] and this should be taken as the correct one as it is also in conformity with the computed value given in Table E.2.

Correcting the 234th *vākya*

Since the desired *vākya* number $i = 234$, the number of the complementary *vākya* $j = 248 - 234 = 14$. The *vākyas* below (V_{j+1}) and above (V_{j-1}) the complementary *vākya* are:

$$V_{15} = 6^r 19°01'30'' \quad \text{and} \quad V_{13} = 5^r 20°24'31''.$$

Their difference is given by

$$\Delta V_{14} = 28°36'59''.$$

By subtracting twice the first *vākya* ($2 \times V_1 = 0^r 24°05'10''$), from the above and dividing the result obtained by 225, we obtain $\delta_{14} = 1'\ 13''$. Subtracting this from the complementary *vākya* (V_{14}), we have

$$\begin{aligned} V'_{14} &= V_{14} - \delta_{14} \\ &= 6^r 04°43'25'' - 1'13'' = 6^r 04°42'12''. \end{aligned}$$

Now the desired *vākya* V_{234} is given by

$$\begin{aligned} V_{234} &= V_{248} - V'_{14} \\ &= 12^r 27°43'29'' - 6^r 04°42'12'' \\ &= 6^r 23°01'17''. \end{aligned} \tag{E.6}$$

The *vākya* number 234 found in the editions of *Veṇvāroha* and *Sphuṭacandrāpti* is "*ṣaṭkāvyajño'mbarīṣaḥ*". This when decoded corresponds to the value $6^r 23°01'16''$ which varies from the value obtained above by 01″. The *vākya*

[10] {VR 1956}, p. 25; {SC 1973}, p. 54.

"*satkāvyajño 'mbarīṣaḥ*" which corresponds to the value given by (E.6) is actually given as a variant reading in the editions of *Veṇvāroha* and *Sphuṭacandrāpti*,[11] and this should be taken as the correct one as it is also in conformity with the computed value given in Table E.2.

Correcting the 242th *vākya*

Since the desired *vākya* number $i = 242$, the number of the complementary *vākya* $j = 248 - 242 = 6$. The *vākyas* below (V_{j+1}) and above (V_{j-1}) the complementary *vākya* are:

$$V_7 = 2^r 27°12'59'' \quad \text{and} \quad V_5 = 2^r 01°19'17''.$$

Their difference is given by

$$\Delta V_6 = 25°53'42''.$$

By subtracting twice the first *vākya* ($2 \times V_1 = 0^r 24°05'10''$), from the above and dividing the result obtained by 225, we obtain $\delta_6 = 28''$. Subtracting this from the complementary *vākya* (V_6), we have

$$\begin{aligned} V_6' &= V_6 - \delta_6 \\ &= 2^r 14°08'28'' - 28'' = 2^r 14°08'00''. \end{aligned}$$

Now the desired *vākya* V_{242} is given by

$$\begin{aligned} V_{242} &= V_{248} - V_6' \\ &= 12^r 27°43'29'' - 2^r 14°08'00'' \\ &= 10^r 13°35'29''. \end{aligned} \tag{E.7}$$

The *vākya* number 242 found in the editions of *Veṇvāroha* and *Sphuṭacandrāpti* is "*taruṇo balīyānāḍhyaḥ*". This when decoded corresponds to the value $10^r 13°35'26''$ which varies from the value obtained above by $03''$. The *vākya* "*dharaṇo balīyānāḍhyaḥ*" which corresponds to the value given by (E.7) is actually given as a variant reading in the editions of *Veṇvāroha* and *Sphuṭacandrāpti*,[12] and this should be taken as the correct one as it is also in conformity with the computed value given in Table E.2.

[11] {VR 1956}, p. 26; {SC 1973}, p. 56.

[12] {VR 1956}, p. 26; {SC 1973}, p. 56.

E.3 Rationale behind the *vākyaśodhana* procedure

We shall explain the rationale behind the *vākyaśodhana* procedure outlined in the previous section.[13] The Moon's true longitude is obtained by applying the 'equation of centre' to the mean longitude. The equation of centre at any instant depends upon the Moon's 'anomaly' which is the angular separation between the 'mean Moon' and the 'apogee' of the Moon. The *khaṇḍadina* is the day at the sunrise of which the Moon's anomaly is zero. The *candravākyas* are based on the following formula for the change in the true longitude of the Moon, i days after the *khaṇḍadina*:

$$V_i = R_1.360.i - \sin^{-1}\left[\frac{7}{80} \times \sin\left(R_2 \times 360.i\right)\right], \tag{E.8}$$

where R_1 and R_2 are the rates of motion of the Moon and its anomaly respectively, in revolutions per day. The second term represents the equation of centre of the Moon. As it stands, V_i is in degrees. The *candravākyas* are essentally the values of V_i, after converting them to *rāśis* (zodiacal signs), degrees, minutes and seconds, and expressed in the *kaṭapayādi* system.

The mean rate of motion of the Moon (R_1) is taken to be $\frac{4909031760}{134122987500} = \frac{1}{27.32167852}$ revolution per day (see (E.1)). It will be seen that the value of R_1 does not play any role in the *vākyaśodhana* procedure. For finding the *Vararuci-vākyas*, R_2 may be taken to be $\frac{9}{248}$ revolution per day. For the *Mādhava-vākyas*, we should take the more accurate value $R_2 = \frac{6845}{188611}$ revolution per day (used in *Veṇvāroha* and *Sphuṭacanrāpti*).

E.3.1 Vararucivākyas

Substituting the value of $R_2 = \frac{9}{248}$ in (E.8), we have

$$V_i = R_1.360.i - \sin^{-1}\left[\frac{7}{80} \times \sin\left(\frac{9}{248} \times 360.i\right)\right].$$

Hence,

[13] The material in this section is based on Sriram (2017).

$$\begin{aligned} V_{248-i} &= R_1.360.(248-i) - \sin^{-1}\left[\frac{7}{80} \times \sin\left(\frac{9}{248} \times 360.(248-i)\right)\right] \\ &= R_1.360.(248-i) + \sin^{-1}\left[\frac{7}{80} \times \sin\left(\frac{9}{248} \times 360.i\right)\right]. \\ V_{248} &= R_1.360.248 + \sin^{-1}\left[\frac{7}{80} \times \sin\left(\frac{9}{248} \times 360.248\right)\right] \\ &= R_1.360.248, \end{aligned}$$

as the last time in the RHS of the equation for V_{248} is 0. Clearly,

$$V_i + V_{248-i} = V_{248}, \quad (\text{modulo } 360^\circ), \tag{E.9}$$

as noted in the previous section.

E.3.2 Mādhava-vākyas

In this case, as $R_2 \neq \frac{9}{248}$, the relation (E.9) clearly does not hold. For the *Mādhava-vākyas*, the *vākyaśodhana* procedure, as explained in the preceding section, is as follows:

Suppose one is in doubt about V_i. Let $j = 248 - i$. Then, V_j is the complementary *vākya*. If V_j, and the *vākyas* above and below it are known, find:

$$V_j - \frac{(V_{j+1} - V_{j-1} - 2V_1)}{225}.$$

Then,

$$V_i = V_{248} - \left[V_j - \frac{(V_{j+1} - V_{j-1} - 2V_1)}{225}\right], \quad j = 248 - i. \tag{E.10}$$

Note that $i = 248 - j$. We rewrite the above equation in the form :

$$\delta_j \equiv V_j + V_{248-j} - V_{248} = \frac{V_{j+1} - V_{j-1} - 2V_1}{225}. \tag{E.11}$$

We now show that the above relation is valid to a very good approximation, using the ubiquitous Indian principle of *trairāśika* or the "rule of three".

E.3.3 Explanation of the vākyaśodhana expression for δ_j

We denote the Mādhava value $\frac{6845}{188611}$ for R_2 by α. Then,

$$\begin{aligned}\delta_j &= V_j + V_{248-j} - V_{248}\\ &= R_1.360.j - \sin^{-1}\left[\frac{7}{80}\ \sin(\alpha.360.j)\right]\\ &+ R_1.360.(248-j) - \sin^{-1}\left[\frac{7}{80}\ \sin(\alpha.360.(248-j))\right]\\ &- R_1.360.248 + \sin^{-1}\left[\frac{7}{80}\ \sin(\alpha.360.248)\right].\end{aligned}$$

Therefore,

$$\begin{aligned}-\delta_j &= \sin^{-1}\left[\frac{7}{80}\ \sin(\alpha.360.j)\right]\\ &+ \sin^{-1}\left[\frac{7}{80}\ \sin(\alpha.360.(248-j))\right]\\ &- \sin^{-1}\left[\frac{7}{80}\ \sin(\alpha.360.248)\right].\end{aligned}$$

We split α as $\alpha = (\alpha - \frac{9}{248}) + \frac{9}{248}$. Hence,

$$\begin{aligned}-\delta_j = \sin^{-1}&\left[\frac{7}{80}\ \sin((\alpha - \frac{9}{248}).360.j + \frac{9}{248}.360.j)\right]\\ + \sin^{-1}&\left[\frac{7}{80}\ \sin((\alpha - \frac{9}{248}).360.(248-j) + \frac{9}{248}.360.(248-j))\right]\\ - \sin^{-1}&\left[\frac{7}{80}\ \sin((\alpha - \frac{9}{248}).360.248 + \frac{9}{248}.360.248)\right].\end{aligned}$$

Let $\epsilon = (\alpha - \frac{9}{248}) \times 360 = 4.6948 \times 10^{-4}$. Using this notation in the above equation we have,

$$\begin{aligned}-\delta_j &= \sin^{-1}\left[\frac{7}{80}\ \sin(\frac{9}{248}.360.j + \epsilon.j)\right]\\ &- \sin^{-1}\left[\frac{7}{80}\ \sin(\frac{9}{248}.360.j - \epsilon.(248-j))\right]\\ &- \sin^{-1}\left[\frac{7}{80}\ \sin(\epsilon.248)\right].\end{aligned}$$

Let f be the function representing the equation of centre, $\sin^{-1}\left[\frac{7}{80}\ \sin(\ )\right]$, where ( ) is the anomaly. Hence,

$$\begin{aligned}-\delta_j &= f\left(\frac{9}{248}.360.j + \epsilon.j\right) - f\left(\frac{9}{248}.360.j - \epsilon.(248-j)\right) - f(\epsilon.248)\\ &= y_1 - f(\epsilon.248), \qquad \text{(E.12)}\end{aligned}$$

where y_1 is the difference in the equation of centre corresponding to a change in the anomaly (which is the argument) equal to $\left(\frac{9}{248}.360.j + \epsilon.j\right) - \left(\frac{9}{248}.360.j - \epsilon.(248 - j)\right) = \epsilon.248 \equiv x_1$, around a value of anomaly equal to $\frac{9}{248} \times 360.j$. Note that the change in the anomaly which is proportional to ϵ is resulting from the departure of $\alpha = R_2$ from $\frac{9}{248}$.

Now consider a different kind of difference:

$$\begin{aligned} V_{j+1} - V_{j-1} &= R_1.360.(j+1) - f\left(\alpha.360.(j+1)\right) \\ &\quad - \left[R_1.360.(j-1) - f\left(\alpha.360.(j-1)\right)\right]. \end{aligned}$$

$$\text{Hence,}\quad V_{j+1} - V_{j-1} = 2.R_1.360 - y_2, \tag{E.13}$$

where y_2 is the difference in the equation of centre corresponding to a change in the anomaly equal to $(\alpha.360.(j+1)) - (\alpha.360.(j-1)) = 2.\alpha.360 \equiv x_2$, around a value of anomaly equal to $\alpha.360.j$. Here, the change in the anomaly is due to the fact we are considering the *vākyas* for two different days, corresponding to $j+1$ and $j-1$.

y_1 and y_2 are the changes in the equation of centre corresponding to changes in the anomaly equal to x_1 and x_2 respectively. Now, we use the *trairāśika* (the rule of three), or the law of proportions,[14] which plays such an important role in Indian mathematics and astronomy:[15]

$$y_1 : x_1 = y_2 : x_2,$$

$$\text{or,}\quad y_1 = \frac{y_2}{x_2}.x_1. \tag{E.14}$$

Using equations (5), (6) and (7), and the values of $x_1 = \epsilon.248$ and $x_2 = 2.\alpha.360$, we have,

$$\begin{aligned} \delta_j &= -y_1 + f(\epsilon.248) \\ &= \frac{(V_{j+1} - V_{j-1}) - 2.R_1.360}{2.\alpha.360} \times \epsilon.248 + f(\epsilon.248). \end{aligned}$$

Now,

$$\begin{aligned} \frac{\epsilon.248}{2.\alpha.360} &= \frac{4.6948.248.188611}{2.360.6845} \times 10^{-4} \\ &= 4.4558 \times 10^{-3} = \frac{1}{224.4244}. \end{aligned}$$

[14] Actually, x_1 is the change in the anomaly around $\frac{9}{248}.360.j$, whereas x_2 is the change in the anomaly around $\alpha.360.j$. As $\alpha = \frac{6845}{18861} \approx \frac{9}{248}$, we ignore this difference, which will lead to changes of higher order in ϵ.

[15] In his explanation of verse 246 in his *Līlāvatī*, Bhāskara remarks that, just as this universe is pervaded by Lord Nārāyaṇa in all his manifestations, "so is all this collections of instructions for computations pervaded by the rule of three terms" ({LV, 1993}, p. 166).

This is approximated as $\frac{1}{225}$. Therefore,

$$\delta_j \approx \frac{V_{j+1} - V_{j-1} - 2X}{225}, \tag{E.15}$$

$$\text{where,}\ X = R_1.360 - \frac{f(\epsilon.248)}{2} \times 225. \tag{E.16}$$

Now, $\frac{f(\epsilon.248)}{2} \times 225 = \frac{1}{2}\sin^{-1}\left(\frac{7}{80}\sin(248.\epsilon)\right) \times 225 = 1.1449$. Hence,

$$X = R_1.360 - 1.1449. \tag{E.17}$$

From equation (1),

$$\begin{aligned} V_1 &= R_1.360 - \sin^{-1}\left(\frac{7}{80}\sin\left(\frac{6845}{188611} \times 360\right)\right) \\ &= R_1.360 - 1.1334. \end{aligned} \tag{E.18}$$

Comparing equations (10) and (11), we find:

$$X \approx V_1. \tag{E.19}$$

Substituting this in equation (8), we have:

$$\delta_j \equiv V_j + V_{248-j} - V_{248} \approx \frac{V_{j+1} - V_{j-1} - 2V_1}{225}, \tag{E.20}$$

which is the same as (E.11).

E.4 Table of *candravākyas* of Vararuci and Mādhava

vākya number	*Candravākyas* of Vararuci in *kaṭapayādi*		in numerals	*Candravākyas* of Mādhava in *kaṭapayādi*		in numerals
1	गीर्नः श्रेयः	*gīrnaḥ śreyaḥ*	$12^{\circ}03'$	शीलं राज्ञः श्रिये	*śīlaṃ rājñaḥ śriye*	$12^{\circ}02'35''$
2	धेनवः श्रीः	*dhenavaḥ śrīḥ*	$24^{\circ}09'$	धिगिदं नश्वरम्	*dhigidaṃ naśvaram*	$24^{\circ}08'39''$
3	रुद्रस्तु नम्यः	*rudrastu namyaḥ*	$1^{r}06^{\circ}22'$	लोलः पुरुषो नार्याम्	*lolaḥ puruṣo nāryām*	$1^{r}06^{\circ}21'33''$
4	भवो हि याज्यः	*bhavo hi yājyaḥ*	$1^{r}18^{\circ}44'$	तपस्वी वैदिकः स्यात्	*tapasvī vaidikaḥ syāt*	$1^{r}18^{\circ}44'16''$
5	धन्येयं नारी	*dhanyeyaṃ nārī*	$2^{r}01^{\circ}19'$	सेव्याळका किन्नरैः	*sevyāḷakā kinnaraiḥ*	$2^{r}01^{\circ}19'17''$
6	धनवान् पुत्रः	*dhanavān putraḥ*	$2^{r}14^{\circ}09'$	दीप्रो दिने भास्करः	*dīpro dine bhāskaraḥ*	$2^{r}14^{\circ}08'28''$
7	गृह्या सुरा राज्ञा[16]	*gṛhyā surā rājñā*	$2^{r}27^{\circ}13'$	धर्मरम्यं सुराष्ट्रम्	*dharmaramyaṃ surāṣṭram*	$2^{r}27^{\circ}12'59''$
8	बालेन कुलम्	*bālena kulam*	$3^{r}10^{\circ}33'$	स्तनौ लीलानुकूलौ	*stanau līlānukūlau*	$3^{r}10^{\circ}33'06''$
9	धनुर्भिः खलैः	*dhanurbhiḥ khalaiḥ*	$3^{r}24^{\circ}09'$	पुत्रादौ न विरागः	*putrādau na virāgaḥ*	$3^{r}24^{\circ}08'21''$
10	दश सूनवः	*daśa sūnavaḥ*	$4^{r}07^{\circ}58'$	शौरिः समर्थ एव	*śauriḥ samartha eva*	$4^{r}07^{\circ}57'25''$
11	होमस्य स्रुवः	*homasya sruvaḥ*	$4^{r}21^{\circ}58'$	दुष्टैर्देशोपद्रवः	*duṣṭairdeśopadravaḥ*	$4^{r}21^{\circ}58'18''$
12	दीनास्ते नृणाम्	*dīnāste nṛṇām*	$5^{r}06^{\circ}08'$	व्यग्रो जनः क्षुन्नाशे	*vyagro janaḥ kṣunnāśe*	$5^{r}06^{\circ}08'21''$
13	मुखं नारीणाम्	*mukhaṃ nārīṇām*	$5^{r}20^{\circ}25'$	योगीश्वरो निराशः	*yogīśvaro nirāśaḥ*	$5^{r}20^{\circ}24'31''$
14	भवभग्नास्ते	*bhavabhagnāste*	$6^{r}04^{\circ}44'$	शिखण्डी भवनेषु	*śikhaṇḍī bhavaneṣu*	$6^{r}04^{\circ}43'25''$
15	श्रीर्निधीयते	*śrīrnidhīyate*	$6^{r}19^{\circ}02'$	नागो यानाधिपतिः	*nāgo yānādhipatiḥ*	$6^{r}19^{\circ}01'30''$
16	शं किल नाथः	*śaṃ kila nāthaḥ*	$7^{r}03^{\circ}15'$	परिणयेऽङ्गनेच्छा	*pariṇaye'ṅganecchā*	$7^{r}03^{\circ}15'21''$
17	श्रेष्ठा सा कथा	*śreṣṭhā sā kathā*	$7^{r}17^{\circ}22'$	कविकण्ठस्था कथा	*kavikaṇṭhasthā kathā*	$7^{r}17^{\circ}21'41''$

[16] This word *gṛhyā* should not be confused with a somewhat similar sounding, and more commonly used word *grāhyā*. While the latter means 'to be accepted or consumed', the word *gṛhyā* in this *vākya* has been employed to convey the special meaning 'that which is to be kept out or avoided'. This is explained in the *Aṣṭādhyāyī* by the *sūtra* पदास्वैरिबाह्यापक्ष्येषु च (*padāsvairibāhyāpakṣyeṣu ca*) [3.1.119].

18	सौख्यस्यानन्दः	*saukhyasyānandaḥ*	$8^{r}01^{\circ}17'$	शीलसम्पद्यानन्दः	*śīlasampadyānandaḥ*	$8^{r}01^{\circ}17'35''$
19	ध्यानं मान्यं हि	*dhyānaṃ mānyaṃ hi*	$8^{r}15^{\circ}01'$	श्रीर्विना न मुकुन्दात्	*śrīrvinā na mukundāt*	$8^{r}15^{\circ}00'42''$
20	धीरो हि राजा	*dhīro hi rājā*	$8^{r}28^{\circ}29'$	निराधारोऽहिराजः	*nirādhāro'hirājaḥ*	$8^{r}28^{\circ}29'20''$
21	श्रुत्वास्य युद्धम्	*śrutvāsya yuddham*	$9^{r}11^{\circ}42'$	कुबेरो विकटधीः	*kubero vikaṭadhīḥ*	$9^{r}11^{\circ}42'31''$
22	अभवच्छ्राद्धम्	*abhavacchrāddham*	$9^{r}24^{\circ}40'$	स्तेनानां श्वा विरोद्धा	*stenānāṃ śvā viroddhā*	$9^{r}24^{\circ}40'06''$
23	गोरसो ननु स्यात्	*goraso nanu syāt*	$10^{r}07^{\circ}23'$	दीर्घरिरंसुर्ना नाके	*dīrghariraṃsurnā nāke*	$10^{r}07^{\circ}22'48''$
24	द्रुमा धन्या नये	*drumā dhanyā naye*	$10^{r}19^{\circ}52'$	वानरो मधुपानाढ्यः	*vānaro madhupānāḍhyaḥ*	$10^{r}19^{\circ}52'04''$
25	इष्टं राज्ञः कुर्यात्	*iṣṭaṃ rājñaḥ kuryāt*	$11^{r}02^{\circ}10'$	घननिकरो निर्ययौ[17]	*ghananikaro niryayau*	$11^{r}02^{\circ}10'04''$
26	धन्या विद्येयं स्यात्	*dhanyā vidyeyaṃ syāt*	$11^{r}14^{\circ}19'$	रोगे धैर्यविपर्ययः	*roge dhairyaviparyayaḥ*	$11^{r}14^{\circ}19'32''$
27	त्वं रक्षा राज्यस्य	*tvaṃ rakṣā rājyasya*	$11^{r}26^{\circ}24'$	स्थूलो गिरिश्चित्रकूटः	*sthūlo giriścitrakūṭaḥ*	$11^{r}26^{\circ}23'37''$
28	क्षेत्रजः	*kṣetrajaḥ*	$8^{\circ}26'$	स्तम्भमात्रो हि	*stambhamātro hi*	$8^{\circ}25'46''$
29	नीले नेत्रे	*nīle netre*	$20^{\circ}30'$	धीरधीस्त्रिनेत्रः	*dhīradhīstrinetraḥ*	$20^{\circ}29'29''$
30	जलं प्राज्ञाय	*jalaṃ prājñāya*	$1^{r}02^{\circ}38'$	प्रपदौ गुरोर्नम्यौ	*prapadau gurornamyau*	$1^{r}02^{\circ}38'12''$
31	शशी वन्द्यः स्यात्	*śaśī vandyaḥ syāt*	$1^{r}14^{\circ}55'$	छन्नो माणवकः किम्	*channo māṇavakaḥ kim*	$1^{r}14^{\circ}55'07''$
32	गोरसप्रियः	*gorasapriyaḥ*	$1^{r}27^{\circ}23'$	गानगोष्ठी सुखाय	*gānagoṣṭhī sukhāya*	$1^{r}27^{\circ}23'03''$
33	वनानि यत्र	*vanāni yatra*	$2^{r}10^{\circ}04'$	काकुध्वनिर्नकारात्	*kākudhvanirnakārāt*	$2^{r}10^{\circ}04'11''$
34	अन्नं गोत्रश्रीः	*annaṃ gotraśrīḥ*	$2^{r}23^{\circ}00'$	तनुर्न नगरे श्रीः	*tanurna nagare śrīḥ*	$2^{r}23^{\circ}00'06''$
35	रुष्टास्ते नागाः	*ruṣṭāste nāgāḥ*	$3^{r}06^{\circ}12'$	शैलाः पुष्पितनगाः	*śailāḥ puṣpitanagāḥ*	$3^{r}06^{\circ}11'35''$
36	धिगन्धः किल	*dhigandhaḥ kila*	$3^{r}19^{\circ}39'$	लोलो जलधिः किल	*lolo jaladhiḥ kila*	$3^{r}19^{\circ}38'33''$
37	पुरोगा अभीः	*purogā abhīḥ*	$4^{r}03^{\circ}21'$	धनी नरोऽङ्गनावान्	*dhanī naro'ṅganāvān*	$4^{r}03^{\circ}20'09''$
38	मान्यः स कविः	*mānyaḥ sa kaviḥ*	$4^{r}17^{\circ}15'$	सर्वविद् व्यासः कविः	*sarvavid vyāsaḥ kaviḥ*	$4^{r}17^{\circ}14'47''$
39	अरिष्टनाशम्	*ariṣṭanāśam*	$5^{r}01^{\circ}20'$	स्तेनेन द्रव्यनाशः[18]	*stenena dravyanāśaḥ*	$5^{r}01^{\circ}20'06''$
40	बालो मे केशः	*bālo me keśaḥ*	$5^{r}15^{\circ}33'$	सूर्यो बलमाकाशे	*sūryo balamākāśe*	$5^{r}15^{\circ}33'17''$
41	कुशधारिणः	*kuśadhāriṇaḥ*	$5^{r}29^{\circ}51'$	मनुष्यो मधुरात्मा	*manuṣyo madhurātmā*	$5^{r}29^{\circ}51'05''$

[17] The version found in the editions of *Veṇvāroha* ({VR 1956}, p. 23.) and *Sphuṭacandrāpti* ({SC 1973}, p. 47) is "*dhananikaro niryayau*" which encodes the value $11^{r}02^{\circ}10'09''$.

[18] The reading above is from *Veṇvāroha* ({VR 1956}, p. 23). The word *'dravśa'* is found in the place of *'dravya'* in *Sphuṭacandrāpti* ({SC 1973}, p. 48) which is incorrect.

42	इष्टिर्विद्यते	*iṣṭirvidyate*	$6^{r}14^{\circ}10'$	गानं नेष्टं विपत्तौ	*gānaṃ neṣṭaṃ vipattau*	$6^{r}14^{\circ}10'03''$
43	स राजा प्रीतः	*sa rājā prītaḥ*	$6^{r}28^{\circ}27'$	पर्वचन्द्रोऽहिग्रस्तः	*parvacandro'higrastaḥ*	$6^{r}28^{\circ}26'41''$
44	सुगुप्रायोऽसौ	*suguprāyo'sau*	$7^{r}12^{\circ}37'$	भोगेच्छालं प्रियेऽर्थे	*bhogecchālaṃ priye'rthe*	$7^{r}12^{\circ}37'34''$
45	धिगस्तु ह्रासः	*dhigastu hrāsaḥ*	$7^{r}26^{\circ}39'$	मागधो गीतरसः	*māgadho gītarasaḥ*	$7^{r}26^{\circ}39'35''$
46	अङ्गानि यदा	*aṅgāni yadā*	$8^{r}10^{\circ}30'$	लीनो नागो निकुञ्जे	*līno nāgo nikuñje*	$8^{r}10^{\circ}30'03''$
47	सेनावान् राजा	*senāvān rājā*	$8^{r}24^{\circ}07'$	गामुक्षा न विरेजे	*gāmukṣā na vireje*	$8^{r}24^{\circ}06'53''$
48	धीराः सन्नद्धाः	*dhīrāḥ sannaddhāḥ*	$9^{r}07^{\circ}29'$	रवौ हरेः सन्निधिः	*ravau hareḥ sannidhiḥ*	$9^{r}07^{\circ}28'42''$
49	शालीनं प्रधानम्	*śālīnaṃ pradhānam*	$9^{r}20^{\circ}35'$	वर्णान् वागनुरुन्धे	*varṇān vāganurundhe*	$9^{r}20^{\circ}34'54''$
50	क्षीरं गोर्नो नयेत्	*kṣīraṃ gorno nayet*	$10^{r}03^{\circ}26'$	भावे स्मरोऽङ्गनानां स्यात्	*bhāve smaro'ṅganānāṃ syāt*	$10^{r}03^{\circ}25'44''$
51	रत्नचयो नृपः	*ratnacayo nṛpaḥ*	$10^{r}16^{\circ}02'$	गायत्री नास्तिकैर्निन्द्या	*gāyatrī nāstikairnindyā*	$10^{r}16^{\circ}02'13''$
52	ताः प्रजाः प्राज्ञाः स्युः	*tāḥ prajāḥ prājñāḥ syuḥ*	$10^{r}28^{\circ}26'$	धनं चोरो हरेन्नित्यम्	*dhanaṃ coro harennityam*	$10^{r}28^{\circ}26'09''$
53	अश्वानां को योग्यः	*aśvānāṃ ko yogyaḥ*	$11^{r}10^{\circ}40'$	धर्मं धिगनपायस्य	*dharmaṃ dhiganapāyasya*	$11^{r}10^{\circ}39'59''$
54	तद्वैरं प्रियायाः	*tadvairaṃ priyāyāḥ*	$11^{r}22^{\circ}46'$	धीगतिर्भद्ररूपेयम्	*dhīgatirbhadrarūpeyam*	$11^{r}22^{\circ}46'39''$
55	धवस्त्वम्	*dhavastvam*	$4^{\circ}49'$	क्षीराब्धौ विभुः	*kṣīrābdhau vibhuḥ*	$4^{\circ}49'26''$
56	ग्रामस्तस्य	*grāmastasya*	$16^{\circ}52'$	यमोऽयमन्तिके	*yamo'yamantike*	$16^{\circ}51'51''$
57	जन्मजरा	*janmajarā*	$28^{\circ}58'$	गौरी स्थाणोर्दाराः	*gaurī sthāṇordārāḥ*	$28^{\circ}57'23''$
58	इष्टका कार्या	*iṣṭakā kāryā*	$1^{r}11^{\circ}10'$	गरळं नोपयुञ्ज्यात्	*garaḷaṃ nopayuñjyāt*	$1^{r}11^{\circ}09'23''$
59	कुलगुरुः स्यात्	*kulaguruḥ syāt*	$1^{r}23^{\circ}31'$	गोमानलं गरीयान्	*gomānalaṃ garīyān*	$1^{r}23^{\circ}30'53''$
60	मुनिस्तु उग्रः	*munistu ugraḥ*	$2^{r}06^{\circ}05'$	सुग्रीवोऽनन्तनिष्ठः	*sugrīvo'nantaniṣṭhaḥ*	$2^{r}06^{\circ}04'27''$
61	प्रमोदकरः	*pramodakaraḥ*	$2^{r}18^{\circ}52'$	प्राज्ञो रामो दैत्यारिः	*prājño rāmo daityāriḥ*	$2^{r}18^{\circ}52'02''$
62	शशाङ्कानुगः	*śaśāṅkānugaḥ*	$3^{r}01^{\circ}55'$	अशुभाशया नागाः	*aśubhāśayā nāgāḥ*	$3^{r}01^{\circ}54'50''$
63	वक्ष्यामि कालम्	*vakṣyāmi kālam*	$3^{r}15^{\circ}14'$	चपलः कामपालः	*capalaḥ kāmapālaḥ*	$3^{r}15^{\circ}13'16''$
64	सम्भेदः खलैः	*sambhedaḥ khalaiḥ*	$3^{r}28^{\circ}47'$	वाग्मी तु वादरागी	*vāgmī tu vādarāgī*	$3^{r}28^{\circ}46'54''$
65	शीलप्रियस्त्वम्	*śīlapriyastvam*	$4^{r}12^{\circ}35'$	गङ्गा भागीरथ्यभूत्	*gaṅgā bhāgīrathyabhūt*	$4^{r}12^{\circ}34'33''$
66	वेलातरवः	*velātaravaḥ*	$4^{r}26^{\circ}34'$	तपस्वीगतिरूर्ध्वम्	*tapasvīgatirūrdhvam*	$4^{r}26^{\circ}34'16''$
67	विभिन्नं कर्म	*vibhinnaṃ karma*	$5^{r}10^{\circ}44'$	जरद्गवोऽनुद्यमः	*jaradgavo'nudyamaḥ*	$5^{r}10^{\circ}43'28''$
68	धर्मवान् रामः	*dharmavān rāmaḥ*	$5^{r}24^{\circ}59'$	सूनुर्धामाभरणम्	*sūnurdhāmābharaṇam*	$5^{r}24^{\circ}59'07''$
69	दिग्व्याळो नास्ति	*digvyāḷo nāsti*	$6^{r}09^{\circ}18'$	गुणोऽसूया धनिषु	*guṇo'sūyā dhaniṣu*	$6^{r}09^{\circ}17'53''$
70	ते बाला भ्रान्ताः	*te bālā bhrāntāḥ*	$6^{r}23^{\circ}36'$	विकृता गौडरीतिः	*vikṛtā gauḍarītiḥ*	$6^{r}23^{\circ}36'14''$

71	कामासन्नः सः	*kāmāsannaḥ saḥ*	$7^r07°51'$	लघुर्न मैथुनेच्छा[19]	*laghurna maithunecchā*	$7^r07°50'43''$
72	होमं पुत्रार्थम्	*homaṃ putrārtham*	$7^r21°58'$	आनन्दमयो रसः	*ānandamayo rasaḥ*	$7^r21°58'00''$
73	मणिर्मानदः	*maṇirmānadaḥ*	$8^r05°55'$	कल्यः शिशुर्मनुजः	*kalyaḥ śiśurmanujaḥ*	$8^r05°55'11''$
74	नाविद्धः पादे	*nāviddhaḥ pāde*	$8^r19°40'$	दृढधीर्लब्धपदः	*dṛḍhadhīrlabdhapadaḥ*	$8^r19°39'48''$
75	उत्पलं निधिः	*utpalaṃ nidhiḥ*	$9^r03°10'$	लीना आपोऽम्बुनिधौ	*līnā āpo'mbunidhau*	$9^r03°10'03''$
76	शूद्रस्तु योद्धा	*śūdrastu yoddhā*	$9^r16°25'$	क्षामवारिस्तोयधिः	*kṣāmavāristoyadhiḥ*	$9^r16°24'56''$
77	विरुद्धं स्त्रीधनम्	*viruddhaṃ strīdhanam*	$9^r29°24'$	भाग्यविरोधः क्रोधात्	*bhāgyavirodhaḥ krodhāt*	$9^r29°24'14''$
78	हीनप्रायो नटः	*hīnaprāyo naṭaḥ*	$10^r12°08'$	धरा हीनाश्रया नित्यम्	*dharā hīnāśrayā nityam*	$10^r12°08'29''$
79	धिगश्वः खिन्नोऽयम्	*dhigaśvaḥ khinno'yam*	$10^r24°39'$	जनोऽन्धो गत्वरो नश्येत्	*jano'ndho gatvaro naśyet*	$10^r24°39'08''$
80	दिशतु नः पथ्यम्	*diśatu naḥ pathyam*	$11^r06°58'$	मुकुन्दान्मोक्ष उपेयः	*mukundānmokṣa upeyaḥ*	$11^r06°58'15''$
81	जनोऽन्धः पापकः	*jano'ndhaḥ pāpakaḥ*	$11^r19°08'$	आगोहीनोऽधिकः पटुः	*āgohīno'dhikaḥ paṭuḥ*	$11^r19°08'30''$
82	गृह्या स्यात्	*gṛhyā syāt*	$1°13'$	ज्ञानी गार्ग्याय	*jñānī gārgyāya*	$1°13'00''$
83	मान्यं लोके	*mānyaṃ loke*	$13°15'$	कृपणः कौण्डिन्यः	*kṛpaṇaḥ kauṇḍinyaḥ*	$13°15'11''$
84	धन्यः शरैः	*dhanyaḥ śaraiḥ*	$25°19'$	लोला दीपशिखा	*lolā dīpaśikhā*	$25°18'33''$
85	सुखी स नित्यम्	*sukhī sa nityam*	$1^r07°27'$	स्वर्गस्तु प्रार्थनीयः	*svargastu prārthanīyaḥ*	$1^r07°26'34''$
86	लाभो धान्यस्य	*lābho dhānyasya*	$1^r19°43'$	सौभ्रात्रं वाधिकं स्यात्	*saubhrātraṃ vādhikaṃ syāt*	$1^r19°42'27''$
87	अङ्कुरं नीरे	*aṅkuraṃ nīre*	$2^r02°10'$	लीनोऽळीनो त्रिनेत्रः	*līno'ḷīno trinetraḥ*	$2^r02°09'03''$
88	धावद्वैद्योऽत्र	*dhāvadvaidyo'tra*	$2^r14°49'$	धिगाहवविकारः	*dhigāhavavikāraḥ*	$2^r14°48'39''$
89	गत्वा सुराष्ट्रम्	*gatvā surāṣṭram*	$2^r27°43'$	शिशिरा वासरश्रीः	*śiśirā vāsaraśrīḥ*	$2^r27°42'55''$
90	गमनकालम्	*gamanakālam*	$3^r10°53'$	अभिरामा नकुली	*abhirāmā nakulī*	$3^r10°52'40''$
91	दयावान् रोगी	*dayāvān rogī*	$3^r24°18'$	धर्मोऽर्थः पूर्वरङ्गः	*dharmo'rthaḥ pūrvaraṅgaḥ*	$3^r24°17'59''$
92	होमस्थानं वनम्	*homasthānaṃ vanam*	$4^r07°58'$	भानुजो मिथुनेऽभूत्	*bhānujo mithune'bhūt*	$4^r07°58'04''$
93	श्रीमान् पुत्रो वा	*śrīmān putro vā*	$4^r21°52'$	रौद्रो यमस्यारम्भः	*raudro yamasyārambhaḥ*	$4^r21°51'22''$
94	तन्मम नाम	*tanmama nāma*	$5^r05°56'$	धिगाशामशनेऽस्मिन्	*dhigāśāmaśane'smin*	$5^r05°55'39''$
95	दानानां क्रमः	*dānānāṃ kramaḥ*	$5^r20°08'$	जनार्दनो नरेशः	*janārdano nareśaḥ*	$5^r20°08'08''$

[19] In this *vākya*, there could be a doubt regarding the usage of the word *laghu* in its masculine form, as it is qualifying a feminine gendered compound ending with *icchā*. However, since the use of feminine suffix is optional, as per the *sūtra* वोतो गुणवचनात् (*voto guṇavacanāt*) [4.1.44] in the *Aṣṭādhyāyī*, it is a perfectly valid usage.

96	क्षेत्रवानस्तु	*kṣetravānastu*	$6^r04°26'$	मृगाः शूरा वनान्ते	*mṛgāḥ śūrā vanānte*	$6^r04°25'35''$
97	शम्भुर्जयति	*śambhurjayati*	$6^r18°45'$	चण्डो वै भोजपतिः	*caṇḍo vai bhojapatiḥ*	$6^r18°44'36''$
98	रत्नाङ्गनार्था	*ratnāṅganārthā*	$7^r03°02'$	धीगम्योऽनङ्ग आसीत्[20]	*dhīgamyo'naṅga āsīt*	$7^r03°01'39''$
99	लक्ष्योऽसौ पार्थः	*lakṣyo'sau pārthaḥ*	$7^r17°13'$	धैर्यालयः सन्यासी	*dhairyālayaḥ sanyāsī*	$7^r17°13'19''$
100	सापत्यनिन्दा	*sāpatyanindā*	$8^r01°17'$	चन्द्रात् तापापनोदः	*candrāt tāpāpanodaḥ*	$8^r01°16'26''$
101	जनो मान्यो हि	*jano mānyo hi*	$8^r15°08'$	सेव्यो जनैर्मुकुन्दः	*sevyo janairmukundaḥ*	$8^r15°08'17''$
102	स वादी राजा	*sa vādī rājā*	$8^r28°47'$	अभिषवोऽहरहः	*abhiṣavo'harahaḥ*	$8^r28°46'40''$
103	आकारो युद्धम्	*ākāro yuddham*	$9^r12°10'$	कार्योऽनार्यैरुपधिः	*kāryo'nāryairupadhiḥ*	$9^r12°10'11''$
104	दास्यामि श्राद्धम्	*dāsyāmi śrāddham*	$9^r25°18'$	मानदेयं मुरळी	*mānadeyaṃ muraḷī*	$9^r25°18'05''$
105	कार्यहानिर्नार्या	*kāryahānirnāryā*	$10^r08°11'$	ज्वलनो यजने नम्यः	*jvalano yajane namyaḥ*	$10^r08°10'34''$
106	दम्भान्नरा नष्टाः	*dambhānnarā naṣṭāḥ*	$10^r20°48'$	क्रीडा दृढा नरनार्योः	*krīḍā dṛḍhā naranāryoḥ*	$10^r20°48'32''$
107	विकलानां कार्याः	*vikalānāṃ kāryāḥ*	$11^r03°14'$	विश्वं गोपाल एकाकी	*viśvaṃ gopāla ekākī*	$11^r03°13'44''$
108	हरणं पाद्यस्य	*haraṇaṃ pādyasya*	$11^r15°28'$	फलाहारो मुख्यकल्पः[21]	*phalāhāro mukhyakalpaḥ*	$11^r15°28'32''$
109	तुला सम्प्रत्यया	*tulā sampratyayā*	$11^r27°36'$	धन्वी शूली सुरैः पूज्यः	*dhanvī śūlī suraiḥ pūjyaḥ*	$11^r27°35'49''$
110	धिगन्धः	*dhigandhaḥ*	$9°39'$	गोमदो गळी	*gomado gaḷī*	$9°38'53''$
111	कविः पुत्रः	*kaviḥ putraḥ*	$21°41'$	धनुर्ज्या विपाठा	*dhanurjyā vipāṭhā*	$21°41'09''$
112	तत्त्वाङ्गनेयम्	*tattvāṅganeyam*	$1^r03°46'$	आढ्यः षड्भागैर्नृपः	*āḍhyaḥ ṣaḍbhāgairnṛpaḥ*	$1^r03°46'10''$
113	जीर्णो मे कायः	*jīrṇo me kāyaḥ*	$1^r15°58'$	धन्यः स्थाणुमुपेयात्	*dhanyaḥ sthāṇumupeyāt*	$1^r15°57'19''$
114	दया हरस्य	*dayā harasya*	$1^r28°18'$	धिगसौख्यं हिरण्यात्	*dhigasaukhyaṃ hiraṇyāt*	$1^r28°17'39''$
115	अशनपरः	*aśanaparaḥ*	$2^r10°50'$	देवो धावन्नैकत्र	*devo dhāvannaikatra*	$2^r10°49'48''$
116	तालुलेखोऽत्र	*tālulekho'tra*	$2^r23°36'$	तन्वी शीलगरिष्ठा	*tanvī śīlagariṣṭhā*	$2^r23°35'46''$
117	सङ्गतो नागः	*saṅgato nāgaḥ*	$3^r06°37'$	लक्ष्मीस्तुङ्गस्तनाङ्गी	*lakṣmīstuṅgastanāṅgī*	$3^r06°36'53''$
118	विशुद्धो योगी	*viśuddho yogī*	$3^r19°54'$	चला लक्ष्मीर्धन्यगा	*calā lakṣmīrdhanyagā*	$3^r19°53'36''$
119	ताराङ्गं नभः	*tārāṅgaṃ nabhaḥ*	$4^r03°26'$	धिगशीघ्रगा नावः	*dhigaśīghragā nāvaḥ*	$4^r03°25'39''$

[20] The reading above is from *Veṇvāroha* ({VR 1956}, p. 24). The word *'eva'* is found in the place of *'āsīt'* in *Sphuṭacandrāpti* ({SC 1973}, p. 50) which is incorrect.

[21] The reading above is from *Veṇvāroha* ({VR 1956}, p. 24). The word *'mūkhya'* is found in the place of *'mukhya'* in *Sphuṭacandrāpti* ({SC 1973}, p. 51) which is incorrect.

120	प्रियार्थं कविः	*priyārthaṃ kaviḥ*	$4^r17^\circ12'$	पूर्णः पयसा कुम्भः	*pūrṇaḥ payasā kumbhaḥ*	$4^r17^\circ11'51''$
121	पापोऽयं निशि	*pāpo'yaṃ niśi*	$5^r01^\circ11'$	कठिनोऽयं कीनाशः	*kaṭhino'yaṃ kīnāśaḥ*	$5^r01^\circ10'21''$
122	धन्यो मान्योऽम्शे	*dhanyo mānyo'mśe*	$5^r15^\circ19'$	धिगहंयुमकस्मात्[22]	*dhigahaṃyumakasmāt*	$5^r15^\circ18'39''$
123	भोगार्धं रामा	*bhogārdhaṃ rāmā*	$5^r29^\circ34'$	षड्भागबन्धुरीशः	*ṣaḍbhāgabandhurīśaḥ*	$5^r29^\circ33'46''$
124	रामा गीयते	*rāmā gīyate*	$6^r13^\circ52'$	कठोरो मृगपतिः	*kaṭhoro mṛgapatiḥ*	$6^r13^\circ52'21''$
125	अत्याहारस्तु	*atyāhārastu*	$6^r28^\circ10'$	क्षीणो न व्याहरति	*kṣīṇo na vyāharati*	$6^r28^\circ10'56''$
126	शारीरकोऽसौ	*śārīrako'sau*	$7^r12^\circ25'$	धर्मशास्त्रं श्रेयसे	*dharmaśāstraṃ śreyase*	$7^r12^\circ25'59''$
127	लोलचक्रस्थः	*lolacakrasthaḥ*	$7^r26^\circ33'$	लोकोऽभिलाषी रसे	*loko'bhilāṣī rase*	$7^r26^\circ34'13''$
128	प्रागनिष्पदम्	*prāganiṣpadam*	$8^r10^\circ32'$	सागरो गोर्न पदम्	*sāgaro gorna padam*	$8^r10^\circ32'37''$
129	दिव्यवान् राजा	*divyavān rājā*	$8^r24^\circ18'$	रविजुष्टं वारिजे	*ravijuṣṭaṃ vārije*	$8^r24^\circ18'42''$
130	अंशार्थिनोर्धीः	*aṃśārthinordhīḥ*	$9^r07^\circ50'$	साम्बोऽनिशं सन्नद्धः	*sāmbo'niśaṃ sannaddhaḥ*	$9^r07^\circ50'37''$
131	सेनायाः क्रोधः	*senāyāḥ krodhaḥ*	$9^r21^\circ07'$	प्रत्यासन्नः पुरोधाः	*pratyāsannaḥ purodhāḥ*	$9^r21^\circ07'12''$
132	दानं भानोर्नष्टम्	*dānaṃ bhānornaṣṭam*	$10^r04^\circ08'$	इष्टिर्दानं विना नेष्टा	*iṣṭirdānaṃ vinā neṣṭā*	$10^r04^\circ08'10''$
133	भूमिस्तस्य नित्यम्	*bhūmistasya nityam*	$10^r16^\circ54'$	नानाभिमतं कनकम्	*nānābhimataṃ kanakam*	$10^r16^\circ54'00''$
134	चक्रार्धं प्राज्ञाय	*cakrārdhaṃ prājñāya*	$10^r29^\circ26'$	श्रीनतः श्रीधरो नित्यम्	*śrīnataḥ śrīdharo nityam*	$10^r29^\circ26'02''$
135	ता भार्याः पापोऽयम्	*tā bhāryāḥ pāpo'yam*	$11^r11^\circ46'$	सुकृतिः स्वयं पाककृत्	*sukṛtiḥ svayaṃ pākakṛt*	$11^r11^\circ46'17''$
136	दिशोऽम्बराण्यस्य	*diśo'mbarāṇyasya*	$11^r23^\circ58'$	काष्ठसमा गात्रयष्टिः	*kāṣṭhasamā gātrayaṣṭiḥ*	$11^r23^\circ57'21''$
137	ग्लौर्नास्ति	*glaurnāsti*	$6^\circ03'$	अरिरनाप्तः	*ariranāptaḥ*	$6^\circ02'20''$
138	मीनजेयम्	*mīnajeyam*	$18^\circ05'$	चण्डभानुर्जयी	*caṇḍabhānurjayī*	$18^\circ04'36''$
139	दानानि नित्यम्	*dānāni nityam*	$1^r00^\circ08'$	व्यवच्छिन्नोऽनुनयः	*vyavacchinno'nunayaḥ*	$1^r00^\circ07'41''$
140	तपः श्रेयः स्यात्	*tapaḥ śreyaḥ syāt*	$1^r12^\circ16'$	कीनाशो व्याघ्रकल्पः	*kīnāśo vyāghrakalpaḥ*	$1^r12^\circ15'01''$
141	अम्बुभिरिष्टैः	*ambubhiriṣṭaiḥ*	$1^r24^\circ30'$	वाङ्माधुरी वरेण्या	*vāṅmādhurī vareṇyā*	$1^r24^\circ29'54''$
142	क्षमास्तु नरैः	*kṣamāstu naraiḥ*	$2^r06^\circ56'$	श्रीकृष्णो मोक्षनिष्ठः	*śrīkṛṣṇo mokṣaniṣṭhaḥ*	$2^r06^\circ55'12''$
143	लोलधीः पुत्रः	*loladhīḥ putraḥ*	$2^r19^\circ33'$	हृष्टो लीलाधिकारी	*hṛṣṭo līlādhikārī*	$2^r19^\circ33'18''$
144	ते रौद्रा नागाः	*te raudrā nāgāḥ*	$3^r02^\circ26'$	स्वामी शरीरेऽनिलः	*svāmī śarīre'nilaḥ*	$3^r02^\circ25'54''$
145	विलोमकुलम्	*vilomakulam*	$3^r15^\circ34'$	स्थाणुर्गङ्गाशयालुः	*sthāṇurgaṅgāśayāluḥ*	$3^r15^\circ33'57''$

[22] The word *ahaṃyu* appearing in this *vākya* (in the second case singular form), is a special *taddhita* usage and means 'one who possesses ego' (*ahaṅkāravān*). The formation of this word is explained by the *sūtra* अहंशुभयोर्युस् (*ahaṃśubhayoryus*) [5.2.140] in the *Aṣṭādhyāyī*.

146	स मन्दो रागी	*sa mando rāgī*	$3^{r}28^{\circ}57'$	तैलार्थी मन्दरोगी	*tailārthī mandarogī*	$3^{r}28^{\circ}57'36''$
147	तैलप्रियस्त्वम्	*tailapriyastvam*	$4^{r}12^{\circ}36'$	स्थानाच्चला रिपवः	*sthānāccalā ripavaḥ*	$4^{r}12^{\circ}36'07''$
148	साम्प्रतं रविः	*sāmprataṃ raviḥ*	$4^{r}26^{\circ}17'$	विनोदरुचिः प्रभुः	*vinodaruciḥ prabhuḥ*	$4^{r}26^{\circ}28'04''$
149	कुलानां कर्म	*kulānāṃ karma*	$5^{r}10^{\circ}31'$	साध्यो योगो नियमात्	*sādhyo yogo niyamāt*	$5^{r}10^{\circ}31'17''$
150	श्रुत्वा स्वराणि	*śrutvā svarāṇi*	$5^{r}24^{\circ}42'$	पत्नी गर्भाभरणा	*patnī garbhābharaṇā*	$5^{r}24^{\circ}43'01''$
151	धर्मो दानं तु	*dharmo dānaṃ tu*	$6^{r}08^{\circ}59'$	स्तेनो न निर्धनैषी	*steno na nirdhanaiṣī*	$6^{r}09^{\circ}00'06''$
152	दूष्यं गोत्रं ते	*dūṣyaṃ gotraṃ te*	$6^{r}23^{\circ}18'$	सेनाधिकाङ्गरक्षा	*senādhikāṅgarakṣā*	$6^{r}23^{\circ}19'07''$
153	तुलार्थिनोऽर्थी	*tulārthino'rthī*	$7^{r}07^{\circ}36'$	श्रीर्गीतालसा नासीत्	*śrīrgītālasā nāsīt*	$7^{r}07^{\circ}36'32''$
154	जित्वास्य रथः	*jitvāsya rathaḥ*	$7^{r}21^{\circ}48'$	धर्माज्जीवेत् परासुः	*dharmājjīvet parāsuḥ*	$7^{r}21^{\circ}48'59''$
155	श्रमणो निन्दा	*śramaṇo nindā*	$8^{r}05^{\circ}52'$	धनी गुणी मनुजः	*dhanī guṇī manujaḥ*	$8^{r}05^{\circ}53'09''$
156	षड्विधान्याहुः	*ṣaḍvidhānyāhuḥ*	$8^{r}19^{\circ}46'$	परिषत्स्वधिकेहा	*pariṣatsvadhikehā*	$8^{r}19^{\circ}46'21''$
157	तत्र गोर्निधिः	*tatra gornidhiḥ*	$9^{r}03^{\circ}26'$	दिव्यः क्षीराम्बुनिधिः	*divyaḥ kṣīrāmbunidhiḥ*	$9^{r}03^{\circ}26'18''$
158	केशास्ते काळाः	*keśāste kāḷāḥ*	$9^{r}16^{\circ}51'$	धीरः कर्णस्तु योद्धा	*dhīraḥ karṇastu yoddhā*	$9^{r}16^{\circ}51'29''$
159	यानानि नो नयेत्	*yānāni no nayet*	$10^{r}00^{\circ}01'$	तनयो ज्ञानिनां नम्यः	*tanayo jñānināṃ namyaḥ*	$10^{r}00^{\circ}01'06''$
160	शिशिरे पानीयम्	*śiśire pānīyam*	$10^{r}12^{\circ}55'$	गोकर्णमित्रं पिनाकी	*gokarṇamitraṃ pinākī*	$10^{r}12^{\circ}55'13''$
161	भोगमात्रं नित्यम्	*bhogamātraṃ nityam*	$10^{r}25^{\circ}34'$	प्रभवो गुणरत्नाढ्याः	*prabhavo guṇaratnāḍhyāḥ*	$10^{r}25^{\circ}34'42''$
162	यूनां दानं पथ्यम्	*yūnāṃ dānaṃ pathyam*	$11^{r}08^{\circ}01'$	प्रकृत्याऽऽनन्द उत्पाद्यः	*prakṛtyā"nanda utpādyaḥ*	$11^{r}08^{\circ}01'12''$
163	सत्येन श्रेयः स्यात्	*satyena śreyaḥ syāt*	$11^{r}20^{\circ}17'$	अनसूया निरपाया	*anasūyā nirapāyā*	$11^{r}20^{\circ}17'00''$
164	मुखे श्रीः	*mukhe śrīḥ*	$2^{\circ}25'$	जिष्णुर्वरिष्ठः	*jiṣṇurvariṣṭhaḥ*	$2^{\circ}24'58''$
165	धारावृष्टिः	*dhārāvṛṣṭiḥ*	$14^{\circ}29'$	दिश्यादिन्द्रो भाग्यम्	*diśyādindro bhāgyam*	$14^{\circ}28'18''$
166	पलितं राज्ञः	*palitaṃ rājñaḥ*	$0^{r}26^{\circ}31'$	नगो नगोऽचरत्[23]	*nago nago'carat*	$26^{\circ}30'30''$
167	तैलजा नार्यः	*tailajā nāryaḥ*	$1^{r}08^{\circ}36'$	गानशीलो जनोऽयम्	*gānaśīlo jano'yam*	$1^{r}08^{\circ}35'03''$
168	ताभिर्नराः स्युः	*tābhirnarāḥ syuḥ*	$1^{r}20^{\circ}46'$	पुराणो भानुरीड्यः	*purāṇo bhānurīḍyaḥ*	$1^{r}20^{\circ}45'21''$
169	मीनलग्नेऽत्र	*mīnalagne'tra*	$2^{r}03^{\circ}05'$	बालोऽभून्नीलनेत्रः	*bālo'bhūnnīlanetraḥ*	$2^{r}03^{\circ}04'33''$
170	तालुमध्ये श्रीः	*tālumadhye śrīḥ*	$2^{r}15^{\circ}36'$	जटी शूली शङ्करः	*jaṭī śūlī śaṅkaraḥ*	$2^{r}15^{\circ}35'18''$
171	नोग्रा दारा राज्ञः	*nogrā dārā rājñaḥ*	$2^{r}28^{\circ}20'$	प्रवृद्धोऽयं जाठरः	*pravṛddho'yaṃ jāṭharaḥ*	$2^{r}28^{\circ}19'42''$

[23] The version given above is from the edition of *Veṇvāroha* ({VR 1956}, p. 25), whereas the one found in *Sphuṭacandrāpti* ({SC 1973}, p. 53) नगो न गोचरत् (*nago na gocarat*), is erroneous as it doesn't convey any proper meaning, besides being grammatically incorrect.

172	धन्यः स्यात् कालः	*dhanyaḥ syāt kālaḥ*	$3^{r}11^{\circ}19'$	सन्निधौ स्यात् कपाली	*sannidhau syāt kapālī*	$3^{r}11^{\circ}19'07''$
173	वर्गे त्वं खलैः	*varge tvaṃ khalaiḥ*	$3^{r}24^{\circ}34'$	दीनेष्वीडा विफला	*dīneṣvīḍā viphalā*	$3^{r}24^{\circ}34'08''$
174	श्वानो दीनो वा	*śvāno dīno vā*	$4^{r}08^{\circ}04'$	खाण्डवघ्नो जनो वै[24]	*khāṇḍavaghno jano vai*	$4^{r}08^{\circ}04'32''$
175	धवाः कारवः	*dhavāḥ kāravaḥ*	$4^{r}21^{\circ}49'$	सौम्यधीः स्वयंप्रभुः	*saumyadhīḥ svayaṃprabhuḥ*	$4^{r}21^{\circ}49'17''$
176	क्षोभः शनैः शनैः	*kṣobhaḥ śanaiḥ śanaiḥ*	$5^{r}05^{\circ}46'$	बालास्तु वाग्मिनोऽमी	*bālāstu vāgmino'mī*	$5^{r}05^{\circ}46'33''$
177	गोशुद्धिकामः	*gośuddhikāmaḥ*	$5^{r}19^{\circ}53'$	शशलक्ष्माधिकांशुः	*śaśalakṣmādhikāṃśuḥ*	$5^{r}19^{\circ}53'55''$
178	दीनो वो ज्ञातिः	*dīno vo jñātiḥ*	$6^{r}04^{\circ}08'$	सुखदं नवनीतम्	*sukhadaṃ navanītam*	$6^{r}04^{\circ}08'27''$
179	तत्र दीयते	*tatra dīyate*	$6^{r}18^{\circ}26'$	धन्वन्तरिर्जयति	*dhanvantarirjayati*	$6^{r}18^{\circ}26'49''$
180	शोभा राज्ञः सेना	*śobhā rājñaḥ senā*	$7^{r}02^{\circ}45'$	वागीशो वारनाथः	*vāgīśo vāranāthaḥ*	$7^{r}02^{\circ}45'34''$
181	आज्ञा साध्या सा	*ājñā sādhyā sā*	$7^{r}17^{\circ}00'$	पयस्यनिच्छा कथम्[25]	*payasyanicchā katham*	$7^{r}17^{\circ}01'11''$
182	नटस्यानन्दः	*naṭasyānandaḥ*	$8^{r}01^{\circ}10'$	धान्ये न कस्यानन्दः	*dhānye na kasyānandaḥ*	$8^{r}01^{\circ}10'19''$
183	धनेशोऽयं जनः	*dhaneśo'yaṃ janaḥ*	$8^{r}15^{\circ}09'$	शमधनाः शापदाः	*śamadhanāḥ śāpadāḥ*	$8^{r}15^{\circ}09'55''$
184	स मन्दो ह्रदः	*sa mando hradaḥ*	$8^{r}28^{\circ}57'$	चारार्थी महाराजः	*cārārthī mahārājaḥ*	$8^{r}28^{\circ}57'26''$
185	नागरो युद्धः	*nāgaro yuddhaḥ*	$9^{r}12^{\circ}30'$	सोमोऽनङ्गारिर्व्याधः[26]	*somo'naṅgārirvyādhaḥ*	$9^{r}12^{\circ}30'57''$
186	धीवशः क्रोधः	*dhīvaśaḥ krodhaḥ*	$9^{r}25^{\circ}49'$	मर्त्योऽधन्वा शरधीः[27]	*martyo'dhanvā śaradhīḥ*	$9^{r}25^{\circ}49'15''$
187	श्रमो दीनो नित्यम्	*śramo dīno nityam*	$10^{r}08^{\circ}52'$	शशी कुमुदिनीनम्यः	*śaśī kumudinīnamyaḥ*	$10^{r}08^{\circ}51'55''$
188	धूली स्याद्राज्ञोऽयम्	*dhūlī syādrājño'yam*	$10^{r}21^{\circ}39'$	रुद्रो धीगम्यः प्राज्ञैः स्यात्	*rudro dhīgamyaḥ prājñaiḥ syāt*	$10^{r}21^{\circ}39'22''$
189	बाह्यवने योग्यम्	*bāhyavane yogyam*	$11^{r}04^{\circ}13'$	ईशप्रियो विनायकः	*īśapriyo vināyakaḥ*	$11^{r}04^{\circ}12'50''$

[24] The version found in the editions of *Veṇvāroha* ({VR 1956}, p. 25) and *Sphuṭacandrāpti* ({SC 1973}, p. 53) is "*bālye'vajño jano vai*" which encodes the value $4^{r}08^{\circ}04'13''$.

[25] The version found in the editions of *Veṇvāroha* ({VR 1956}, p. 25) and *Sphuṭacandrāpti* ({SC 1973}, p. 54) is "*phalajñānecchā katham*" which encodes the value $7^{r}17^{\circ}00'32$.

[26] The versions found in both the editions *Veṇvāroha* ({VR 1956}, p. 26) and *Sphuṭacandrāpti* ({SC 1973}, p. 54) carry a misprint in this *vākya*. The former omits the *avagraha*, whereas the latter omits the *repha*.

[27] We have supplied an *avagraha* in this *vākya*, which is found missing in both the editions *Veṇvāroha* ({VR 1956}, p. 26) and *Sphuṭacandrāpti* ({SC 1973}, p. 54). This *avagraha* is crucial, in order to avoid grammatical error. The form of the word *dhanvā* can appear only in compounded form such as शार्ङ्गधन्वा (*śārṅgadhanvā*), and not in its isolated form. Hence, to get the word right, as well as the meaning of the sentence – 'without a bow, merely thinking of arrow, [makes the warrior] bound to die' – correctly, *avagraha* needs to be there.

190	विगतपापोऽयम्	*vigatapāpo'yam*	$11^{r}16^{\circ}34'$	विद्योज्ज्वला तार्किकस्य	*vidyojjvalā tārkikasya*	$11^{r}16^{\circ}34'14''$
191	तावदत्र कार्यः	*tāvadatra kāryaḥ*	$11^{r}28^{\circ}46'$	धनाप्ताभूदद्रिकन्या	*dhanāptābhūdadrikanyā*	$11^{r}28^{\circ}46'09''$
192	ग्रामो नष्टः	*grāmo naṣṭaḥ*	$10^{\circ}52'$	दुर्गेयमनिन्द्या	*durgeyamanindyā*	$10^{\circ}51'38''$
193	शशी रात्रौ	*śaśī rātrau*	$22^{\circ}55'$	प्राज्ञौ विष्णुरुद्रौ	*prājñau viṣṇurudrau*	$22^{\circ}54'02''$
194	दुःशुभा नष्टाः	*duḥśubhā naṣṭāḥ*	$1^{r}04^{\circ}58'$	पद्माक्षी शोभनास्या	*padmākṣī śobhanāsyā*	$1^{r}04^{\circ}56'51''$
195	भानुः सद्यः स्यात्	*bhānuḥ sadyaḥ syāt*	$1^{r}17^{\circ}04'$	भर्गो गोनाथः पूज्यः	*bhargo gonāthaḥ pūjyaḥ*	$1^{r}17^{\circ}03'34''$
196	दयार्धं श्रेयः	*dayārdhaṃ śreyaḥ*	$1^{r}29^{\circ}18'$	हरिः सेव्यो धरया	*hariḥ sevyo dharayā*	$1^{r}29^{\circ}17'28''$
197	प्रभायाः पुत्रः	*prabhāyāḥ putraḥ*	$2^{r}11^{\circ}42'$	पौलस्त्यो भयङ्करः	*paulastyo bhayaṅkaraḥ*	$2^{r}11^{\circ}41'31''$
198	हर्यश्वः श्रेष्ठः	*haryaśvaḥ śreṣṭhaḥ*	$2^{r}24^{\circ}18'$	स्थाने जयो वरिष्ठः	*sthāne jayo variṣṭhaḥ*	$2^{r}24^{\circ}18'07''$
199	धनुः सेनाङ्गम्	*dhanuḥ senāṅgam*	$3^{r}07^{\circ}09'$	मानधनः सानुगः	*mānadhanaḥ sānugaḥ*	$3^{r}07^{\circ}09'05''$
200	शाक्यज्ञोऽरागी[28]	*śākyajño'rāgī*	$3^{r}20^{\circ}15'$	तरुणः को न रागी	*taruṇaḥ ko na rāgī*	$3^{r}20^{\circ}15'26''$
201	सलिलं नवम्	*salilaṃ navam*	$4^{r}03^{\circ}37'$	गौरी सलीला न वा	*gaurī salīlā na vā*	$4^{r}03^{\circ}37'23''$
202	वैद्यः स कविः	*vaidyaḥ sa kaviḥ*	$4^{r}17^{\circ}14'$	करीभव्योऽसौ युवा	*karībhavyo'sau yuvā*	$4^{r}17^{\circ}14'21''$
203	मेनका नाम	*menakā nāma*	$5^{r}01^{\circ}05'$	तमस्विनीयं निशा	*tamasvinīyaṃ niśā*	$5^{r}01^{\circ}04'56''$
204	सेना मध्यमा	*senā madhyamā*	$5^{r}15^{\circ}07'$	रत्नासनमुपेमः	*ratnāsanamupemaḥ*	$5^{r}15^{\circ}07'02''$
205	संयुद्धक्रमः	*saṃyuddhakramaḥ*	$5^{r}29^{\circ}17'$	हेम सम्पद्धारिणाम्	*hema sampaddhāriṇām*	$5^{r}29^{\circ}17'58''$
206	स्वर्गलोकोऽस्ति	*svargaloko'sti*	$6^{r}13^{\circ}34'$	जडो विडम्बयति	*jaḍo viḍambayati*	$6^{r}13^{\circ}34'38''$
207	गुणार्थी रतिः	*guṇārthī ratiḥ*	$6^{r}27^{\circ}53'$	सलिलाशा सम्प्रति	*salilāśā samprati*	$6^{r}27^{\circ}53'37''$
208	काव्यप्रियोऽसौ	*kāvyapriyo'sau*	$7^{r}12^{\circ}11'$	गुरुकार्ये प्रयासः	*gurukārye prayāsaḥ*	$7^{r}12^{\circ}11'23''$
209	भद्रतरोऽर्थी	*bhadrataro'rthī*	$7^{r}26^{\circ}24'$	पाण्डवाः प्राप्तरथाः	*pāṇḍavāḥ prāptarathāḥ*	$7^{r}26^{\circ}24'31''$
210	धू राज्ञः पादे	*dhū rājñaḥ pāde*	$8^{r}10^{\circ}29'$	शिवधीरनापदे	*śivadhīranāpade*	$8^{r}10^{\circ}29'45''$
211	गुरुर्वरदः	*gururvaradaḥ*	$8^{r}24^{\circ}23'$	चापी वीरो द्विरदे	*cāpī vīro dvirade*	$8^{r}24^{\circ}24'16''$
212	मानदो निधिः	*mānado nidhiḥ*	$9^{r}08^{\circ}05'$	शश्वन्मौनीजनोऽन्धः	*śaśvanmaunījano'ndhaḥ*	$9^{r}08^{\circ}05'45''$
213	रङ्गस्य श्रद्धा	*raṅgasya śraddhā*	$9^{r}21^{\circ}32'$	मूलं फलाढ्यश्राद्धः	*mūlaṃ phalāḍhyaśrāddhaḥ*	$9^{r}21^{\circ}32'35''$
214	स्वभावो ज्ञानस्य	*svabhāvo jñānasya*	$10^{r}04^{\circ}44'$	तमालाभो घनो नित्यम्	*tamālābho ghano nityam*	$10^{r}04^{\circ}43'56''$
215	अवस्थेयं नार्याः	*avastheyaṃ nāryāḥ*	$10^{r}17^{\circ}40'$	प्रभोर्धिगच्छकनकम्	*prabhordhigacchakanakam*	$10^{r}17^{\circ}39'42''$

[28] Here again an *avagraha* has been introduced, as in the *vākya* 186, in order to get the meaning – 'the one who has completely understood [the teachings of] Buddha would be dispassionate'. Otherwise, the *vākya* will have an offensive import.

216	पुत्रो ज्ञानाढ्योऽयम्	*putro jñānāḍhyo'yam*	11^{r}00°21′	वैश्वानरं नानुयायात्	*vaiśvānaraṃ nānuyāyāt*	11^{r}00°20′44″
217	धवः श्रेयः पथ्यम्	*dhavaḥ śreyaḥ pathyam*	11^{r}12°49′	रागादिवैराग्यं पथ्यम्	*rāgādivairāgyaṃ pathyam*	11^{r}12°48′32″
218	तेन शरैः पटुः	*tena śaraiḥ paṭuḥ*	11^{r}25°06′	श्रीरामनाम रम्याग्र्यम्	*śrīrāmanāma ramyāgryam*	11^{r}25°05′22″
219	वैद्योऽसौ	*vaidyo'sau*	7°14′	प्रज्ञावान् पार्थः	*prajñāvān pārthaḥ*	7°14′02″
220	हयो धन्यः	*hayo dhanyaḥ*	19°18′	बीभत्सुर्योधोऽयम्	*bībhatsuryodho'yam*	19°17′43″
221	अप्रियो नये	*apriyo naye*	1^{r}01°20′	ग्रामाधिपः कनकी	*grāmādhipaḥ kanakī*	1^{r}01°19′52″
222	शास्त्रबाह्योऽयम्	*śāstrabāhyo'yam*	1^{r}13°25′	धम्मिल्ले फुल्लपुष्पम्	*dhammille phullapuṣpam*	1^{r}13°23′59″
223	भोगमात्रस्य	*bhogamātrasya*	1^{r}25°34′	कालो बली शरण्यः	*kālo balī śaraṇyaḥ*	1^{r}25°33′31″
224	ग्रामार्थी नरः	*grāmārthī naraḥ*	2^{r}07°52′	तुङ्गयशसो नराः	*tuṅgayaśaso narāḥ*	2^{r}07°51′36″
225	यात्रान्नं श्रेष्ठम्	*yātrānnaṃ śreṣṭham*	2^{r}20°21′	धर्मज्ञाः प्राङ्नरेन्द्राः	*dharmajñāḥ prāṅnarendrāḥ*	2^{r}20°20′59″
226	भिन्नाङ्गो नागः	*bhinnāṅgo nāgaḥ*	3^{r}03°04′	दीर्घाङ्गो नीलनागः	*dīrghāṅgo nīlanāgaḥ*	3^{r}03°03′48″
227	प्रज्ञातो योगी	*prajñāto yogī*	3^{r}16°02′	फलाढ्यो ऋतुकालः	*phalāḍhyo ṛtukālaḥ*	3^{r}16°01′32″
228	मुख्यो धीरो लीनः	*mukhyo dhīro līnaḥ*	3^{r}29°15′	केशवो योद्धा रङ्गे	*keśavo yoddhā raṅge*	3^{r}29°14′51″
229	गावः प्रिया वः	*gāvaḥ priyā vaḥ*	4^{r}12°43′	सौबलो वरयुवा	*saubalo varayuvā*	4^{r}12°43′37″
230	सुरतन्त्रिभिः	*suratantribhiḥ*	4^{r}26°27′	पद्मेषु रन्ता रविः	*padmeṣu rantā raviḥ*	4^{r}26°26′51″
231	त्रिराज्ञाङ्कुशाः	*trirājñāṅkuśāḥ*	5^{r}10°22′	भ्रमरश्रीर्निकामम्	*bhramaraśrīrnikāmam*	5^{r}10°22′52″
232	धाराभिः श्रमः	*dhārābhiḥ śramaḥ*	5^{r}24°29′	तपोधराः स्वैरिणः	*tapodharāḥ svairiṇaḥ*	5^{r}24°29′16″
233	त्रिभिर्हानिस्ते	*tribhirhāniste*	6^{r}08°42′	आद्यो गोविन्द एषः	*ādyo govinda eṣaḥ*	6^{r}08°43′10″
234	अनङ्गाश्रिता	*anaṅgāśritā*	6^{r}23°00′	सत्काव्यज्ञोऽम्बरीषः[29]	*satkāvyajño'mbarīṣaḥ*	6^{r}23°01′17″
235	धन्यः स नाथः	*dhanyaḥ sa nāthaḥ*	7^{r}07°19′	उद्यानं स्त्रीसनाथम्	*udyānaṃ strīsanātham*	7^{r}07°20′10″
236	तिलस्य रसः	*tilasya rasaḥ*	7^{r}21°36′	दीपतैलं पात्रस्थम्	*dīpatailaṃ pātrastham*	7^{r}21°36′18″
237	तव मानदः	*tava mānadaḥ*	8^{r}05°46′	सूर्योऽस्तु वो मानदः	*sūryo'stu vo mānadaḥ*	8^{r}05°46′17″
238	षड्विधं पदम्	*ṣaḍvidhaṃ padam*	8^{r}19°46′	भानुः सर्वाधिको हि	*bhānuḥ sarvādhiko hi*	8^{r}19°47′04″
239	मङ्गलं नीळम्	*maṅgalaṃ nīḷam*	9^{r}03°35′	पीनोत्तुङ्गाङ्गो नळः	*pīnottuṅgāṅgo naḷaḥ*	9^{r}03°36′01″
240	योग्यः संयुद्धे	*yogyaḥ saṃyuddhe*	9^{r}17°11′	हीनपापः सुयोद्धा	*hīnapāpaḥ suyoddhā*	9^{r}17°11′08″
241	योगो ज्ञानिनः स्यात्	*yogo jñāninaḥ syāt*	10^{r}00°31′	सूनुः कुलीनोऽनुनेयः	*sūnuḥ kulīno'nuneyaḥ*	10^{r}00°31′07″

[29] The version found in the editions of *Veṇvāroha* ({VR 1956}, p.26) and *Sphuṭacandrāpti* ({SC 1973}, p. 56) is "*ṣaṭkāvyajño'mbarīṣaḥ*" which encodes the value 6^{r}23°01′16″.

242	शैलालयो नम्यः	*śailālayo namyaḥ*	$10^{r}13^{\circ}35'$	धरणो बलीयानाढ्यः[30]	*dharaṇo balīyānāḍhyaḥ*	$10^{r}13^{\circ}35'29''$
243	मन्त्रितं प्राज्ञाय	*mantritaṃ prājñāya*	$10^{r}26^{\circ}25'$	बालेऽभिरुचिरनल्पा	*bāle'bhiruciranalpā*	$10^{r}26^{\circ}24'33''$
244	अनिधानं कपेः	*anidhānaṃ kapeḥ*	$11^{r}09^{\circ}00'$	स्थिरधीर्महीनायकः	*sthiradhīrmahīnāyakaḥ*	$11^{r}08^{\circ}59'27''$
245	श्रोत्रियः प्रियस्य	*śrotriyaḥ priyasya*	$11^{r}21^{\circ}22'$	स्वर्नारी रम्यरूपाढ्या	*svarnārī ramyarūpāḍhyā*	$11^{r}21^{\circ}22'04''$
246	मङ्गलम्	*maṅgalam*	$3^{\circ}35'$	भीमो वललः	*bhīmo valalaḥ*	$3^{\circ}34'54''$
247	कवेः शक्यम्	*kaveḥ śakyam*	$15^{\circ}41'$	शशी नभोमध्ये	*śaśī nabhomadhye*	$15^{\circ}40'55''$
248	भवेत् सुखम्	*bhavet sukham*	$27^{\circ}44'$	धीरगीर्भासुरा	*dhīragīrbhāsurā*	$27^{\circ}43'29''$

Table E.1: The *candravākyas* of Vararuci and Mādhava.

[30] The version found in the editions of *Veṇvāroha* ({VR 1956}, p.26) and *Sphuṭacandrāpti* ({SC 1973}, p. 56) is "*taruṇobalīyānāḍhyaḥ*" which encodes the value $10^{r}13^{\circ}35'26''$.

E.5 Table of computed *candravākyas*

The following Table E.2 lists the 248 values of the longitudes as encoded by the *candravākyas* of Mādhava as amended (in the case of *vākya* numbers 25, 174, 181, 234 and 242) and listed in Table E.1 (column 2), and compares them with the computed values of the longitudes (columns 3–4) obtained by using the exact value as well as the approximate value $\frac{9}{248}$ for the rate of motion of the Moon's anomaly. The mean daily rate of motion of the Moon is taken to be $\frac{4909031760}{134122987500}$.

We can clearly see that the computed values with the rate of motion of the anomaly being approximated by $\frac{9}{248}$, differ significantly from those obtained by using the exact value for the rate of motion of anomaly as given in *Karaṇa-paddhati*. However, it can be checked that if we make use of the approximate rate of motion $\frac{6845}{188611}$ used in the *Sphuṭacandrāpti* of Mādhava, the resulting value of longitude are the same upto the thirds with those computed using the exact rate of motion. More importantly the computed true longitudes coincide with the values encoded by *Mādhava-vākyas* (as emended by us in case of *vākya* numbers 25, 174, 181, 234 and 242) except for a difference of $1''$ in a few places.

Vākya No.	Values encoded in *Mādhava-vākyas*				Computed true longitudes of the Moon on each day for different rates of motion of Moon's anomaly									
					$\frac{G}{H} = \frac{45742113404287O9}{1260404054365475OO}$					$\frac{G}{H} \approx \frac{9}{248}$				
	r	$\circ$	$'$	$''$	r	$\circ$	$'$	$''$	$'''$	r	$\circ$	$'$	$''$	$'''$
1	0	12	2	35	0	12	2	34	41	0	12	2	34	50
2	0	24	8	39	0	24	8	39	10	0	24	8	39	26
3	1	6	21	33	1	6	21	33	1	1	6	21	33	21
4	1	18	44	16	1	18	44	15	36	1	18	44	15	58
5	2	1	19	17	2	1	19	16	46	2	1	19	17	5
6	2	14	8	29	2	14	8	28	32	2	14	8	28	43
7	2	27	12	59	2	27	12	58	34	2	27	12	58	33
8	3	10	33	6	3	10	33	6	11	3	10	33	5	53
9	3	24	8	21	3	24	8	21	6	3	24	8	20	29
10	4	7	57	25	4	7	57	25	17	4	7	57	24	19
11	4	21	58	18	4	21	58	17	41	4	21	58	16	22
12	5	6	8	21	5	6	8	21	8	5	6	8	19	30
13	5	20	24	31	5	20	24	30	53	5	20	24	29	0
14	6	4	43	24	6	4	43	24	14	6	4	43	22	10
15	6	19	1	30	6	19	1	30	29	6	19	1	28	21
16	7	3	15	21	7	3	15	21	16	7	3	15	19	12
17	7	17	21	41	7	17	21	40	39	7	17	21	38	47
18	8	1	17	35	8	1	17	35	2	8	1	17	33	31
19	8	15	0	42	8	15	0	42	22	8	15	0	41	19

20	8	28	29	20	8	28	29	20	0	8	28	29	19	33
21	9	11	42	31	9	11	42	30	48	9	11	42	31	2
22	9	24	40	6	9	24	40	6	28	9	24	40	7	27
23	10	7	22	48	10	7	22	48	14	10	7	22	49	58
24	10	19	52	4	10	19	52	4	9	10	19	52	6	36
25	11	2	10	4	11	2	10	4	1	11	2	10	7	7
26	11	14	19	32	11	14	19	31	56	11	14	19	35	33
27	11	26	23	37	11	26	23	37	29	11	26	23	41	27
28	0	8	25	46	0	8	25	46	2	0	8	25	50	9
29	0	20	29	29	0	20	29	28	38	0	20	29	32	42
30	1	2	38	12	1	2	38	11	47	1	2	38	15	33
31	1	14	55	7	1	14	55	7	16	1	14	55	10	31
32	1	27	23	2	1	27	23	2	25	1	27	23	4	56
33	2	10	4	11	2	10	4	11	9	2	10	4	12	44
34	2	23	0	6	2	23	0	6	24	2	23	0	6	55
35	3	6	11	35	3	6	11	34	39	3	6	11	33	59
36	3	19	38	33	3	19	38	33	3	3	19	38	31	12
37	4	3	20	10	4	3	20	9	41	4	3	20	6	40
38	4	17	14	47	4	17	14	46	37	4	17	14	42	33
39	5	1	20	6	5	1	20	5	45	5	1	20	0	46
40	5	15	33	16	5	15	33	16	28	5	15	33	10	49
41	5	29	51	5	5	29	51	4	51	5	29	50	58	48
42	6	14	10	3	6	14	10	3	20	6	14	9	57	12
43	6	28	26	41	6	28	26	41	2	6	28	26	35	7
44	7	12	37	34	7	12	37	33	50	7	12	37	28	29
45	7	26	39	35	7	26	39	34	35	7	26	39	30	7
46	8	10	30	3	8	10	30	2	40	8	10	29	59	21
47	8	24	6	53	8	24	6	52	41	8	24	6	50	46
48	9	7	28	42	9	7	28	41	40	9	7	28	41	18
49	9	20	34	54	9	20	34	53	59	9	20	34	55	16
50	10	3	25	44	10	3	25	43	30	10	3	25	46	26
51	10	16	2	13	10	16	2	12	48	10	16	2	17	17
52	10	28	26	9	10	28	26	9	20	10	28	26	15	11
53	11	10	39	59	11	10	39	59	9	11	10	40	6	7
54	11	22	46	39	11	22	46	38	53	11	22	46	46	37
55	0	4	49	26	0	4	49	26	23	0	4	49	34	31
56	0	16	51	51	0	16	51	50	48	0	16	51	58	54
57	0	28	57	22	0	28	57	22	24	0	28	57	30	4
58	1	11	9	22	1	11	9	22	25	1	11	9	29	12
59	1	23	30	53	1	23	30	52	52	1	23	30	58	23
60	2	6	4	27	2	6	4	27	15	2	6	4	31	10
61	2	18	52	2	2	18	52	2	3	2	18	52	4	5
62	3	1	54	50	3	1	54	50	3	3	1	54	50	3
63	3	15	13	16	3	15	13	15	56	3	15	13	13	49
64	3	28	46	55	3	28	46	54	47	3	28	46	50	36
65	4	12	34	34	4	12	34	33	38	4	12	34	27	33
66	4	26	34	16	4	26	34	15	49	4	26	34	8	5
67	5	10	43	28	5	10	43	27	41	5	10	43	18	40
68	5	24	59	7	5	24	59	7	5	5	24	58	57	14

69	6	9	17	53	6	9	17	52	45	6	9	17	42	33
70	6	23	36	14	6	23	36	14	25	6	23	36	4	23
71	7	7	50	43	7	7	50	42	54	7	7	50	33	36
72	7	21	58	0	7	21	58	0	28	7	21	57	52	23
73	8	5	55	11	8	5	55	10	37	8	5	55	4	13
74	8	19	39	47	8	19	39	47	24	8	19	39	43	5
75	9	3	10	4	9	3	10	3	35	9	3	10	1	38
76	9	16	24	57	9	16	24	56	49	9	16	24	57	24
77	9	29	24	13	9	29	24	13	25	9	29	24	16	34
78	10	12	8	29	10	12	8	29	16	10	12	8	34	53
79	10	24	39	8	10	24	39	7	31	10	24	39	15	22
80	11	6	58	14	11	6	58	13	44	11	6	58	23	27
81	11	19	8	29	11	19	8	28	40	11	19	8	39	48
82	0	1	12	60	0	1	12	59	37	0	1	13	11	36
83	0	13	15	11	0	13	15	10	41	0	13	15	22	55
84	0	25	18	33	0	25	18	32	49	0	25	18	44	40
85	1	7	26	34	1	7	26	33	30	1	7	26	44	20
86	1	19	42	27	1	19	42	26	36	1	19	42	35	50
87	2	2	9	3	2	2	9	2	35	2	2	9	9	41
88	2	14	48	39	2	14	48	39	24	2	14	48	43	56
89	2	27	42	55	2	27	42	54	46	2	27	42	56	26
90	3	10	52	40	3	10	52	40	26	3	10	52	39	5
91	3	24	17	59	3	24	17	59	0	3	24	17	54	38
92	4	7	58	4	4	7	58	3	46	4	7	57	56	33
93	4	21	51	22	4	21	51	21	36	4	21	51	11	51
94	5	5	55	38	5	5	55	38	25	5	5	55	26	36
95	5	20	8	7	5	20	8	6	46	5	20	7	53	28
96	6	4	25	35	6	4	25	34	46	6	4	25	20	39
97	6	18	44	36	6	18	44	35	53	6	18	44	21	39
98	7	3	1	39	7	3	1	39	5	7	3	1	25	29
99	7	17	13	19	7	17	13	19	0	7	17	13	6	46
100	8	1	16	26	8	1	16	26	11	8	1	16	15	59
101	8	15	8	17	8	15	8	16	39	8	15	8	9	3
102	8	28	46	41	8	28	46	40	48	8	28	46	36	16
103	9	12	10	11	9	12	10	10	43	9	12	10	9	34
104	9	25	18	5	9	25	18	5	28	9	25	18	7	49
105	10	8	10	34	10	8	10	33	31	10	8	10	39	18
106	10	20	48	32	10	20	48	32	14	10	20	48	41	13
107	11	3	13	44	11	3	13	44	26	11	3	13	56	12
108	11	15	28	32	11	15	28	32	22	11	15	28	46	20
109	11	27	35	50	11	27	35	49	43	11	27	36	5	13
110	0	9	38	53	0	9	38	52	30	0	9	39	8	44
111	0	21	41	9	0	21	41	9	8	0	21	41	25	18
112	1	3	46	10	1	3	46	10	18	1	3	46	25	31
113	1	15	57	19	1	15	57	18	38	1	15	57	32	7
114	1	28	17	39	1	28	17	38	47	1	28	17	49	47
115	2	10	49	48	2	10	49	47	46	2	10	49	55	39
116	2	23	35	46	2	23	35	46	29	2	23	35	50	47
117	3	6	36	53	3	6	36	52	44	3	6	36	53	10

118	3	19	53	37	3	19	53	36	35	3	19	53	33	3
119	4	3	25	39	4	3	25	38	31	4	3	25	31	9
120	4	17	11	51	4	17	11	50	39	4	17	11	39	45
121	5	1	10	21	5	1	10	20	45	5	1	10	6	51
122	5	15	18	39	5	15	18	38	52	5	15	18	22	39
123	5	29	33	46	5	29	33	45	30	5	29	33	27	47
124	6	13	52	21	6	13	52	21	0	6	13	52	2	39
125	6	28	10	56	6	28	10	55	33	6	28	10	37	32
126	7	12	25	59	7	12	25	59	23	7	12	25	42	39
127	7	26	34	13	7	26	34	13	2	7	26	33	58	27
128	8	10	32	37	8	10	32	37	10	8	10	32	25	34
129	8	24	18	42	8	24	18	42	9	8	24	18	34	10
130	9	7	50	36	9	7	50	36	8	9	7	50	32	15
131	9	21	7	12	9	21	7	11	38	9	21	7	12	9
132	10	4	8	10	10	4	8	9	36	10	4	8	14	31
133	10	16	54	1	10	16	54	0	31	10	16	54	9	39
134	10	29	26	3	10	29	26	2	35	10	29	26	15	32
135	11	11	46	17	11	11	46	17	5	11	11	46	33	12
136	11	23	57	21	11	23	57	21	18	11	23	57	39	48
137	0	6	2	20	0	6	2	20	4	0	6	2	40	1
138	0	18	4	36	0	18	4	36	11	0	18	4	56	34
139	1	0	7	40	1	0	7	40	20	1	0	8	0	6
140	1	12	15	1	1	12	15	0	52	1	12	15	18	58
141	1	24	29	54	1	24	29	53	37	1	24	30	9	7
142	2	6	55	12	2	6	55	12	5	2	6	55	24	6
143	2	19	33	18	2	19	33	18	9	2	19	33	26	1
144	3	2	25	54	3	2	25	54	16	3	2	25	57	30
145	3	15	33	57	3	15	33	57	24	3	15	33	55	45
146	3	28	57	36	3	28	57	35	33	3	28	57	29	3
147	4	12	36	7	4	12	36	7	20	4	12	35	56	15
148	4	26	28	4	4	26	28	4	27	4	26	27	49	20
149	5	10	31	17	5	10	31	16	58	5	10	30	58	33
150	5	24	43	1	5	24	43	0	39	5	24	42	39	50
151	6	9	0	6	6	9	0	5	49	6	8	59	43	39
152	6	23	19	7	6	23	19	7	2	6	23	18	44	40
153	7	7	36	33	7	7	36	33	16	7	7	36	11	51
154	7	21	48	58	7	21	48	58	3	7	21	48	38	42
155	8	5	53	10	8	5	53	9	41	8	5	52	53	28
156	8	19	46	21	8	19	46	21	0	8	19	46	8	46
157	9	3	26	18	9	3	26	18	12	9	3	26	10	41
158	9	16	51	29	9	16	51	28	35	9	16	51	26	14
159	10	0	1	6	10	0	1	5	53	10	0	1	8	52
160	10	12	55	13	10	12	55	13	7	10	12	55	21	23
161	10	25	34	42	10	25	34	42	29	10	25	34	55	38
162	11	8	1	12	11	8	1	12	4	11	8	1	29	29
163	11	20	17	0	11	20	17	0	10	11	20	17	20	58
164	0	2	24	57	0	2	24	57	29	0	2	25	20	38
165	0	14	28	18	0	14	28	18	3	0	14	28	42	23
166	0	26	30	29	0	26	30	29	27	0	26	30	53	43

167	1	8	35	3	1	8	35	2	34	1	8	35	25	30
168	1	20	45	21	1	20	45	21	27	1	20	45	41	52
169	2	3	4	33	2	3	4	33	10	2	3	4	49	57
170	2	15	35	18	2	15	35	18	13	2	15	35	30	26
171	2	28	19	42	2	28	19	41	47	2	28	19	48	44
172	3	11	19	7	3	11	19	6	39	3	11	19	7	55
173	3	24	34	8	3	24	34	8	14	3	24	34	3	41
174	4	8	4	32	4	8	4	32	26	4	8	4	22	14
175	4	21	49	16	4	21	49	16	28	4	21	49	1	6
176	5	5	46	33	5	5	46	32	43	5	5	46	12	56
177	5	19	53	55	5	19	53	54	57	5	19	53	31	43
178	6	4	8	26	6	4	8	26	25	6	4	8	0	55
179	6	18	26	49	6	18	26	49	13	6	18	26	22	45
180	7	2	45	34	7	2	45	34	10	7	2	45	8	5
181	7	17	1	11	7	17	1	10	58	7	17	0	46	38
182	8	1	10	19	8	1	10	18	32	8	1	9	57	14
183	8	15	9	55	8	15	9	54	53	8	15	9	37	46
184	8	28	57	27	8	28	57	26	43	8	28	57	14	43
185	9	12	30	58	9	12	30	57	41	9	12	30	51	30
186	9	25	49	15	9	25	49	15	15	9	25	49	15	16
187	10	8	51	55	10	8	51	54	56	10	8	52	1	13
188	10	21	39	22	10	21	39	21	51	10	21	39	34	9
189	11	4	12	49	11	4	12	49	12	11	4	13	6	56
190	11	16	34	14	11	16	34	13	51	11	16	34	36	7
191	11	28	46	10	11	28	46	9	33	11	28	46	35	14
192	0	10	51	39	0	10	51	38	35	0	10	52	6	24
193	0	22	54	2	0	22	54	2	16	0	22	54	30	48
194	1	4	56	51	1	4	56	50	55	1	4	57	18	41
195	1	17	3	34	1	17	3	33	38	1	17	3	59	12
196	1	29	17	28	1	29	17	28	7	1	29	17	50	8
197	2	11	41	31	2	11	41	30	46	2	11	41	48	2
198	2	24	18	7	2	24	18	7	19	2	24	18	18	53
199	3	7	9	5	3	7	9	4	52	3	7	9	10	3
200	3	20	15	26	3	20	15	25	32	3	20	15	24	0
201	4	3	37	23	4	3	37	22	47	4	3	37	14	33
202	4	17	14	21	4	17	14	20	31	4	17	14	5	58
203	5	1	4	55	5	1	4	55	23	5	1	4	35	12
204	5	15	7	2	5	15	7	1	38	5	15	6	36	50
205	5	29	17	58	5	29	17	58	22	5	29	17	30	12
206	6	13	34	38	6	13	34	38	14	6	13	34	8	7
207	6	27	53	37	6	27	53	37	2	6	27	53	6	31
208	7	12	11	24	7	12	11	23	50	7	12	10	54	29
209	7	26	24	31	7	26	24	31	11	7	26	24	4	33
210	8	10	29	45	8	10	29	45	18	8	10	29	22	46
211	8	24	24	16	8	24	24	15	50	8	24	23	58	39
212	9	8	5	45	9	8	5	45	1	9	8	5	34	7
213	9	21	32	35	9	21	32	35	18	9	21	32	31	19
214	10	4	43	55	10	4	43	55	9	10	4	43	58	24
215	10	17	39	42	10	17	39	42	14	10	17	39	52	35

216	11	0	20	43	11	0	20	43	25	11	0	21	0	23
217	11	12	48	32	11	12	48	32	2	11	12	48	54	48
218	11	25	5	22	11	25	5	22	21	11	25	5	49	46
219	0	7	14	2	0	7	14	1	57	0	7	14	32	37
220	0	19	17	43	0	19	17	42	47	0	19	18	15	10
221	1	1	19	51	1	1	19	51	27	1	1	20	23	52
222	1	13	23	59	1	13	23	58	58	1	13	24	29	46
223	1	25	33	31	1	25	33	30	39	1	25	33	58	12
224	2	7	51	36	2	7	51	35	52	2	7	51	58	42
225	2	20	20	58	2	20	20	58	27	2	20	21	15	21
226	3	3	3	48	3	3	3	47	51	3	3	3	57	52
227	3	16	1	32	3	16	1	31	45	3	16	1	34	17
228	3	29	14	51	3	29	14	50	55	3	29	14	45	45
229	4	12	43	37	4	12	43	36	37	4	12	43	23	59
230	4	26	26	51	4	26	26	51	18	4	26	26	31	48
231	5	10	22	52	5	10	22	51	56	5	10	22	26	32
232	5	24	29	16	5	24	29	16	9	5	24	28	46	7
233	6	8	43	10	6	8	43	10	6	6	8	42	36	57
234	6	23	1	18	6	23	1	17	43	6	23	0	43	9
235	7	7	20	11	7	7	20	10	31	7	7	19	36	19
236	7	21	36	18	7	21	36	17	53	7	21	35	45	49
237	8	5	46	17	8	5	46	17	12	8	5	45	48	56
238	8	19	47	4	8	19	47	3	57	8	19	46	41	0
239	9	3	36	1	9	3	36	1	13	9	3	35	44	50
240	9	17	11	8	9	17	11	8	19	9	17	10	59	25
241	10	0	31	8	10	0	31	7	38	10	0	31	6	46
242	10	13	35	29	10	13	35	29	20	10	13	35	36	36
243	10	26	24	33	10	26	24	33	9	10	26	24	48	14
244	11	8	59	27	11	8	59	27	11	11	8	59	49	21
245	11	21	22	4	11	21	22	3	49	11	21	22	31	57
246	0	3	34	53	0	3	34	53	10	0	3	35	25	53
247	0	15	40	55	0	15	40	54	53	0	15	41	30	29
248	0	27	43	29	0	27	43	28	39	0	27	44	5	19

Table E.2: Comparison of *candravākyas* of Mādhava with the computed values.

Appendix F
Literal meanings of selected *vākyas*

In this Appendix we shall present the literal meaning of selected set of *vākyas*. This is just to highlight the fact that the *vākyas* often carry beautiful literal meanings too.

vākyas		Literal meaning
in *Devanāgarī*	in transliterated form	
नानाज्ञानप्रगल्भः	*nānājñānapragalbhaḥ*	Illustrious with various kinds of knowledge
तिलबलमसुसूक्ष्मम्	*tilabalamasusūkṣmam*	Power of sesamum [seed] is not subtle
धयेद्राजदम्भः	*dhayedrājaddambhaḥ*	Hypocrisy of the king*
भद्रोदन्तोधरेन्द्रः	*bhadrodantodharendraḥ*	King about whom the narratives are good
निरनुसृगधिसौख्यम्	*niranusṛgadhisaukhyam*	Comfort without repeated inarticulate sound
वरिष्ठोऽभिषङ्गः	*variṣṭhobhiṣaṅgaḥ*	Humiliated senior
दोर्दण्डाग्रेऽद्रिनाथः	*ddorddaṇḍāgreddrināthaḥ*	Lord of mountains, upholding the pole by [his] arm
विषमितविपिनम्	*viṣamitavipinam*	Impassable forest
चन्द्ररेखाम्बुखिन्ना	*candrarekhāmbukhinnā*	Ray of the Moon distressed by water
अनृशंसः कळार्थीसमर्त्यः	*anṛśaṃsaḥ kaḷārthīsamartyaḥ*	Along with a man who is not cruel*

*Literal meaning of the words "*dhayed*" and "*kaḷārthī*" is not clear.

Table F.1 *Vākyas* presenting the revolutions made by the planets in a *mahāyuga*. (Refer Table 1.1)

V. Pai et al., *Karaṇapaddhati of Putumana Somayājī*, Sources and Studies in the History of Mathematics and Physical Sciences, https://doi.org/10.1007/978-981-10-6814-0

vākyas		Literal meaning
in *Devanāgarī*	in transliterated form	
गोपाज्ञया दिनधामाः	*gopājñayā dinadhāmāḥ*	We have been accomodated for a day by the order of *Gopa*
चण्डिकेशो भर्गस्निग्धोऽसौ	*caṇḍikeśo bhargasnigdhosau*	*Caṇḍikeśa* has close affinity towards effulgence
गूढवाशोऽनुभूतिः	*gūḍhavāśo'nubhātiḥ*	This is an experience of deep sense of roaring
प्रभुर्धराचक्रपालः	*prabhur dharācakrapālaḥ*	Lord is the monarch of the earth
रागी तुम्बुरुर्गणेश्वरः	*rāgī tumbururgaṇeśvaraḥ*	The leader of the group who carries a special fruit is desirous
प्रज्ञासन्नो धर्मवान्	*prajñāsanno dharmavān*	One who possesses *dharma* is attracted to the intelligent
काशी साम्बसन्नचोळः	*kāśī sāmbasannacoḷaḥ*	The *Sāmbasannacoḷa* shines
प्रबलः प्राज्ञो नरः	*prabalaḥ prājño naraḥ*	A wise man is powerful
कविर्विभुर्नकुलः	*kavirvibhurnakulaḥ*	The poet *Nakula* is around everywhere

Table F.2 *Vākyas* presenting the corrected mean motion of planets. (Refer Table 1.19)

vākyas		Literal meaning
in *Devanāgarī*	in transliterated form	
श्रीर्गुणमित्रा	*śrīrguṇamitrā*	Wealth is the friend of the virtuous
भूर्विधिपक्षा	*bhūrvidhipakṣā*	The earth is on side of the creator
स्त्री रतिशूरा	*strī ratiśūrā*	Woman skilful in amorous acts
भोगवरा ते	*bhogavarā te*	She is well suited for enjoyment
भावचरोऽरिः	*bhāvacaroriḥ*	The enemy wanders over thoughts
तेन वशत्त्वं	*tena vaśattvaṃ*	Entangled by that
लोकजभीतिः	*lokajabhītiḥ*	Fear born out of the world
स्थूलहयोऽयम्	*sthūlahayo'yam*	This is a stout horse
अङ्गधिगारः	*aṅgadhigāraḥ*	O [friend] wretched is the planet
स्तम्भितनाभिः	*stambhitanābhiḥ*	One who has arrested his navel
नित्यशशीशो	*nityaśaśīśo*	Eternal lord of the Moon
यागमयोऽयम्	*yāgamayo'yam*	A [place is] filled with sacrifices

Table F.3 *Vākyas* used in computing the instance of entry of the Sun into different zodiacal signs. (Refer Table D.2)

vākyas		Literal meaning
in *Devanāgarī*	in transliterated form	
श्रेष्ठं हि रत्नम्	*śreṣṭhaṃ hi ratnam*	This indeed is a precious jewel
धान्यधरोऽयम्	*dhānyadharo'yam*	He is a possesser of grains
सुखी अनिलः	*sukhī anilaḥ*	A happy wind
धरण्यां नभः	*dharaṇyāṃ nabhaḥ*	The sky on the earth
वानरा अमी	*vānarā amī*	These are monkeys
मुनीन्द्रोऽनन्तः	*munīndro'nantaḥ*	[Lord] Ananta is foremost amongst the seers
बलाढ्यो नाथः	*balāḍhyo nāthaḥ*	Lord endowed with power
जले निनादः	*jale ninādaḥ*	The humming noise in the river
शूलधरो हि	*śūladharo hi*	He certainly possesses a trident
साम्बो हि प्रधानः	*sāmbo hi pradhānaḥ*	Parameśvara is indeed the foremost
धर्मसूक्ष्मं नित्यम्	*dharmasukhaṃ nityam*	Eternal are the secrets of *dharma*
लक्ष्मी सुरपूज्या	*lakṣmī is surapūjyā*	Lakṣmī venerable among the divine beings

Table F.4 *Vākyas* presenting the mean longitude of the Sun at the zodiacal transits of the true Sun. (Refer Table D.3)

vākyas		Literal meaning
in *Devanāgarī*	in transliterated form	
योग्यो वैद्यः	*yogyo vaidyaḥ*	A qualified doctor
तपः सत्यम्	*tapaḥ satyam*	[Speaking] truth [by itself] is austerity
धन्यः पुत्रः	*dhanyaḥ putraḥ*	A blessed son
खरो वरः	*kharo varaḥ*	A donkey is better
वीरः शूरः	*vīraḥ śūraḥ*	A skilful warrior
शरो वज्री	*śaro vajrī*	Indra's arrow
भद्रं गोत्रो	*bhadraṃ gotro*	This clan is safe
रुरुः करी	*ruruḥ karī*	The antelope and elephant
धन्यः सेव्यो मया लोके	*dhanyaḥ sevyo mayā loke*	In the world only the blessed are to be served by me
कायी दीनः	*kāyī dīnaḥ*	The one with a bulky body is pitiable
स्तनाङ्गना	*stanāṅganā*	A lady with big breasts
याज्ञी यज्ञाङ्गना शूना	*yājñī yajñāṅganā śūnā*	The wife of the *Yajamāna* and the performer of the sacrifice is swollen
स्तेनो दीनो	*steno dīno*	The thief is miserable
धुनी नतः	*dhunī naṭaḥ*	The river is a dancer
आपः पापः	*āpaḥ pāpaḥ*	The water is the culprit
पयः पथ्यं	*payaḥ pathyaṃ*	Milk is good
पूज्या धेनुर्दिनेऽर्थिनः	*pūjyā dhenurdine'rthinaḥ*	Cow is to be worshiped during the day by those desirous of becoming wealthy
तनुर्भिन्ना	*tanurbhinnā*	The body has been split
खनी ज्ञानी	*khanī jñānī*	The wise is like a mine
रत्नं भानुः	*ratnaṃ bhānuḥ*	The Sun is a pearl
सुनिर्नयः	*sunirnayaḥ*	The one who is completely unscrupulous

Table F.5 *Vākyas* used in computing the longitude of the Sun at any given instance. (Refer Table D.8)

Appendix G
The *Maṇḍalas* and *dhruvas* of the planets

In this Appendix, we first present the *maṇḍalas* and *dhruvas* for the various planets, as specified in the text *Vākyakaraṇa*,[1] in Tables G.1 to G.5. Verse 4 of Chapter 4 of *Karaṇapaddhati* presents an algorithm for computing the *maṇḍalas* and *dhruvas* of the planets. These have been computed and presented in Tables G.6 to G.10. A comparison of these tables clearly shows that indeed a large set of *maṇḍala* and *dhruva* value, the procedure outlined in *Karaṇapaddhati* which include all the *maṇḍala* and *dhruva* values given in *Vākyakaraṇa*, provided we ignore differences of the order of a few *nāḍikās* (in the case of *maṇḍalas*) or a few minutes (in the case of *dhruvas*).

Maṇḍalas (M_i)			*Dhruvas* (D_i)	
	in numerals			in numerals
in *kaṭapayādi*	days	*nāḍikās*	in *kaṭapayādi*	(in minutes)
dhanadhīdānabhaṅgārtaiḥ धनधीदानभङ्गार्तैः	634089	09	*vanānām* वनानाम्	+004
putradhīhāmarālayaiḥ पुत्रधीहामरालयैः	132589	21	*satrajñāḥ* सत्रज्ञाः	+027
kavissammodahāraiḥ कविस्सम्मोदहारैः	28857	41	*gaṅgārcyā* गङ्गार्च्या	+133
saṅgajanmapathikaiḥ सङ्गजन्मपथिकैः	17158	37	*vaneśaḥ* वनेशः	−504
vanadhīdhṛtapuṇyaiḥ वनधीधृतपुण्यैः	11699	04	*jalārtā* जलार्ता	+638

Table G.1 The *maṇḍalas* and *dhruvas* of Mars.

[1] {VK 1962}, p. 51–55.

V. Pai et al., *Karaṇapaddhati of Putumana Somayājī*, Sources and Studies in the History of Mathematics and Physical Sciences, https://doi.org/10.1007/978-981-10-6814-0

Maṇḍalas (M_i)			*Dhruvas* (D_i)	
	in numerals			in numerals
in *kaṭapayādi*	days	*nāḍikās*	in *kaṭapayādi*	(in minutes)
viṣṇuryajñajatuṣṭaiḥ विष्णुर्यज्ञाजतुष्टैः	16801	54	*yajñajñaḥ* यज्ञज्ञः	−001
guṇanāmotsavaiḥ गुणनामोत्सवैः	4750	53	*dhīvedyā* धीवेद्या	+149
śokadhīviśikhaiḥ शोकधीविशिखैः	2549	15	*sattvavit* सत्त्ववित्	−447

Table G.2 The *maṇḍalas* and *dhruvas* of Mercury.

Maṇḍalas (M_i)			*Dhruvas* (D_i)	
	in numerals			in numerals
in *kaṭapayādi*	days	*nāḍikās*	in *kaṭapayādi*	(in minutes)
saraṇisthadevotsavaiḥ सरणिस्थदेवोत्सवैः	474875	27	*anenāḥ* अनेनाः	000
narmadāvartamātrakaiḥ नर्मदावर्तमात्रकैः	125648	50	*dhanāni* धनानि	−009
satyodayanimittaiḥ सत्योदयनिमित्तैः	65018	17	*balāḍhyā* बलाढ्या	+133
saṅkīrṇakulanīlaiḥ सङ्कीर्तनकुलनीलैः	30315	17	*kathājñaḥ* कथाज्ञः	−071
deveddhaguṇiputraiḥ देवेद्धगुणिपुत्रैः	21539	48	*dhīkṛtaḥ* धीकृतः	−619
viśvotsāhagarvaiḥ विश्वोत्साहगर्वैः	4387	44	*bhūsurāḥ* भूसुराः	+274

Table G.3 The *maṇḍalas* and *dhruvas* of Jupiter.

Maṇḍalas (M_i)			*Dhruvas* (D_i)	
	in numerals			in numerals
in *kaṭapayādi*	days	*nāḍikās*	in *kaṭapayādi*	(in minutes)
dhanāśāvṛddhisaṅgāḍhaiḥ धनाशावृद्धिसङ्गाढैः	487945	09	*jñāninaḥ* ज्ञानिनः	000
dānavṛddhiśubhārthakaiḥ दानवृद्धिशुभार्थकैः	174594	08	*dhīprājñāḥ* धीप्राज्ञाः	+029
guṇitātmā sudehaiḥ गुणितात्मा सुदेहैः	88756	53	*damajñaḥ* दमज्ञः	−058
gururakṣādhibhāvaiḥ गुरुरक्षाधिभावैः	44962	23	*lūnaḥ potrī* लूनः पोत्री	+2103
jaleddhakandharaiḥ जलेद्धकन्धरैः	2919	38	*viśveḍyaḥ* विश्वेड्यः	−144

Table G.4 The *maṇḍalas* and *dhruvas* of Venus.

Maṇḍalas (M_i)			*Dhruvas* (D_i)	
	in numerals			in numerals
in *kaṭapayādi*	days	*nāḍikās*	in *kaṭapayādi*	(in minutes)
hīnabhogamunisamaiḥ हीनभोगमुनिसमैः	570534	08	*māninī* मानिनी	+005
guruvṛddhidharājayaiḥ गुरुवृद्धिधराजयैः	182994	23	*lokajñaḥ* लोकज्ञः	−013
jñānayamīśaparaiḥ ज्ञानयमीशपरैः	21551	00	*gurvājñā* गुर्वाज्ञा	+043
raṅgabhaktadhanikaiḥ रङ्गभक्तधनिकैः	10964	32	*yajñavān* यज्ञवान्	+401

Table G.5 The *maṇḍalas* and *dhruvas* of Saturn.

hāraka	*śeṣa*	*maṇḍala*		*dhruva*
		day	*nāḍikās*	(minutes)
7	88224523260	5459	33.03	−1143.08
15	49106475712	11699	2.21	636.25
22	39118047548	17158	35.24	−506.83
37	9988428164	28857	37.45	129.42
133	9152763056	103731	27.60	−118.59
170	835665108	132589	5.05	10.83
1833	796111976	1429622	18.08	−10.31
2003	39553132	1562211	23.12	0.51
41893	5049336	32673850	0.56	−0.07
295254	4207780	230279161	27.03	0.05
337147	841556	262953011	27.59	-0.01
1980989	0	1545044218	45.00	0.00

Table G.6 Computed values of the *maṇḍalas* and *dhruvas* of Mars.

hāraka	*śeṣa*	*maṇḍala*		*dhruva*
		day	*nāḍikās*	(minutes)
3	325016711593	347	37.55	−1042.74
19	185768858442	2201	37.83	595.99
22	139247853151	2549	15.38	−446.74
41	46521005291	4750	53.20	149.25
104	46205842569	12051	1.78	−148.24
145	315162722	16801	54.98	1.01
21274	192085157	2465130	49.48	−0.62
21419	123077565	2481932	44.46	0.39
42693	69007592	4947063	33.94	−0.22
64112	54069973	7428996	18.41	0.17
106805	14937619	12376059	52.35	−0.05
384527	9257116	44557175	55.45	0.03
491332	5680503	56933235	47.79	−0.02
875859	3576613	101490411	43.24	0.01
1367191	2103890	158423647	31.04	−0.01
2243050	1472723	259914059	14.28	0.004
3610241	631167	418337706	45.31	−0.002
9463532	210389	1096589472	44.90	0.001
32000837	0	3708106125	00.00	−0.00

Table G.7 Computed values of the *maṇḍalas* and *dhruvas* of Mercury.

hāraka	*śeṣa*	*maṇḍala*		*dhruva*
		day	*nāḍikās*	(minutes)
10	16514274166	3988	51.28	−1714.41
11	2640802728	4387	44.41	274.15
76	669457798	30315	17.72	−69.50
239	632429334	95333	37.55	65.65
315	37028464	125648	55.27	−3.84
5594	2945446	2231365	17.08	0.31
67443	1683112	26902032	20.24	−0.17
73037	1262334	29133397	37.32	0.13
140480	420778	56035429	57.56	−0.04
494477	0	197239687	30.0	0

Table G.8 Computed values of the *maṇḍalas* and *dhruvas* of Jupiter.

hāraka	*śeṣa*	*maṇḍala*		*dhruva*
		day	*nāḍikās*	(minutes)
1	893666409854	583	55.54	−8668.88
2	439391325219	1167	51.07	4262.25
5	14883759416	2919	37.68	−144.38
147	7762302155	85837	3.79	75.30
152	7121457261	88756	41.47	−69.08
299	640844894	174593	45.27	6.22
3441	72163427	2009287	59.39	−0.70
27827	63537478	16248897	40.42	0.62
31268	8625949	18258185	39.81	−0.08
246703	3155835	144056197	19.09	0.03
524674	2314279	306370580	18.00	−0.02
771377	841556	450426777	37.09	0.01
2067428	631167	1207224135	32.18	−0.01
2838805	210389	1657650913	9.27	0
10583843	0	6180176875	0	0

Table G.9 Computed values of the *maṇḍalas* and *dhruvas* of Venus.

hāraka	*śeṣa*	*maṇḍala*		*dhruva*
		day	*nāḍikās*	(minutes)
28	56981546371	10586	26.98	−357.90
57	6823125659	21550	59.21	42.86
484	2396541099	182994	20.67	−15.05
1025	2030043461	387539	40.54	12.75
1509	366497638	570534	1.21	−2.30
8570	197555271	3240209	46.57	1.24
10079	168942367	3810743	47.78	−1.06
18649	28612904	7050953	34.35	0.18
103324	25877847	39065511	39.52	−0.16
121973	2735057	46116465	13.87	0.02
1201081	1262334	454113698	44.37	−0.01
2524135	210389	954343862	42.60	0
16345891	0	6180176875	0	0

Table G.10 Computed values of the *maṇḍalas* and *dhruvas* of Saturn.

Appendix H
The *guṇakāras* and *hārakas* of the *Agaṇita* system

In this Appendix, we first present the *hārakas* of various planets, which are used in the calculation of the associated *śodhyabdas* in the *Agaṇita* system, as noted in the Commentary I of *Karaṇapaddhati*.[1] These are compiled in Tables H.1 to H.5. The verse 14 of Chapter 4 of *Karaṇapaddhati* gives an algorithm for computing these *hārakas* of the *Agaṇita* system. Following this algorithm, the *agaṇitahārakas* of various planets have been computed and presented in Tables H.6 to H.10. A comparison of these tables clearly shows that the *hārakas* given in Commentary I are indeed a subset of those which have been computed following the algorithm given in *Karaṇapaddhati*.

hārakas	
in *kaṭapayādi*	in numerals
yatnam	1
ratnam	2
mānyam	15
rāgam	32
dhṛtsu	79
rāghavaḥ	284

Table H.1 The *agaṇita-hārakas* of Mars.

hārakas	
in *kaṭapayādi*	in numerals
tajñaḥ	6
sthānam	7
gopaḥ	13
gaṃgā	33
tatvajñaḥ	46
dhanvī sutaḥ	6749

Table H.2 The *agaṇita-hārakas* of Mercury.

[1] ({KP 1956}, p. 145–146.

V. Pai et al., *Karaṇapaddhati of Putumana Somayājī*, Sources and Studies in the History of Mathematics and Physical Sciences, https://doi.org/10.1007/978-981-10-6814-0

hārakas	
in *kaṭapayādi*	in numerals
puṇyam	11
priyam	12
gajaḥ	83
pakṣīndraḥ	261
vibhaṃgaḥ	344

Table H.3 The *agaṇita-hārakas* of Jupiter.

hārakas	
in *kaṭapayādi*	in numerals
dhyānam	1
khinnam	2
gānam	3
dānam	8
mallāriḥ	235
gobharam	243
haṃsaugham	478

Table H.4 The *agaṇita-hārakas* of Venus.

hārakas	
in *kaṭapayādi*	in numerals
dhātrī	29
dharmam	59
yajñeśaḥ	501

Table H.5 The *agaṇita-hārakas* of Saturn.

guṇakāra	*hāraka*
1	1
1	2
8	15
17	32
25	47
42	79
151	284
193	363
2081	3914
2274	4277
47561	89454
335201	630455
382762	719909
2249011	4230000
8996044	16920000

Table H.6 Computed values of the *agaṇita-guṇakāras* and *agaṇita-hārakas* of Mars.

guṇakāra	*hāraka*
4	1
25	6
29	7
54	13
137	33
191	46
28023	6749
28214	6795
56237	13544
84451	20339
140688	33883
506515	121988
647203	155871
1153718	277859
1800921	433730
2954639	711589
4755560	1145319
12465759	3002227
42152837	10152000

Table H.7 Computed values of the *agaṇita-guṇakāras* and *agaṇita-hārakas* of Mercury.

guṇakāra	*hāraka*
1	11
1	12
7	83
22	261
29	344
515	6109
6209	73652
6724	79761
12933	153413
45523	540000
91046	1080000

Table H.8 Computed values of the *agaṇita-guṇakāras* and *agaṇita-hārakas* of Jupiter.

guṇakāra	*hāraka*
1	1
2	1
3	2
5	3
13	8
382	235
395	243
777	478
8942	5501
72313	44486
81255	49987
641098	394395
1363451	838777
2004549	1233172
5372549	3305121
7377098	4538293
27503843	16920000

Table H.9 Computed values of the *agaṇita-guṇakāras* and *agaṇita-hārakas* of Venus.

guṇakāra	*hāraka*
1	29
2	59
17	501
36	1061
53	1562
301	8871
354	10433
655	19304
3629	106953
4284	126257
42185	1243266
88654	2612789
574109	16920000

Table H.10 Computed values of the *agaṇita-guṇakāras* and *agaṇita-hārakas* of Saturn.

Glossary

abhimata	Desired.
abhyastā	Having multiplied (same as *hatvā*).
adhika	Excess; additive.
adhikamāsa, *adhi-māsa*	Intercalary month: a lunar month in which no *saṅkrānti* (solar transit across zodiacal signs) occurs – it is not counted as a part of the lunar year.
adhimāsakhaṇḍa	The number of civil days elapsed since the beginning of the *kaliyuga* till the beginning of an *adhimāsa* which is close to the beginning of a *Meṣādi*.
āḍhya	Quantity that is to be added.
ādi	Beginning, starting point.
ādityamadhyama	(1) The mean Sun. (2) The mean longitude of the Sun.
agrā	Amplitude at rising, that is, the perpendicular distance of the rising point from the east–west line.
agrāṅgula	*agrā* specified in *aṅgulas*.
ahargaṇa	Count of days. Number of civil days elapsed since the commencement of a chosen epoch.
āhatya	Having multiplied (same as *hatvā*).
ākāśakakṣyā	Boundary circle of the sky, the circumference of which is the linear distance traversed by a planet in a *yuga*.
akṣa	(1) Terrestrial latitude (see also *vikṣepa*). (2) Rsine of terrestrial latitude.
ākṣa	Relating to (terrestrial) latitude.
akṣajīvā, *akṣajyā*	Rsine of the terrestrial latitude.

V. Pai et al., *Karaṇapaddhati of Putumana Somayājī*, Sources and Studies in the History of Mathematics and Physical Sciences, https://doi.org/10.1007/978-981-10-6814-0

akṣakarṇa	Hypotenuse of the equinoctial shadow.
akṣamaurvikā	Same as *akṣajyā*.
akṣavalana	Deflection due to the latitude of the observer. Part of the inclination of the ecliptic to the local vertical, due to the observer's latitude.
alpaguṇakāra	Smaller multiplier.
alpahāraka	Smaller divisor.
aṁśa	(1) Part. (2) Numerator. (3) Degree.
aṅgula	A unit of measurement used to measure linear distances.
antyaphala	Epicycle radius.
antarāla	(1) Difference. (2) The perpendicular distance from a point to a straight line or plane. (3) Divergence. (4) Intervening.
antyakarṇa	The last hypotenuse in the iterative process for the computation of the *manda*-hypotenuse K, such that the relation $\frac{r_m}{R} = \frac{r}{K}$ is satisfied.
antyakrānti	Maximum declination, taken to be 24 degrees, which is the same as the inclination of the ecliptic to the celestial equator.
anvita	That which is added.
apahṛta	That which is subtracted.
apakrama	(1) Declination of a celestial body measured along the meridian circle from the equator towards the north/south pole. (2) Rsine of the declination.
apakramajyā	Rsine of the declination.
apakramamaṇḍala, *apakramavṛtta*, *apamaṇḍala*	Ecliptic: the great circle in the celestial sphere along which the Sun moves in the background of stars, during the course of a year.
apavartana	The process of reduction of a fraction to its lowest terms by division.
apavartita	That which is divided by a common factor without remainder.
apavartya	That which is to be reduced [by division] to the smallest quantity.
āpta	That which is obtained by [division].

asu	21600th part of a sidereal day, or 4 sidereal seconds, which is said to correspond to the time taken by a healthy person to inhale and exhale.
ardhajyā	Rsine of an arc, which is half of the chord.
arka	Sun
arkabhagaṇa	Number of revolutions made by a planet in the course of a *Mahāyuga* (4320000 years).
avama	Omitted/lapsed *tithi*: a *tithi* that commences after sunrise and ends before the next sunrise.
avasāna	End.
ayanacalana	Motion of the equinoxes as well as solsticial points.
bāhu	(1) Rsine. (2) Number two in the *Bhūtasaṅkhyā* system. (3) Side of a geometrical figure.
bāhujyā	Rsine.
bāṇa	(1) Literally, arrow. (2) Rversed sine: $R(1 - \cos\theta)$. (3) Number five in the *Bhūtasaṅkhyā*.
bha	(1) Star. (2) Asterism.
bhacakra	Circle of asterisms.
bhacakraliptā	Minutes of arc contained in a circle which is equal to $360 \times 60 = 21600$.
bhakakṣyā	Orbit of the asterisms.
bhāga	See *aṁśa*.
bhagaṇa	Number of revolutions [of the planets].
bhājaka	Divisor.
bhājya	Dividend.
bhūdina, *bhūdivasa*	(1) Terrestrial/civil day, the average time interval between two successive sunrises. (2) The number of civil days in a *yuga*/*kalpa*.
bhūguṇa	Literally earthsine. It is the Rsine of the arc of a diurnal circle intercepted between local horizon and 6 o'clock circle. If δ is the declination of the celestial body (usually the Sun) and ϕ is the local latitude, then the earthsine is given by $\frac{R\sin\phi . R\sin\delta}{R\cos\phi}$.
bhujā	(1) Opposite side of a right-angled triangle. (2) The *bhujā* of an angle is obtained from the degrees gone in the odd quadrants and to go in the even quadrants.

bhujājyā	Rsine of an angle, or the usual sine multiplied by the radius.
bhujāphala	Equation of centre.
bhukti	(1) Motion. (2) Daily motion of a celestial body.
bimba	Disc of a planet.
bimbavyāsa	Diameter of the disc of a planet.
cakra	(1) Circle. (2) Cycle.
cakrakalā, *cakraliptā*	Minutes of arc contained in a circle which is equal to $360 \times 60 = 21600$.
calakendrahāra	Divisor pertaining to *śīghra* anomaly.
candravākyas	True longitudes of the Moon for 248 days represented in 248 phrases.
cāndramāsa	Lunar month. The time interval between two successive new moons whose average value is ≈ 29.54 civil days.
cāpa	(1) Arc of a circle. (2) Constellation *Dhanus*.
cara	Ascensional difference: equal to the arc of the celestial equator lying between the 6 o'clock circle for a place with a specified latitude, and the horizon. Usually expressed in *nāḍikās*.
carajyā	Rsine of ascessional difference.
caramaphala	See *antyaphala*.
carāsava	*cara* (ascensional difference) expressed in *prāṇas* (sidereal seconds).
dala	Half.
dṛḍhaguṇakāra	Steady [state] multiplier.
dṛḍhahāra	Steady [state] divisor.
dhana	(1) Positive. (2) Additive.
dhanus	Arc of a circle.
dharādina/ *dhātrīdina*	Civil days.
dhruva	(1) Celestial pole (north or south). (2) Fixed initial positions or longitudes of planets at a chosen epoch. (3) Change in the longitude over a specified interval of time.
dhruva-saṁskāra-hāra	Divisor which is used for correcting the *dhruva*.

dinagati	Mean daily motion.
dinakara	Sun.
dinayojanagati	Mean daily motion in *yojanas*.
doḥ	Literally, hand. See *bhujā/bāhu*.
doḥphala	Opposite side of a right–angled triangle conceived inside an epicycle of specified radius with one of the vertices coinciding with the centre of the epicycle, and the angle subtended at that vertex being the *mandakendra* or *śīghrakendra*.
dorjyā	Rsine.
dvitīyahāra	Second order divisor.
dyugaṇa	See *Ahargaṇa*.
dyujyā	Day-radius: Radius of the diurnal circle, whose magnitude is $R\cos\delta$, δ being the declination of the celestial body
dyuvṛtta	See *Ahorātravṛtta*.
eka	(1) Unit. (2) Unit's position. (3) One.
gati	(1) Motion. (2) Rate of motion (of celestial bodies).
ghana	(1) Cube of a number. (2) A solid object.
ghanamūla	Cube root.
ghāta	Product.
ghna	That which is multiplied.
graha-bhramaṇa-vṛtta	Literally, circular orbit of the motion of a planet. This is generally identified with the *pratimaṇḍala* or the eccentric circle.
grahabhukti/ grahagati	Daily motion of a planet.
grāhaka/ grāhakabimba	Eclipsing body.
grahaṇa	Eclipse.
grahaṇa-kāla	Time or duration of an eclipse.
grahaṇa-madhya	Middle of an eclipse.
grahaṇa-parilekhana	Geometrical or graphical representation of the course of an eclipse.

graha-sphuṭa	True longitude of a planet.
grahakakṣyā	Orbit of a planet.
guṇa	(1) Multiplication. (2) Multiplier. (3) Rsine.
guṇaka, guṇakāra	Multiplier.
guṇana	Multiplication.
guṇita	That which is multiplied.
guṇya	Multiplicand.
hanana	Multiplication.
hara, hāra, hāraka	Divisor.
haraṇa	Division.
haraṇaphala	Result of division, quotient.
hata	That which is multiplied.
hṛta	That which is divided.
indu	Moon.
iṣṭa	Desired quantity.
iṣṭabhujācāpa	Arc corresponding to the desired Rsine.
iṣṭadigvṛtta	Vertical circle passing through the zenith and the given celestial body.
iṣṭadyujyā	Desired *dyujyā* (Rcosine of declination).
iṣṭajyā	Rsine at the desired arc.
jīvā	Rsine of an arc. $R\sin\theta$ where θ is the angle corresponding to the arc and R is the *trijyā*, which is the radius of the circle.
jīve-paraspara-nyāya	Rule for obtaining the Rsine of the sum or difference of two angles, according to which the Rsine of one angle is multiplied by the Rcosine of the other and vice-versa and the results are added or subtracted from each other. i.e., $R\sin(A\pm B)=\dfrac{R\sin A.R\cos B \pm R\cos A.R\sin B}{R}$.
jyā	See *jīvā*.
jyācāpāntara	Difference between an arc and its Rsine.
kakṣyā	Orbit of a planet.
kakṣyāmaṇḍala, kakṣyāvṛtta	Deferent or concentric circle, on which the mean planet moves.
kakṣyā-vyāsārdha	Mean radius of the planetary orbit.

kalā	Minute of an arc (angular measure), $\frac{1}{21600}$th part of the circumference of a circle.
kalāgati	Daily motion [of the planets] in minutes.
kālajīva	Rsine of right ascension.
kālalagna	(1) Time elapsed after the rise of the vernal equinox. (2) Time interval between the rise of the vernal equinox and the sunrise.
kalidinagaṇa	Number of civil days elapsed since the beginning of the *Kaliyuga.*
kalpādidhruva	Initial positions (longitudes) of planets at the beginning of the *kalpa.*
kalyādi	Beginning of the *Kaliyuga.*
kalyādi-dhruva	Initial positions (longitudes) of planets at the beginning of the *Kaliyuga.*
kalyahargaṇa	Number of civil days elapsed since the beginning of the *Kaliyuga.*
karaṇa	(1) Construction. (2) A genre of astronomical work (3) Period corresponding to half a *tithi.*
karka, karki	The sign of Cancer.
karkyādi	Six signs commencing from the sign of Cancer.
karṇa	Hypotenuse.
kendrabhukti	Daily motion of the anomaly.
kendragrahāntara	The distance of the planets from the center of the celestial sphere (earth).
kendrahāra/ kendrahāraka	Divisor giving the successive approximations to the rate of anomalistic motion.
khaṇḍa	(1) A part. (2) An epoch closer to *Ahargaṇa.* (3) A day on which the instant of conjunction of Moon and it's apogee is close to the sunrise. (4) A day satisfying specified conditions, close to given *Ahargaṇa.*
khecara, kheṭa	That which wanders in space (planet).
koṭi	(1) Adjacent side of a right-angled triangle. complement of *bhujā* that is Rcosine. (2) 10^7 (both number and place value).
koṭicāpa	90 degrees minus *cāpa.*
koṭijyā	Rcosine of an arc.

krānti	See *apakrama*.
krāntijyā	Rsine of the declination.
krāntikoṭi	Rcosine of the declination.
krāntimaṇḍala	See *apakramamaṇḍala*.
kṣepa	(1) Celestial latitude. (2) Additive quantity.
kṣitidina/ kṣitivāsara	Number of civil days in a *Mahāyuga*.
kṣitijyā	See *bhūjyā*
kudina	Number of civil days in a *Mahāyuga*.
Kulīra	See *Karki*.
lagna	Orient ecliptic point. The longitude of the ecliptic point at the eastern horizon.
lagnasamamaṇḍala	Vertical circle passing through the orient ecliptic point.
lambaka, lambana	(1) Plumb-line. (2) Rsine of co-latitude, i.e., Rcosine of latitude (3) Parallax. (4) Parallax in longitude.
lambana-nāḍikā	Parallax in longitude in *nāḍikās* (24 sidereal minutes).
liptā	Minute of an arc (angular measure).
madhyabhukti/ madhyagati	The mean rate of motion of a planet.
madhyabhuktikalā	The mean rate of motion of a planet in minutes.
madhyagraha	Mean longitude of the planet.
madhyagrahaṇa	Mid-eclipse.
madhyāhna	Midday.
madhyāhnacchāyā	Midday-shadow.
madhyāhnāgrāṅgula	Measure of amplitude at noon in terms of *aṅgula*.
madhyajyā	Meridian sine, i.e. Rsine of the zenith distance when the planet crosses the prime meridian.
madhyakāla	(1) Mean time. (2) Middle of an eclipse etc.
madhyāhna-kālalagna	Meridian ecliptic point—the point of the ecliptic on the prime meridian.
madhyama	Mean longitude of a planet.
madhyānayana-hāraka	Divisor employed in obtaining the mean longitudes of the planets.

mahācchāyā	Literally, great shadow. The distance from the foot of the *mahāśaṅku* to the centre of the Earth. Rsine zenith distance.
mahāguṇakāra	Largest multiplier.
mahāhāra	Largest divisor.
mahājyā	The 24 Rsines used for computation.
mahāśaṅku	(1) Great gnomon. (2) The perpendicular dropped from the Sun to the horizon (when the radius of the celestial sphere is taken to be R), which is equal to Rsine altitude or Rcosine of zenith distance.
makara	Capricorn.
makarādi	The six signs commencing from *Makara* (Capricorn).
māna	(1) Measure. (2) An arbitrary unit of measurement.
manda	(1) Slow. (2) Associated with the equation of centre. (3) Saturn.
manda-karṇa	Hypotenuse associated with *manda* correction.
manda-karṇa-vṛtta	Circle with a radius equal to *manda-karṇa*.
mandakendra	*manda* anomaly, that is the difference in the longitude between the *mandocca* (apogee or apsis) and the mean planet. Mean anomaly.
mandakendrahāra	Divisors pertaining to *manda* anomaly.
maṇḍala	(1) Circle. (2) Orb.
maṇḍalahāraka	Divisors employed in the process of obtaining *maṇḍalas*.
manda-paridhi	Circumference of the epicycle associated with the equation of centre.
mandaphala	The equation of centre.
manda-saṁskāra	*manda* correction in planetary computation. Procedure for obtaining the equation of centre.
manda-sphuṭa, *manda-sphuṭa-graha*	The longitude of a planet obtained after applying the *manda* correction (equation of centre) to the mean longitude (known as *madhyagraha*).
manda-vṛtta	*manda* epicycle, that is, the epicycle associated with the equation of centre.
mandocca	Uppermost point in the *manda* epicycle. Apogee. Apsis.
maṅgalācaraṇa	Invocation.
māsa	Month.

māsavākyas	A set of 12 *vākyas* which gives the cumulative number of civil days at the end of each solar month.
mauḍhya	Invisibility of a planet due to its direction/longitude being close to that of the Sun.
maurvikā	See *jyā*.
meṣa	Aries.
meṣādi	(1) First point of Aries. (2) Commencing point of the ecliptic. (3) Six signs beginning with *meṣa*.
mīna	Pisces.
mithuna	Gemini.
mṛga	The 10th sign: *Makara* (Capricorn).
mṛgādi	The six signs beginning with Capricorn.
mūla	(1) The base or starting point of a line or arc. (2) Square root, cube root etc.
nakra	Capricorn (generally refered to as *Makara*).
nakṣatra	(1) Star. (2) Asterism. (3) Constellation.
nākṣatradina	Sidereal day, which is equal to the time interval between two successive transits of a particular star across the horizon or the meridian ($\approx 23^h 56^m$ of a civil day).
nakṣatrakakṣyā	Orbit of the asterisms.
nakṣatra-saṅkrāntivākyas	A set of 27 *vākyas* giving the times when Sun crosses from one *nakṣatra* ($13°20'$ of ecliptic) to another.
natakāla	Right ascension.
nighna	That which is multiplied.
oja	Odd.
ojapada	Odd quadrants (the first and the third) of a circle.
pada	(1) Square root. (2) Terms of a series. (3) Quarter. (4) Quadrant of a circle.
padīkṛta	Having obtained the square root.
palabhā/palaprabhā	Equinoctial shadow.
palajyā	Rsine of the latitude.
palaśruti	The hypotenuse of the gnomon.
parama	Maximum.
paramakrānti	Maximum declination.

paramakrānti-maurvikā	Rsine of maximum declination.
paramakṣepa	Maximum latitude.
paramāpama/ paramāpakrama	Maximum declination.
paridhi	Circumference.
paryaya	(1) Count of a certain repeated process. (2) Number of revolutions of a planet.
prāṇa	4 sidereal seconds. See *asu*.
pāta	Moon's node.
prāṇakalāntara	Difference between the longitude and right ascension of the Sun in *prāṇas*.
pratatparā	Angular measure corresponding to one-sixtieth of thirds (*tatparā*), or fourths.
rāhu	The ascending node of the Moon.
rāśi	Literally, a group. It refers to: (1) A number (which is a member of a group). (2) A zodiacal sign equal to 30 degrees in angular measure.
rāśicakra	Ecliptic.
rūpa	Unity or number one in the *Bhūtasaṅkhyā* system (literally, form, which is unique to every entity).
ṛṇa	Negative or quantity to be subtracted.
samamaṇḍala	Prime vertical (circle passing through the zenith and the east and west points of the horizon).
samamaṇḍalachāyā	Rsine of zenith distance of a celestial body when it is on the prime vertical.
sama-maṇḍala-śaṅku, sama-śaṅku	Rsine of altitude of a celestial body when it lies on the prime vertical.
samasaṅkhyā	Even number.
saṁguṇa	Multiplied.
saṁvardhita	That which is multiplied.
saṁhṛta	That which is divided.
saṅkramaṇa, saṅkrānti	Sun's transit from one *rāśi* to the next (refers to both the instant as well as the transit).
saṅkramaṇadhruva	The weekday at the beginning of the *kalpa*.

saṅkrāntivākyas	A set of 12 *vākyas* using which the time of transit of the Sun from one sign to another can be computed.
śaṅku	(1) Gnomon (usually of 12 units). (2) Sometimes *mahāśaṅku* (great gnomon), the perpendicular dropped from the Sun to the horizon (= Rsine of altitude). (3) The number 10^{13}.
śaṅkucchāyā	Shadow of the gnomon.
śaṅkukoṭi	Compliment of altitude or zenith distance.
śara	(1) Arrow. (2) Rversed sine, $R(1 - \cos\theta)$.
saura	Related to Sun, solar.
sauramāsa	Solar month.
sāvanadina	Civil days.
saṁskṛtaparyaya	Corrected revolutions of the planets.
saṁvardhita	Multiplied.
sannikarṣa	Proximity.
śakābdasaṁskāra	A correction applied to the mean planet based on the elapsed number of *śaka* years.
śaśāṅka	Moon.
śaśī	Moon.
śaśituṅgayoga	Conjunction of the Moon and its apogee.
śaśyucca	Moon's apogee.
śeṣa	Remainder in an operation.
śīghra-karṇa	(1) Hypotenuse associated with *śīghra* correction. (2) Geocentric radius vector.
śīghrakendra	Anomaly of conjunction; Angular separation between *śīghrocca* and *manda-sphuṭa* (planet corrected for equation of centre) of a planet.
śīghra-kendrahāra	Divisor pertaining to *śīghra* anomaly.
śīghra-paridhi	Circumference of the *śighra* epicycle.
śīghraphala	The correction to be applied to the *manda-sphuṭa* (a planet corrected for the equation of centre) to obtain the geocentric longitude of the planet.
śīghra-saṃskāra	*Śīghra* correction or the application of the *śīghraphala*.
śīghra-sphuṭa	The longitude of a planet obtained after applying the *śīghraphala*.

śīghra-vṛtta	The *śīghra* epicycle, that is, the epicycle associated with the anomaly of conjunction.
śīghrocca	(1) Higher apsis (or the uppermost point) of the epicycle employed in the *śīghra* correction. (2) Apex of the planet moving faster.
śīghroccayoga-samaya	The instant of conjunction (*yoga*) of the *śīghrocca* with the planet.
śiñjinī	See *jyā*.
śiṣṭa	Remainder in an operation.
śiṣṭacāpa	The difference between the given *cāpa* and the nearest *mahājyācāpa* (arc whose Rsine is tabulated).
śodhya	*Śodhya* corresponds to a time interval (not necessarily an integer) close to the desired *ahargaṇa*, at which the *śīghrakendra* is zero, and the longitudes of the mean planet and the *śīghrocca* (which are equal) are close to the longitude of *mandocca*.
śodhyabda	The number of years elapsed since the beginning of the *Śaka* era to the beginning of that year when the mean planet and the *śīghrocca* are in conjunction at the *Meṣādi*.
sphuṭa-vikṣepa	Corrected celestial latitude.
sphuṭayojanakarṇa	The actual physical distance of the planet from the earth.
sphuṭendu	True Moon or True longitude of the Moon.
śruti	See *karṇa*.
śuddha	Subtraction.
sūrya	Sun
svadeśahāraka	Divisor used to calculate the time difference between two places on the same latitudinal circle.
svalpacāpa	An arc having a small magnitude.
sva(m)	(1) Addition. (2) Additive quantity.
tāḍana	Multiplication.
tatparā	Angular measure corresponding to one-sixtieth of a second (*vikalā*).
tṛtīyahāra	Third order divisor.

tithi	Lunar day, a thirtieth part of a synodic lunar month, or the time interval during which the difference in the longitudes of the Moon and the Sun increases by 12°.
tithikṣaya	See *avama*.
tithyanta	End of a *tithi*.
trairāśika	(1) Rule of three. (2) Direct proportion.
tribhajyā	Rsine of three *rāśis*, same as *trijyā* or radius.
tribhuja	A three-sided figure, triangle.
trijyā, trirāśijyā	Rsine 90 degrees. The radius of the circle.
trimaurvikā	See *trijyā*.
triśarādi	Set of odd numbers 3, 5, 7, etc.
tuhinagu	Moon.
tulā	Libra.
tulādi	The six signs commencing from *Tulā*.
tuṅga	Apogee or aphelion (literally, 'peak', *ucca*).
ucca	Higher apsis pertaining to the epicycle (*manda* or *śīghra*). Equivalently, the farthest point in the *pratimaṇḍala* from the centre of the *kakṣyāmaṇḍala*. The apogee of the Sun and the Moon, and the aphelion of the planets.
uccavāra	Friday.
udaya	(1) Rising. (2) Heliacal rising. (3) Rising point of a star or constellation at the horizon.
udayakāla	The moment of rising of a celestial body.
udayalagna	Rising sign. The orient ecliptic point.
uddhṛta	That which is divided.
ūna	(1) Less. (2) That which is subtracted.
upāntya	(1) Close to the end. (2) Penultimate (term).
upāntyajyā	Penultimate Rsine.
uparāga	Eclipse.
utkramajyā	Rversed sine ($R(1 - \cos\theta)$, where θ is the angle corresponding to the arc).
vadha	Multiplication.
vākyakhaṇḍa/ *vākyoktakhaṇḍa*	*Khaṇḍa* mentioned in *vākya* texts.

vallī	An array of numbers.
vallyupasaṁhāra	A mathematical process for obtaining successive approximations to a ratio of integers.
varga	Square.
vargaikya	Sum of the squares.
vargaikyamūla	Square-root of sum of the squares.
vargamūla	Square root.
vāyukāla	Right ascension.
vibhajya	Having divided.
vidhūcca	Moon's apogee
vidhuntuda	Moon's node.
vihaṅgama	Mean planet.
vihṛta	That which is divided.
vikṣepa	(1) Latitudinal deflection (Rsine of celestial latitude). (2) Celestial latitude. (3) Polar latitude.
vikṣepacalana	Related to deflection of Moon's obit from the ecliptic.
vināḍī	$\frac{24}{60}$ of a sidereal second.
viparītakarṇa	Reverse or inverse hypotenuse: $\frac{R^2}{K}$, where K is the *aviśiṣṭa-mandakarṇa* (iterated *manda* hypotenuse).
viparyaya	Inverse or reverse, also called *viparyāsa.*
viṣama	(1) Odd number or quadrant. (2) Difficult.
viśeṣa	Speciality, Difference.
viṣkambha	(1) Diameter. (2) The first of 27 daily *yogas.*
viṣkambhadala	Semi-diameter.
viśleṣa	(1) Subtraction. (2) Difference.
viśodhya	Having subtracted.
vistarārdha	Semi-diameter or radius.
vistṛtidala	Semi-diameter (*vistṛti* is diameter).
viśuddha	That which is subtracted.
viṣuvacchāyā/ viṣuvadbhā	Equinoctial midday shadow, that is, the shadow of a gnomon measured at the meridian transit, of the Sun when at the equinox.
viṣuvadbhāgra	Tip of the shadow on the equinoctial day.

viṣuvanmaṇḍala	Celestial equator, which is the same as the path traced by a star rising exactly in the east and setting exactly in the west.
viṣuvat	Vernal or autumnal equinox.
viṣuvatkarṇa	Hypotenuse of equinoctial shadow.
vitribhalagna	Central ecliptic point or nonagesimal, i.e., the point on the ecliptic whose longitude is less than that of the *lagna* (ascendant or the ecliptic point on the eastern horizon) by 90 degrees.
vivara	(1) Difference. (2) Gap. (3) Space in between.
viyoga	Subtraction.
vṛtta	Circle.
vṛttakendra	Centre of a circle.
vṛttaparidhi	Circumference of a circle.
vyāsa	Diameter of a circle.
vyāsa-dala/ardha	Semi-diameter, radius.
vyasta-karṇa	See *viparīta-karṇa.*
yogadhruvas	Corrections applied to the longitudes of Moon and its apogee.
yogodayāntara	Difference between the instances of conjunction of Moon and its apogee and the sunrise.
yogyādivākyas	A set of 48 *vākyas* which are used to find the true longitude of the Sun at any given instant.
yojanagati	Daily motion in terms of *yojanas* .
yuga	Aeon. A large unit of time, for instance, *Kaliyuga* whose duration is 432000 years or *Mahāyuga* whose duration is 4320000 years.
yugādhimāsa	Number of *adhimāsas* in a *yuga.*
yugabhagaṇa	Number of revolutions made by a planet in the course of a *Mahāyuga.*
yugma	Even.
yukta, yuta	That which is added.

Bibliography

Source Works

{AB 1874} *Āryabhaṭīyaṃ* of Āryabhaṭa with the commentary *Bhaṭadīpikā* of Parameśvara, ed. by B. Kern, Leiden 1874 (repr. 1906, 1973).

{AB 1976} *Āryabhaṭīya* of Āryabhaṭa, ed. with tr. and notes by K. S. Shukla and K. V. Sarma, INSA, New Delhi 1976.

{ABB 1930} *Āryabhaṭīya* of Āryabhaṭācārya with the *Mahābhāṣya* of Nīlakaṇṭha Somasutvan, Part I, *Gaṇitapāda*, ed. by Sāmbaśiva Śāstrī, Trivandrum Sanskrit Series 101, Trivandrum 1930.

{ABB 1931} *Āryabhaṭīya* of Āryabhaṭācārya with the *Mahābhāṣya* of Nīlakaṇṭha Somasutvan, Part II, *Kālakriyāpāda*, ed. by Sāmbaśiva Śāstrī, Trivandrum Sanskrit Series 110, Trivandrum 1931.

{ABB 1957} *Āryabhaṭīya* of Āryabhaṭācārya with the *Mahābhāṣya* of Nīlakaṇṭha Somasutvan, Part III, *Golapāda*, ed. by Śūranāḍ Kuñjan Pillai, Trivandrum Sanskrit Series 185, Trivandrum 1957.

{ABB 1976} *Āryabhaṭīya* of Āryabhaṭa with the commentary of Bhāskara I and Someśvara, ed. by K. S. Shukla, INSA, New Delhi 1976.

{CCG 1976} *Candracchāyāgaṇita* of Nīlakaṇṭha Somayājī, with auto-commentary, ed. and tr. by K. V. Sarma, VVRI, Hoshiarpur 1976.

{CV 1948} *Candravākyas* of Vararuci, C. Kunhan Raja, Adyar Library, Madras 1948.

{DG 1963} *Dṛggaṇita* of Parameśvara, ed. by K. V. Sarma, VVRI, Hoshiarpur 1963.

{GCN 1954} *Grahacāranibandhana* of Haridatta, ed. by K. V. Sarma, KSRI, Madras 1954.

V. Pai et al., *Karaṇapaddhati of Putumana Somayājī*, Sources and Studies in the History of Mathematics and Physical Sciences, https://doi.org/10.1007/978-981-10-6814-0

{GD 1916} *Goladīpikā* of Parameśvara, ed. by T. Ganapati Sastri, Trivandrum 1916.

{GD 1957} *Goladīpikā* of Parameśvara, with auto-commentary *Vivṛti*, ed. by K. V. Sarma, Adyar Library, Madras 1957.

{GS 1970} *Golasāra* of Nīlakaṇṭha Somayājī, ed. and tr. by K. V. Sarma, VVRI, Hoshiarpur 1970.

{GVV 1979} *Grahasphuṭānayane-vikṣepavāsanā* of Nīlakaṇṭha Somayājī, in *Gaṇitayuktayaḥ*, ed. by K. V. Sarma, VVRI, Hoshiarpur 1979.

{GYB 1948} *Gaṇita-yukti-bhāṣā*, of Jyeṣṭhadeva, *Gaṇitādhyāya* ed. with notes in Malayalam by Ramvarma Maru Tampuran and A. R. Akhileswara Aiyar, Trichur 1948.

{GYB 1953} *Gaṇita-yukti-bhāṣā* of Jyeṣṭhadeva, ed. by K. Chandrasekharan, Government Oriental Manuscripts Library, Madras 1953.

{GYB 2008} *Gaṇita-yukti-bhāṣā* of Jyeṣṭhadeva, ed. and tr. by K. V. Sarma with Explanatory Notes by K. Ramasubramanian, M. D. Srinivas and M. S. Sriram, 2 Volumes, Hindustan Book Agency, New Delhi 2008 (repr. Springer 2009).

{HC 1948} *Haricarita* of Parameśvara Bhaṭṭa, ed., Krishnamacharya, GOML, Madras 1948.

{JAM 1930} *Jātakādeśamārga* of Putumana Somayājī, ed. by V. N. Sankunni Nayar, Kunnamkulam 1930.

{JAM 1942} *Jātakādeśamārga* of Putumana Somayājī, ed. and tr., by V. Subrahmanya Sastri and M. M. Bhat, Bangalore 1942.

{JAM 1960} *Jātakādeśamārga* of Putumana Somayājī, ed. with a Malayalam *Ṭīkā* by Kanippayur Sankaran Nambudiripad, Kunnamkulam 1960.

{JAM 1971} *Jātakādeśamārga* of Putumana Somayājī, ed. with Hindi tr. by Gopesh Kumar Ojha, Delhi 1971.

{JM 1977} *Jyotirmīmāṃsā* of Nīlakaṇṭha Somayājī, ed. by K. V. Sarma, VVRI, Hoshiarpur 1977.

{KA 1975} *Karaṇāmṛta* of Citrabhānu, with commentary of Śaṅkara Vārīyar, ed. by V. Narayanan Nambudiri, Trivandrum 1975.

{KP 1937} *Karaṇapaddhati* of Putumana Somayājī, ed. by K. Sāmbaśiva Śāstrī, Trivandrum 1937.

{KP 1953} *Karaṇapaddhati* of Putumana Somayājī, ed. by P. K. Koru, Astro Printing and Publishing Co., Cherp 1953.

{KP 1956} *Karaṇapaddhati* of Putumana Somayājī, ed. with two Malayalam commentaries by S. K. Nayar, GOML, Madras 1956.

{KU 1964} *Karaṇottama* of Acyuta Piṣāraṭi, with auto-commentary *Vivaraṇa*, ed. by K. Raghavan Pillai, Trivandrum 1964.

{LB 1963} *Laghubhāskarīya*, of Bhāskara I, ed. and tr. with notes by K. S. Shukla, Lucknow 1963.

{LM 1952} *Laghumānasa* of Muñjāla with *Vyākhyā* of Parameśvara, ed. by B. D. Apte, Pune 1944; 2nd edn 1952.

{LM 1990} *Laghumānasa* of Muñjāla, ed. with tr. and notes by K. S. Shukla, INSA, New Delhi 1990.

{MB 1957} *Mahābhāskarīya* of Bhāskara I, ed., with commentary of Govindasvāmin and *Siddhāntadīpikā* of Parameśvara, by T. S. Kuppanna Sastri, KSRI, Madras 1957.

{MB 1960} *Mahābhāskarīya* of Bhāskara I, ed. and tr. with notes by K. S. Shukla, Lucknow 1960.

{PS 1889} *Pañcasiddhāntikā* of Varāhamihira, ed. and tr. by G. Thibaut and Sudhakara Dvivedi, Varanasi 1889.

{PS 1971} *Pañcasiddhāntikā* of Varāhamihira, ed. and tr. by O. Neugebauer and D. Pingree, Copenhagen 1971.

{PS 1993} *Pañcasiddhāntikā* of Varāhamihira, ed. and tr. by T. S. Kuppanna Sastri and K. V. Sarma, PPST Foundation, Chennai 1993.

{RS 1977} *Rāśigolasphuṭanīti* of Acyuta Piśāraṭi, ed. and tr. by K. V. Sarma, VVRI, Hoshiarpur 1977.

{SC 1973} *Sphuṭacandrāpti* of Mādhava, ed. and tr. by K. V. Sarma, VVRI, Hoshiarpur 1973.

{SDA 1976} *Siddhāntadarpaṇa* of Nīlakaṇṭha Somayājī, ed. with auto-commentary and tr. by K. V. Sarma, VVRI, Hoshiarpur 1976.

{SDV 1981} *Śiṣyadhīvṛddhidatantra* of Lalla, ed. and tr. by Bina Chatterjee, 2 vols, INSA, New Delhi 1981.

{SNT 1974} *Sphuṭanirṇayatantra* of Acyuta Piśāraṭi, ed. with auto-commentary by K. V. Sarma, VVRI, Hoshiarpur 1974.

{SRM 2011} *Sadratnamālā* of Śaṅkaravarman, tr. by S. Madhavan, KSRI, Chennai 2011.

{SSR 1981} *Siddhānta-śiromaṇi* of Bhāskara II, with Bhāskara's *Vāsanā* and Nṛsiṃha Daivajña's *Vāsanāvārtika*, ed. by Muralidhara Chaturvedi, Varanasi 1981.

{SSR 2000} *Siddhānta-śiromaṇi* of Bhāskara II, *Grahagaṇitādhyāya* tr. by D. Arkasomayaji, Rastriya Sanskrit Vidyapeetha, Tirupati 2000.

{TS 1958} *Tantrasaṅgraha* of Nīlakaṇṭha Somayājī, with *Laghuvivṛti*, ed. by S. K. Pillai, Trivandrum 1958.

{TS 1977} *Tantrasaṅgraha* of Nīlakaṇṭha Somayājī, with *Yuktidīpikā* (for chapters I–IV) and *Laghu-vivṛti* (for chapters V–VIII) of Śaṅkara Vārīyar ed. by K. V. Sarma, VVRI, Hoshiarpur 1977.

{TS 1999} *Tantrasaṅgraha* of Nīlakaṇṭha Somayājī, tr. by V. S. Narasimhan, *Indian Journal History of Science*, INSA, New Delhi 1998–99.

{TS 2011} *Tantrasaṅgraha* of Nīlakaṇṭha Somayājī, tr. with mathematical notes by K. Ramasubramanian and M. S. Sriram, HBA, Delhi and Springer, London 2011.

{VK 1962} *Vākyakaraṇa* with the commentary by Sundararāja, ed. by T. S. Kuppanna Sastri and K. V. Sarma, KSRI, Madras, 1962.

{VP 1980} *Vākyapadīya* of Bhartṛhari, *Vākyakhaṇḍam* ed. with Puṇyarāja Commentary, by Raghunatha Sharma, Sampurnanand Sanskrit University, Varanasi 1980.

{VR 1956} *Veṇvāroha* by Mādhava, ed. with Malayalam commentary of Acyuta Piṣāraṭi by K. V. Sarma, The Sanskrit College Committee, Tripunithura, Kerala 1956.

Secondary Works

Amma T. A. Sarasvati, *Geometry in Ancient and Medieval India*, Motilal Banarsidass, Delhi 1979.

Datta B and Singh A. N., *History of Hindu Mathematics*, 2 vols, Lahore 1935, 1938 (repr. Asia Publishing House, Bombay 1962; repr. Bharatiya Kala Prakashan, Delhi 2004).

Emch G., Srinivas M. D. and Sridharan R (eds), *Contribution to the History of Mathematics in India*, Hindustan Book Agency, New Delhi 2005.

Evans James, *The History and Practice of Ancient Astronomy*, Oxford University Press, New York 1998.

Gupta R. C., Second order interpolation in Indian mathematics up to the fifteenth century, *Indian Journal of History of Science*, pp. 86-98, **4**, 1969.

Gupta R. C., Solution of the Astronomical Triangles as found in the *Tantrasaṅgraha*, *Indian Journal of History of Science*, pp. 86-99, **9**, 1974.

Gupta R. C., Mādhava's Rule for finding angle between the ecliptic and the horizon and Āryabhaṭa's knowledge of it, in *History of Oriental Astronomy*, eds. G. Swarup, A. K. Bag and K. S. Shukla, Cambridge 1987, pp. 197-202.

Hariharan S., Declination in Indian Astronomy and the Approach of Kerala Astronomers, *Jour. Asiatic Soc.*, pp. 39-49, **30**, 1988.

Hari K. Chandra, *Vākyakaraṇa*: A study, *Indian Journal of History of Science*, pp. 127-149, **36**, 2001.

Hari K. Chandra, Computation of the true moon by Mādhava of Saṅgamagrāma, *Indian Journal of History of Science*, pp. 231-253, **38**, 2003.

Jones A., Adaptation of Babylonian Methods in Greek Numerical Astronomy, *Arch. Hist. Exact. Sci.*, pp. 1-36, **29.1**, 1983.

Kak. S., The Vararuci Cipher, *Cryptologia*, pp. 79-82, **14**, 1990.

Khinchin A. Ya., *Continued fractions*, University of Chicago Press, 1964.

Madhavan S., *Veṇvāroha* from a modern perspective, *Indian Journal of History of Science*, pp. 699-717, **49**, 2012.

Mahesh K., Pai R. Venketeswara and Ramasubramanian K., 'Turning an algebraic identity into an infinite series', in *History of Mathematical Science II*, eds. B. S. Yadav and S. L. Singh, Cambridge Scientific Publishers, UK 2010, pp. 61-81.

Mahesh K., Pai R. Venketeswara and Ramasubramanian K., Mādhava series for π and its fast convergent approximations, *Astronomy and Mathematics in Ancient India*, ed. J. M. Delire, Peeters Publishers, Leuven 2012, pp. 175-198.

Pai R. Venketeswara, *A Critical Study of Karaṇapaddhati of Putumana Somayājī*, Ph.D. Thesis (Unpublished), IIT Bombay, 2011.

Pai R. Venketeswara, Obtaining the longitude of the moon by *Veṇvāroha* in *Addressing History: Science in Medieval Kerala* ed. Vinod Kumar, Maharajas College, Ernakulam 2013, pp. 44-68.

Pai R. Venketeswara, Joshi Dinesh Mohan and Ramasubramanian K., The *Vākya* method of finding the moon's longitude, *Gaṇita Bhāratī*, pp. 39-64, **31**, No. 1-2, 2009.

Pai R. Venketeswara, Mahesh K. and Ramasubramanian K., *Kriyākalāpa*: A Commentary of *Tantrasaṅgraha* in Keralabhāṣā, *Indian Journal of History of Science*, pp. T1-T47, **45**, No. 2, 2010.

Pai R. Venketeswara, Ramasubramanian K., and Sriram M. S., Rationale for *Vākyas* pertaining to the Sun in *Karaṇapaddhati*, *Indian Journal of History of Science*, pp. 245-258, **50**, No. 2, 2015.

Pai R. Venketeswara, Ramasubramanian K., Srinivas M. D. and Sriram M. S., The *Candravākyas* of Mādhava, *Gaṇita Bhāratī* (in press 2017).

Parameswaran S., *The Golden Age of Indian Mathematics*, Swadeshi Science Movement, Kochi 1998.

Pingree D., *Jyotiśśāstra: Astral and Mathematical Literature*, Wiesbaden 1981.

Pingree D., *Census of the Exact Sciences in Sanskrit,* American Philosophical Society, Philadelphia, Series A, Vol 4, 1981.

Plofker Kim, *History of Mathematics in India: From 500 BCE to 1800 CE*, Princeton, 2009.

Raju C. K., *Cultural Foundations of Mathematics: The Nature of Mathematical Proof and the Transmission of the Calculus from India to Europe in the 16th c. CE*, Pearson Education, Delhi 2007.

Ramasubramanian K., Aniket Sule and Mayank Vahia, *History of Indian Astronomy A Handbook*, SandHI, IIT Bombay and TIFR Bombay 2017.

Ramasubramanian K., and Pai R. Venketeswara, *Infinite Series Inlaid in Verses*, ICRTGC-2010, A Satellite Conference of the International Congress of Mathematicians (ICM) 2010.

Ramasubramanian K. and Srinivas M. D., Development of calculus in India, in *Studies in the History of Indian Mathematics* ed. C. S. Seshadri, Hindustan Book Agency, New Delhi 2010, pp. 201-286.

Ramasubramanian K. and Sriram M. S., Correction to the Terrestrial Latitude in *Tantrasaṅgraha*, *Indian Journal of History of Science*, pp. 129-144, **38**, 2003.

Ramasubramanian K., Srinivas M. D. and Sriram M. S, Modification of the earlier Indian planetary theory by the Kerala astronomers (c. 1500 AD) and the implied heliocentric picture of the planetary motion, *Current Science*, pp. 784-790, **66**, 1994.

Rao S. Balachandra, *Indian Mathematics and Astronomy: Some Landmarks*, Bangalore 1998.

Rao S. Balachandra, *Indian Astronomy: An Introduction*, Hyderabad 2000.

Sarma K. V., *A Bibliography of Kerala and Kerala-based Astronomy and Astrology*, Hoshiarpur 1972.

Sarma K. V., *A History of the Kerala School of Hindu Astronomy*, Hoshiarpur 1972.

Sarma K. V., Putumana Somayājī, in *Encyclopedia of Science Technology and Medicine in Non-Western Cultures*, ed. H. Selin, Springer New York 2008, pp. 1856-57.

Sarma K. V. and Hariharan S., *Yuktibhāṣā* of Jyeṣṭhadeva: A book of rationale in Indian mathematics and astronomy: An analytical appraisal, *Indian Journal of History of Science*, pp. 185-207, **26**, 1991.

Sarma U. K. V., Pai Venketeswara and Ramasubramanian K., '*Madhyamñayanaprakāraḥ*: A Hitherto Unknown Manuscript Ascribed to Mādhava', *Indian Journal of History of Science*, pp. T1-T29, **46**, No. 1, 2011.

Sarma U. K. V., Bhat Vanishri, Pai Venketeswara and Ramasubramanian K., The Discovery of Mādhava Series by Whish: An Episode in Historiography of Science, *Gaṇita Bhāratī*, pp. 115-126, **32**, No. 1-2 (2010).

Sastri T. S. Kuppanna, A Brief History of Tamil Astronomy, Madras Univ Jl, Section C 41(ii), 1971, Reprinted in *Selected papers on Jyotisha*, Rashtriya Sanskrit Vidyapeetha, Tirupati, 1989, pp. 329-344.

Sen S. N. and Shukla K. S. (eds), *A History of Indian Astronomy*, New Delhi 1985 (rev. edn 2000).

Shukla K. S., Use of hypotenuse in the computation of the equation of center under the epicyclic theory in the school of Āryabhaṭa, *Indian Journal of History of Science*, 43-57, **8**, 1973.

Somayaji D. A., *A Critical Study of Ancient Hindu Astronomy*, Karnatak University, Dharwad 1972.

Srinivas M. D., Indian approach to science: The case of *Jyotiśśāstra*, *PPST Bulletin*, Madras, pp. 69-76, **19-20**, 1990.

Sriram M. S., Planetary and lunar models in *Tantrasaṅgraha* and *Gaṇitayuktibhāṣā*, in *Studies in the History of Indian Mathematics*, ed. C. S. Seshadri, Hindustan Book Agency, New Delhi 2010, pp. 353-389.

Sriram M. S., *Vākya* system of astronomy, in *Handbook of Archeoastronomy and Ethnoastronomy*, ed. C. L. N. Ruggles, Springer Science, 2015, pp. 1991-2000.

Sriram M. S., Explanation of *Vākyaśodhana* procedure for *Candravākyas*, *Gaṇita Bhāratī* (in press 2017).

Sriram M. S. and Pai R. Venketeswara, Use of Continued fractions in *Karaṇapaddhati*, *Gaṇitabhāratī*, pp. 137-160, Vol. 34, No. 1-2, 2012.

Sriram M. S., Ramasubramanian K. and Pai R. Venketeswara, On the relation between the arc (*jyā*) and the Rsine (*bhujā*) in *Tantrasaṅgraha* and other Kerala works. *Indian Journal of History of Science*, pp. 719-736, **47.4**, 2012.

Sriram M. S., Ramasubramanian K. and Srinivas M. D. (eds), *500 Years of Tantrasangraha: A Landmark in the History of Astronomy*, Shimla 2002.

Subbarayappa B. V. and Sarma K. V., (eds), *Indian Astronomy: A Source Book*, Bombay 1985.

Subbarayappa B. V., *The Tradition of Astronomy in India: Jyotiśśāstra*, PHISPC vol. IV, part 4, Center for Studies in Civilizations, New Delhi 2008.

Whish C. M., On the Hindu quadrature of the circle and the infinite series of the proportion of the circumference to the diameter exhibited in the four Shastras, the Tantrasangraham, Yucti Bhasa, Carana Paddhati and Sadratnamala, *Trans. R. Asiatic. Soc.* (GB), **3**, pp. 509-523, 1834.

Index

V. Pai et al., *Karaṇapaddhati of Putumana Somayājī*, Sources and Studies in the History of Mathematics and Physical Sciences, https://doi.org/10.1007/978-981-10-6814-0

Index of Half-verses

V. Pai et al., *Karaṇapaddhati of Putumana Somayājī*, Sources and Studies in the History of Mathematics and Physical Sciences, https://doi.org/10.1007/978-981-10-6814-0

Printed in the United States
By Bookmasters